metric edition

Purpose-made Joinery

Frank Hilton

Lately Senior Lecturer Building Department Bolton Technical College
With drawings by the author

Longman London & New York

Longman Group Limited
Longman House, Burnt Mill, Harlow
Essex CM20 2JE, England
Associated Companies throughout the world

Published in the United States of America
by Longman Inc., New York

© Longman Group Limited 1978

First published 1978
Third impression 1983

Library of Congress Cataloging in Publication Data

Hilton, Frank
Purpose-made joinery.

  1. Joinery. I. Title
TH5663.H54   694'.4   77-21884
ISBN 0-582-41142-4

Type photoset in 10 point Times
by Woolaston Parker Ltd, Leicester
Printed in Singapore by Selector Printing Co Pte Ltd

By the same author

Metric editions
Advanced carpentry and joinery

Building geometry and drawing

Craft technology for carpenters and joiners

Preface

A recent review of courses of technical education designed for craftsmen in construction crafts has produced a new approach to the educational requirements of building craft apprentices and of those craftsmen who aim to become craft or general foremen.

The important considerations which have brought about this review of further education have been:

1. The rapid changes in technology where it has become increasingly important that the craftsman should be educated in the scientific and technological principles employed in industry if he is to be able to understand, appreciate and apply new techniques as they are introduced.

2. The importance of the 'industrial relations' factor.
   (a) changing technology may increase the 'overlap' between craftsmen's activities without necessarily removing the need for specialist and traditional craft skills and knowledge: the individual craftsman needs more than ever to study his own craft not only to understand his own function but also to be able to cooperate with other craftsmen in the industry;
   (b) understanding the job makes job satisfaction more probable;
   (c) the ability to communicate enables the craftsman to do his job, to talk about his job, to explain his ambitions and be satisfied that he has been understood.

3. The rapid change in the pattern of social life: the attainment of a satisfactory personal status within his community and a satisfaction in personal life and an opportunity to engage in worthwhile leisure pursuits are as important to job performance as they are to self-respect.

The aims of the further education given in the various levels of these courses are to:

1. Provide the knowledge and appreciation of techniques and materials which a craftsman will need to do his job with efficiency and understanding.

2. Provide a broad understanding of relevant science and technology, with background industrial studies so that the student craftsman

   (a) acquires an understanding of the principles of his craft;
   (b) appreciates the work and problems of craftsmen engaged in associated occupations and the relationship of his work to theirs;
   (c) is better equipped to adjust to changes in the nature of his work caused by technological development, changes in industrial conditions, change of job within his own industry or transfer to a similar occupation in another.

3. Provide opportunity for continued study in preparation for advancement in the industry.

4. Widen the student craftsman's understanding of the industry in which he works and the society in which he lives.

5. Provide opportunity for the development of responsible attitudes to quality of work and to costs.

6. Introduce a study of elements of supervision and job organisation.

7. Develop the student craftsman as a person, so as to encourage the growth of mature attitudes in industry and society in general, of powers of thought, reasoning, and communication, and of his appreciation of the value of learning.

Selected option studies in Carpentry and Joinery at Advanced Level include purpose-made joinery. An endeavour has been made in this book to provide, for the first time, a fairly complete course of instruction in purpose-made joinery, and is intended to cover a range of good-quality joinery work associated with housing, flats, schools, industrial buildings, offices, hospitals, laboratories, hotels, churches, banks, and public buildings generally.

A special feature has been made of the illustrations used to convey as much information as possible, facing in most cases the relevant text which has been kept to a minimum. It will be seen that a unique feature has been introduced, again for the first time in a technical work, that of project work. Certain special situations have been chosen to introduce the student to all the various stages of work involved and their objectives, from the design and planning stages, to the construction and completion of the various joinery components. These particular projects, will therefore, allow scope for the treatment of a series of different specialised joinery items of such a variety that, the student has the opportunity of experiencing techniques and practices which will assist him to understand the fundamental principles of a far wider range of items that would normally be the case, in his specialism.

While primarily intended for students preparing for the City and Guilds examinations leading to an Advanced Craft Certificate in Carpentry and Joinery, students preparing for the Institute of Builders examinations and TEC technician subjects in construction, teachers in colleges and certain personnel in builders' and architects' offices should, it is hoped, find this work useful.

In conclusion the author wishes to express his gratitude to his publishers for their help and cooperation and trusts his efforts will have an equally friendly and appreciative reception as accorded to his earlier books and prove to be equally useful.

Frank Hilton
1977

Acknowledgements

Sincere thanks are tendered to the following firms and organisations for their courtesy and help in supplying valuable technical information, and for their permission to include in this book extracts from publications, etc. Arborite Ltd, Bakelite, Black and Decker Ltd, British Equipment Co. Ltd, Celcure Ltd, CIBA (ARL) Ltd, Commercial Secretary, High Commissioner for Canada, Controller Her Majesty's Stationery Office, G. F. Wells Ltd, GKN Ltd, Lawtons of Liverpool, MacAndrews and Forbes Ltd, Plywood Manufacturers Association of British Columbia, Rawlplug Co. Ltd, Spotnails Ltd, Stanley Works (Gt Britain) Ltd, Timber Research and Development Association, Twinaplate, H. Arnold and Co. Ltd, Wadkin Ltd, Weyroc Ltd, Whitehill Spindle Tools Ltd, Robert Adams Ltd, A. H. Anderson Ltd, Austins Ltd, S. N. Bridges & Co. Ltd, British Equipment Co. Ltd, Centec Machine Tools Ltd, P. G. Henderson Ltd, William Kay (Bolton) Ltd, Linden Doors Ltd, Tomo Trading Co. Ltd, William Newman Ltd, Westland Engineers Ltd.

F.H.

Contents

List of illustrations

Metrication tables

Basic SI units

Quantity	Unit	Symbol
length	metre	m
mass	kilogram	kg
time	second	s
temperature	degree	°C or degC*
luminous intensity	candela	cd
electric current	ampere	A

*The basic SI unit of temperature is 'degree kelvin', in practice 'degree Celsius' is used.

The symbol °C denotes customary temperature, degC denotes a temperature interval.

'Celsius' is the unit of temperature formerly called 'centigrade'.

Derived SI units with complex names

Quantity	Unit	Symbol
area	square metre	m^2
volume	cubic metre	m^3
density	kilogram per cubic metre	kg/m^3
frequency	cycle per second	c/s
velocity	metre per second	m/s
pressure, stress	newton per square metre	N/m^2
thermal conductivity	watt per metre degree Celsius	W/mdegC

Derived SI units with special names

Quantity	Unit	Symbol
force	newton	N
work, energy	joule	J
power	watt	W
illuminance	lux	lx
electric potential	volt	V

Conversion factors

Length

Metric unit to British unit		British unit to metric unit	
1 kilometre	0·621 mile	1 mile	1·609 kilometre
1 metre	3·281 feet	1 foot	0·305 metre
1 millimetre	0·0394 inch	1 inch	25·4 mm

Area

1 square kilometre	0·386 square mile	1 square mile	2·590 km^2
1 square metre	10·764 square feet	1 square foot	0·093 m^2
1 square millimetre	$1·55 \times 10^{-3} in^2$	1 square inch	645·16 mm^2

Mass

1 kilogram	2·205 lb	1 pound	0·454 kg
1 gram	0·305 ounce	1 ounce	28·35 grams

Volume

1 cubic metre	35·315 cubic feet	1 cubic foot	0·028 m^3

Force

newton (1 kgf=9·806 N)	0·225 lbf	1 pound force	4·448 N

Pressure

newton per square metre	$1·4504 \times 10^{-4} lbf/in^2$	1 lb force per square inch	6894·8 N/m^2

Density

1 kilogram per cubic metre	0·062 lb/ft^2	1 lb per cubic foot	16·019 kg/m^3

Chapter 1

Timber

Timber is divided into two classes: 1. the coniferous trees (Fig. 1.1.) known as *softwoods;* and 2. the deciduous trees, (Fig. 2.1.) known as *hardwoods.* Softwoods are usually evergreen with needle-pointed leaves and are cone-bearing, (Fig. 1.2). Hardwood trees have broad leaves, which in most cases are shed at the end of the growing season (Fig. 2.3.). There are certain exceptions, one example being the holly tree which is evergreen throughout the year.

A tree consists of three main parts: the roots, the stem, and the crown. The root fixes the tree in the ground and takes in moisture from the soil. The stem or trunk stores foodstuffs, conducts these to the leaves and provides strength and rigidity to the tree. The timber which man has used since the earliest ages is, of course, cut from the trunk. The crown consists of branches, twigs, and leaves in which the chemical process essential to growth takes place.

Hardwoods bear fruit in which the seeds are to be found (Fig. 2.2.) the chestnut of the horse-chestnut, the acorn of the oak, and the berries of the holly tree are examples.

The terms 'softwood' and 'hardwood' are by no means accurate in every case; they are, however, generally descriptive and established terms in the trade. Some hardwoods are as soft as, or even softer than, the nominally softwood, while some softwoods are harder than many hardwoods.

Most of the timber used by the carpenter and joiner in the construction of buildings is softwood. This is mainly of the pine and fir class, and though they are to be found in many parts of the world, the chief sources of pine are the forests of Canada, North America, Scandinavia, and Russia. The forest belts providing the hardwoods are to be found in the tropical zones, namely: Central and South America, West Africa, regions of India, Burma, and Malaysia and eastern Australia.

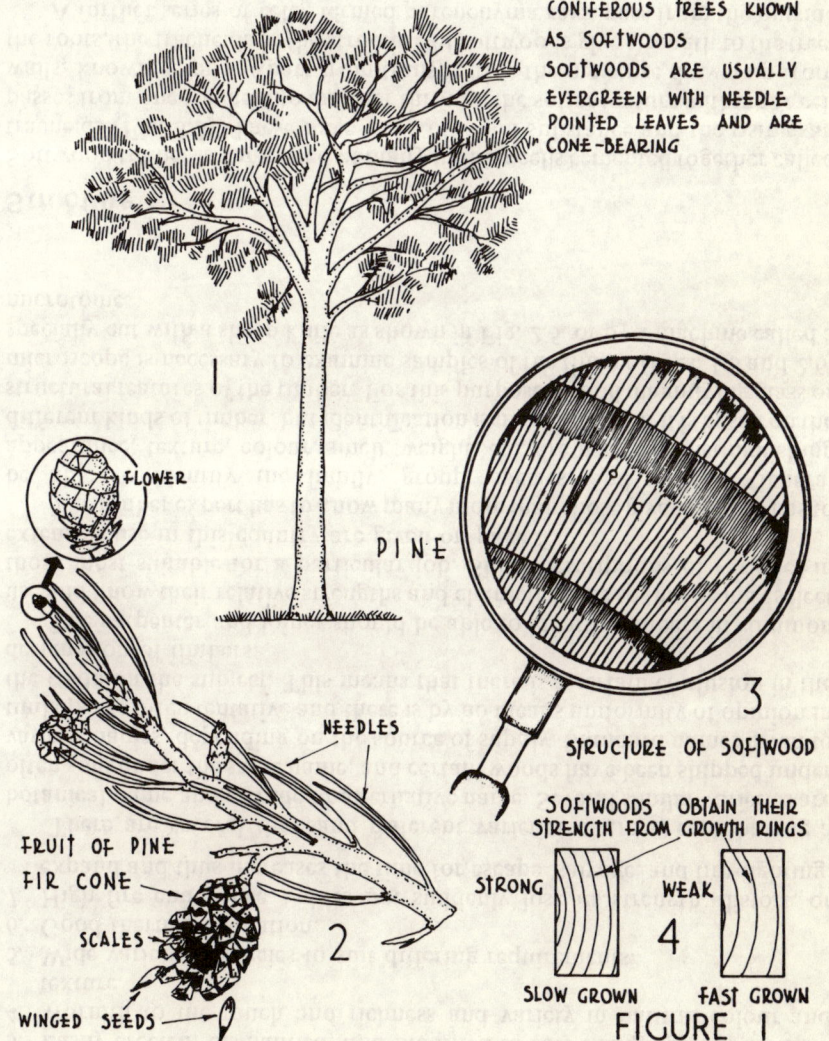

SOFTWOODS

CONIFEROUS TREES KNOWN AS SOFTWOODS
SOFTWOODS ARE USUALLY EVERGREEN WITH NEEDLE POINTED LEAVES AND ARE CONE-BEARING

FLOWER

PINE

NEEDLES

FRUIT OF PINE
FIR CONE
SCALES
WINGED SEEDS

STRUCTURE OF SOFTWOOD

SOFTWOODS OBTAIN THEIR STRENGTH FROM GROWTH RINGS

STRONG   WEAK

SLOW GROWN   FAST GROWN

FIGURE I

H A R D W O O D S

MOST HARDWOODS SHED THEIR LEAVES AT THE END OF THE GROWING SEASON

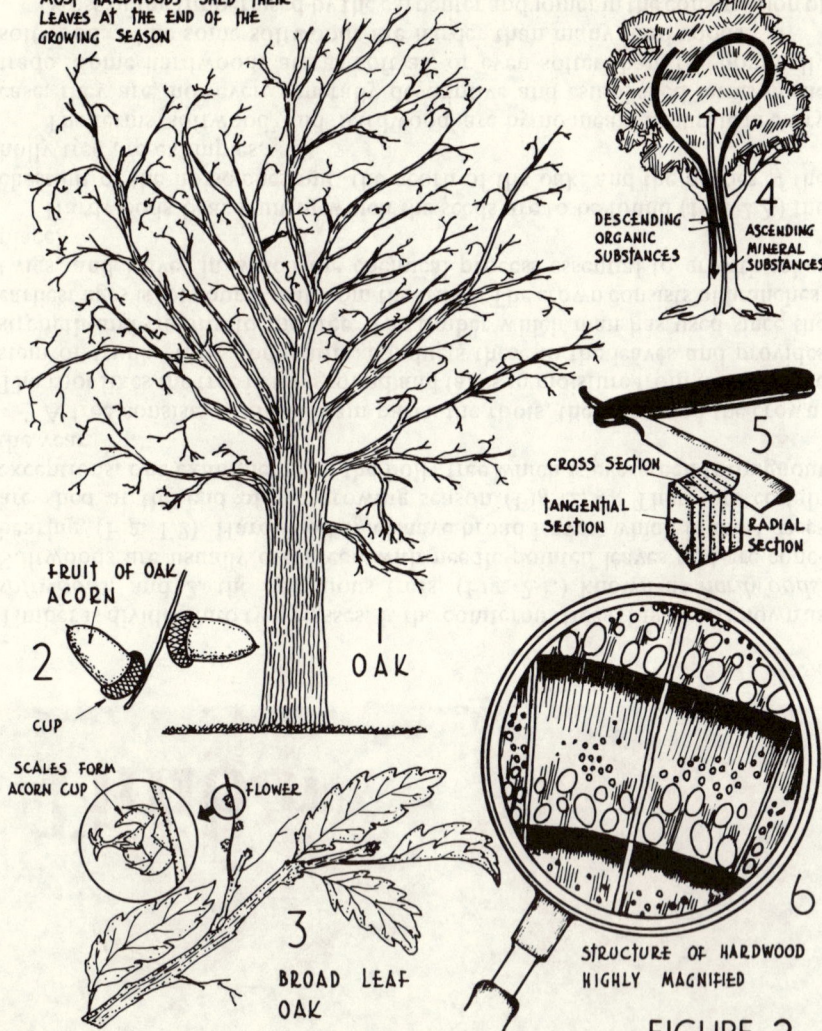

4
DESCENDING ORGANIC SUBSTANCES
ASCENDING MINERAL SUBSTANCES

5
CROSS SECTION
TANGENTIAL SECTION
RADIAL SECTION

1 OAK

2 FRUIT OF OAK ACORN
CUP

SCALES FORM ACORN CUP
FLOWER

3 BROAD LEAF OAK

6 STRUCTURE OF HARDWOOD HIGHLY MAGNIFIED

FIGURE 2

Advantages of wood

Wood has the following advantages:

1. Very high strength compared to its weight.
2. Easily worked and shaped.
3. Easily erected, dismantled, and modified to suit changing conditions.
4. Warmth to the touch and richness and variety in natural colour and texture.
5. Wide variety of species to suit differing requirements.
6. Good thermal insulation.
7. High fire endurance – does not suddenly lose its strength, distort, or expand and thus increases the time for escape, salvage, and fire-fighting.

There are several thousand different varieties of trees, each having a botanical name and a trade or alternative name. Several similar varieties are often sold under the same name, and certain woods have been shipped under various names, depending on the source of supply. Standard names given to timbers are often tentative and there is by no means uniformity of opinion in the trade on the subject. This means that there is a certain confusion in the designation of timbers.

The carpenter and joiner should be able to identify timbers in common use and know their relative strengths and characteristics so that he may select those most suitable for a particular job. Short descriptions of varieties in extensive use in this country are given on page 5.

The timber expert has to know many more than those given, and he has to be able to identify the family, group, species, and variety. General appearance, texture, colour, smell, weight, etc. are useful in distinguishing different kinds of timber, but identification is more reliable if it is based on the structural features of the timber. For this purpose a hand magnifying glass or microscope is necessary to examine samples of the timber (Figs. 1.3 and 2.6) specially cut with a sharp knife as shown in Fig. 2.5, or by a machine called a microtome.

Structure

Softwood timber is composed of many tubular cells cemented together called tracheids (Fig. 3.3). These have walls of wood substance and the rising sap passes from one tracheid to another through the softer texture within the cell walls, known as pits. Apart from transporting the moisture drawn up from the roots, the tracheids in the structure of softwoods give strength to the tree.

A further series of cells, termed parenchyma rays, pass from the outside of the tree towards the pith and are formed of a pithy substance. These cells, which are shorter than the tracheids, store reserves of food which can be passed to any part of the tree which requires them. The rays are often used as a means of identifying timbers.

Resin canals sometimes occur in softwoods. These are placed in a horizontal and vertical direction and are shown in Fig. 3.3.

The structure of hardwoods (Fig 3.2) is more complicated than that of softwoods. The main feature of the structure is the presence of large cells or

S T R U C T U R E

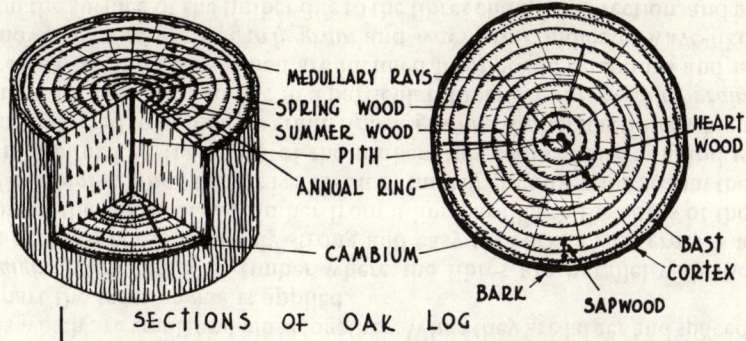

SECTIONS OF OAK LOG

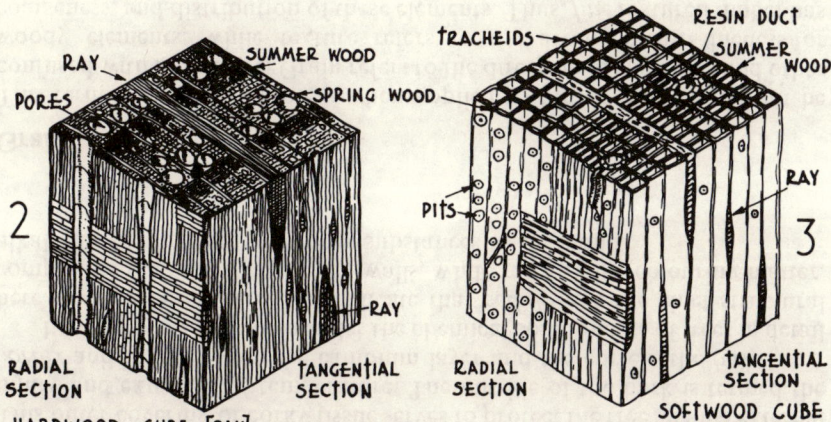

Labels: MEDULLARY RAYS, SPRING WOOD, SUMMER WOOD, PITH, ANNUAL RING, CAMBIUM, HEART WOOD, BAST, CORTEX, BARK, SAPWOOD

2 — **HARDWOOD CUBE [OAK]**

Labels: RAY, PORES, SUMMER WOOD, SPRING WOOD, RAY, RADIAL SECTION, TANGENTIAL SECTION

3 — **SOFTWOOD CUBE**

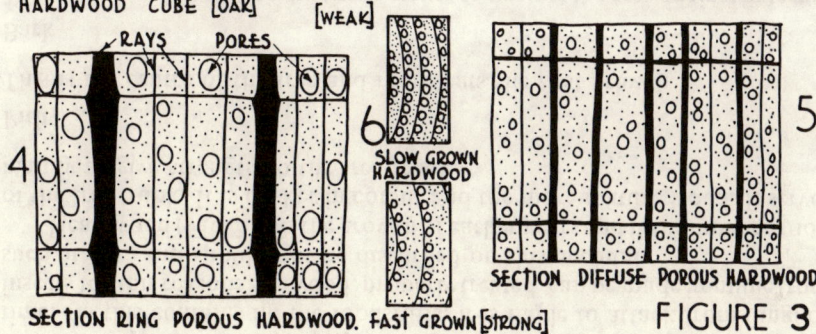

Labels: TRACHEIDS, RESIN DUCT, SUMMER WOOD, PITS, RAY, RADIAL SECTION, TANGENTIAL SECTION

4 — **SECTION RING POROUS HARDWOOD. FAST GROWN [STRONG]**

Labels: RAYS, PORES

6 — **SLOW GROWN HARDWOOD** [WEAK]

5 — **SECTION DIFFUSE POROUS HARDWOOD**

FIGURE 3

---

vessels which pass the moisture up the tree from the roots to the leaves. Along with the large vessels are rays of parenchyma cells and fibres. The latter serve to give strength to the tree.

There are two types of hardwoods: (1) ring-porous (Fig. 3.4); and (2) diffuse-porous (Fig. 3.5). In ring-porous timbers large cells are produced during the early part of the growth ring and these become smaller in size as the season progresses. In diffuse-porous timbers the cells are generally the same size within the growth ring (Fig. 3.5).

Growth

The season when growth takes place in this country is from April to September when new wood is formed in the zone called the *cambium*. This is a very thin layer of cells beneath the *bark* (Fig. 3.1). This layer cuts off the cells on each side of it, those nearest the *pith* become wood and the ones on the outside become bark. The cambium passes the food-stuffs we call sap from the leaves down the trunk of the tree causing growth to take place. The wood produced in the earlier part of the growing season, called *spring-wood*, has cells which are more open and with thinner walls than the later wood called *summer-wood*. This is usually heavier, harder, stronger, and darker in colour than the earlier spring-wood.

The proportion of summer-wood is a measure of the *density* or specific gravity of the wood. The higher the density the greater the strength, and a visual means of selecting timber of superior strength is by noting the percentage of summer-wood.

The principle of tree growth is that moisture absorbed by the roots passes up the trunk through the sapwood to the leaves. Here, sugars are formed from the carbon dioxide of the air and the moisture from the roots, through the action of the sun. These sugars are returned to the stem to act as food for the growing tree, as shown in Fig. 2.4.

Annual or growth rings

These are formed by the early spring-wood and are arranged in roughly concentric formation round the pith, as shown in Fig. 3.1. Each growing season an additional sheath of tissue is produced around the tree, increasing the diameter and pushing the bark outwards.

In softwoods which have been grown slowly, the timber will have more growth rings than one which has been grown quickly, resulting in much stronger timber, (Fig. 1.4). The age of a tree can be determined by counting these rings of annual growth.

There are many more large cells and fewer fibres in a slow-grown ring-porous hardwood. This means that a weaker timber is produced than is the case with hardwood which is fast-grown (Fig. 3.6).

Medullary rays

These exist in all woods. They are seen as lines on the transverse section, radiating from the pith to the bark and running with the grain of the tree. Generally, these rays are not easy to see without the use of a lens or microscope, except in certain hardwoods, particularly oak. It is these rays which give many hardwoods their rich decorative figuring.

Sapwood and heartwood

New wood formed on the outside, next to the bark, is called sapwood (Fig. 3.1). Every part of wood in any tree has, therefore, been sapwood at some time. As this contains all the foodstuffs, it is liable to attack from fungi or insects for this reason. Sapwood, properly treated, can be made immune from such attack, and should not be discarded on this account.

Heartwood (Fig. 3.1) is the growth of earlier years and is the inner portion of the tree trunk. It is darker in colour and the more mature wood. It serves mainly to give strength to the tree trunk.

Pith

This is the centre of the tree and represents the first growth.

Bark

This outer covering of corky tissue serves to protect the tree against external injury and extremes of temperature. The outside of the bark is termed the *cortex* and that between the cambium layer and the cortex, the *bast*.

It is not proposed to consider the chemical components of trees in detail here. It may be sufficient to indicate that *cellulose* is the chief structural component contained in the cell walls, while resins, oils, colouring matter, alkaloids, tannins, etc. are other substances to be found.

Grain

This term is used very loosely when applied to timber and should not be confused with its texture. Grain refers to the direction of the fibres and other woody elements, while texture refers to the arrangement, fineness or coarseness, and distribution of these elements. Thus, *fine* textured timber has elements which are small and close together. When they are larger and spaced wider apart the term *coarse* is applied.

Straight grain refers to timber where the fibres are parallel with the surface; such timber is relatively strong and easy to work. *Cross grain* is a deviation of the fibres of the timber from a line parallel to the edges of the wood. *Diagonal-grained* timber is a result of improper conversion so that the fibres are inclined to the edges of the timber; this reduces strength and is sometimes referred to as oblique grain. *Spiral-grained* timber has fibres which take a more or less spiral course in a particular direction. *Interlocking grain* has fibres partly spiralling which are inclined in opposite directions and is often known as *wild grain*. *Curly grain* and *wavy grain* indicates wave-like stripes on the surface of the timber due to the fibres changing direction, and is valued because of its highly decorative appearance. *Short grain* indicates that the timber may fracture due to fibres lying in a certain direction. *End grain* refers to the section of a cross-cut surface, showing the arrangement of the exposed fibres.

Figure is the pattern on the surface of the timber and is due entirely to the structure of the wood. Straight-grained timber has only a plain figure, whereas wavy or interlocked-grained timber produces a finely marked and

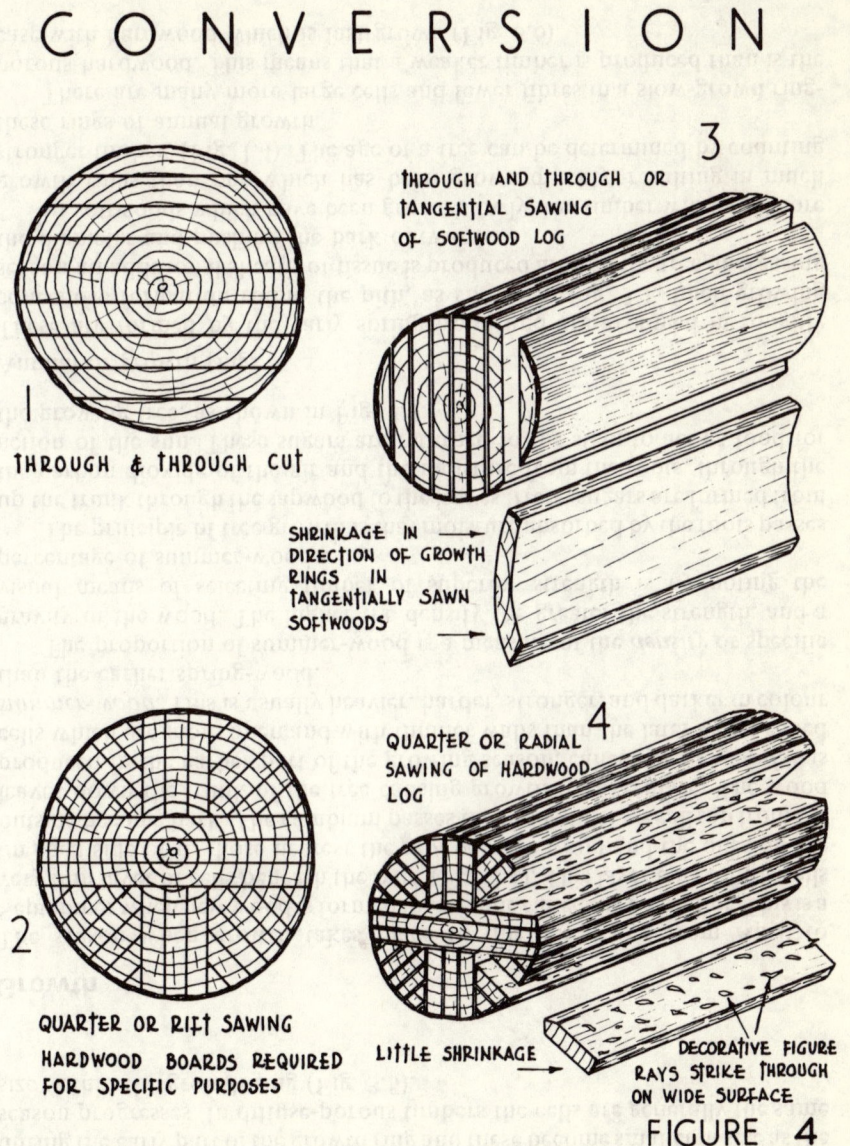

CONVERSION

1 THROUGH & THROUGH CUT

3 THROUGH AND THROUGH OR TANGENTIAL SAWING OF SOFTWOOD LOG

SHRINKAGE IN DIRECTION OF GROWTH RINGS IN TANGENTIALLY SAWN SOFTWOODS

2 QUARTER OR RIFT SAWING HARDWOOD BOARDS REQUIRED FOR SPECIFIC PURPOSES

4 QUARTER OR RADIAL SAWING OF HARDWOOD LOG

LITTLE SHRINKAGE

DECORATIVE FIGURE RAYS STRIKE THROUGH ON WIDE SURFACE

FIGURE 4

attractive figure.

The method of conversion affects the nature of the figure. Quarter sawing in the case of oak used for such purposes as panelling and furniture, where appearance is most important, discloses on the surface the medullary rays which gives the 'silver grain' or rich figure, as shown in Fig. 4.4.

A complete list of commercially used timbers is outside the scope of this book, but the following short descriptions are of varieties in extensive use. The standard name of the timber is given first, followed by alternatives.

Softwoods

Douglas fir (British Columbian pine, Columbian pine, Oregon pine) Average weight 528 kg/m³. Available in long lengths and large sections; straight-grained and resilient; easy to work by hand or machine. Reddish brown to pinkish brown in colour. Being one of the hardest softwoods it can take heavy, continuous wear. The strongest, for its weight, of any softwood in the world, with a high resistance to acids and decay, has good gluing and high insulation qualities. Used for first-class joinery, light and heavy structural work, glued laminated work and timber buildings. Large quantities of plywood are made from Douglas fir.

Hemlock, Western (Pacific hemlock, British Columbian hemlock) Average weight 480 kg/m³. It ranks high in strength and durability and has a fine uniform texture. Straight-grained, stiff, yet easily worked, and light brown in colour. Easy to work by hand or machine and good for gluing. Its smooth clear surface takes stain, paint, and varnish without difficulty. Used for interior joinery work, built-in furniture, agricultural and timber buildings.

Larch, European Average weight 592 kg/m³. One of the most valuable and most used home-grown timbers. Reddish brown in colour, very strong and durable; resinous; straight-grained. The larch grows to a height of 30–48 m or more, with a girth of 4·5 m in some trees. Used for all kinds of carpentry work, fencing, gates, posts, garden furniture, flooring, and railway sleepers.

Parana pine Average weight 544 kg/m³. This South American softwood has an even texture and is straight-grained. It is unsuitable for exterior work, being brittle and not durable. The colour is from light to darkish brown with some reddishness. Suitable for all classes of interior joinery but is inclined to split on nailing. Takes screws, glue, and paint well.

Pitch pine (Gulf Coast pitch pine, longleaf pitch pine) Average weight 672 kg/m³. From southern USA, this wood is highly resinous and provides a large proportion of the world's supply of resin. It is very durable, strong and heavy, being used in ship-building, heavy construction, church, and school work. Light red in colour, it is mild to work but the excessive resin affects the ease of working in that machine saws, cutters, and hand tools, soon become clogged.

Red pine (Canadian pine, Quebec red pine) Average weight 528 kg/m³.

Similar to Baltic redwood and used for similar purposes; internal carpentry and cabinet work. Fairly strong and durable, easy to work but is known for an excessive number of large knots. Colour: light red to a reddish yellow. Takes glue, screws, and nails well.

Redwood (Scots pine, red pine, red deal, red, yellow deal) Average weight 528 kg/m³. This is the most used softwood in this country and is imported in large quantities from the Baltic countries and north Russia; it is also grown in Scotland. Used extensively for all classes of work in building, being very durable, straight-grained and easily worked, although knots may be troublesome. Depending on quality, it is used for interior and exterior work, joists, flooring, beams, windows, rafters, doors, fitments, etc. Colour: pale reddish brown.

Spruce, European (Whitewood, white deal, Baltic whitewood, northern whitewood) Average weight 432 kg/m³. From northern and central Europe and the British Isles. It is less durable than redwood but with similar strength properties. Pale yellow or pinkish white in colour. The timber is easily worked by hand or machine and is used for a large variety of purposes. Flooring, blockboard core, packing cases, internal carpentry, pit-props, and temporary work are some of its uses.

Spruce, Sitka (Silver spruce) Average weight 464 kg/m³. From British Columbia, western USA. Average heights of this variety range from 30 to 45 m, some reaching far beyond this, while the diameters range from 900 mm to 3 m. It is fairly durable and easily worked, being straight-grained with a satiny finish. Strong in proportion to its weight. Used extensively for interior joinery and takes glue, nails, and screws well.

Sugar pine Average weight 448 kg/m³. From Oregon and California, this wood has a similar appearance to yellow pine (see below) and is used as a substitute. Used for general internal joinery.

Western red cedar (British Columbian red cedar, red cedar) Average weight 384 kg/m³. This is the largest of the American cedars and grows to a great size, with the trunk often clear of branches for between 15 m and 18 m from the ground. Large clear boards are thus obtainable. As with all cedars, it contains an aromatic oil which renders the wood free from insect attack. The timber is not suitable for rough usage as its weight implies, but it is easy to work and straight-grained.

The colour varies from a reddish pink to dark brown and when exposed to the weather the colour turns to a silver grey. It is used for general carpentry and joinery, roof shingles, and decorative work including panelling. Complete houses are constructed in red cedar, the outside being left unpainted as there is no deterioration in the wood under the severest climatic conditions. Nails used for outside fixing should be galvanised.

Western white pine (Soft pine, Idaho white pine) Average weight 448 kg/m³. From British Columbia and the north-western USA. Its properties are compared to those of yellow pine (see below), but it is harder and stiffer and more resistant to shock loads. It is straight-grained and easy to work; it neither warps or twists and is thus a suitable material for the pattern-maker and a substitute for yellow pine. Colour: light brown. It takes glue, nails, screws, paints, stains, etc. well.

Yellow pine *(Northern pine, Quebec yellow pine, white pine, soft pine, Weymouth pine)* Average weight 416 kg/m³. Colour: pale straw to light reddish brown. Yellow pine is regarded as Canada's most valuable softwood. It is easy to work by hand or machine being even-grained and soft. This makes it a most suitable timber for the pattern-maker as it is most reliable against warp or twist. Further uses are for interior fittings, panelling doors, drawing-boards, and textile and agricultural work.

Hardwoods

Afrormosia (Kokrodua) Average weight 688 kg/m³. This wood resembles teak in appearance, but has a finer grain. It is very durable and works well. It is suitable for high-class joinery, ship, and carriage work. Care should be taken when used on outside work in direct contact with ironwork to avoid staining.

Agba (Nigerian cedar, pink mahogany) Average weight 480 kg/m³. This West African timber grows up to 60 m in height.

American whitewood *(Canary whitewood, yellow poplar)* Average weight 528 kg/m³. Essentially a wood for interior work to be painted. It takes glue, nails, and screws well and is easy to work. The tree grows to a height of 45–60 m and up to 3 m in diameter. The colour varies according to the age of the wood, between quite yellow and grey. It is not particularly strong and will deteriorate quickly in damp situations. Botanically, whitewood is a hardwood, but in workability it is to all intents and purposes a softwood.

Beech, European Average weight 720 kg/m³. This is one of the most used hardwoods in this country, large quantities being imported from central and southern Europe. The timber is hard, close-grained, and durable, with a fine texture. It is used extensively for furniture, particularly chair-making, wooden planes, handles of the woodworkers' saws and other tools, block and parquet flooring. It shows silver grain and is used for veneers on this account. Colour reddish yellow or light brown.

Birch Average weight 672 kg/m³. From Europe generally, also Canada and other regions of North America. European birch is used principally for plywood. Large quantities from Finland and Sweden are imported into this country. Colour white to light brown. Straight-grained and medium texture. Similar to beech in many ways but is more inclined to warp.

Black bean Average weight 720 kg/m³. Chocolate brown with greyish-brown streaks giving an attractive rich appearance to the wood. Similar to French walnut in colour, hard to work. From New South Wales and Queensland, the timber is excellent for veneers, high-class joinery, panelling, and furniture.

Ebony, African *(Gaboon ebony, Nigerian ebony)* Average weight 1041 kg/m³.

Ebony, Ceylon (Indian ebony) Average weight 1233 kg/m³. The Ceylon and African varieties are similar and will be considered together. Both are jet black in colour with some brown streaking. The restriction in supplies of this timber limits its use to turnery, decorative inlay, certain tool handles, fancy goods, and good-quality tee-square blade edges, and golf clubs. Very hard, close-grained, and slow working by hand and machine.

Elm, British (Common elm, red elm) Average weight 560 kg/m³. Grows up to 45 m in height with diameters between 900 mm and 1·5 m. Used for panelling and interior work. Light brown in colour, tough, and durable under water. Tendency to warp and twist. Used for coffins, dock work, turnery, barge- and boat-building.

Elm, Rock (Canadian rock elm, white elm) Average weight 720 kg/m³. From eastern Canada and USA. Pale brown in colour, is the strongest and finest grained of the elms. Used principally in ship- and boat-building, agricultural work, and wagon and coach work. Rock elm works well and is very durable in wet conditions. It takes nails and screws without splitting and is very tough and elastic.

Greenheart Average weight 1040 kg/m³. From British Guiana, the wood is shipped in long straight lengths up to 24 m, for length is necessary in pile work. The timber is very strong and durable and does not suffer when in contact with iron. Where great strength is required in lock gates, dock and harbour work, bridges and the like, greenheart is unexcelled. It is also highly resistant to acids and is, therefore, used in chemical laboratory work. Its colour ranges from green to dark brown and black with brown streaking. Greenheart is well known for the manufacture of fishing rods.

Gurjun Average weight 720 kg/m³. From India, Ceylon, Burma, Siam, and China. Used for general construction work, bridge decking, wagon-building, and flooring. This dense hardwood is reddish brown in colour with much oil in its composition. It is not easy to work but is straight-grained with a pleasant odour. It is often used as a substitute for teak.

Idigbo (Black Afara, Emeri, Framire) Average weight 560 kg/m³. The timber is pale yellow to light brown in colour and comes from western tropical Africa, French Guiana, the Ivory Coast, Ghana, and Nigeria. For its weight, Idigbo has excellent strength properties, is durable, and fairly resistant to fungi and insects. Being highly resistant to penetration, it is not an easy timber to treat with preservatives. It works well by hand and machine and takes glue, nails, screws, stains and polishes fairly well. It is used for all types of joinery work, panelling, and fitments.

Iroko (African teak, Nigerian teak, tule, Odoum) Average weight 672 kg/m³. Iroko has the appearance (but not the qualities) of teak. It is very hard, strong, very durable, and well figured. The colour varies from pale yellowish brown to dark chocolate. It works fairly well with hand and machine, though with some dulling effect on the cutting edges. It takes nails and screws well and is used for good-quality interior and exterior joinery, doors, windows and laboratory furniture. It is also suitable for dock and wharf work.

Jarrah Average weight 913 kg/m³. This Western Australian timber is easily recognised by its deep rich red colour, its hard and dense grain, and heavy weight; occasionally it is figured. It is hard to work by hand and machine, and is resistant to fire. It is very strong and durable under all conditions, straight-grained, and obtainable in long lengths and good dimensions. It is used extensively on dock and harbour work, sleepers, flooring, and road blocks end-grain up.

Mahogany, African (Axim, Benin, Cape Lopez, Lagos, Grand Bassam mahogany) Average weight 520–720 kg/m³. From West Africa and Uganda. Average heights of this variety range between 30 and 45 m, with a trunk clear of branches up to 27 m. The diameter varies from 900 mm to 1·8 m and considerably more at the base. African mahogany is perhaps the most widely used of the mahoganies, being the cheapest and most readily obtained in all sizes. The timber is more difficult to work than the other varieties of mahoganies owing to interlocked grain, but it is still fairly easy to work with hand and machine. It is reasonably strong for its weight and moderately resistant to decay. It is used for panelling, furniture, veneered work, ships' cabins, and high-class joiners' work. Colour light brown to deep red.

Mahogany, Cuban (Spanish, West Indian, Porto Rico, Jamaica Mahogany) Average weight 786 kg/m³. Deep rich red or brown in colour, this wood possesses a high golden lustre and is often beautifully figured, is hard, and fine to medium textured. It has excellent working qualities and is unsurpassed for the clean finish and polish effects obtained. Used for furniture, high-class joinery, shop fitting, turnery, and ship and yacht work.

Mahogany Honduras (Central American, Baywood) Average weight 544 kg/m³. The colour of Honduras mahogany varies with the country of origin from light reddish to yellowish brown to rich red. Much figure such as blister, curl, fiddle-back, mottle, and stripe is available. Honduras is generally softer and lighter than Cuban mahogany with either straight or interlocked grain. It is used for all types of high-class joinery, pattern-making, ship-building, small boats, and propellors. It is excellent to work by hand and machine, takes glue, nails, screws, and stains well.

Maple Rock (Hard maple, bird's-eye maple, sugar maple) Average weight 720 kg/m³. From Canada, USA, and Europe this fine-grained wood has excellent finishing qualities. Imported maple is almost entirely confined to converted planks, boards, and flooring tongued and grooved both at the ends and sides. The grain is dense giving a fine finished surface. It is used in the manufacture of joinery work, panelling, furniture, veneers, letter blocks in the printing trade, textile rollers, turnery, and billiard cues.

From rock maple is produced the beautiful figured bird's-eye maple. This figure is said to be the result of attack by woodpeckers and other birds on the surface in search of the sugar content of the sap. These injured parts are covered over by a healing growth and the tree develops with this mass of marks and dimples year by year with the resulting figure known as bird's-eye maple. The colour is light yellowish brown.

Oak, English Average weight 720 kg/m³. English oak is perhaps the principal timber of this country and for strength and durability no other oak is comparable; the colour is a light yellowish brown with areas of darker brown. Strength covers hardness, toughness, stiffness, elasticity, and resistance to bending, shearing, compression, etc. These qualities in English oak have been evident over the centuries, whether for roof-trusses, ship-building, wagon-building, and almost every conceivable type of woodwork. It is susceptible to fungi attack. Ironwork should not be placed in contact with oak which is only partially seasoned otherwise disfiguration in the form of bluish stains will result. The reason for this chemical contamination is a result of the tannin content in the timber combining with the iron.

Oak, Slavonian (Austrian, Hungarian, Polish oak) Average weight 670 kg/m³. Perhaps the mildest and most easily worked of all the oaks. It is straighter grained than English, with a bold silver flash and uniform colour. It is sometimes described as 'wainscot oak', being cut for figure and used for high-class joinery.

Oak, American white Average weight 640 kg/m³. From eastern Canada and USA; the wood varies in colour from pale yellow brown to pale brown. It is easier to work than European oak and is used for cabinet work, fittings, panelling, and ship-building. Much furniture in this country, finished in light oak, is surfaced in selected American oak veneers. It is also used in drawer sides, rails, backs, etc. in furniture construction.

Oak, Japanese Average weight 670 kg/m³. From Japan, this is a much milder wood than European. It has a fine and attractive rich grain and flash obtained by careful sawing up of the log. It is especially suitable for panelling, bank work, church work, and shop fittings.

Obeche (Nigerian whitewood, whitewood, arere, African whitewood) Average weight 320–384 kg/m³. This is a soft light wood with a rather coarse texture. Its colour is near-white to pale straw. The straight-grained variety is easy to work but spongy when cutting across the grain as in dovetailing. It is used in fittings and furniture for drawer sides, backs, bottoms, partitions. When stained the grain resembles mahogany, but as it is an open-pored wood, special preparation using wood-filler paste is necessary before polishing.

Sapele (Sapele mahogany, scented mahogany) Average weight 640 kg/m³. From both East and West Africa, this close-textured wood is similar to African mahogany, although there is much variation in the grain. It is dark reddish brown in colour, marked by a regular stripe, and almost as hard as oak. It works fairly well with hand and machine and takes nails and screws well. It is used extensively in the furniture trade for cabinet and veneer work, shop fitting, and fittings where it polishes well.

Sycamore (Plane, great maple) Average weight 624 kg/m³. From Europe and the British Isles the timber is almost white with a slight tinge of yellow. It is close-grained, even and fine, with a lustrous silky finish. It is used extensively in the textile industry for rollers, for dairy utensils, brush handles, general turnery, and in building for wall panelling, fitments, and flooring. The highly figured veneers of sycamore used in bedroom furniture and the like are often dyed *greywood*, or harewood so fashionable these days. It is comparable to oak for strength but is more resistant to splitting as it is brittle. Because of its even grain, it is a useful carver's wood.

Teak Average weight 720 kg/m³. From India and Burma, this tree grows to heights exceeding 39 m, and up to 2·4 m in diameter. It is a very strong and durable timber under all conditions and when seasoned, shrinkage is slight and it does not warp or twist. It is a golden brown in colour with a surface which is greasy to the touch. It is this chemical which helps towards its exceptional qualities. It offers a great resistance to insect attack and is fire-resistant. Although the timber is mild, it is one of the most difficult to work, tools soon become dull. Teak is inclined to splinter, and these can sometimes be a source of danger causing blood-poisoning. It is used for high-class joinery, external doors and windows, chemical laboratory furniture, drainer boards, and ship-building.

Utile Average weight 640 kg/m³. From both East and West Africa, this close-textured wood is closely related to sapele. In general the timber works quite well with only a slight blunting effect on the cutting edges of tools. Utile is a good timber for steam bending purposes and for turnery. Glues adhere well and the nail- and screw-holding properties are satisfactory. Chiefly used for furniture, interior joinery, cabinet work, and flooring, but is also used in plywood manufacture and provides a certain amount of decorative veneer. Sapwood is palish brown in colour, while the heartwood is reddish to a purple-brown.

Conversion

The sawing of logs, or breaking down into various shaped and sized pieces for specific purposes, is known as conversion. There are very few firms which take the logs in the round state. Timber is purchased mostly cut through and through, quartered or squared and carried out at the place of shipment in the case of softwoods. Purchasing timber in this way does mean that the buyer is able to select his material with the finished product in mind, noting the quality, texture, grain, colour, etc. thus avoiding having to handle unsuitable materials often occupying valuable yard or floor space.

The types of saws used for conversion will include one or more of the following: band re-saw; roller feed sawbench; crosscut; straight-line edger, conversion being carried out with the minimum of waste.

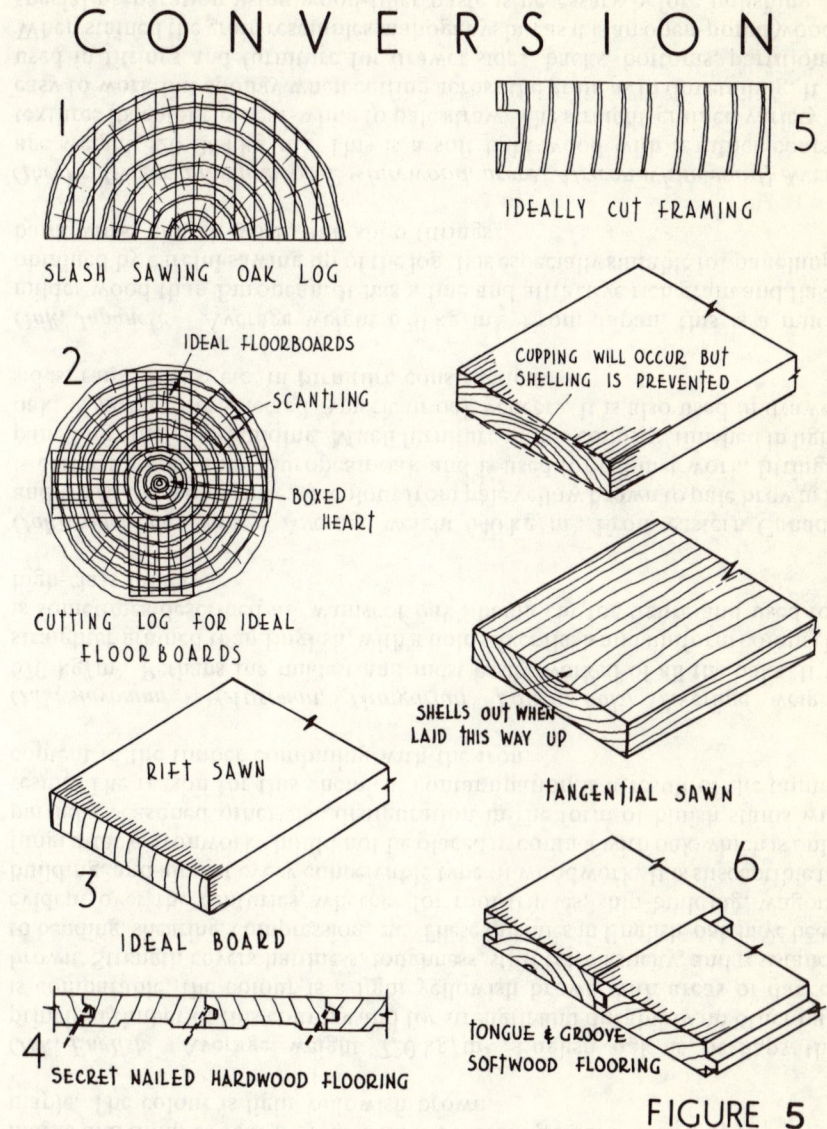

CONVERSION

1 SLASH SAWING OAK LOG

2 CUTTING LOG FOR IDEAL FLOORBOARDS

IDEAL FLOORBOARDS
SCANTLING
BOXED HEART

3 RIFT SAWN
IDEAL BOARD

4 SECRET NAILED HARDWOOD FLOORING

5 IDEALLY CUT FRAMING

CUPPING WILL OCCUR BUT SHELLING IS PREVENTED

SHELLS OUT WHEN LAID THIS WAY UP
TANGENTIAL SAWN

6 TONGUE & GROOVE SOFTWOOD FLOORING

FIGURE 5

CONVERSION

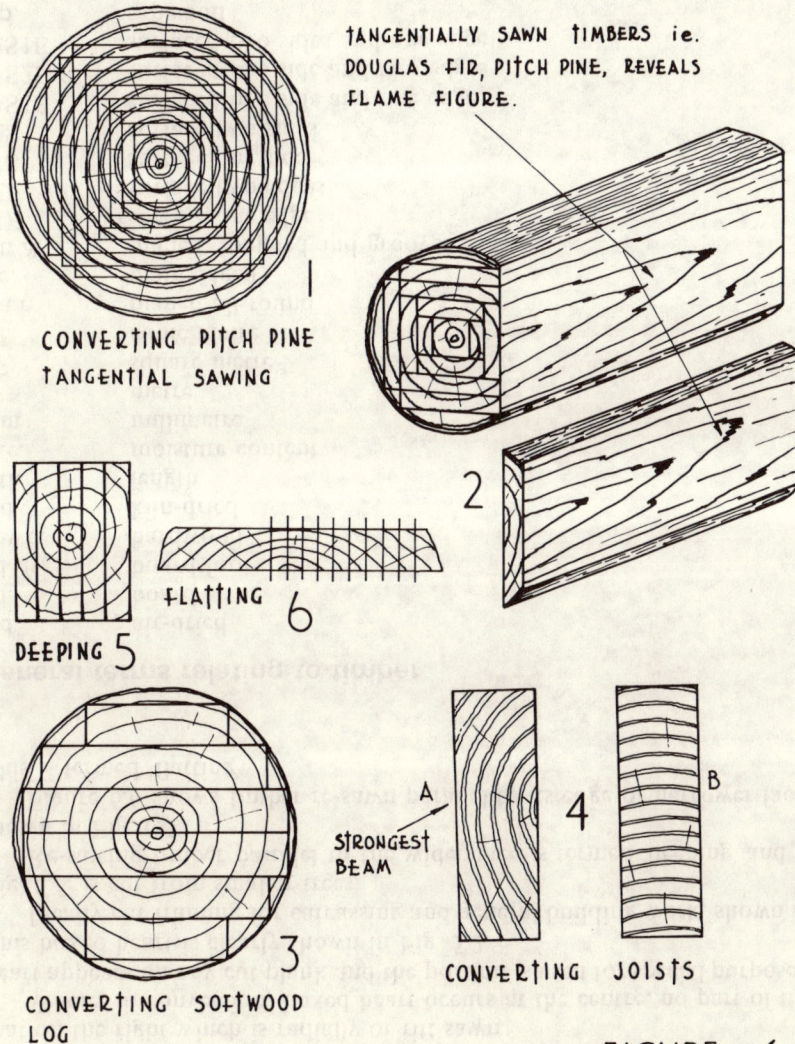

CONVERTING PITCH PINE
TANGENTIAL SAWING

TANGENTIALLY SAWN TIMBERS i.e.
DOUGLAS FIR, PITCH PINE, REVEALS
FLAME FIGURE.

DEEPING 5

FLATTING 6

CONVERTING SOFTWOOD LOG

STRONGEST BEAM

CONVERTING JOISTS

FIGURE 6

There are several ways of converting a log, for example: (1) through and through or tangential sawing shown in Fig. 4.1; (2) quarter or rift sawing shown in Fig. 4.2.

The terms 'radial' and 'tangential' refer to the surfaces secured by the cut of the saw in relation to the growth rings of the tree.

Quarter or rift sawing

Figure 4.2 shows a hardwood log quartered or rift sawn, producing boards which are required for specific purposes. In quarter sawing the growth rings meet the face of the board at an angle of not less than 45 degrees. In timbers having clearly defined medullary rays, as in oak, this method of conversion produces silver-grain, which is highly valued in good-class joinery, as the medullary rays are exposed on the face of the boards as shown in Fig. 4.4.

Quarter-sawn boards shrink less than flat-sawn material and generally give more even wear, as, for example, in flooring. To convert an entire log by quarter sawing is more expensive than the trough and through method, as there is much more waste. This accounts for the fact that figured oak, sawn as in Fig. 4.2, is more expensive than plain oak sawn as in Fig. 5.1.

In this method, the log is halved instead of quartering it, and then cut into boards. The outside boards are plain with the boards towards the centre being of better quality, and containing increasing amounts of figure.

Floorboards and thin boards should be rift or radial sawn as shown in Fig. 5.3. Although this method is expensive, boards cut in this way have better wearing qualities and shrink less. For special floors, i.e. dance floors, boards are laid and secret nailed as shown in Fig. 5.4. It will be seen that the nails are driven through the splayed top edge of the tongue. This method of nailing maintains an unblemished surface to the floor. Although the specially shaped tongue is more wasteful in machining as against the normal square tongue in softwood flooring (Fig. 5.6) the boards are brought together easily without the need for cramping, and also there is less danger of splitting the tongue or damaging the board edge when nailing. Nails are usually driven into boards at an angle to assist in bringing the board joints together.

Tangential sawing

This method is shown in Fig. 6.1, and is adopted with timbers such as pitch-pine and Douglas fir having clearly defined growth rings. As the boards have their faces tangential to the growth rings the grain shows up to the best advantage (Fig. 6.2). When converting softwoods for floor joists and the like, where strength is most important, the lengths of timber must be tangentially sawn.

Through and through sawing

This method, shown in Fig. 4.1, is straightforward cutting of the log into boards of any required thickness, without regard for showing particular grain and with a minimum of waste. This plain sawing is usually the cheapest form of conversion. In this method most of the boards will be flat sawn which means that the growth rings meet the face of the board in any part of an angle of less than 45 degrees.

The method of conversion has a bearing upon the resultant figure or pattern obtainable. Flat-sawn timber from softwoods has a more decorative appearance than quarter or rift sawn, but this does not apply to hardwoods.

Softwoods, generally, and especially inferior timbers, are converted to obtain the maximum amount of timber. Ordinary planks are cut through and through as shown in Fig. 6.3.

The method of cutting joists to obtain the maximum strength is shown in Fig. 6.4. The section on the left, tangential sawn, will give a stronger joist than that on the right which is radially or rift sawn.

Where, in conversion, boxed heart occurs at the centre, no part of the heart appears on any cut plank and the portion is used for special purposes. This boxed heart is clearly shown in Fig. 5.2.

Ideally cut framing for carcassing and general building work, shown in Fig. 5.5, is cut from smaller trees.

Re-sawing timber parallel to the wider face is termed 'deeping' and is shown in Fig. 6.5.

Figure 6.6 shows timber re-sawn parallel to its edge or narrower face. This is termed 'flatting'.

General terms relating to timber

a.d.	air-dried
bd.	board
bd. ft.	board foot
hdwd.	hardwood
k.d.	kiln-dried
lgth.	length
m.c.	moisture content
mm	millimetre
m	metre
m²	square metre
m³	cubic metre
p.a.r.	planed all round
p.e.	plain edged
p.t. & g.	planed, tongued and grooved
S1E	surface one edge
S2E	surface two edges
S1S	surface one side
S2S	surface two sides
S1S1E	surfaced one side and one edge
S1S2E	surfaced one side and two edges
S2S1E	surfaced two sides and one edge
sap.	sapwood
s.e.	square edged
sftwd.	softwood
sqr.	a square (approx 9 m²)
sup.ft	superficial foot
t.&g.	tongued and grooved
t.g.b.	tongued, grooved, and beaded
t.g.v.	tongued, grooved, and V-jointed
u/s	unsorted

Sizes of timber

The woodworking industry in this country receives its timber in various forms: (1) logs and round trees ready for conversion; (2) square-edged boards; (3) sawn unedged boards; and (4) imported boards, deals, battens, etc.

The sizes of timber are limited to the dimensions of the trees from which they have been converted and the carpenter and joiner should familiarise himself with the everyday names applied to cut timber. These are illustrated in Fig. 7.

At the present time, imported sawn hardwoods are being used to a greater extent in building, replacing the shortage of suitable softwoods, although softwood imports account for the majority of the timbers imported in this way. The main sources of supply of softwoods are the European countries such as Sweden, Finland, and the USSR. Smaller quantities come from other countries. As the trees in these European countries are smaller than those in western Canada, the sawn timber is, therefore, of smaller dimensions.

Of the species of imported sawn softwoods, redwood and whitewood are the two species mainly used in building. These are widely used; the lower grades for carcassing and the better for joinery.

The softwood species from western Canada are of large dimensions and include Douglas fir, western hemlock, western red cedar, sitka spruce, and western white pine.

Spruce is the timber mainly imported from eastern Canada and is very like European whitewood in appearance and properties.

Pitch pine is imported from southern USA and from Belize, while Parana pine is imported from Brazil.

Hardwood sizes are dependent upon the tree from which the material is to be cut. Sawn hardwoods are normally imported in random widths. This is the direct opposite to the position with softwoods where it is common to have a whole parcel of boards of one width. Therefore, large quantities of sawn hardwoods in one width may be difficult to obtain and likely to be expensive in widths over 150 mm.

West Africa is one of the few areas from which logs are usually converted in Britain. Much of this conversion is through-and-through cutting (see page 9), although quarter sawing can be undertaken if logs are large enough.

Hardwoods from Malaya are imported as sawn stock; the principal timbers concerned are keruing and red meranti.

From Japan, oak is the principal timber imported, with some qualities of such species as elm, maple, and beech.

European countries provide sawn oak and beech.

Hardwoods from Canada and the USA comprise: American oak, Canadian yellow birch, and rock maple.

All home-grown timber is, of course, available in log form in this country; much of it is, however, cut into 25 mm and thicker boards, and the sizes vary considerably with species. The trade also tends to be somewhat specialised; one firm dealing mainly in timber for fencing, gates, etc. another in wood for pit-props, a third in oak for church work, etc.

Imported sawn timber usually has shipping marks either on the end, in the case of softwood, or on the face and ends on imported hardwoods. These marks show the country of origin or the district or mill in which the timber was produced, and also the quality. As there are many hundreds of such marks grouped to identify the quality, a given grade for one country may not be equal in quality to the same grade from another.

The grading is usually done by taking into consideration the lengths, number, and size of allowable knots and their spacing, spacing of growth rings, waney edge, and other defects. In connection with softwood imports, which are received in a prepared state for use in the manufacture of joinery and in carpenter's work, abbreviations are used. The general terms relating to timber are listed on page 10.

Measurement of timber

Softwood timber is bought and sold in one of four ways:

1. Per cubic metre (m³).

The majority of softwoods are sold and marketed at a price per cubic metre. This form of measurement replaces the *standard*. (The standard is equal to 4·672 m³.)

2. Per square metre (m²).

This is simply length multiplied by width. All planed interlocking boards, i.e. flooring and matchboarding, should be charged on a net surface measure (laid) basis. This means that the buyer will be charged for the visible face though obviously there must be, in any costing, a reservation of 10 mm for the invisible tongue. (The square is approximately 9 m².)

3. Per lineal metre (m).

Small quantities sold from the timber yard are bought in this way. This may cover narrow tongued and grooved boards, door and window stuff, and ready machined sections, mouldings, etc.

4. By cross-sections in millimetres (mm).

This is simply width by thickness, i.e. 150 mm × 50 mm. Hardwoods are generally sold in quantities containing a specific number of cubic metres (m³) – that is a cube measuring 1·000 m × 1·000 m × 1·000 m.

Density of various timbers

Earlier the characteristics of timber in determining its use and identification were discussed. Perhaps the most important factor concerned with the strength of timber is its weight or density. The weight of timber is expressed in terms of a standard volume (kg/m³) and this figure is called the density.

Density can be measured by finding the weight of one unit of volume and may be expressed as:

$$\text{Density} = \frac{\text{weight}}{\text{volume}}$$

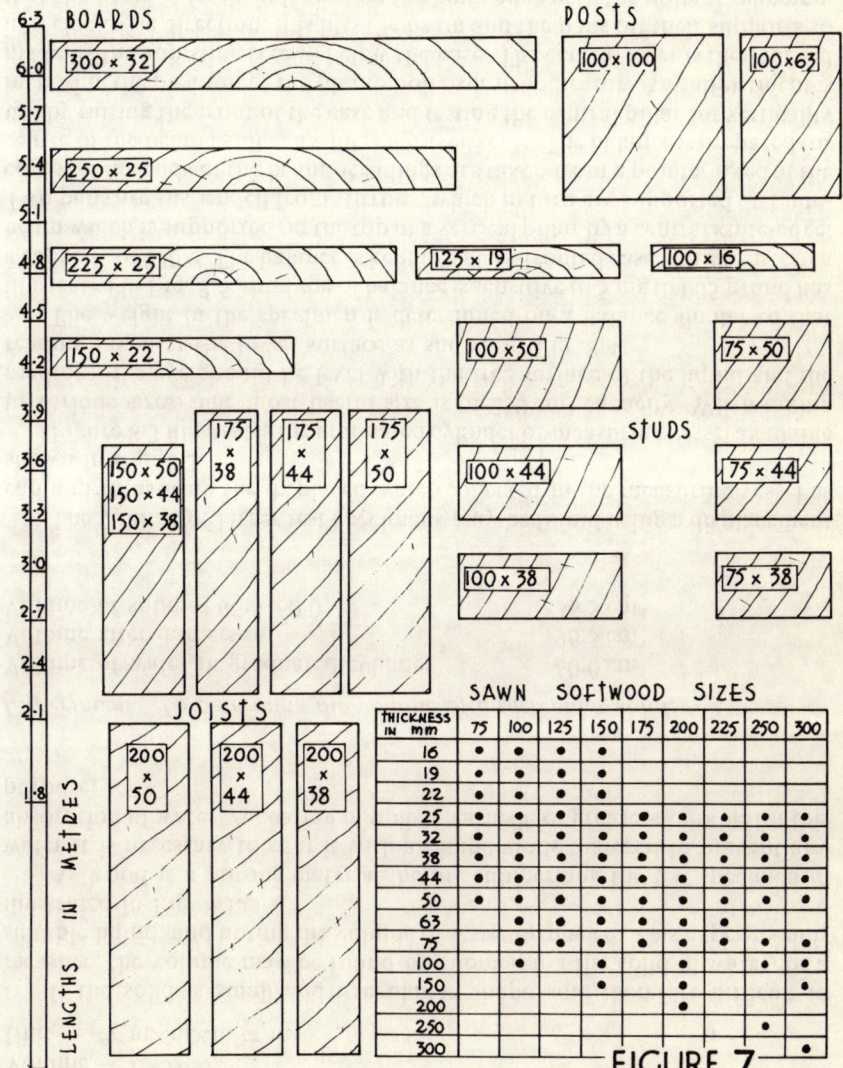

FIGURE 7

PHYSICAL MEASUREMENT

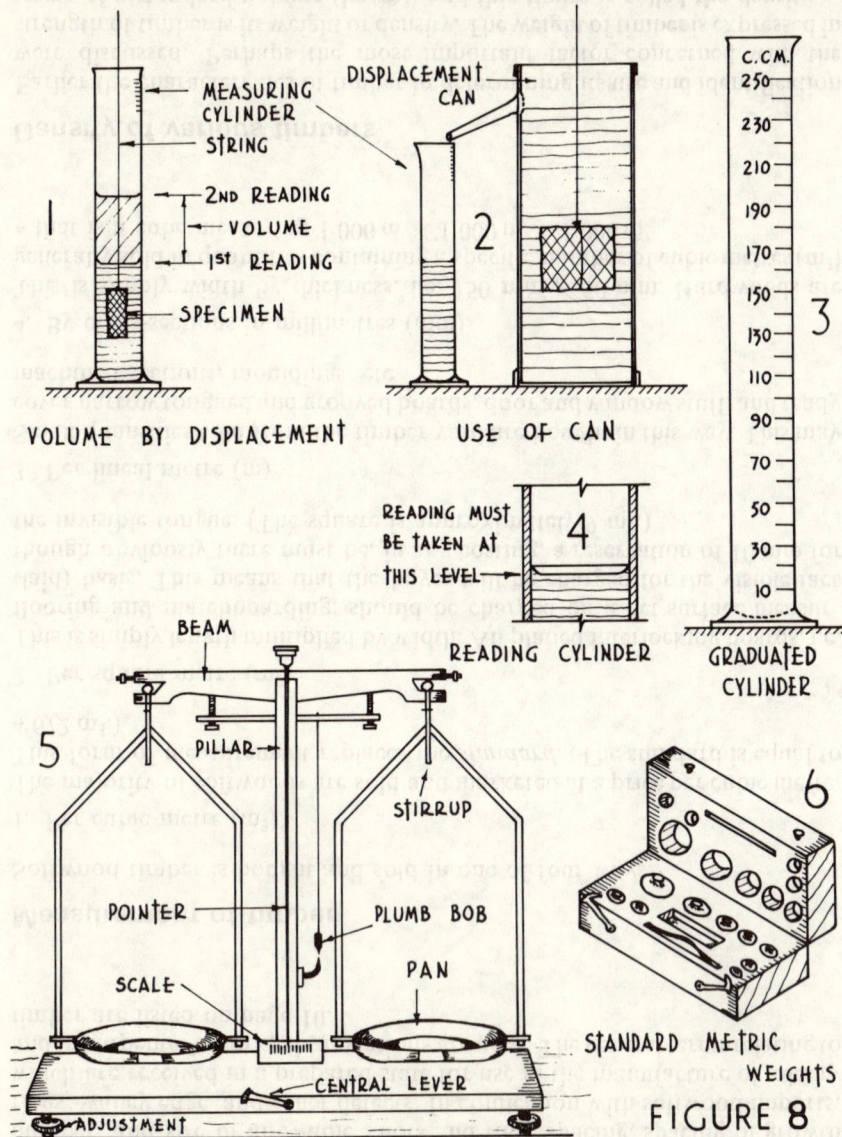

1 VOLUME BY DISPLACEMENT

- MEASURING CYLINDER
- STRING
- 2ND READING
- VOLUME
- 1st READING
- SPECIMEN

2 USE OF CAN

- DISPLACEMENT CAN

3 GRADUATED CYLINDER

C.CM.
250
230
210
190
170
150
130
110
90
70
50
30
10

4 READING CYLINDER

READING MUST BE TAKEN AT THIS LEVEL

5

- BEAM
- PILLAR
- STIRRUP
- POINTER
- PLUMB BOB
- PAN
- SCALE
- CENTRAL LEVER
- ADJUSTMENT

6 STANDARD METRIC WEIGHTS

FIGURE 8

The weight is determined on a balance and the volume by one of several ways. The simplest is by calculation, taking the length (*L*), breadth (*B*), and thickness (*T*) (or height) and from these data calculating the volume of the solid:

Volume $= L \times B \times T$
Unit $= m \times m \times m = m^3$

If the solid is small and irregular in shape, and therefore difficult to measure, the volume may be found by submerging the solid in water, or a suitable liquid, and noting the volume of water displaced. This experiment is illustrated in Fig. 8.1.

As wood is a porous material, before submerging the test specimen in water it is necessary to coat it with an impervious material to prevent any absorption of water. A coating of polish, varnish, or grease is suitable for this purpose.

Experiment   *To determine the volume of an irregular solid:*

Volume of water in graduated cylinder	$60 \cdot 0$ cm³
Volume after immersion	$96 \cdot 5$ cm³
Volume of solid $= 96 \cdot 5 - 60 \cdot 0$	$36 \cdot 5$ cm³

The volume of larger test specimens may be found using a displacement can and measuring the displaced water collected in the measuring vessel as shown in Fig. 8.2.

Figure 8.3 illustrates the graduated cylinder or measuring vessel available in various sizes. The most useful size is of 250 cm³ capacity. When taking readings, the eye should be level with the free surface of the liquid and the reading taken at the lower surface as shown in Fig. 8.4.

The weight of the specimen is determined on a balance similar to that illustrated in Fig. 8.5; this type of balance is sensitive to 5 mg (0·005 g) and has a capacity of 250 g. The balance is kept in a glass-fronted case and consists of a beam which is supported on the top of a vertical pillar by a central knife-edge. Two pans are suspended from stirrups, which in turn are supported on knife-edges, at the ends of the beam. Readings are taken from a pointer fixed at the centre of the beam against a white engraved scale. The balance is prepared for use by raising the front of the case and testing the central pillar for verticality by noting the position of the plumb bob over its indicator. Any adjustment is made by the adjusting screws below the base. The central lever is then turned in a clockwise direction, this lifts the beam and the pans off their supports so that the beam is freely balanced on the knife-edge with the pointer swinging. The balance is correctly adjusted when the pointer comes to rest in front of the central division on the scale. The balance is then ready for use.

The weights used with the balance are shown in Fig. 8.6 and are standard metric weights supplied in a box with a pair of tweezers for handling the weights. These should always be placed in the right-hand pan with the specimen to be weighed in the left-hand pan. A weight estimated to be too large is placed in the pan. The beam is then raised slightly and the movement of the pointer noted. This will show whether the weight is too much or too little. If it is too large the next smaller weight is selected from the box and the procedure repeated until balance is achieved.

Specific gravity

The specific gravity of a substance is merely the relative density of that substance in comparison with a standard density.

The standard substance used for comparing densities of woods or any solid is water. The weight of 1 cm³ of water is 1 g. Therefore, provided the weight of any given volume of water is known, the density of the same volume of all other substances can be calculated from their specific gravities.

1 m³ of water weighs 1001 kg

Density in kg/m³ = s.g.×1001

The specific gravity may be found from its density when known.

Example

A solid whose density is 2·4 g/cm³ has a s.g. of 2·4.

$$\text{Density} = \text{s.g.} \times 1001$$
$$= 2402 \text{ kg/m}^3$$
$$\text{s.g.} = \frac{\text{density}}{1001} = 2 \cdot 4$$

The specific gravity, or density, of the cell walls and cavities in perfectly dry wood is approximately $1\frac{1}{2}$ in all timbers. This means that the cell walls are approximately $1\frac{1}{2}$ times as heavy as water. There is great variation in the weight in kilograms per cubic metre of different timbers as a result of the differences in ratio of cell wall to air space. The composition of these cell walls and their thickness are important factors in relation to density and strength.

The weights of various timbers are classified earlier in this chapter to which reference should be made.

Chapter 2

Timber seasoning and storage

The term 'seasoning' refers to the drying out of a certain amount of free water and moisture contained in the cell cavities and cell walls of green timber. Before the converted timber is handled by the carpenter and joiner it is dried or *seasoned*. This is a slow process and is important for several reasons. The presence of water affects the cost and workability of the timber and the amount determines the strength and weight; this in turn affects the cost of handling and transportation. The durability of the wood, by controlling the activities of wood-destroying growths, may be influenced by the amount of moisture. Timber with a low percentage of moisture is immune from any attack by fungus growths or insects, and to ensure this condition the moisture should not exceed 20 per cent of the dry weight of the timber. The method of estimating the amount of moisture is dealt with on page 21. It is important for the student, in understanding seasoning, to realise that the drying out of timber by evaporation takes place only through the surface. When the moisture close to the surface evaporates in drying, moisture from the centre takes its place. As a result of this, the centre of the wood is liable to be damper than at the surface and any tests made with a moisture meter may be misleading. It is, however, most important that the drying out of the surface is not done too rapidly compared with the damper centre, as case-hardening may be set up (see page 17).

There are two methods of removing the excess moisture from wood, namely: (1) air seasoning; and (2) kiln seasoning. A further phrase sometimes used in this connection is the term 'shipping dry'. This is meant to define timber which has been only partially dried or seasoned to prevent mould and stain (see page 43) during shipment.

Air seasoning

This is done by piling the converted timber into stacks, separating the boards by using skids or stickers, so that the moisture is evaporated by the free

TIMBER PILING

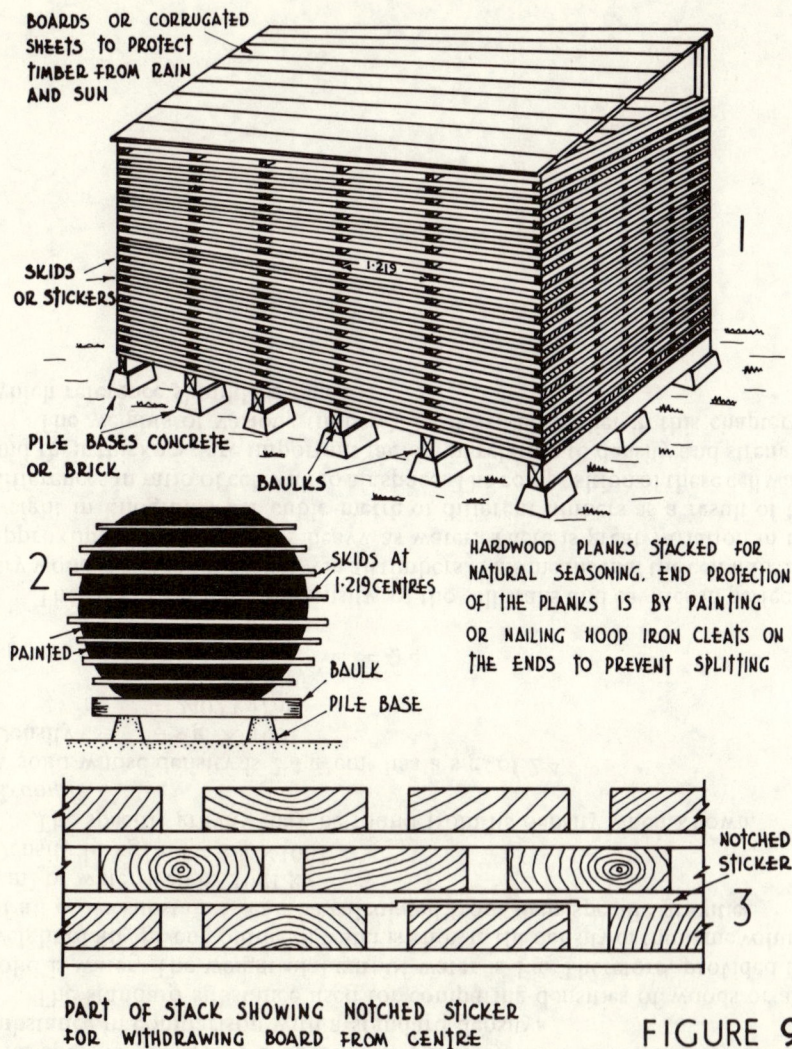

BOARDS OR CORRUGATED SHEETS TO PROTECT TIMBER FROM RAIN AND SUN

SKIDS OR STICKERS

1·219

1

PILE BASES CONCRETE OR BRICK

BAULKS

SKIDS or STICKERS

2

SKIDS AT 1·219 CENTRES

PAINTED

BAULK

PILE BASE

HARDWOOD PLANKS STACKED FOR NATURAL SEASONING. END PROTECTION OF THE PLANKS IS BY PAINTING OR NAILING HOOP IRON CLEATS ON THE ENDS TO PREVENT SPLITTING

NOTCHED STICKER

3

PART OF STACK SHOWING NOTCHED STICKER FOR WITHDRAWING BOARD FROM CENTRE

FIGURE 9

circulation of air. A good dry site with a firm foundation is necessary. This site or yard should be in an accessible position so that transport for delivery and handling may be used. Some undercover space for storing close-piled timber which is sufficiently dry for use is also necessary. Timber which will not require any further seasoning will also require under-cover space.

Boards are stacked as shown in Fig. 9.1. Hardwood boards are stacked in the same order as they are cut from the log with skids placed vertically one over the other to prevent any distortion to the boards due to the weight of the timber above.

Any timber stacked outside should be sheltered from the prevailing winds and also protected from the sun by a covering of some sort – boards or corrugated iron sheets may be used.

Pile bases or foundations are often used in drying yards, sited with consideration of the prevailing winds and the method of handling in the yard. These piers should be high enough to give a good air circulation and constructed to support the heaviest piles. Figure 9.2 shows sawn logs stacked for air-drying supported on concrete piers.

Usually softwoods which dry quickly are stacked in the spring and hardwoods in the winter when the humidity of the air is high to allow slow drying at first. Drying in this way varies according to timber sizes and species and is a slow process. Softwood 25 mm thick should be reduced to about 30 per cent moisture content in about 12 weeks in spring. In good summer weather in this country, the moisture content of air-dried timber may fall to about 15 per cent with the average taken at nearly 20 per cent.

Hardwoods, 25 mm thick, under similar conditions should take up to 8 or 9 months.

The moisture content chart reproduced by permission of the Controller, Her Majesty's Stationery Office and shown in Fig. 15 gives the moisture content of timber for various purposes, giving average values. The ideal value of this content varies as to the type of timber, its situation, and environment.

Because of the slow speed of drying, particularly during the winter, it is not possible in this country to dry timber outside sufficiently for use in artificially heated buildings.

Kiln seasoning

A kiln is a drying room with heating pipes arranged in the floor and ceiling, fans to circulate the hot air, and jets to allow the introduction of steam. A section through a reversible air circulation kiln is shown in Fig. 10.1.

Most kilns in this country are of the compartment type in which the load of timber remains in the kiln throughout the drying and the air conditions are regulated in accordance with a suitable schedule. Kiln-dried timber will normally be of a lower moisture content than air-dried material and may be between 6 and 15 per cent, depending on the intended use and final position within the building.

Kiln drying implies the use of temperatures in excess of atmospheric temperatures in winter and summer, and the most important feature in any drying kiln design is air circulation. It is a simple matter to provide air in a kiln at a predetermined temperature and humidity, the problem is rapidly to bring

K I L N S

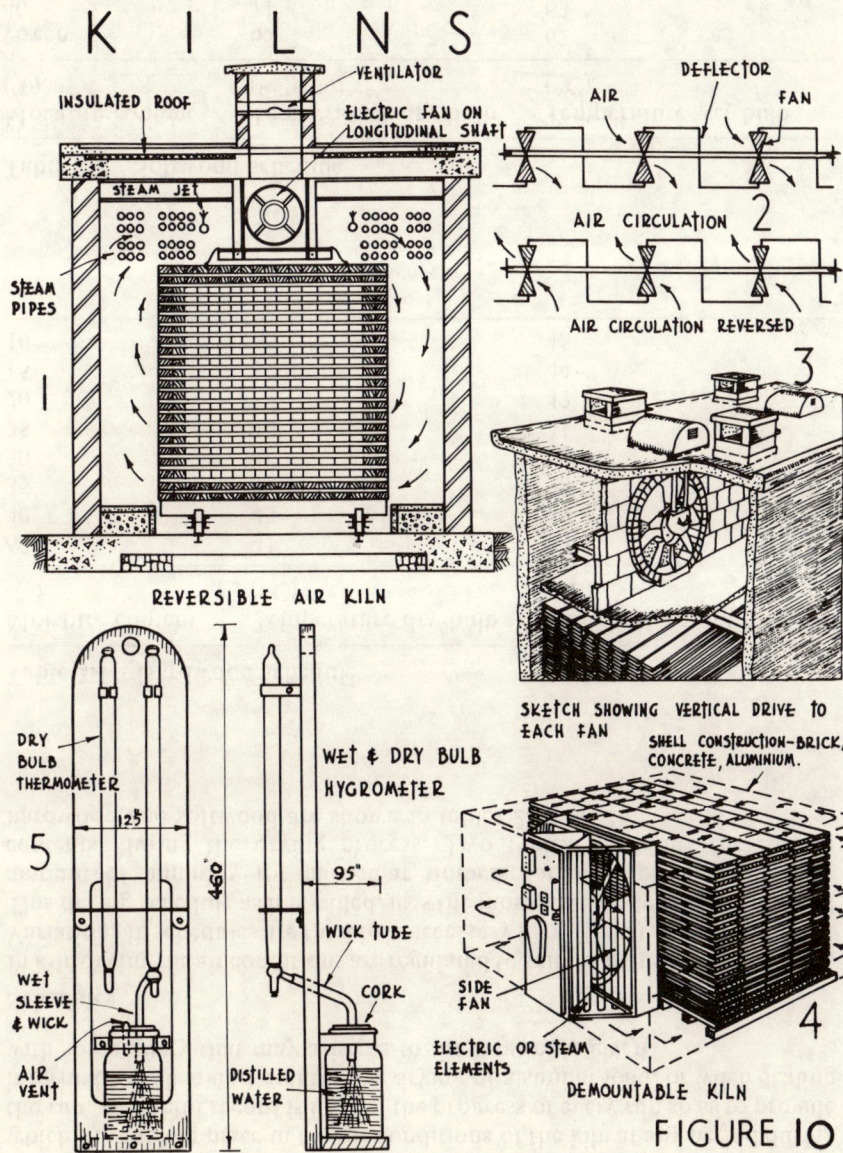

INSULATED ROOF
VENTILATOR
ELECTRIC FAN ON LONGITUDINAL SHAFT
STEAM JET
STEAM PIPES

1

REVERSIBLE AIR KILN

AIR
DEFLECTOR
FAN

AIR CIRCULATION 2

AIR CIRCULATION REVERSED

3

SKETCH SHOWING VERTICAL DRIVE TO EACH FAN

DRY BULB THERMOMETER

WET & DRY BULB HYGROMETER

125

5

460"

95"

WICK TUBE

WET SLEEVE & WICK

CORK

AIR VENT

DISTILLED WATER

SHELL CONSTRUCTION-BRICK, CONCRETE, ALUMINIUM.

SIDE FAN

ELECTRIC OR STEAM ELEMENTS

4

DEMOUNTABLE KILN

FIGURE 10

this conditioned air into close contact with both sides of every board in the stack, including those in the centre of the stack. The fast circulating air ensures a plentiful supply of heat for evaporation and quick removal of the evaporated moisture, which in turn helps to promote drying that is uniform throughout the stack.

However, if the air were dry as well as hot, surface evaporation would exceed the rate that moisture could move or transfuse up to the surface. Wood near the surfaces would dry and shrink, often well in advance of wood more remote from the surfaces, and this shrinkage would clearly become restrained by centre portions where shrinkage tendencies were less at any instant. Stresses then would develop and these might very speedily grow large enough to tear the wood apart, i.e. to split it. The humidity in a kiln, and hence the evaporation rate, can be controlled by regulating, by means of air vents, the amount of moisture permitted to leave the chamber as damp air and, when necessary, by introducing vapour in the form of live steam.

Therefore, no drying must be allowed to commence until each plank or board is heated through to its centre. A board cannot be heated to its centre without drying taking place, unless it is heated by very moist, or saturated air. This is done by blowing live steam into the kiln through jets.

As soon as each board has become heated to the core, drying can be allowed to commence. Drying is commenced by reducing steam from the jets and thus using air of lower humidity.

Most kilns are steam heated and in them steam also provides the means of humidifying the air. The fans for promoting the air movement in a kiln usually are of the propeller type and sometimes situated to one side of the timber stack. In many kilns the fans are placed near the roof above the pile of timber.

Figure 10.1 shows a reversible air-circulation kiln. In this type the side at which the air enters must become the side at which the air leaves, and vice versa, thus equalising the drying rate at both sides of the stack.

The arrangement of the fans is shown in Fig. 10.2. When the shaft is revolving in one direction the air is as shown by the arrows. When the shaft revolves in the opposite direction, the whole air movement is reversed, as shown by the arrows in Fig. 10.2.

It will be noted that only one motor is required and that the fans are inside the kiln instead of outside, thus avoiding heat losses from ductwork.

Figure 10.3 shows a sketch of an overhead fan kiln with a vertical drive. Each fan has a separate motor situated under a cover on the roof and the drive is by vertical chain enclosed in the two vertical tubes.

For the smaller kilns, the air-circulating, heating, humidifying, and ventilating equipment can be condensed into one compact unit as shown in Fig. 10.4. The chamber to enclose the timber can be built in a variety of local materials or in insulated panels. After erection, connections are made to the steam or electric supply, or to both.

In cases where steam is not available at any time, heat and humidity can be provided entirely by electric units in this form of prefabricated kiln. As in the brick and concrete units the fans may be to one side as shown, or above the pile of timber and indeed, apart from the construction of the shell, they are alike in all essentials.

Kiln operation

In controlling the conditions in a kiln correctly the operator is provided with means for reading the temperature and humidity of the circulating air. A suitable instrument for this purpose is the wet-and-dry bulb hygrometer, illustrated in Fig. 10.5. Charts provide a record for the operator of the changes which have taken place in the air conditions of the kiln at any period during the run. A careful record is kept of the progress of every run so as to provide information when undertaking the drying of a similar load, or when dealing with some query that may arise as to the treatment given.

Schedules

In kiln drying the air conditions are regulated to suit the materials being dried. Variations in schedules are therefore necessary for different species of timber. This drying schedule as it is called, lists the most suitable degrees of heat and amount of humidity for particular timbers for corresponding moisture contents, during the drying process. Two typical schedules for 25 mm hardwood and softwood are shown in tables 2.1 and 2.2.

Table 2.1 Hardwood schedule

Moisture content (%)	Temperature dry bulb (°C)	Temperature wet bulb (°C)
60	41	36
40	43	39
35	43	38
30	46	40
25	49	41
20	52	42
15	57	44
10	63	45

Table 2.2 Softwood schedule

Moisture content (%)	Temperature dry bulb (°C)	Temperature wet bulb (°C)
Green	65	62
40	71	63
25	77	65
20	82	66
15	85	64
10	90	64

If a particular hardwood was found to contain 40 per cent of moisture, reference to the table shows that the drying could commence at 43°C. If it contained only 25 per cent of moisture, drying could commence at 49°C.

The correct percentage of humidity in the air is obtained by blowing live steam into the kiln until the 'wet' thermometer shows the specified temperature.

Drying time

The length of time in which the timber must remain in a kiln depends principally upon the moisture content of the wood, the species, the dimensions, and the degree of dryness required.

As a rough guide, a period of 1 week per 25 mm of thickness for softwoods, and 2 weeks per 25 mm of thickness for hardwoods, up to 75 mm thick, may be taken; although in actual practice the drying period for 25 mm softwood is sometimes as short as 2 days. During this period sample pieces are periodically removed from the stack to test the percentage of moisture remaining.

Progressive kilns

These are built in the form of a tunnel, having a dry warm end and a moist cool end. The tunnel is completely filled with a train of trucks carrying timber. When a wet load is pushed in at the cool end, a dry load is ejected at the warm end.

Progressive kilns must be of considerable length in order to obtain, and maintain, a sufficient temperature difference between the cool and warm ends. Once filled, the kiln must not be allowed to become empty, nor must gaps occur in the train of trucks. Every truck in the train must carry timber of the same species and thickness. The usefulness of any progressive kiln is limited to those cases where there is a large and continuous output of material of the same size and species.

Both good air drying and kiln drying have the same effect upon the strength of the wood – the drier the wood is, the stronger it becomes.

Shrinkage

As moisture is withdrawn from the cell walls in the process of drying, the cells contract and shrinkage takes place, and in reverse, if additional moisture is absorbed by the cell walls after seasoning, swelling results.

Shrinkage is greatest in the direction of the annual rings, that is, tangential to the circumference. Radial shrinkage also takes place and is about half tangential shrinkage while longitudinal shrinkage can be ignored. Various forms of defects which may be caused by uneven drying are splits and checks, shown in Figs. 11 and 12. The direction of shrinkage with regard to the position of the rays in cut boards is illustrated in Fig. 4.3. The quartered log shrinks parallel to the growth rings as shown in Fig. 11.2, while boards cut 'through-and-through' curl away from the heart. Consequently it should be possible to foretell how a board will tend to twist or warp in drying, and where if it checks, these are likely to occur.

Most hardwoods, like softwoods, are affected in similar ways so far as

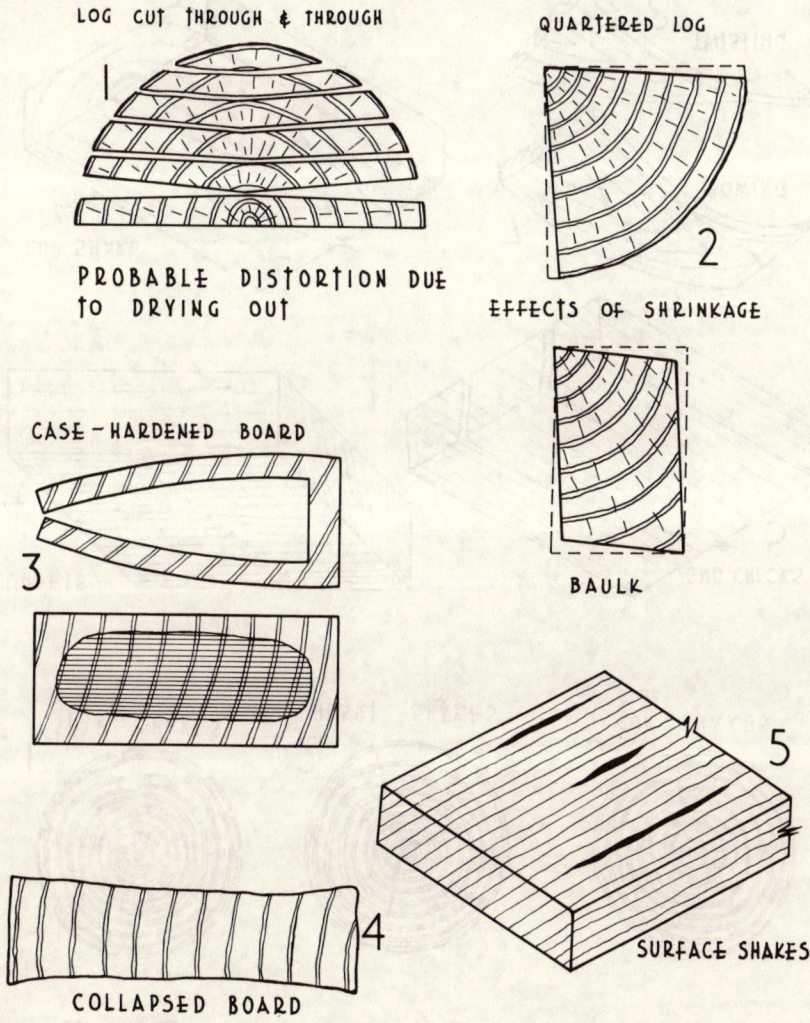

SHRINKAGE

LOG CUT THROUGH & THROUGH

1

PROBABLE DISTORTION DUE TO DRYING OUT

QUARTERED LOG

2

EFFECTS OF SHRINKAGE

CASE-HARDENED BOARD

3

BAULK

COLLAPSED BOARD

4

5

SURFACE SHAKES

FIGURE II

shrinkage, swelling, and distortion are concerned, though not to the same extent. An important safeguard with finished joinery work is to delay its delivery to the site until the building is roofed in and the heating installed. This means that any movement in timber, kiln dried to the required moisture content, is reduced to the minimum, thus reducing any future maintenance work.

As it is impossible to restrain movement in timber, an important aspect of the joiner's work is in the preparation and completion of joinery to allow this movement to take place without damage. Where wide surfaces are essential, the surface must be securely held to prevent any distortion due to twisting, but at the same time, provision must be made to allow for movement due to shrinkage or swelling. This applies in doors constructed with solid panels and some types of external doors filled with matchboarding. The need for the same precautions applies to wide solid counter tops, laboratory benches, and certain tables. The fixing of such surfaces may be carried out using slot-screwed cleats; tapered dovetail keys housed into the top; rebated bearers fixed with wooden buttons, or wooden or metal plate buttons.

Defects caused by shrinkage

Apart from those previously mentioned, these include bowing, cupping, checking, springing, collapse, and case-hardening.

Bowing   This is shown in Fig. 12.7, and is the curve in the direction of the length, due to the board not being cut parallel to the growth rings.

Twisting   This is shown in Fig. 12.8.

Cupping   The curve across the width of a board due to the greater shrinkage on the concave side, shown in Fig. 12.

Checking   This occurs in drying when the exposed ends of boards are unprotected and dry out more quickly than the rest of the board, shown in Fig. 12.5. This is often called end-checking or splitting, and may be prevented by painting with a good moisture-proof coating or nailing hoop iron across the board ends.

Surface shakes   These are shown in Fig. 11.5.

Springing   This is a curvature on the edge of a piece of timber, the face remaining flat. This is shown in Fig. 12.7.

Collapse   This is shown in Fig. 11.4, where affected boards become distorted through being kiln-dried too rapidly. This may be prevented if drying in the earlier stages is done at a low temperature.

Case-hardening   When the timber is dried rapidly the moisture in the centre of the wood cannot be extracted fast enough, the timber is likely, therefore, to be dry at the outer surfaces but wet inside. The innermost cells of the wood, which are saturated, hold their moisture and prevent shrinkage taking place at the outer surfaces. Internal stresses are set up which tend to twist the timber

D E F E C T S

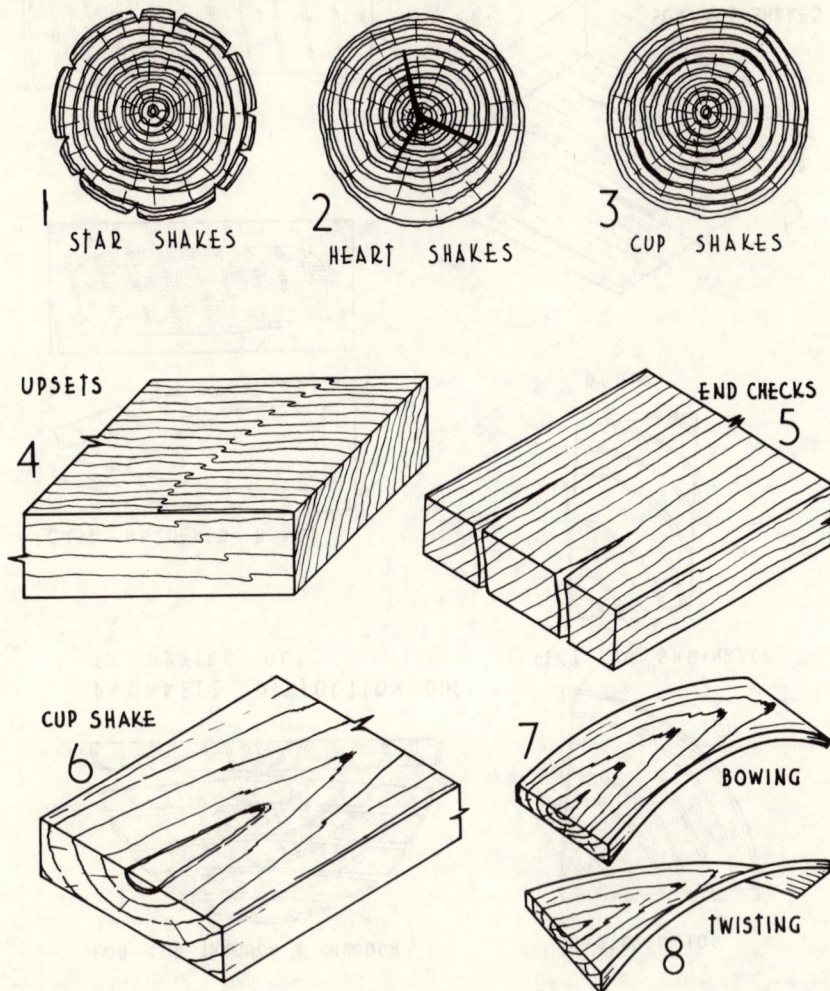

1 STAR SHAKES 2 HEART SHAKES 3 CUP SHAKES

4 UPSETS 5 END CHECKS

6 CUP SHAKE 7 BOWING 8 TWISTING

FIGURE 12

when it is resawn.

A simple test for case hardening is shown in Fig. 11.3. A sample section about 13 mm thick is cut from the plank with the centre portion cut away. Where unequal stresses are present, exposure to the air will cause the two prongs to spring together.

Natural defects in trees

These may be said to fall into two groups: (1) natural defects developed during the growth of the tree, and (2) those occurring after it has been felled. Some of the commoner defects are described below.

Star shakes (Fig. 12.1).
These are caused by too rapid drying on the outside of the tree and are a defect in felled timber.

Heart shakes (Fig. 12.2).
These begin at the heart of the log and are generally due to over-maturity.

Cup shakes (Figs. 12.3 and 12.6).
This defect is often called 'ring-shaking' as the shakes develop between two adjacent annual rings and is caused through lack of nutriment.

Knots

These are of many kinds and are the bases of side branches of the tree. Knots are the most frequent source of weakness in timber and may be classed according to their size, position, and number in the converted timber. Sound knots, which are solid and hard and show no signs of decay, are no serious detriment in timber for work other than for structural purposes. Where walls are panelled in timber, such as 'knotty pine', knots are considered to add beauty and effect in the final appearance. In joinery, dead or loose knots should be cut out and filled with inserts of sound wood. Further reference on this subject should be made to the stress grading of timber on page 81.

Upsets

This is the fracture of the fibres across the grain and may be the result of injury in felling. The defect, common in several species of mahogany, is often not detected until after planing. Upsets are shown in Fig. 12.4.

Waney edge

This defect lies in the sapwood which accompanies wane due to too economical conversion of a log shown in Fig. 7.

Twisted grain

Timber affected in this way is short-grained and unsuitable for structural work. The twisted fibres are usually due to the fact that the particular tree has been continually exposed to strong winds which cause twisting.

Blue stain

This defect, sometimes called blueing or sap-stain, is a discoloration in timber having a high moisture content and is caused by stacking too closely in confined spaces. The stain is the result of harmless fungi which lives on the stored foods in the cell walls; the hyphae of the fungi, which are dark in colour (Fig. 25) give a bluish colour to the timber. This blue appearance of the sapwood is found in many timbers, especially redwood.

Blue stain can be prevented by seasoning the timber immediately after conversion, or where this is not possible, by treating with a preservative (see page 25) which is poisonous to the fungi, and later drying.

As the fungi do not attack the cell walls of the wood there is no appreciable loss in the strength of the timber. Where appearance is not important, it is considered that, for ordinary purposes, stained timber is entirely serviceable.

If logs of susceptible timber are left lying on the ground for any length of time, the whole sapwood may become stained. The rapidity of this development will depend chiefly on environmental temperatures. Infection will occur from cut ends of logs, from branch wounds, damaged bark or may even be introduced by bark- or wood-boring beetles. The entry and exit holes of beetles can provide access for the fungus.

Blue stain is mostly confined to the sapwood of trees that possess a distinctive heartwood. Thus pines are more liable to staining than are spruces. Among the temperate hardwoods, staining is chiefly restricted in occurrence to poplar and ash. Many light-coloured, tropical hardwoods are also susceptible to blue stain, Ramin, obeche, balsa, etc.

Other drying methods

Unorthodox methods of drying timber have received attention from time to time, as for example, the method of drying with radio-frequency heating. In this method, the wood is placed between two metal plates to which is applied an electric current oscillating at a very high frequency. The high-frequency current causes the moistures in the wood to heat up at a more or less uniform rate throughout. Eventually boiling point is reached and, if the steam that then forms can escape freely, the drying rates may become very great indeed, depending upon the power input.

In a wood such as beech, which is very permeable, steam generated within the wood can readily move along the length of a plank and leave via the ends. This is theoretically an ideal method of drying wood since all parts of any cross-section receive equal quantities of heat and hence, dry at equal rates. There is no tendency, therefore, for the surfaces to dry in advance of the core and so to induce stresses leading to case-hardening and splitting. Many woods, such as oak, will not permit the steam to escape readily and, as a result, pressure builds up inside the wood and soon these pressures tear apart the fibres and so ruin the material. Apart from the technical difficulties of application, the cost of producing heat in this manner often proves to be prohibitive.

Temperature gradient method

In this method of drying the core of the wood is heated by radio-frequency and the surface deliberately cooled by moist air, thereby inducing moisture movement from the hotter centre to the colder surfaces. This method is likewise very expensive.

Other methods of drying timber include those in which the heat is introduced by the way of oil. creosote or xylene. The method of boiling off water from the wood by immersing it in heated creosote in a cylinder and drawing a vacuum is called the Boulton process, after the inventor. At the reduced pressure the moisture in the wood tends to boil at temperatures below that of the heated creosote, and some of it, particularly that near the surfaces, is evaporated. The wood is partially dried and rendered fit for the subsequent normal creosote impregnation treatment that is then carried out in the same cylinder.

Chemical drying

In this method the surfaces of the timber are caused to absorb, when in the green state, certain hygroscopic salts such as urea or even common eating salt. The presence of such a salt in the surface tends to keep these damp and inhibits shrinkage while moisture from within is able to diffuse outwards into and out of the surface layers into the atmosphere. The treated wood has still to be air or kiln dried and even though lower humidities may now be employed without damage, the drying times are not necessarily any shorter. The main difficulties in this method are in knowing just how much chemical should be introduced, and to what depth, also in ensuring that the timber is not badly discoloured.

Second seasoning

Joinery for good-class work is subjected to a further conditioning of the timber termed 'second seasoning'. After the material has been machined, the work is framed loosely together without wedges. It is then left in the workshop or convenient room where the moisture content will be similar to that where the work will finally be fixed in the building. Should any members develop faults, such as shakes or excessive drying in, during this second drying period of 2 or 3 months, they are replaced. The work is then ready for final assembly prior to delivery to the site for fixing.

Care and storage on site

The care and storage of materials, joinery, timber, and the like on site are most important. Facilities for the storage of joinery and internal finishings in timber correctly seasoned are rarely adequate except on larger contracts. The use of such materials would, of course, be of little avail if they are eventually left outside in the rain during building operations. A carefully prepared schedule of fabrication, delivery to the site and fixing is therefore necessary to reduce the period of site storage to a minimum. All joinery should be covered in transit as well as on the site, and delivery delayed as long as possible. A priming coat of paint carried out in the shop rather than at the site affords some protection. Various moisture retardants are now available and are discussed on page 80.

Internal finishings in timber should not be delivered to the site for fixing until the heating is installed and in use, with the building properly dried out. This is important in the case of special floors in hardwood.

External finishings and assemblies should be stored under cover, preferably in a dry, ventilated room, stacked flat and skidded to avoid deformation, twisting or accidental damage.

Moisture movement of timber

The variation in size and shape of timbers having varying moisture contents may be studied and compared by experiment.

Apparatus

Vernier calipers (illustrated in Fig. 13.2) tank, and sections of various timbers.

Description

Specimens of various timbers, cut with the grain running as shown in Fig. 13.1, approximately 25 mm × 25 mm in section and 113 mm long, are suitable. The specimens are dried in an oven at 105°C, and measured to the nearest 0:025 mm using the vernier calipers as shown in Fig. 13.2. They are then submerged in a tank of water for a period of say 1 week. On removal, the surplus water is wiped off and the length measured to determine the expansion. The results should then be recorded as follows:

Type of timber   Dry reading(mm)   Wet reading(mm)   Moisture movement(%)

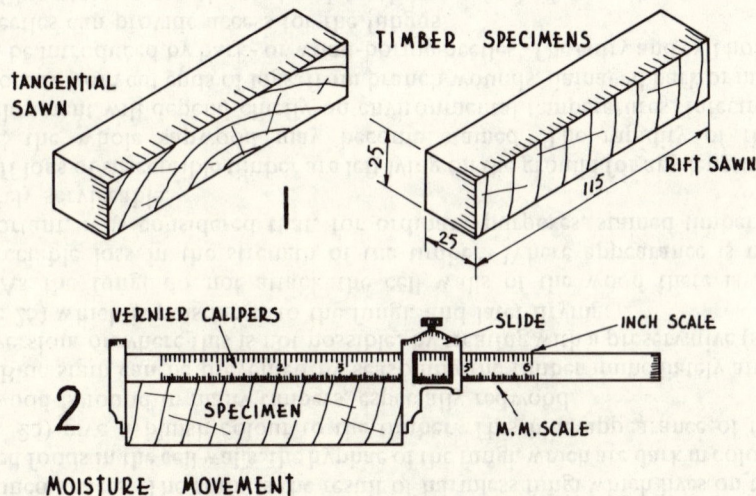

TIMBER SPECIMENS

TANGENTIAL SAWN

RIFT SAWN

25   115   25

1

VERNIER CALIPERS   SLIDE   INCH SCALE

SPECIMEN   M.M. SCALE

2

MOISTURE MOVEMENT

FIGURE 13

MOISTURE CONTENT

CUTTING SAMPLE FOR MOISTURE CONTENT TESTING

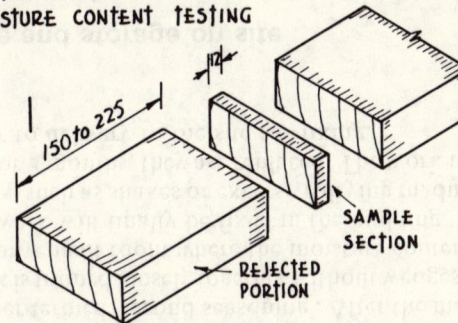

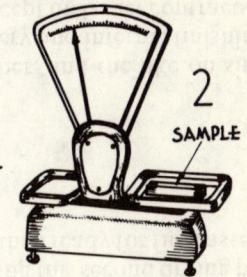

150 to 225

SAMPLE SECTION

REJECTED PORTION

SAMPLE

2

SAMPLE SECTION WEIGHED

4

OVEN FOR DRYING SAMPLES

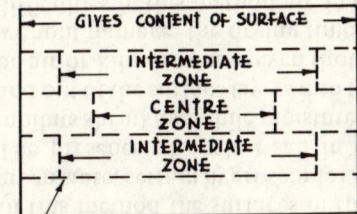

PERCENTAGE MOISTURE CONTENT:

$$= \frac{\text{ORIGINAL WEIGHT} - \text{DRY WEIGHT}}{\text{DRY WEIGHT}} \times 100$$

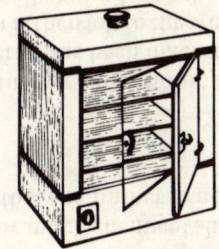

GIVES CONTENT OF SURFACE

INTERMEDIATE ZONE

CENTRE ZONE

INTERMEDIATE ZONE

3

DISTRIBUTION OF MOISTURE

SECTIONS CUT ON DOTTED LINES

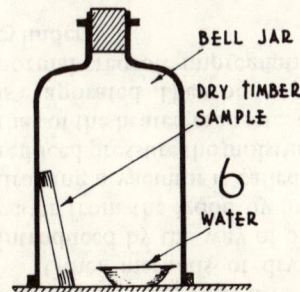

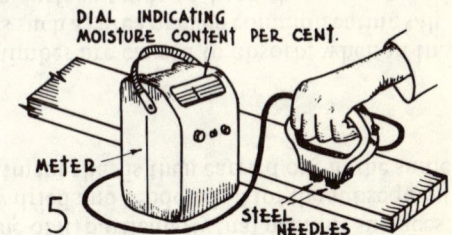

BELL JAR

DRY TIMBER SAMPLE

WATER

6

DIAL INDICATING MOISTURE CONTENT PER CENT.

METER

STEEL NEEDLES

5

MOISTURE METER IN USE

FIGURE 14

The percentage moisture movement may be calculated as follows:

$$\text{Per cent moisture movement} = \frac{\text{increase in length}}{\text{dry reading}} \times 100$$

Where it is possible to leave the specimens submerged for a longer period, the results calculated will give the complete moisture movement.

Moisture content

Reference to the moisture content of timber in connection with seasoning was made on page 13. This refers to the actual weight of moisture expressed as a percentage of the dry weight of the wood.

The normal method of accurately determining the moisture content is by drying in an oven, as shown in Fig. 14.4, at a temperature between 100 and 105°C, a small piece of a typical sample under consideration and noting the loss in weight. The amount of moisture present in the piece expressed then as a percentage of the dry weight of the piece is obtained from the following formula:

$$\text{Percentage moisture content} = \frac{\text{original weight} - \text{dry weight}}{\text{dry weight}} \times 100$$

Example
Original weight of specimen   = 23 g
Dry weight of specimen   = 20 g
Therefore, loss in weight = 23 — 20 =   3 g

This equals weight of moisture driven off, and is expressed as a percentage thus:

$$\text{Moisture content} = \frac{23 - 20}{20} \times 100 = 15 \text{ per cent}$$

This formula can also be expressed as:

$$\text{Percentage moisture content} = \left(\frac{\text{original weight}}{\text{dry weight}} - 1 \right) \times 100$$

In order to obtain the average moisture content of any particular board or plank, the sample should be cut at least 225 mm from the end of the board as shown in Fig. 14.1. The end portion should be rejected for it will usually be found to be drier than sections nearer the centre. The sample cross-section is then weighed on a balance (Fig. 14.2) as soon as possible after cutting before drying in the oven. When the sample loses no further weight this indicates that all the moisture will have been driven off. The moisture content can then be calculated from the formula. By cutting up into strips a further cross-section and weighing the pieces individually before and after drying in the oven, the way in which the moisture is distributed may be determined. Figure 14.3 shows the possible moisture distribution when the sample is cut along the dotted lines as indicated.

A further method of determining moisture content, and often practised, aims at eliminating the waste of material and time involved in checking sample pieces from boards throughout the drying.

The current moisture content of the stack is found by first calculating the dry weight of the board from which the sample piece is cut. This is determined using the following formula, when the dry weight and original weight of a similar piece 13 mm in length are known:

$$\text{Dry weight of board} = \frac{\text{original weight}}{(\text{moisture content}/100) + 1}$$

As the dry weight of the board is now determined, its moisture content can be calculated at any time during the drying process. The equation then becomes:

$$\text{Moisture content of boards in stack} = \left(\frac{\text{current weight}}{\text{dry weight}} - 1 \right) \times 100$$

For making on-the-spot moisture content checks of timber in kilns or stacked in the open, an electric appliance known as a moisture content meter is used. This is a portable meter (shown in Fig. 14.5) with a calibrated dial and is an essential part of the timber merchant's plant. Readings are quickly and easily read directly off the meter's dial. The operator simply has to push a twin electrode into the surface of the timber and read off the moisture content from the dial. Although this method is speedier and less wasteful, it is less accurate than the oven-drying method. The range of reasonable accuracy is from about 6 per cent to about 25 per cent. Above the higher limit the moisture in the wood makes it so conductive that the instrument is, as it were, short circuited and at the lower limit the wood becomes so good an insulator that unusually sensitive instruments would be required to measure the resistance with any accuracy. However, despite their limitations, these instruments are undoubtedly useful and, if sensibly used, are accurate enough for many industrial purposes.

Moisture content of timber in use will vary according to the temperature and humidity conditions of the particular environment. The moisture content to which the timber should be dried should be the average value it will attain in use in the building. For this reason the wood should only be introduced into the building when the air conditions have become reasonably steady.

The chart shown in Fig. 15 indicates the average values of moisture content of timbers for various purposes. These must be regarded as being approximate values due to variations attributable to different species of timber, treatment, exposure, etc.

The following notes will assist in the study of the moisture content of specimens of timber stored under different conditions of humidity:

Apparatus
Balance and weights, bell-jar, six samples of timber, evaporating dish, oven.

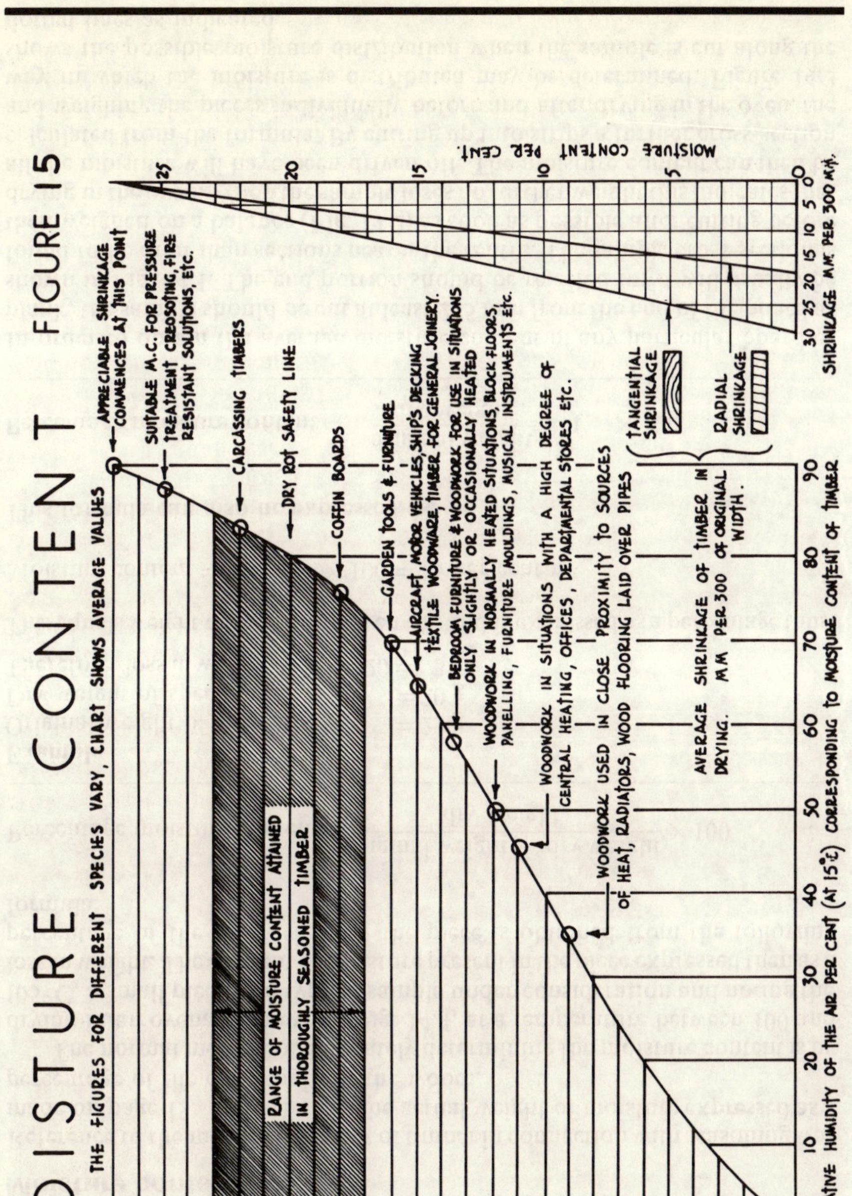

MOISTURE CONTENT — FIGURE 15

THE FIGURES FOR DIFFERENT SPECIES VARY, CHART SHOWS AVERAGE VALUES.

MOISTURE CONTENT PER CENT
SHRINKAGE MM PER 300 MM.

- APPRECIABLE SHRINKAGE COMMENCES AT THIS POINT
- SUITABLE M.C. FOR PRESSURE TREATMENT CREOSOTING, FIRE RESISTANT SOLUTIONS, ETC.
- CARCASSING TIMBERS
- DRY ROT SAFETY LINE
- COFFIN BOARDS
- GARDEN TOOLS & FURNITURE
- AIRCRAFT, MOTOR VEHICLES, SHIP'S DECKING, TEXTILE WOODWARE, TIMBER FOR GENERAL JOINERY.
- BEDROOM FURNITURE & WOODWORK FOR USE IN SITUATIONS ONLY SLIGHTLY OR OCCASIONALLY HEATED
- WOODWORK IN NORMALLY HEATED SITUATIONS, BLOCK FLOORS, PANELLING, FURNITURE, MOULDINGS, MUSICAL INSTRUMENTS ETC.
- WOODWORK IN SITUATIONS WITH A HIGH DEGREE OF CENTRAL HEATING, OFFICES, DEPARTMENTAL STORES ETC.
- WOODWORK USED IN CLOSE PROXIMITY TO SOURCES OF HEAT, RADIATORS, WOOD FLOORING LAID OVER PIPES
- AVERAGE SHRINKAGE OF TIMBER IN DRYING. MM PER 300 OF ORIGINAL WIDTH.

TANGENTIAL SHRINKAGE
RADIAL SHRINKAGE

RANGE OF MOISTURE CONTENT ATTAINED IN THOROUGHLY SEASONED TIMBER

RELATIVE HUMIDITY OF THE AIR PER CENT (AT 15°C) CORRESPONDING TO MOISTURE CONTENT OF TIMBER

MOISTURE CONTENT OF TIMBER PER CENT

ARTIFICIAL HEAT TO SECURE SUFFICIENT DRYING — AIR DRYING SUFFICIENT

Description

Six samples are prepared as in Fig. 14.1. Five are placed in different parts of the building marking each sample with the locality – i.e. on a bench, close to radiator for a period of 1 week. One sample is dried in the oven. All are then weighed accurately.

Place the dry specimen under the bell-jar with a dish of water (illustrated in Fig. 14.6) to maintain a damp atmosphere, and place the other specimens in the oven.

Prepare a table of weights as shown to be used for subsequent weighings to be recorded during the week as the specimens dry. Next week make final check weighing of each specimen.

Determine the moisture content of each specimen using the formula.

Specimen and locality	Original weight (g)	Final weight (g)	Moisture (g)	Percentage moisture

Chapter 3

Timber preservation

Dry rot

This is a general term applied to wood that has been attacked by fungus which feeds upon it. Fungi are plants which produce countless numbers of spores or seeds, these will take root quickly upon any timber which is damp enough to enable them to grow. Having germinated, the spores throw out strands which push their way through the wood developing into a grey mass of interlaced networks. These strands which rot the wood by feeding on the cellulose content, are termed hyphae. As the fungus develops, the timber becomes light brownish in colour, with cracks appearing both across and along the grain dividing it into small cubes. A certain musty smell is associated with the rot.

The fungus responsible for most of the destruction of timber in buildings is *Merulius lacrymans*. This highly infectious disease will attack timber which is allowed to absorb moisture to a level above 20 per cent and flourishes at levels beyond this figure. Timber used in general joinery work, such as door-frames, skirting boards, panelling and flooring, along with constructional timbers, can become affected. In all cases the fungus prefers damp, stagnant air in which to become established. These conditions are often found in buildings where the damp-proof course is faulty, or not inserted, and in badly ventilated areas below ground floors. Floors which may have been wetted by a dripping tap and roof timbers affected by leaking roofs and damaged gutters, are other starting points for rot. In its search for further supplies of timber on which to feed, the fungus will penetrate through brickwork or along steel joists to find it. This often means difficulty at first in determining the extreme limits of the rot.

Cellar fungus

This fungus is usually found in damp cellars and bathrooms and requires wet conditions for an attack to develop. It is responsible for the decay of timber used outside, in fencing posts and the like, and is often found in houses where linoleum has been laid over damp floorboards. *Cellar fungus* cracks the wood

along the grain and not across it, as with *Merulius lacrymans*. It also darkens the wood.

There are other forms of fungi which attack timber out of doors. *Lentinus lepideus* causes much destruction in softwood timbers used outside. *Mine fungus* is frequently found in damp mines and is similar to cellar fungus. Both of these types are often termed 'wet rots'. This is because they will only grow upon wood with a very high moisture content. Once the wood is dried out they die away and are not liable to start up again as in the case of dry rot.

Prevention

Possible starting points for any form of rot have been mentioned. The chief protective measure in the prevention of the disease is to keep the timber dry. Thorough ventilation and special attention to the cause of any damp within the building is necessary. Defective gullies and drains, rain-water pipes, gutters, radiators, and defective plumbing are possible causes of damp and should receive attention.

Eradicating rot

After the source of damp which caused the rot has been traced, the following points should be observed:

1. All infected timber should be removed and burnt. It is also important to allow a margin of safety by cutting apparently sound timber away for a distance of about 600 mm to make sure no fungus strands remain.
2. All brickwork and materials other than timber should be brushed down with a stiff brush and sterilised by heating with a blowlamp and treated with a suitable fungicide. Solid floors, groundwork, and subsoil should also be sprayed freely.
3. Any sound timber remaining, and new timber used for replacement, should be treated with a wood preservative. This may be applied by brush, spray, or by immersing the timber in a bath or tank.

Preservative should be used freely on floor-joists, especially the ends where they are built into exterior walls. The undersides of floorboards, wall plates and any built-in timbers should be treated freely. Where the back of wall-panelling comes into contact with the wall, skirting boards, and fixing grounds should all be treated.

In new joinery, the members should be treated with a preservative before it is put together, so that the joints are thoroughly impregnated. Window-frames, door-frames, sills, and any exposed joinery should be treated.

Properly seasoned timber should always be used, and care should be taken that adequate protection is provided on a building during erection, or on the site, to keep it under cover in the period before fixing.

Insect borers

Timber, especially in old buildings, is frequently subjected to the ravages of different insects, which bore into the wood and eventually destroy it. Insect damage of timber can occur in standing trees, freshly felled logs, and other unseasoned timber, as well as in seasoned timber during storage and use.

Insect larvae (grubs) burrow in the wood to obtain food and shelter and make characteristic tunnels which sometimes are packed with bore dust, termed 'frass'.

The most important group of insects attacking timber in this country are the beetles. A certain amount of damage to softwoods is caused by the wood wasps, and the carpenter moths occasionally excavate large galleries in fruit and other trees. In tropical countries, severe timber deterioration is caused by termites.

Three types of insects, mainly restricted to partially decayed timber, are: the death watch beetle, the weevils, and the wharf borer; three others are predominantly forest pests: wood wasps, ambrosia beetles, and longhorns; the termites are mainly tropical and rare in this country; powder post beetles are chiefly pests of the timber yard, attacking partially seasoned timber; the house longhorn beetle is restricted mainly to certain areas of Surrey; and the common furniture beetle is the only species which is of widespread occurrence in worked timber in buildings.

In order to understand and to diagnose insect attack it is essential to have a knowledge of the life-cycle. This can be divided into four stages for all insects (except termites) causing timber deterioration:

1. *The egg.* These are laid by female insects, usually in cracks, splits, on rough surfaces, in the pores of the wood, on end grain areas, or in tunnels bored into the timber. Eggs are usually very small and may not be discernible to the unaided eye. They usually hatch within a short period of laying.
2. *The larva.* The larva, or grub, hatches from the egg and it is this stage which causes most of the damage to timber. The larva has a completely different appearance to the adult, being worm-like and usually dull white-brown or white in colour. Larvae, when they first hatch, are usually very small, but in some species they may become over 25 mm long when fully grown. The larval stage varies in duration from a few weeks to many years.
3. *The pupa.* During an inactive period the larva ceases to feed, and changes in form and structure occur. The pupa resembles a pale yellow beetle covered with a transparent skin, which darkens in colour.
4. *The insect.* As the adult insect emerges from the pupal skin it bores its way out of the timber, forming a hole which is round or oval in shape and varies in size according to the insect species. As the adults of most wood-boring insects can fly strongly they are responsible for the spread of infestation to fresh timber.

In order to prevent the attack and to undertake the extermination of these insects, it is necessary to know the characteristic features of each. Descriptions of the three common wood-boring types of beetles follow:

Furniture beetle

This is the most common and best known of these destructive pests. In spite of its name it is frequently found in structural timbers of buildings such as rafters and joists, as well as in furniture. The beetles are brown in colour and about 3 mm long. They emerge from the wood during June to August and fly to any suitable timber to lay eggs in any crack or crevice in the surface. As the eggs hatch the larvae bore into the wood. The larva, a white-coloured grub with

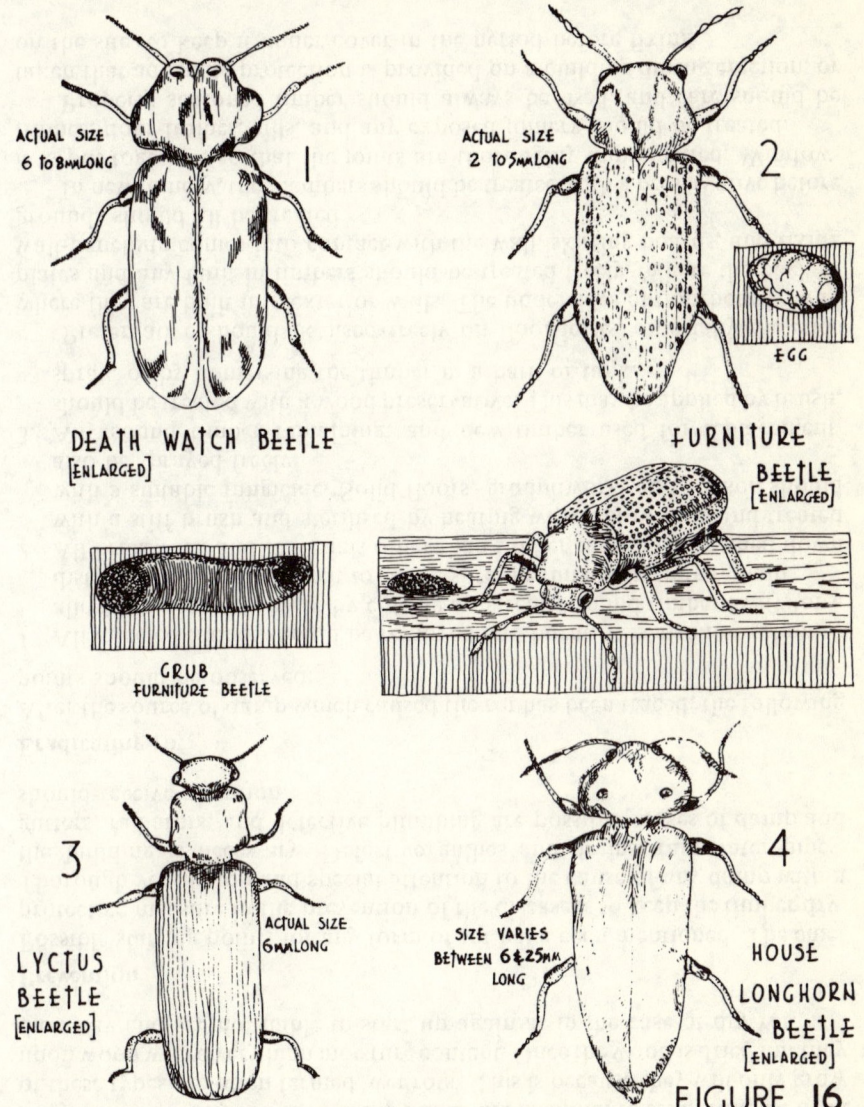

INSECT BORERS

ACTUAL SIZE 6 to 8mm LONG

1

DEATH WATCH BEETLE [ENLARGED]

GRUB FURNITURE BEETLE

ACTUAL SIZE 2 to 5mm LONG

2

EGG

FURNITURE BEETLE [ENLARGED]

3

ACTUAL SIZE 6mm LONG

LYCTUS BEETLE [ENLARGED]

4

SIZE VARIES BETWEEN 6 & 25mm LONG

HOUSE LONGHORN BEETLE [ENLARGED]

FIGURE 16

strong jaws, bores into the wood first in the line of the grain and then in every direction. All the time it is feeding on the cellulose in the wood and fattening, so that the tunnels increase in diameter. This burrowing continues for a year or more until the larva approaches the surface of the wood where it hollows out a small cave for itself and sheds its pupal case. Now a full-grown beetle, it emerges from the wood, leaving a small round hole, to continue the havoc of its predecessors. An adult beetle and larva are shown greatly enlarged in the sketch (Fig. 16.2.).

Powder post beetle

There are several varieties of these beetles in this country. All are very similar in both appearance and their habits. As the larvae of the furniture beetles live by digesting the cellulose in the wood, the powder post beetles must have starch for their food. It is for this reason that it is found only in the sapwood of certain hardwoods. It is never found in softwoods and, although ash, elm, and oak suffer, the damage applies solely to the sapwood of these hardwoods. This beetle is illustrated in Fig. 16.3.

Death watch beetle

This insect is perhaps best known for its destruction to roof timbers in churches, and to old and historic buildings.

The beetle belongs to the same family as the furniture beetle. It is, however, about twice the size. The beetle goes through the same course of transformation as the furniture beetle and is well known for the ticking sound made during the mating season. This is caused by the insects knocking their heads against the wood. This beetle is illustrated in Fig. 16.1.

Extermination of wood-boring beetles

It has already been stated that it is the larvae or grubs of the beetles tunnelling in the wood which cause the damage. These tunnels are filled with a very fine dust or 'frass' which is wood that has been digested by the grubs. This dust should be cleared where possible by means of a vacuum cleaner and a preservative flooded into all holes, cracks and crevices until no more can be absorbed. This may be applied by brush or spray, with any new timber used for replacement, immersed in a bath of the preservative.

Structural timbers which are built in, such as the ends of joists, rafters, and wall plates, are often difficult to treat. In such cases holes bored into the heart of the timber should be flooded to saturate the surrounding wood. Timber badly affected should be cut out until sound wood is reached. Two coats of the preservative should be applied, the second coat when the first has sunk in but not dried out.

It should be obvious that it is no easy matter to kill grubs boring within the wood and that once worm holes appear on the surface of the timber treatment should be started immediately. Regular inspections of the affected timbers are necessary with possible further applications of preservative at intervals.

The longhorn beetle is shown in Fig. 16.4. This beetle was considered uncommon in this country until recently. It attacks softwood roof timbers mainly, and destroys the interior of the wood, leaving only a thin surface skin. The damage may go quite unsuspected until actual collapse takes place.

There are two common species of wood-boring weevils in this country. They are usually found in timber which is partially decayed and are considered as secondary timber pests. The adult weevil has a long snout on which the antennae are situated. They have red-black bodies 3–5 mm long. Both adults and larvae bore into the wood, producing tunnels which run in the wood grain direction leaving paper-thin walls between them.

Hardwood and softwood sapwood may be attacked, and heartwood may occasionally be affected. Attack by weevils is reported in plywood and panelling, but it is probably confined to plywood bonded with animal and vegetable-based adhesives.

Preservation

Timbers vary greatly in their durability and resistance to attack by fungi and wood-boring insects. For instance, oak or teak will remain sound for much longer than beech or redwood under adverse conditions. It has already been stated that sapwood, and in some cases heartwood, of many timbers used in building are liable under certain conditions of moisture content of the timber, of temperature and humidity of the air to be attacked by wood-destroying fungi and insects. The conditions under which such attack may occur have also been mentioned. In order to increase the long-term durability of the timber we use, it is therefore desirable to treat it with some form of preservative which will penetrate into the cells and fibres and act as a strong poison against these destroying fungi. At the present time there is a growing demand for preservatives which will provide protection against decay and which will also act as a fire-retardant.

A suitable preservative should be toxic but safe for men to handle, it should penetrate easily, be permanent and not wash off in rain. It should not affect metals, and be easily obtainable in large quantities.

There are three main types of preservatives: (1) tar oils, (2) water-borne preservatives, and (3) organic solvent preservatives.

Creosote

Which is in group (1), is a black or brownish oil, produced by the distillation of coal-tar, it is widely used in the treatment of timbers used in the transmission of electricity, telecommunications, and in railway work. Creosote-treated telegraph poles, railway sleepers, fencing, bridge and piling timbers are examples where a service life is obtained of many times that which would result from the use of untreated timbers. For instance the service life of a treated telegraph pole is 30 years, whereas an untreated pole would decay in 5 years.

Creosote has a fairly strong smell and for this reason it should not be employed for preserving timbers to be used near or in contact with food-stuffs. Because of staining it is unwise to use a tar oil type of preservative such as creosote on wood in contact with plaster.

Water-borne preservatives

Group (2) consist of chemicals dissolved in water. Salts of copper, zinc, mercury, sodium, and chromium are the principal chemicals used today and

PRESERVATION

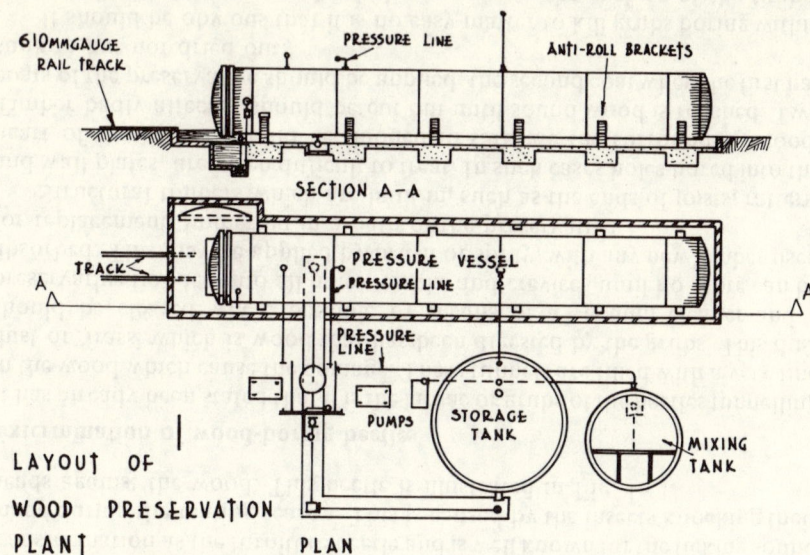

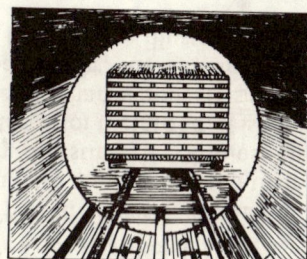

610MM GAUGE RAIL TRACK

PRESSURE LINE

ANTI-ROLL BRACKETS

SECTION A-A

TRACK

PRESSURE VESSEL
PRESSURE LINE

A

A

PRESSURE LINE

PUMPS

STORAGE TANK

MIXING TANK

1

LAYOUT OF WOOD PRESERVATION PLANT

PLAN

2

TIMBER ENTERING PRESSURE VESSEL

SAPWOOD   HEARTWOOD

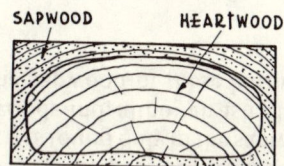

SECTION SHOWING PENETRATION OF PRESERVATIVE BY PRESSURE VESSEL TREATMENT

FIGURE 18

have the following characteristics. Because they are so well fixed and are resistant to leaching and none of the ingredients evaporate quickly, their strong poison value remains constant. Treated timbers can be painted or varnished after redrying. They may also be glued, but advice on this should be sought from the glue manufacturers. Water-borne preservatives usually have no smell and are non-inflammable. Treated timber may be used for food storage containers and linings of refrigerators without risk of contamination. They are suitable for internal and external joinery but older types of preservative may leach out of wood in contact with damp or water.

Organic solvent type

This is among the most recent developments in preservatives, and is in group (3). In this form, the poisonous ingredients are dissolved in organic solvents such as white spirit, naphtha, light petroleum distillates, diesel oil, or gas oil. These solvents evaporate after treatment leaving the preservative in the wood.

Almost all organic solvent preservatives in present use are based on either pentachlorophenol, copper or zinc naphthenates and have characteristics very similar to those described for water-borne preservatives, with the following exceptions: they are usually more expensive and certain types have a strong smell; after treatment, timber can be used straight away as there is no swelling and therefore no need for redrying; penetration of the wood is usually better than other types of preservatives and hence are more suited to treatment by brush, spray, or dipping.

The most effective method of applying preservatives to timber is by pressure impregnation which can only be carried out in yards equipped with a pressure plant. Other ways of applying preservatives are 'hot and cold' treatment in an open tank, steeping, dipping, and brushing or spraying.

Before describing the various methods of preservation it should be realised that the degree of protection obtained by treating timber with a preservative depends more on the depth to which a preservative is forced into the wood than on the type of preservative used.

Pressure impregnation

This method ensures deep penetration and uniform distribution of specified quantities of the preservative, which are necessary to give a long life to the timber, with a minimum of maintenance costs. The process requires a steel cylinder similar to that shown in Figs. 18.1 and 18.2, which can be opened at one or both ends. This may be between 6 and 45 m in length and between 900 mm and 2·7 m in diameter so that the timber to be treated can be loaded on trolleys and run into the cylinders on tracks. Storage tanks for the preservative with pumps, gauges, etc. are also required.

Pressure treatment is of two kinds: (1) *full cell*, and (2) *empty cell*.

Full cell

The timber to be treated is placed in the cylinder which is then sealed. The vacuum pump then creates a vacuum to remove air from the cells of the timber and without breaking this, the preservative is introduced at a temperature of 60–82°C. As the cylinder is filled, pressure is built up by a pump to between 690 and 1 380 kN/m². This pressure is maintained until the right amount of preservative is forced into the timber. This time will vary

PRESERVATION

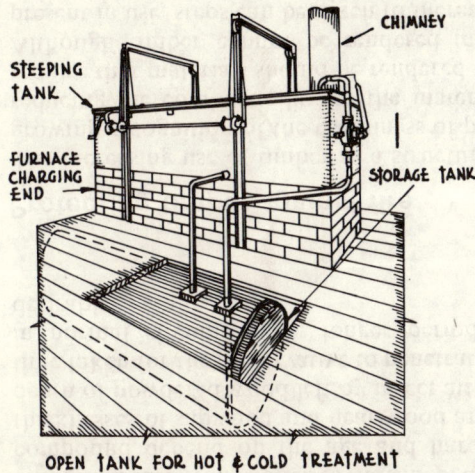

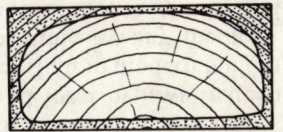

STEEPING TANK

CHIMNEY

FURNACE CHARGING END

STORAGE TANK

OPEN TANK FOR HOT & COLD TREATMENT

SECTION SHOWING PENETRATION OF SAPWOOD AND SOME HEARTWOOD BY HOT AND COLD TREATMENT

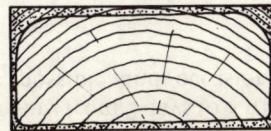

SECTION SHOWING LIMITED PENETRATION OF SAPWOOD DEPENDING ON LENGTH OF STEEPING TIME

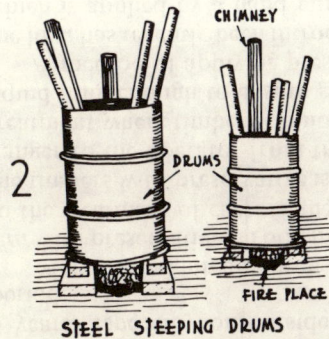

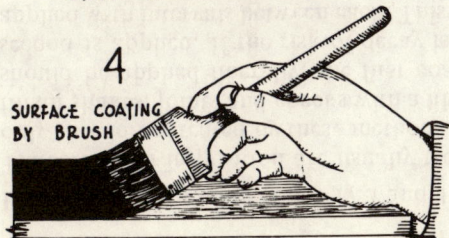

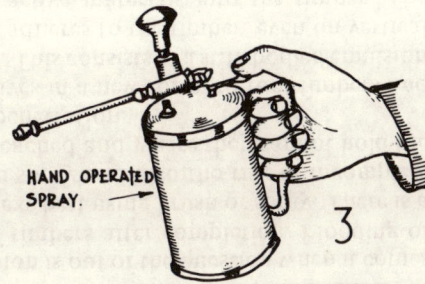

CHIMNEY

DRUMS

FIRE PLACE

STEEL STEEPING DRUMS

HAND OPERATED SPRAY.

SURFACE COATING BY BRUSH

SECTION SHOWING VARIABLE PENETRATION INTO SAPWOOD BY SPRAY OR BRUSH GIVING SURFACE PROTECTION

FIGURE 17

according to the variety and sizes of the material under treatment. At the end of the pressure treatment, the pressure is released and the surplus preservative is run back into the storage tank leaving the cells of the timber full.

This form of full cell treatment, known as the *Bethel* process, is given to railway sleepers and timber to be submerged in water or subject to alternate wetting and drying as found in dock work, piers, jetties, etc.

Empty cell

With this method the timber in the cylinder is subjected to a pressure of about 276 kN/m² of compressed air. The pressure is kept constant for about half an hour after which time the preservative is forced into the cylinder at a pressure exceeding 690 kN/m² and at a temperature of 65–93°C. This pressure is maintained until the required absorption is obtained. The surplus preservative is then pumped back into the storage tank. During this time the compressed air within the wood expands and expels much of the preservative from the cells. More preservative is removed from the cells as the pump reduces the air pressure. This leaves the cell walls well coated with preservative yet the cells themselves are free or empty. Telegraph poles and the like are usually treated by this method called the *Reuping* process.

Timbers used for building purposes are also treated very effectively by this method and it is comparatively cheap as much less preservative is used compared with the full cell method.

In areas where it is not practicable to obtain pressure-treated timber easily, simpler methods of treatment such as steeping, dipping, brushing, and spraying are used. Generally, these methods are not as effective as pressure impregnation but are often quite suitable for small quantities of timber.

Steeping

The timber is submerged in cold preservative in a tank or bath for a period from days to weeks, depending on the type and dimensions of the timber. Where the tank is not big enough to take the timber, the ends of the posts or planks can be immersed in turn in a bucket of preservative and left to soak, the remainder of the timber being brush treated.

Dipping

This consists of submerging timber in a tank for very short periods and is the best of the surface-treatment methods. All types of preservatives may be used, the longer the immersion the better, but to obtain maximum penetration the preservative should be heated.

Hot and cold treatment

In this method shown in Fig. 17.1, the timber is submerged and fastened down in a tank of preservative which is heated to a temperature of 93°C for a period of up to 2 hours. This is then allowed to cool, the preservative pumped back to the storage tank and the timber removed. During the heating period the hot liquid heats the wood which causes some air to be expelled from the cells. On cooling, the air left in the cells contracts and the partial vacuum created causes the preservative to be forced into the wood. This absorption of the liquid by the timber occurs during the cooling period.

A simple form of hot and cold open tank treatment for fencing posts, gate posts and the like can be effected using an open-topped 225 litre drum, shown in Fig. 17.2. Part-filled with preservative, the posts are stood on end in the drum and a fire lit underneath to heat the liquid. After about an hour the fire is put out to allow the liquid to cool. The process is then repeated for the opposite ends of the props or for further timbers to be treated as required.

Brushing and spraying (Figs. 17.3 and 17.4)

Timbers fixed in position are usually treated in this way. Seasoned timber only should be treated by these methods and special care should be taken to fill all shakes, joints and checks with a liberal application. At least two coats should be applied liberally, the first coat being allowed to dry before the second is applied. If the risk of decay is high then further coats should be applied with intervals between each. This treatment should be renewed every 3 years, especially on outside work, to preserve the timbers over a long period.

Mention has already been made of the difficulties of treating timbers *in situ* with preservatives. Total immersion is out of the question when it comes to the treatment of carcass and roof timbers after completion. Flooding of the timbers with preservative is the next best using brush or spray. There is a danger to the operator from the fine spray and also the risk of incomplete treatment where timbers cannot be reached and where they cannot hold the liquid long enough to ensure good penetration.

A method of applying preservatives in a new form, to roof timbers and the like, has recently been introduced. This consists of a stiff-bodied emulsion which is applied by a hand gun and adheres to the timber, even on vertical surfaces, and slowly discharges its active materials into the timber. The amount of emulsion applied to the timber and the thickness and spacing of the compound depend on the age and hardness of the timber. The relative thicknesses of sapwood and heartwood and, by inspection, the amount and depth of powdered wood left by insect attack, are other considerations. The time taken for the preservative to penetrate may vary from 2 days to 3 weeks in normal circumstances; longer periods are possible and occasionally desirable.

Protecting timber against fire

The increasing use of timber as a structural material has brought with it a growing recognition of the usefulness of protective treatment as a means of reducing the combustibility of the material, and local by-laws frequently require that materials should be rendered fire resistant in modern buildings. Although timber cannot be rendered incombustible by any methods at present in use, steps can be taken to increase its resistance to ignition and to delay or even prevent its active participation in a fire. Such resistance can be given to timber by the application of fire-retardants.

Modern fire-retardants are of two types:

1. Those applied to the timber by pressure impregnation:
2. Those for surface application.

The first method is normally used when there is a risk of intense fire in other adjoining materials such as structural timbers in warehouses, garages, hangars, etc. Surface coatings are generally used for lightweight and decorative wood, plywood, and wallboards where the risk is of ignition from a fairly small source and may be applied either by brush, spray, or dipping.

The protection of timber against fire is distinct from its preservation from attack by destructive insects and fungi described elsewhere in this chapter, but a combined treatment is now available.

Among the large number of chemicals which have been suggested for the protection of timber against fire are monoammonium dihydrogen ortho-phosphate, ammonium sulphate, boric acid, zinc chloride, and ammonium chloride. These substances, and compositions containing one or more of them, are the principal water-soluble fire-retardants used in pressure impregnation today, although some are also used in surface coatings.

A fire-retardant surface coating will protect the surface of the timber against the spread of flame and ignition in one or more ways:

1. The coating may swell to form an insulating layer which retards the access of heat.
2. The fire-retardant may change the thermal decomposition of the timber.
3. The coating may liquefy by the heat to produce glassy substances which coat the surface and prevent access of oxygen.
4. The coating may be thick, heavy, and inactive, keeping out oxygen and absorbing or dissipating the heat so that the timber remains cool.

Timber which has been treated and is to be used in the open should be protected from the weather by painting it with a paint of low combustibility and, if timber which has been treated with a fire-retardant is to be glued, the makers of the particular adhesive used and the manufacturers of the fire-retardant should be consulted.

The effectiveness of fire-retardant treatment is determined by testing treated timbers in controlled conditions and comparing the results with similar untreated timbers in the same conditions. The accepted tests in this country are those to BS 476:1953.

Fire retardance of plywood

Plywood panels have definite fire-retardant properties. This is particularly true of phenolic-bonded plywood. Tests have shown that although the face veneer of the phenolic-bonded plywood was completely destroyed, charred flakes of this veneer still adhered to the panel affording a degree of protection to the underlying layers. Furthermore, the lack of joints does prevent the penetration of the flames until the panel is completely burnt through.

Highly fire-retardant panels are also available. In some, the cores are made with incombustible materials, such as asbestos sheets, and in others the plywood has received treatment by pressure impregnation with a fire-retardant process.

Chapter 4

Plywood and manufactured boards

Plywood

Plywood is an assembled product made up of plies, or veneers, and adhesives. It is manufactured in various thicknesses from 3 mm upwards, in dimensions up to 3 m × 1·220 m sheet sizes.

Plywood is produced in two main types:

1. Interior: for most interior joinery, flush doors, panels, wall-panelling, fitments, balustrades, sub-floors, and in any work where resistance to moisture is not required.

2. Resin bonded: for similar uses as the interior type along with exterior work in flush doors, formwork and shuttering for concrete, signs, and for any purpose where it may be exposed to moisture.

Three-ply

Figure 20.1 shows three-ply construction consisting of a face and back veneer and a core or inner-ply. The grain of the veneer in the core runs at right angles to the veneers on either side of it.

Multi-ply

The construction of multi-ply is similar to three-ply except that the inner core consists of three of more veneers, as shown in Fig. 20.3.

Plywood construction may be balanced or unbalanced. Balanced construction reduces the tendency of the finished board to distort as the construction consists of an odd number of veneers arranged symmetrically.

Properties of plywood

Plywood has certain outstanding advantages over solid timber. It is more suitable for large uninterrupted surfaces and has greater uniformity of strength and dimensional stability. This is due to the cross-banding of the veneers as a means of distributing the high longitudinal strength of the wood in all directions.

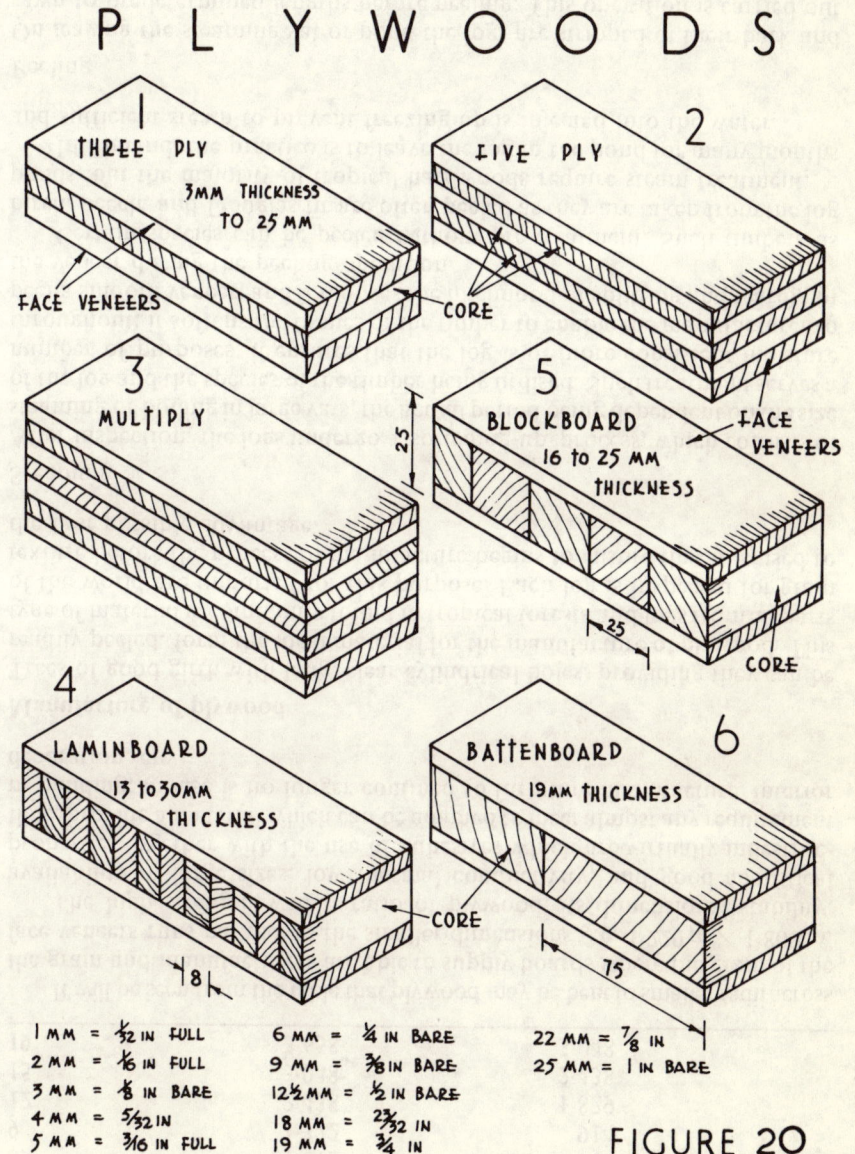

PLYWOODS

1 THREE PLY — 3MM THICKNESS TO 25 MM — FACE VENEERS

2 FIVE PLY — CORE

3 MULTIPLY

5 BLOCKBOARD — 16 to 25 MM THICKNESS — FACE VENEERS — CORE — 25

4 LAMINBOARD — 13 to 30 MM THICKNESS — CORE — 8

6 BATTENBOARD — 19MM THICKNESS — 75

1 MM = 1/32 IN FULL	6 MM = 1/4 IN BARE	22 MM = 7/8 IN
2 MM = 1/16 IN FULL	9 MM = 3/8 IN BARE	25 MM = 1 IN BARE
3 MM = 1/8 IN BARE	12½ MM = ½ IN BARE	
4 MM = 5/32 IN	18 MM = 23/32 IN	
5 MM = 3/16 IN FULL	19 MM = 3/4 IN	

FIGURE 20

Resistance to splitting

Straight-grained timber splits readily. This is due to the fact that when a cleavage force is applied it tends to pull the timber apart across the grain, in which direction its strength is the least.

Plywood has no line of cleavage and cannot therefore split, since the grain of each alternate layer is opposed to the direction of the force. It can be nailed or screwed close to the edges of the panel when fixing, without the danger of splitting, in addition to which the criss-cross arrangement of the fibres offers considerable resistance to the pull through of nail or screw heads.

Impact resistance

The same property, the absence of a line of cleavage, also has its effect on the impact strength of plywood. Tests have shown that an impact force greater than the tensile strength of the timber is required to cause failure. The multiple-layer construction prevents splitting by distributing the pressure at the point of impact: the longitudinal fibres fracture on the underside of the panel with successive shattering of the other plies, but no complete break, as in the case of solid wood subject to similar loading, will occur.

Thermal insulation

Plywood as an insulating material possesses the same properties as the wood of which it is composed. The thermal conductivity value of timber varies according to the species but an average value of 1·44 W/mdegC is normally accepted for timber in general.

Its dimensional stability allows the use of large sheets of plywood which offers these advantages: by reducing the number of joints necessary to clad a wall, plywood – when well nailed or glued to timber studding – minimises heat loss and virtually eliminates draughts which might otherwise occur through cracks or badly fitting joints, and thereby increases the insulation value of the construction.

Bending properties

Although plywood is normally used as a flat material its resilient properties enable it to be bent to a reasonably small radius of curvature without suffering damage. The safe minimum radius of curvature of any plywood varies according to the overall thickness and the individual thickness of the veneers used in the construction of the panel as well as the species of wood, or woods, of which it is composed. Three-ply composed of veneers of equal thickness, will bend more readily than a panel consisting of a substantial core and thin outer veneers.

Table 4.1 has been reproduced to give some idea of the radii of curvature obtainable in different thicknesses of plywood.

The radii shown in this table are minimum radii, and may have to be increased when plywood has substantial end grain exposure.

The figures shown have been prepared principally for use in connection with Douglas fir plywood, but it is reasonable to assume that it offers some guidance in the application of hardwood plywoods. The figures were obtained from plywood bent in a dry state at ordinary room temperature, but plywood bonded with suitable adhesives can be soaked or steamed to enable it to be bent to smaller radii.

Table 4.1. Radii of curvature

Thickness (mm)	Along grain (mm)	Across grain (mm)
6	610	380
9	1·372	914
12	2·438	1·829
15	3·048	2·438
19	3·658	3·048

It will be seen from the table that plywood may be bent to smaller radii across the grain and manufacturers are able to supply boards where the grain of the face veneers runs parallel to the smaller dimensions, i.e. 1·220 m × 1·800 m.

The high strength/weight ratio of plywood, its dimensional stability, availability in large sizes, low thermal conductivity, and good acoustical properties together with the use of adhesives which are virtually indestructible, provide a material which can be adapted to meet almost any requirement in building; its use is no longer confined to furniture manufacture, interior decoration, etc.

Manufacture of plywood

Trees of good girth with long, clear cylindrical boles, providing they can be readily peeled, form the ideal material for the manufacture of plywood. This type of material is commonly found in tropical forests and logs from all parts of the world are imported for this purpose. Each log is inspected for grain texture before the process of manufacture begins to ensure that it is used to the best possible advantage.

Steaming

After inspection, the logs undergo a 'softening-up' process, which consists of steaming or boiling in large vats, the actual period being dependent on the size of the log and the species of the timber being utilised. Such treatment serves a number of purposes; it ensures that the log is of more consistent moisture throughout; it softens the tissues of the timber to enable the manufacturer to peel a smooth veneer; and it reduces the likelihood of splitting and tearing of the veneer during the peeling operation.

Certain species can be peeled without pre-treatment. Such timbers as birch, beech, and Douglas fir are often peeled as they are taken from the log ponds, but the majority of tropical hardwoods require steam treatment.

In Finland, the practice is to leave the log in the pond for many months and sufficient steam to prevent freezing up is injected into the water.

Peeling

On leaving the steaming vat or pond the logs are stripped of their bark and sawn to predetermined lengths before peeling. This operation is carried out on a large rotary lathe or peeler, as shown in Fig. 19.1. The lathe rotates the log against the knife which runs the full length of the log and a continuous ribbon of veneer is thus produced. Gears feed the knife towards the log as its diameter is reduced so that a constant thickness of veneer is maintained. This operation is one of the most important stages in the manufacturing process

PLYWOODS

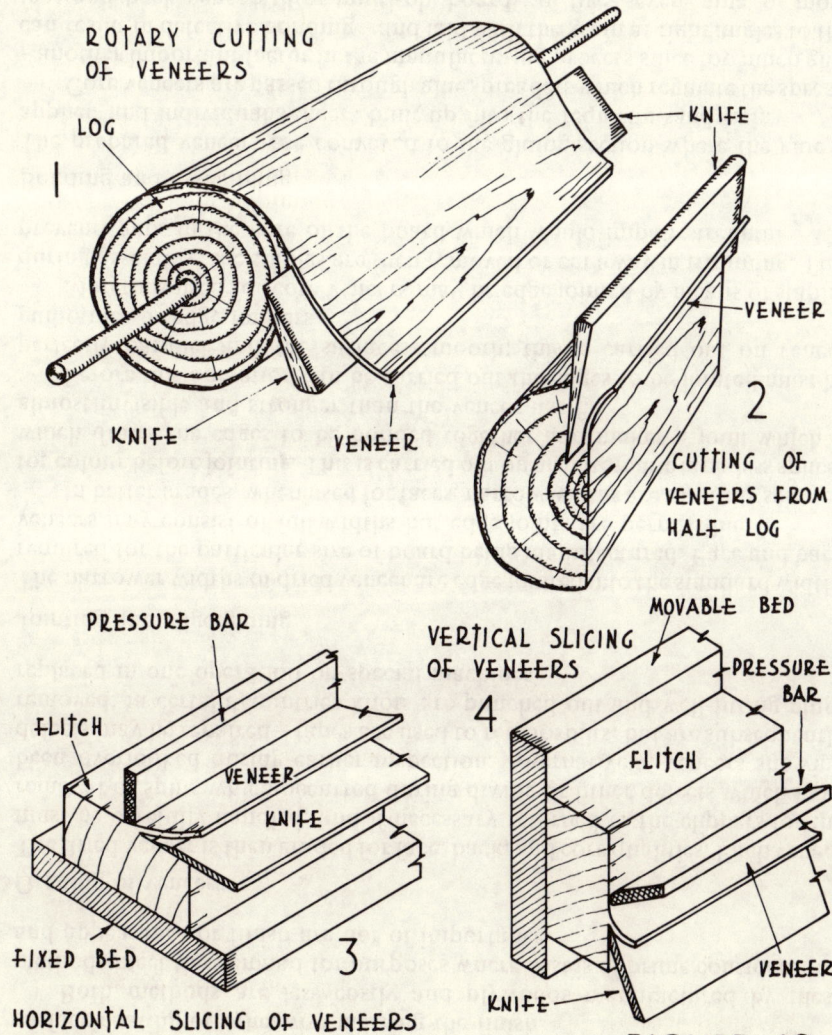

ROTARY CUTTING OF VENEERS

1

LOG

KNIFE

KNIFE

VENEER

VENEER

2

CUTTING OF VENEERS FROM HALF LOG

PRESSURE BAR

VERTICAL SLICING OF VENEERS

MOVABLE BED

PRESSURE BAR

4

FLITCH

VENEER

KNIFE

FLITCH

FIXED BED

3

VENEER

KNIFE

HORIZONTAL SLICING OF VENEERS

FIGURE 19

and checks have to be made by the lathe operator to ensure that a smooth veneer of even thickness throughout is produced.

Unless such control is kept the quality of the finished product will suffer. For this reason the rate of feed, angle of knife, etc. is adjusted to the best advantage for individual logs and/or species.

Plywood manufacturers peel veneers for commercial plywood varying in thickness from about 1 mm to about 4 mm. Boards of the same nominal thickness from different sources of supply may differ in regard to the number of plies and the thickness of individual veneers used in their construction.

This factor, in combination with the species of wood used in the make-up, is of importance in assessing the strength properties of a particular board, in addition to having a direct bearing on the limiting radius of curvature of the material.

Clipping

As the veneer is peeled it is either carried on to conveyor tables or wound on to spindles. It is then fed either through automatic clippers which cut the veneer to predetermined widths, or through manually operated clippers which remove strips of veneer with large defects unsuitable for the manufacture of plywood.

Drying

To ensure that the plywood stays flat it is imperative that the veneers used in its construction are of equal moisture content, otherwise stresses are likely to be set up within the board due to uneven moisture distribution, resulting in the twisting or warping of the finished article.

For this reason, the still wet veneer is fed through a continuous drier, a long heated chamber through which the veneer is carried on continuous belts or rollers. The rate of feed and temperatures are controlled according to thickness and species, so that each veneer emerges at the other end at a predetermined moisture content. The actual figure depends largely on the type of adhesive employed as a bonding medium, since certain adhesives are sensitive to moisture and maximum adhesion may only be achieved within certain limits. The general range is from 5 to 14 per cent moisture content. Excessive moisture in the veneers has a detrimental effect on phenolic adhesives and can cause faulty adhesion.

A recent development is the drying of veneers in a continuous ribbon. After peeling, but before clipping, the veneer is fed into the drier, straight off the reel. It is clipped into required widths when dry after passing through the drier. The advantages are fewer splits, a saving in labour costs, and a reduction in waste.

Wet and semi-dry cementing

Alternative methods, less frequently employed in the manufacture of plywood, are the wet and semi-dry cementing processes. 'Wet-cemented' plywood is manufactured from veneers that are not dried after peeling, but are spread with glue and pressed in a wet condition.

Plywood produced by the 'semi-dry' process may be composed either of veneers that are partially dried or of face and back veneers that have been dried, and core veneers that are still wet prior to assembly and pressing.

Wet or semi-dry cemented plywoods are inferior to those produced by the 'dry-cemented' process described above. They are more inclined to warp and twist and are unsuitable for veneering, painting, varnishing, etc. due to the numerous tiny surface checks caused by stresses set up in the board by the heat and pressure applied during the pressing operation. These checks vary in width under fluctuating humidity conditions, resulting in a breakdown of the surface coating or otherwise marring the finish.

Both methods are less costly and plywoods manufactured by these methods meet the demand for purposes where cost is of prime consideration and appearance or finish are not of importance.

Grading of veneers

The dried veneer is then graded for face, back, and core qualities. Each veneer must be carefully handled, and if necessary returned to the clippers for the removal of splits which occurred during drying or other defects which have been overlooked during earlier inspection. Alternatively, veneers showing defects may be repaired – tapes are used to repair splits, but are subsequently removed; in certain countries knots are punched out and well-fitting plugs replaced in one operation on special machines.

Jointing and edge gluing

The narrower widths of dried veneer are edge jointed into the standard widths required for the particular size of board being manufactured. Face and back veneers may consist of full widths but edge joints are permissible.

In better grades, when used for faces, narrow widths are carefully selected for colour before jointing. This is carried out on an automatic tape-less splicer which draws the edges to be bonded together and makes a joint which is almost invisible and stronger than the veneer itself.

Before this operation can be carried out the edges to be jointed must be perfectly parallel and the surfaces smooth; this is carried out on veneer guillotines or edge jointers.

Materials used for core veneers may be edge jointed by means of staples during assembly; the staples are then removed or cut away in trimming. This prevents gaps in the core of the board which would impair strength.

Bonding and assembling

The prepared veneers are conveyed to the gluing section where the glue is applied and individual veneers built up into the requisite sandwich.

Core veneers are passed through glue spreaders which regulate the spread – another important factor in the manufacturing process since too much glue can result in defective bonding – and laid with the grain at right angles to the face and back veneers. For multi-ply boards, of five, seven, nine, or more veneers, alternate veneers are bonded and laid at right angles to their neighbours until the requisite number of veneers needed for the thickness to be pressed is obtained.

An alternative form of adhesive which is occasionally employed is film glue, a thin phenolic resin-impregnated paper, which is simply cut to size and laid between the veneers during assembly.

The veneers forming any one ply and the corresponding ply on the opposite side of the central plane of the board should be of the same thickness and species and should be cut by the same method, i.e. either all rotary-cut or all sliced, as shown in Fig. 19.

Pressing

The veneers are assembled in packs and are sometimes prepressed cold before hot pressing. This latter operation is carried out in hydraulic presses incorporating multiple heated platens between which the packs are loaded by hand or by automatic feeders. The latter speeds up production by placing the new assemblies between the platens, at the same time ejecting the pressed boards.

Platen temperatures, moisture content, pressure, and time are all controlled to suit the particular batch being pressed.

Conditioning and finishing

Pressing at high temperatures and the various operations carried out during the manufacturing process make it necessary for the boards to be reconditioned, since they may have been left with uneven moisture contents – an undesirable condition where balance is all important. This is effected by stacking or some other suitable treatment until the moisture content of the panels is uniform throughout, after which the panels are cut to size by double trimming saws.

Automatic sanding machines complete the process and ensure that both faces are perfectly smooth and ready for final inspection.

Adhesives

The bonding media used in the manufacture of plywood are of the utmost importance since, among other things, their properties determine the characteristics and end-usage of the final product.

For example, plywood used for internal applications such as furniture needs to be well bonded but does not necessarily require resistance to moisture or water. Conversely, plywood used externally, exposed to all climates, requires an adhesive capable of withstanding the full vagaries of the weather without fear of breakdown.

It is essential, therefore, that the user knows the properties of the various types of adhesives used in the manufacture of the different plywoods available before being able to specify a certain plywood for a particular purpose.

The adhesives most commonly used in the manufacture of plywood may be divided into the following groups for durability:

Interior – animal glues

Manufactured from hide, fleshing, bone, and fish offal, such glues provide an excellent bond under dry conditions of service. They are not water-resistant and if exposed to high relative humidities they are liable to destruction by micro-organisms (moulds and bacteria).

Blood albumen

These glues are prepared from fresh blood obtained from slaughtered animals or from dried soluble blood albumen. Such glues give a moderately strong bond with a high resistance even to boiling water. They are, however, liable to attack by micro-organisms which rapidly cause a breakdown of the glue line under damp conditions. Mixes containing paraformaldehyde are less rapidly attacked but are by no means immune to bacteria.

Casein

Casein glues consist of a mixture of the curds of milk, hydrated lime, and certain other chemicals. As a plywood adhesive, casein provides a strong bond in the dry state, but has only a short life when subjected to even moderately severe conditions of exposure. If exposed for long periods to wet or damp conditions, it is attacked and destroyed by micro-organisms (other chemicals are sometimes incorporated to increase its resistance to bacteriological attack). Exposure to chemical action is also likely to cause a breakdown in the glue line, but casein has good resistance to reasonably high temperatures under dry conditions.

Urea-formaldehyde

Urea resins are widely used in the manufacture of plywood and provide a high bonding strength in the dry state and even after prolonged soaking in water at normal temperatures test, but they can withstand immersion in boiling water for minutes. They are suitable for use under most normal conditions but will break down under continuous exposure to extreme weather conditions. They are immune to micro-organism attack and possess high resistance to acids and alkalis.

For use in plywood manufacture, urea resins are often extended by the addition of cereal flour or blood albumen, but this is accompanied by a reduction in the strength of the bond and in its resistance to micro-organism attack. Conversely, they can be fortified by the inclusion of other resins to increase their resistance to boiling water.

Urea-melamine formaldehyde

By the addition of melamine to urea-formaldehyde resins their resistance to immersion in boiling water is greatly improved. Their chemical and physical properties are also superior and they are immune to attack by micro-organisms; their dry strength is the same. Their durability is also greatly improved, but over long periods they do not compare with phenol-formaldehyde.

Phenol-formaldehyde

Phenolic resins have high strength properties under all conditions of exposure. Long-term weathering tests have indicated their ability to withstand the most severe conditions without deterioration. They are ideal adhesives for conditions of maximum durability and are immune from micro-organism attack. They are very resistant to common solvents, wood preservatives, fire-retardant chemicals, and most acids.

Resorcinal formaldehyde

These resins are principally used as special-purpose glues since they are more expensive than those already mentioned and are therefore not normally employed in their natural state for the manufacture of plywood, but may be

used as fortifiers. They are easy to use and are capable of gluing substances other than wood. They are boil-proof, immune to micro-organism attack and have an excellent record for durability under the severest conditions of exposure.

All British-made, and most imported, plywood is manufactured with synthetic resin adhesives in one form or another.

Some manufacturing countries use soya, or other vegetable proteins, casein, casein/blood albumen mixes, or synthetic resin adhesives.

Production of plywood

Plywood is imported into the UK from numerous producing countries. Finland, Sweden, America, and Canada are examples. It can be obtained with a wide variety of finish veneers such as Douglas fir, Parana pine, and most of the decorative hardwoods, oak, mahogany, walnut, etc.

Unlike a number of other plywood-producing countries, the British manufacturer has to rely on imported timbers for his supplies. Logs imported from West Africa and Asia form the bulk of woods used and consist of such hardwoods as agba, gaboon, khaya, makore, sapele, seraya, and utile.

Sizes

Commercial plywood is normally obtained in thicknesses ranging from 3 to 25 mm but thinner and thicker plywood can be obtained for special purposes. Sheet sizes range from $1·219\,m \times 1·219\,m$ up to as much as $1·829\,m \times 1·219\,m$. The maximum width available is $1·524\,m$; the other standard sizes are 914 mm, $1·066\,m$, $1·219\,m$ and $1·372\,m$. The first dimension quoted when ordering plywood indicates the length of the board.

Decorative veneers

The wood veneer, in both its constructional and decorative forms, is of great importance today. In the former it is the raw material of plywood in its many varieties; in the latter it is the source of an almost limitless range of decorative effects.

Veneers can display decorative features even in less exotic woods which often remain hidden in the conversion of solid timber.

Decorative effects are obtained in a number of ways; by cutting veneers to reveal certain structural features, and from irregularities in growth and defects, colour variations and the arrangement and jointing of veneers.

Two important structural features, growth rings and rays (bands of tissue radiating from the centre of the tree to the bark), contribute to the figuring of veneers and may dictate the method of cutting.

Slicing is a method used to produce decorative veneers for use in furniture, ship-building, etc. but not for standard plywood production. The method is as follows: the logs are first converted to flitches which are normally boiled or steamed and are then clamped to the bed of the slicing machine, of which there are two main types. These are the horizontal and vertical slicers. The horizontal slicer is shown in Fig. 19.3, where the bed is stationary and the knife moves across the flitch. In the vertical slicer, shown in Fig. 19.4, the timber is fed against the knife by means of a movable bed. Accurate matching of the figure when laying requires the veneers to be kept in sequence as they are cut. Slicing is used for the production of nearly all decorative veneers.

Peeling is the only true way of cutting tangentially to the growth rings, and some hardwoods yield a more decorative figure from the rotary cutter than from any other method of conversion. Growth abnormalities such as burrs, which are wart-like growths distorting the annual rings, produce an attractive figure. Such abnormalities are repeated in the veneer with each revolution of the peeler-log. Thus bird's-eye figure in maple arises from repeatedly cutting through small irregularities in the annual rings. The width of the annual rings and the contrast between early wood and late wood are other factors which affect the pattern.

Decorative veneered plywood may be obtained, either with a decorative surface as an integral part of the board construction, or alternatively as a plywood board which has had the decorative veneer added.

The decorative veneer should be laid at right angles to the grain of the face whether based on a plywood, blockboard, or laminboard. The moisture content of the board and veneer should be approximately the same.

Plywood, blockboard, and laminboard are of balanced construction and therefore stable. If a decorative veneer is added to such boards, a compensating veneer should be applied to the other face to prevent distortion.

Shaped plywood

Plywood may be bent to conform to simple curvatures, as shown in Fig. 21.2, the limiting radius being governed by the thickness and construction of the plywood and the direction in which it is bent. The radius of curvature of any plywood can be improved by steaming, and various other means such as a saw kerf or kerfs at the apex of the curve, or the reduction of the number of veneers for use with shaped or flexible mouldings. These methods are quite satisfactory for simple curves in joinery, cabinet-making and other purposes but cannot easily be applied where compound curves are required.

Moulded plywood

Plywood may be manufactured into pre-formed shapes in which the grain of the veneers may be orientated so that maximum strength is obtained in any given direction. This fact, plus the durability of present-day glues, opens up a wide field of possible uses.

Moulded plywood, shown in Fig. 21.1, is produced in three ways by using:

1. Male and female forms into which the veneers are placed and bonded under pressure, usually by high-frequency electrical heating.
2. Vacuum process in which a male form only and a rubber bag or sheet is required.
3. By specially constructed shaping or moulding presses or pressure cylinders, whereby fluid pressure is exerted together with heat.

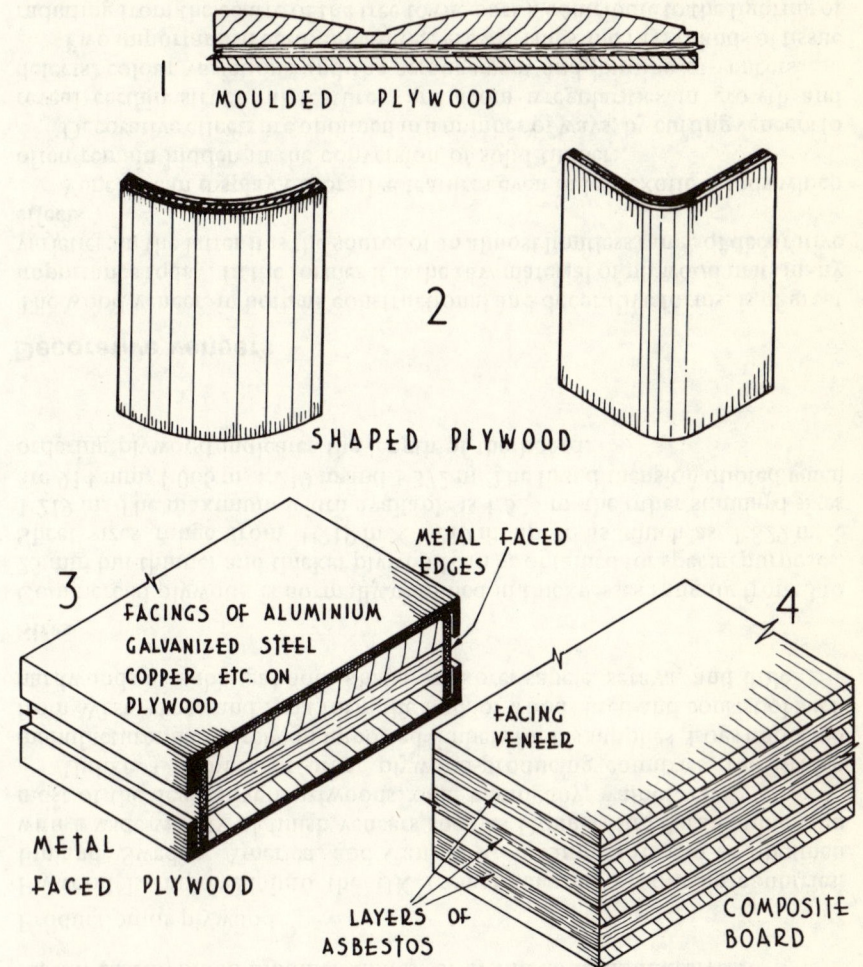

PLYWOODS

FIGURE 21

Rubber bag presses

For certain curved work, vacuum and dome rubber bag presses have been widely adopted. The press consists essentially of a heat-resisting rubber blanket, over which are electrical heating elements covered by a dome. The work is placed on a table, which forms the base of the machine, and the rubber blanket presses down on the work. The pressure is brought about by means either of compressed air blown into the dome, or as a result of the air being evacuated from between the blanket and the table, or both. Heating elements are also fitted below the table. Rubber bag presses are commonly used for applying veneers to shaped cores and for making curved laminated panels. In the latter case, the shape is obtained from solid moulds lying on the table.

Strip heating

Apart from steam-heated presses and bag presses there are two forms of industrial heating especially applicable to the woodworking industries. These are strip heating and radio-frequency heating. Each has its special uses, but both are versatile in that they are not applicable only to flat work, and the heat is generated only where it is required.

Strip heating – or low-voltage heating – involves heating the joint with an electrically heated bare metal strip. The current is often about 6 V, and does not usually exceed 12 V.

Apart from a step-down transformer capable of handling a large current, the only other requirement is a jig suitable for the work in hand.

The working temperature of the strip will vary according to the work in hand, but as an indication it can be said that for the application of wood veneers a temperature of about 94°C will be suitable. This is well below the temperature at which the wood will scorch, but it is high enough to set the glue rapidly. Using Aerolite KL glue and powder hardener L58, the basic setting time will be about 25 seconds. If laminated plastic is being applied, the temperature should be kept down to about 70°C. Using the same glue and hardener, the basic setting time will now be 2 or 3 minutes. If there are glue lines more remote from the heating element, longer time must be given.

Jigs can usually be made with the skill and resources available in an average joinery shop or can be bought from a firm specialising in this work. Most parts are of wood. The design should be such that a pressure of at least 172 kN/m² can be brought to bear on the glue line. This can be done by means of screw clamps, levers attached to eccentric cams, or rams operated by compressed air or hydraulic pressure.

Radio-frequency heating

This form of heating glue lines depends on the principle that, when two metal electrodes are connected to a source of electric current alternating at radio frequency, heat is generated in non-metallic materials situated between the plates.

The advantages of radio-frequency heating are that the material, if of the same kind and uniform in section, is heated uniformly throughout its mass; the quantity of heat generated in the work can be accurately controlled; there is no waste of power during periods of rest or shut-down; and thermal efficiency is high since it can usually be arranged that most of the heat is generated only where it is wanted. The most important advantage is, however, that the rate of heating is usually greater than in methods depending on conduction, and productivity can, therefore, be greatly increased. The speeding of gluing processes depends upon the fact that synthetic resin adhesives, such as Aerolite, set much more rapidly when they are heated. Since they have a higher power factor than wood, they heat under radio-frequency more readily than the wood.

In designing a radio-frequency jig, the disposition of the glue line or lines relative to the electrodes is a main consideration. There are three types of arrangement.

In glue-line heating, the electrodes are so placed that the glue is in a plane approximately between them.

In through heating, the two electrodes are placed on either side of the joint, so that in a simple joint there are three layers between the electrodes – wood, glue, wood. In laminated assemblies, there will be numerous layers of wood and glue.

In stray-field heating, the electrodes are brought into the neighbourhood of the glue line, and because of the glue's greater power factor, the radio-frequency energy deviates into the glue line, heating and thus accelerating the cure.

Tongued and grooved plywood

Tongued and grooved panels with accurately matched joints are available for uses which include flooring, roof cladding, partitions, wall panels, ceilings, as well as shuttering and heavily loaded ships' decking. Their advantage lies in speedy assembly and the result is perfect alignment of surfaces and adjacent panels.

Standard panels are up to 2·438 m × 1·219 m; 12 mm, 16 mm, 19 mm and 25 mm in thickness; and grooved and tongued along alternate long edges. Short edges are square and should be supported on framing. Some panels are grooved on both long edges and fitted with a loose plywood tongue. Other panels are also produced in standard sizes 1·524 m/406 mm/508 mm, and 610 mm, and 12 mm and 18 mm in thickness. These panels, tongued on two adjacent edges and grooved on the remaining edges, provide load-carrying and matched joints which may be unsupported, and panel edges need not, therefore, coincide with joists on timber framing. The above types of panel have been designed to carry domestic floor loading at 146·40 kg/m² on joists or supports spaced at nominal centres ranging from 406 mm to 610 mm, according to the thickness of the plywood. Heavier loads are possible with closer spacing of supports, or by supporting the panels on their long edges or on all four edges.

Large panels, scarf jointed from standard size components, are regularly produced. In this well-established method the scarf joint is as strong and durable as the plywood itself. The size is virtually limited only by the facilities for physically handling the panels. Lengths of 6·096 m – 9·144 m are common.

STRUCTURAL PLYWOOD

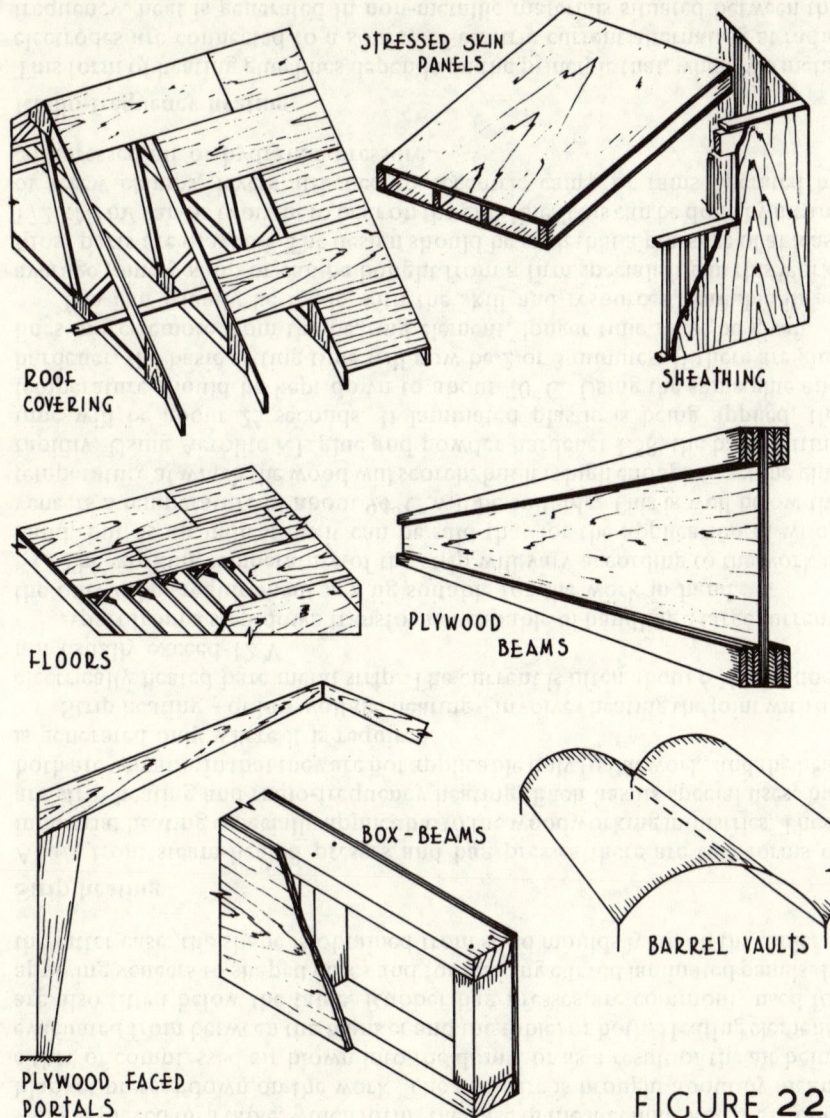

ROOF COVERING

STRESSED SKIN PANELS

SHEATHING

FLOORS

PLYWOOD BEAMS

BOX-BEAMS

BARREL VAULTS

PLYWOOD FACED PORTALS

FIGURE 22

Protective treatments

The term 'weatherproof' when applied to plywood refers only to the bonding media employed in its manufacture. No fear of delamination due to dampness or exposure to weather need be entertained in plywood of this type, but the wood of which the plywood is comprised is nevertheless liable to destruction by fungi attack. This can only be prevented by the following:

1. The use of plywood constructed throughout with a species known to be durable.
2. Affording a suitable means of protection to the face, edges and back of the material.
3. Treating the plywood with a suitable preservative.

There are numerous exterior grade paints and varnishes available, but it should be noted that a surface application is not sufficient to withstand continuous exposure to weather. The face should first be primed with a suitable primer and edges of boards sealed with paint and/or lead paste to prevent the infiltration of moisture through the edges of the boards. The backs of boards should be treated with a red or white lead paint if used under damp conditions.

Plywood bonded with a suitable adhesive may be impregnated with certain preservatives.

Various examples of the use of plywood in timber structures are shown in Fig. 22.

Laminboard

This is built-up board and often included in the more general sense under the heading of plywood. The construction consists of narrow strips of softwood faced on both sides with face and back veneers. These face veneers may be either single or double with the grain running at right angles to the core. Thicknesses of laminboard range from 12 mm to 25 mm. It is most suitable for a base for veneering. Figure 20.4 shows the construction of laminboard.

Blockboard

Blockboard consists of a core of wood made from strips up to 25 mm wide placed together, with or without glue between each strip, to form a slab which is sandwiched between outer veneers of 2–3·5 mm with their grain direction at right angles to the grain of the core. Cross-grained blockboards may have single and double outer plies on each side of the core, but the grain of all veneers runs at right angles to the grain of the core.

When the length exceeds the width, the blockboard should be five-ply construction: a core of the same construction as that mentioned above, a veneer on each side of about 2 mm thickness running at right angles to the core, and a thinner outer ply running parallel to the core.

Boards where the strips have not been glued together to form a slab are sometimes known as stripboards.

Blockboard is shown in Fig. 20.5.

Battenboard

This is a variation on blockboard construction in which the core is built up of strips usually not exceeding 75 mm wide, as shown in Fig. 20.6.

Metal and plastic-faced plywood

Plywood of almost any thickness may be obtained faced with a number of metals – stainless or galvanised steel, copper, aluminium; or, faced with a decorative laminated plastic veneer. Such panels are used for counter-tops and wall cladding where surfaces are required to be both decorative and hygienic and easy to clean. Sheets may be sealed at the edges and rendered waterproof and vermin-proof.

Plastic-faced plywood with high abrasion resistance is also produced for use in concrete formwork to give a smoother surface finish and a greater number of re-uses.

Metal-faced panels are used in places where great rigidity is required, for example in showers and shopfitting work where the plywood core stiffens the panel and provides fixing for the metal facings.

Metal-faced plywood is shown in Fig. 21.3.

Composite boards

Plywood with cores of insulating materials such as cork, asbestos fibre, and foam rubber are generally made specially. They are used for cladding where heat insulation is important, such as in cold storage rooms, and sometimes for sound damping in studios and telephone booths. Composite board is shown in Fig. 21.4.

Manufactured boards

The manufacture of 'synthetic wood' from wood chips or sawdust has over recent years developed on a large scale commercially.

The process requires the main constituents: wood chips or sawdust, a bonding agent, and a powerful heated press. The bonding agent normally used is a synthetic resin urea-formaldehyde greatly used in plywood manufacture. Boards may be produced by this method in any thickness from 6 mm upwards in dimensions up to 4·8 m × 1·220 m, the latter being governed by the size of press used.

Insulation board and wood fibre hardboard are other examples of wood waste boards.

Particle boards

These consist basically of wood chips or flax shives (i.e. the woody part of the flax plant after the fibres have been removed) bonded together with a synthetic adhesive. Their most important advantages lie in the large flat panels which can be obtained and their freedom from variability in physical properties.

The boards are produced in a press by the application of heat and pressure by a process similar to that used in veneering. There are some wood chipboards where the resin and chip-mix is forced between parallel metal platen, the distance between which corresponds to the thickness of the board produced. These core-boards do not have the uniform structural strength properties obtained in the flat pressed boards.

The material nails well and will hold them well except near the edges. It is comparatively stable under fluctuating conditions of temperature and humidity and can be worked equally well with hand or machine tools as in normal woodworking practice. Particle boards are suitable for use in most constructional situations, such as under roofing, flooring, and panelling where they are not subject to the action of the weather.

'Weyroc' chipboards now firmly established as a building material are produced in various grades. The standard grade is used for built-in fitments, roof lining, partitioning, panelling, and many other applications in building where a rigid and stable board is required. It is available in board sizes up to 5 m × 1·5 m from 10 mm thickness to 36 mm.

Other grades in the medium-density range of chipboard, but of graded density construction, are available for furniture manufacture. They may be obtained veneered with oak or mahogany with a plain compensating veneer on the reverse side, or decorative both sides.

Chipboards are widely used for industrialised building.

Flooring

Chipboard for flooring is slightly heavier though of similar construction to the standard grade, available in similar panel sizes, but of one thickness, 19 mm. It can be used for suspended flooring or laid on a solid sub-floor. For floors with joists at 400 mm centres, with superimposed loading up to 480 kg/m², the flooring grade of chipboard has adequate bending strength and impact resistance, and will sustain normal pedestrian traffic and dead loading occasioned by furniture and equipment.

Stressed skin flooring panels

The advent of industrialised building methods has brought increasing interest in stressed skin construction. The prime objects are to reduce timber size and therefore cost, or alternatively, increase the length of span that a given timber joist can cover. An efficient stressed skin construction using 19 mm Weyroc nailed and glued to one side of timber joists will give a stiffness and load-bearing capacity of some 2–2½ times the values for a normal joisted floor. This means that with such a construction, the joist depth 3/5 may be reduced by approximately 25 per cent, or alternatively the span could be increased by about 12 per cent.

Roofing

Chipboard may be used in low pitch or flat roof of timber shell construction. Being available in large board sizes, an area can be covered rapidly effecting a saving in labour. The smooth surface, even thickness, freedom from warping and uniform properties enable the final cladding material to be applied. Used purely as a structural support to the roof covering, insulation may be obtained by fibreglass quilts or similar, or by using low-density boards it can give insulation by itself.

Partitions

Chipboard used for the construction of internal partitions may be applied to on-site studding or prefabricated with equal facility. Panels may be veneered with any available wood veneer or faced with laminated plastic, PVC, or other decorative sheet materials used in the building industry.

Wallboards

There are many proprietary brands of wallboards used mainly for covering walls, ceilings, and partitions. A number are essentially the province of the plasterer but those made from fibre and pulp are fixed by the joiner.

Insulating board

In the manufacture of insulating board most common species of timber are used, including both hardwoods and softwoods or sugar-cane fibre.

Timber is converted to pulp either by wet grinding on abrasive stones rotating at high speed, or by first chipping the timber and then reducing the chips to fibres.

The pulp is mixed with a small percentage of sizing additives and is then fed on to a forming machine, consisting of an endless belt, or a cylinder of meshwire. Here, most of the water is drained, sucked, and pressed out of the pulp. The continuous strip of wet board so made is rolled to thickness, cut to lengths, dried and finally trimmed to size.

The finishes of insulation board range from cream to grey and brown in colour with smooth, rough textured, or moulded surfaces.

Sizes are from 12 to 25 mm in thickness and widths up to 1·2 m and lengths up to 3·6 m.

Hardboard

The forming of hardboard is similar to insulation board. They are manufactured by two processes: (1) wet, and (2) dry.

In the wet process the board is formed and cut to size then loaded into a multi-day light press operating at high temperature and pressure. A wire screen and plate inserted under each board allows water and steam to escape and accounts for the pattern on one face of all hardboards made by the wet process.

The dry, pressed sheets of hardboard emerging from the press are usually further cured by heat treatment. They are then humidified before being trimmed to size.

In the dry process, a board is fed into the press after drying. Since there is no water and steam, no meshwire is used, resulting in a hardboard sheet with both sides smooth.

In a second dry process method of manufacture, wood is converted to fibres in the dry state. The dry fibre is handled pneumatically and fed through blowers on to the forming machine. The pressed boards have both sides smooth also by this method.

Tempered hardboard is made by impregnating standard hardboard sheets with hot oil or resin.

Decorated hardboards

Boards which are painted, printed, plastic laminated, or otherwise veneered in the factory to provide a finished panelling or surfacing material are termed 'decorated hardboards'.

Moulded hardboard

These boards are made by using figured plates in the press where the pattern is transferred to one side only of the hardboard. Patterns available include: reeded, leather-grained, tiled, bamboo, etched wood grains, linenfold, and cable patterns.

Moulded hardboard is used for panelling furniture, shop and bar fittings, display and exhibition work.

Perforated hardboard

This type of hardboard is used for interior decoration, shopfitting and display work. Ventilation panels and surfacing to acoustic materials are other applications. Perforations are usually round, but different shapes are available including square, diamond and slots.

Laminated plastics

Today, plastics play an ever-increasing part in the building industry which includes, of course, the carpenter and joiner.

Decorative laminates may be described as high-pressure, thermoset decorative laminated plastic veneers. They are built up of a series of synthetic resin impregnated paper sheets which are fused into a homogeneous mass by heat and pressure. The core of the material is built up of impregnated kraft papers which is covered by a plain colour or patterned sheet of paper. An overlay transparent sheet heavily impregnated with melamine resin is then laid over the decorative paper. This top sheet is transformed in the press into a hard and durable surface, impervious to moisture, resistant to bacteria, fungus growths, grease, oil, cigarette burns, alcohol, household solvents, and most chemical substances.

The layer composition of Arborite decorative laminate is shown in Fig. 23.1. It may be used as a surfacing material for walls, ceilings, work surfaces, cabinets, furniture, displays, doors, window-sills, or practically any surface in the interior of a building.

For general-purpose applications mentioned above, the recommended thickness is 1·5 mm standard grade.

Sizes

Standard sheet sizes are 3 m × 1·2 m and 2·4 m × 1·2 m. The largest sheet size available is 3·6 m × 1·5 m. Smaller sizes are also available.

Colours, patterns

The complete range of Arborite decorative laminates contains over 200 patterns and colours including many woodgrains. Edge trims may be obtained in rolls to match all patterns to give a flush finish to edges, sills, etc. shown in Figs. 23.3 and 23.4.

LAMINATED PLASTICS

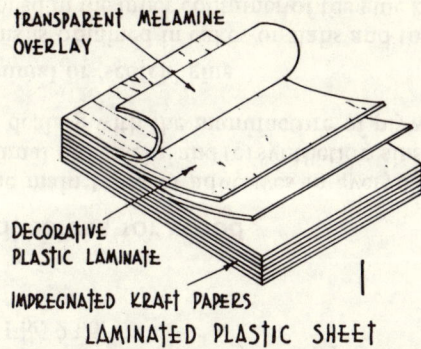

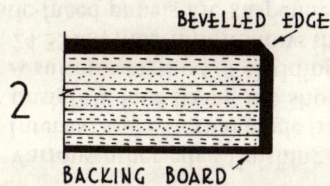

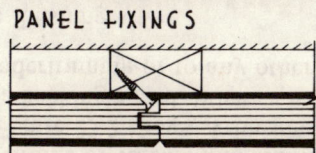

TRANSPARENT MELAMINE OVERLAY

DECORATIVE PLASTIC LAMINATE

IMPREGNATED KRAFT PAPERS

LAMINATED PLASTIC SHEET

1

D.V.C. EDGE TRIM 3

COUNTER NOSING

HARDWOOD EDGE 4

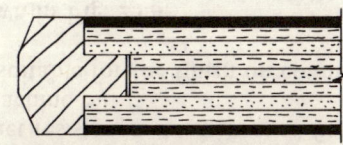

HARDWOOD EDGE

BEVELLED EDGE

2

BACKING BOARD

PANEL FIXINGS

JOINTING PANELS 5

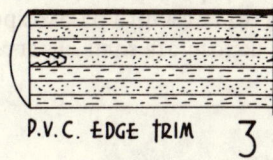

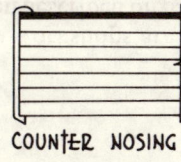

PLASTIC INSERT EDGE

VARIOUS METHODS OF EDGE FINISHINGS

FIGURE 23

Cutting

This may be carried out using a fine-toothed tenon saw, keeping the angle of the cutting stroke acute in relation to the panel. Power tools may also be used quite successfully but it should be pointed out that the material should be placed face down because of the direction of rotation of the cutting blade.

Decorative laminates may also be cut on the circular saw bench. A hollow ground high-speed steel blade, 300 mm in diameter and of 9 gauge, is most suitable, keeping the saw as low as possible in the bed to avoid chipping.

Bonding

The thin plastic veneer requires some form of supporting or backing material for most occasions in actual use, although there are many instances where the surface to which the veneer is to be bonded already exists. For table tops and panels for general use, a base of blockboard, laminated board, or chipboard is employed. The thickness of the backing member or 'corestock' will depend upon the use of the panel, but a safe guide is that when the panel is to be securely fixed to a carcass or underframe, the minimum thickness of 12 mm should be used. An unsupported panel should have a minimum thickness of 18 mm. Exterior grade plywood for the corestock of panels provides a core which has a commercially acceptable standard of flatness.

Adhesives

There are three types of adhesives used in the bonding of laminated plastics. These are:

1. Contact adhesives which are rubber based.
2. Poly-vinyl-acetate (PVA) adhesives.
3. Synthetic resin adhesives.

Contact adhesives The principle involved in the use of contact adhesives is that a synthetic rubber body is carried in a vehicle of solvent, having a base of acetone. The solvent enables the body to be applied to an area uniformly. After application the solvent evaporates, leaving a deposit in the form of a thin rubber film. When two such films are pressed firmly together they will weld. Sustained pressure is not necessary when contact adhesives are used because of the very high initial tack rate.

The contact adhesive should be applied thinly and evenly over both the surfaces to be bonded and left for 15–20 minutes for the solvent to evaporate from the adhesive before the surfaces are brought together. As soon as the two coated surfaces come into contact adhesion takes place, therefore it is necessary for the panel to be accurately positioned before closing the joint.

Because there is no tolerance on a rubber-based glue line as far as location is concerned, it is always advisable to cut the panel a little oversize, thus enabling the panel to be cleaned back to the corestock.

Poly-vinyl-acetate adhesives These adhesives are in the form of a plastic emulsion. They do not have a high initial tack rating, and overall pressure must be applied to the assembled panel during the glue line curing period. This may be accelerated by the application of heat (see strip-heating, page 35).

As this type of adhesive can be applied by brush or roller it is much easier to handle than the contact type, enabling the panel to be positioned accurately after the surfaces have been brought together through the glue line, allowing the panel to slide freely.

Synthetic resin adhesives   (see pages 33 and 41).

Sealing

After laminated panels have been bonded to a base of plywood, blockboard, or chipboard, the back side of the panel must be restrained against movement caused by changes of humidity. This can be done by securing it to a rigid underframe or carcass and sealing it with paint or varnish.

Figure 23.2 shows Arborite-surfaced corestock sealed on the back side with a backing board. Where a panel is to be free-standing and is not fixed to an underframing or to any other structure, it should be backed as shown.

Edge finishing

Figure 23.2 shows a bevelled-edge finish worked on the corners, using a metal plane or second-cut file, to increase impact resistance. An improved finish may be obtained with a smear of light machine oil.

Figure 23.3 shows a PVC edge-trim suitable for the finish on countertops.

Various other edge finishing are shown in Fig. 23.4.

Internal and external angle fixing of laminated plastic panels eliminating any unsightly face fixing are shown in Figs. 24.1 and 24.2.

A suitable method of cladding walls with plastic-faced panels is shown in Fig. 24.3. The illustration shows the grounds, clad with plastic upon which the plastic-faced panels are suspended. Contrasting patterns or colours may be used to pleasing effect in this way.

Various methods of jointing panels and fixing to groundwork are shown in Fig. 23.5.

Adhesives for wood

The main types of adhesives in everyday use in joinery workshops are (a) animal, (b) casein, and (c) synthetic resins. These have been described on page 32 dealing with the manufacture of plywood.

Animal or 'scotch' glue

This is obtained in cakes or slabs and to prepare it the slabs are broken up, placed in the inner container of the glue kettle and covered with water to soak overnight. The water in the outer container is then heated until the glue softens and is of even consistency ready for use. Should the glue be too thick, hot water may be added to bring it to the required consistency.

Casein

This glue is sold in powder form and is mixed by measure with water. It is used cold with pressure applied by cramping. Casein is liable to stain some softwoods and many hardwoods. This staining does not matter so far as wood

LAMINATED PLASTICS

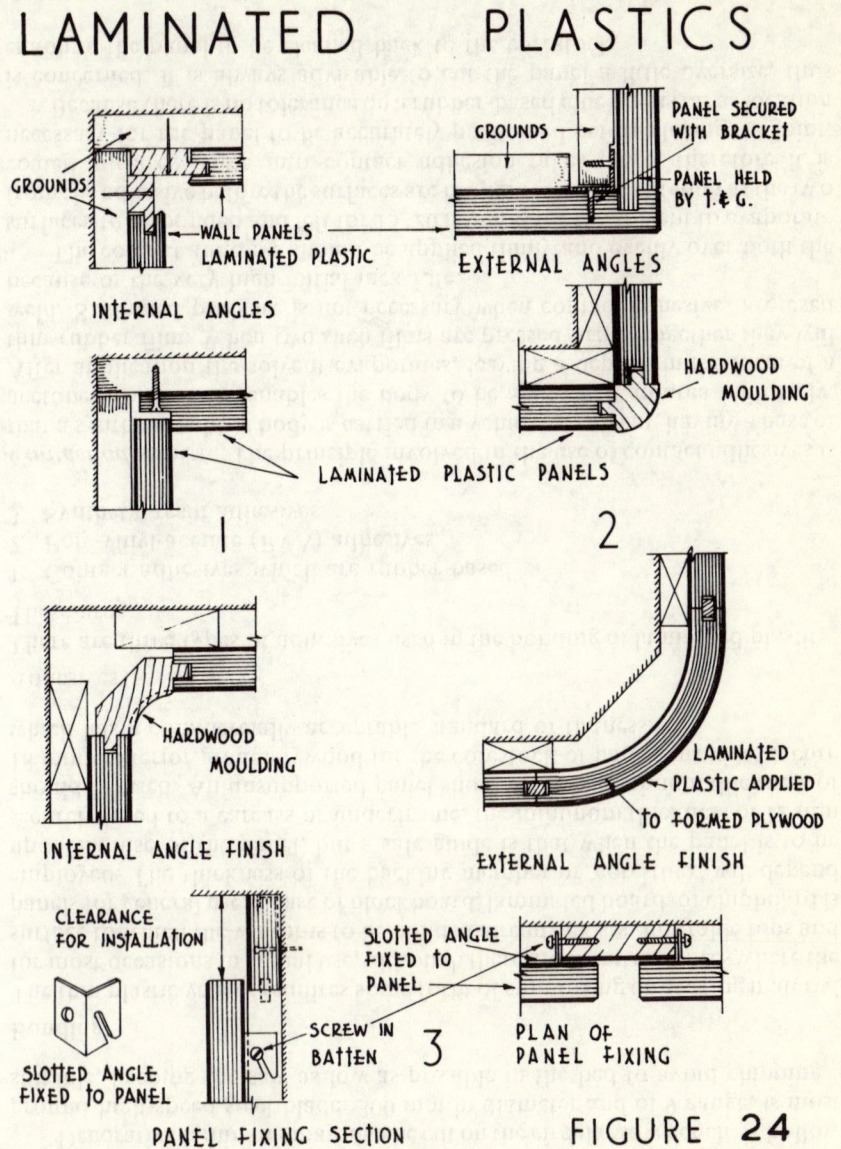

FIGURE 24

for painting is concerned. A further disadvantage of this type of glue is that it has only a short life when subjected to even moderately severe exposed conditions. If exposed for long periods to wet or damp conditions a breakdown in the glue line soon occurs.

Synthetic resins

Cold-setting synthetic glues have now largely replaced animal glues and casein for joinery and constructional purposes, since they have the great advantage that they are proof against damp conditions and are not attacked by fungi or micro-organisms.

Among gap-filling glues, Aerolite 300 has been widely adopted. It has the advantage that it is gap-filling, fulfilling the requirements of BSS 1204 in this respect; that is to say it can make strong permanent joints even when the surfaces are 1·25 mm apart. This compensates for small inaccuracies in the machining of the timber. It also means that only light clamping is required.

Different hardeners can be used to obtain the desired setting time. At ordinary shop temperatures and using a hardener of medium speed the joint can be safely handled after about 3 hours, though the glue takes a considerably longer period to acquire its full strength, and water resistance is not fully developed for a week or two.

Where it is required to speed up production, the setting time of the glue can be greatly accelerated by the application of heat. For example, at 88°C the setting time with a hardener of medium speed is 2½ minutes.

For assembly gluing, the glue has the advantage that it is used by the separate application method, i.e. the glue is applied to one of the two surfaces to be joined and the hardener to the other surface. Setting does not begin until the two parts of the joint are brought into contact, and thereafter it proceeds speedily. For some large structures it may be convenient to use a mixed application of glue with a longer setting time. In such cases Aerolite 303 SM is suitable.

The above type of glue is available in powder form, having a long shelf life. To prepare liquid resin for use, it is only necessary to add water and stir.

There are further types of glues available by Aerolite, intended primarily for use under conditions which do not call for gap-filling properties. They can be used in the cold, but by the use of heat the speed of setting is accelerated. Reference should be made to hot presses, rubber bag presses, strip heating, or radio-frequency heating (see pages 34 and 35).

A gap-filling resin especially suitable for the making of large laminated timber structures is Aerolite 311. Used with one of the hardeners, it has a long closed assembly time. That is to say, plenty of time is available in making up the structure for adjustments before full clamping pressure is applied.

Resorcinol-formaldehyde resins

These glues have outstanding durability under severe conditions of exposure. They are gap-filling and have excellent resistance to moisture and attack by micro-organisms. Aerodux 185 and 500 are very tolerant of moisture in the wood being glued. They cure at normal shop temperatures and are used very extensively for the construction of laminated building structures, trusses, and joinery. They are excellent adhesives for many different materials including cement-asbestos, brick, concrete, unglazed porcelain, rubber, cork, and linoleum.

Epoxy resins

Among plastic materials which are bringing about major changes in the building industry, epoxy resins can be regarded as of prime importance. These resins, since they harden without giving off any volatile matter, can be used for joining even completely non-porous materials such as metals and glass. They are capable of 'welding' concrete, brick, tile, terra-cotta, granite, marble, and ceramics, as well as wood and most plastics.

Bonds made with Araldite are exceptionally strong, and when joints are made properly, any subsequent rupture occurs in the material bonded and not in the resin.

The resin can be cured cold and the cured material is exceptionally strong and resilient, waterproof, and unaffected by acids, alkalis, and solvents. It therefore provides a perfect bonding medium for masonry of all kinds. It is generally used, however, where exceptional properties are desired and it is not suggested as an alternative to conventional materials where their performance is satisfactory.

Epoxy resins are used extensively in the manufacture of paints and coatings. Araldite-based coatings possess outstanding adhesion, chemical, and water-resistant properties, and give films of exceptional toughness and abrasion and corrosion resistance. Applications include, clear resistant finishes for wood surfaces including floors, and heavy-duty protective coatings for concrete, metal, brick, and other surfaces.

Chapter 5

Acid and corrosive effects in timber

Effect of acids and metals on oak and other woods

Oak, along with certain other woods – chestnut, western red cedar, Douglas fir, etc. – contains gallic acid which rapidly corrodes ironwork when these two materials are used together. This results in disfiguration of the timber in the form of an inky-blue stain on the surface of the wood.

Acids fall into two classes: weak and strong. All have a sour taste, as vinegar, lemon juice, etc. and contain hydrogen.

Acetic acid is a weak acid, while sulphuric acid is in the strong group and will cause severe burning should it come into contact with the skin. Hydrochloric acid is also in the latter group.

Tannins also occur in oak in considerable quantities and are characterised by their relation with iron salts with which they form greenish-black or dark blue-black compounds. In the past, use has been made of this property to make inks. Great care should be taken when using woods containing amounts of tannin, like oak and western red cedar, in association with nails and ironmongery, to avoid any such disfiguration of the surface as the timber becomes moist.

Where metallic fixing and fastenings in contact with such timbers are unavoidable, all ironwork, including nails, should be galvanised or sherardised.

Rainwater running off roofs covered with cedar shingles is acidic and liable to corrode some metals. Gutters, flashings, etc. should therefore be galvanised or in copper or plastic.

Various experiments to show the chemical effects of metals on timber may be carried out as follows: steel, brass, and black-japanned screws are put into specimens of various timbers which are then immersed in tap water and salt water. The results are compared and noted.

The staining effect by metals on oak and the breakdown of the metal itself can be observed when specimens of 50 mm × 50 mm section are part covered with lead, copper, zinc, and steel and left exposed to the weather.

The effects of metals on various woods may also be observed in the same way.

Acidity and alkalinity

The presence of acids in oak, Douglas fir, birch, and certain other common woods makes them well known for their corrosive properties, especially if they are stored under warm, damp conditions. An important consideration in the manufacture of wooden packing cases, containers, and crates for the storage and transport of metallic articles, is the moisture content of the timbers to be used. The moisture content of the timber controls the humidity of the air within the container, hence the risk of corrosion to metallic contents is real where unseasoned timber is used.

Many woods contain volatile oils as well as moisture. Turpentine is an example. These oils have a solvent action on coating materials applied to contents for protection.

The volatile substances given off from timber vary from one timber to another when subject to tests under conditions of high temperature and high humidity. Of the acids contained in oak and Douglas fir, the most prominent is acetic acid. This acid vapour is one of the most corrosive vapours known. Cases or containers should be ventilated, or a lining of either a moisture-barrier or aluminium foil used; this helps to counter any possible corrosion in preventing acid vapours reaching the contents.

Before discussing the degrees of acidity in woods, some mention of corrosion is necessary.

Corrosion

This is a process of oxidation (or rusting) and the combination of oxygen with metals produces oxides of those metals. The extent to which corrosion takes place is limited by the formation of scales on the metals' surface due to chemical action. This coating of a film of rust can attain great thickness in layers which tend to fall off, exposing other parts of the metal which in turn rust. Other metals form an oxide on the surface but flaking off does not occur as with iron, so that some protection is afforded the metal from the coating, preventing further oxidation.

The surface film which copper develops serves as protection against corrosion by the action of the atmosphere. Copper goes black and turns green in colour in the course of time. Sheet lead also forms a protective coating when exposed to air and turns a dull grey.

Corrosion of metals normally occurs when the air in contact with the metals has a high relative humidity. Rusting uses up part of the air in the chemical action which takes place and the higher the relative humidity, the higher the rate of corrosion. When water falls on iron one of the two elements in water, oxygen, combines with the iron and produces rust or oxide of iron. Some protection from the atmosphere is usually given to iron by a surface coat of paint.

The scale which peels from hot iron when struck with a hammer is an oxide, and is produced by oxygen of the air combining with the hot surface of the iron.

It has been said that corrosion of metals normally occurs when the air in contact with the metals contains moisture and oxygen. Also, that the presence of acids in timber increases the rate of corrosion in metals, when timber and metal are used in association with each other. It should be pointed out that serious corrosion of metal is not likely to occur when well-seasoned timber is used or when dry conditions prevail.

Degrees of acidity in woods

The measurement of acidity in woods is made by experiment, preparing one part sawdust to five parts of distilled water and measuring the acidity and alkalinity. This value, termed pH value, is a measurement of the suspension, using a pH meter. Comparative values of acidity are given in the table, where a pH value of 7·0 is neutral. Values below 7·0 indicate increases in acidity. Where a pH value drops from 4·0 to 3·5, this equals an increase in acidity of five times.

Hardwoods	pH value	Softwoods	pH value
Beech	4·5	Douglas fir	3·0
Chestnut	3·6	Pine, Scots	4·3
Elm	6·0	Spruce	4·0
Jarrah	3·0	Western red cedar	3·0
Mahogany	4·5	Western hemlock	4·8
Oak	3·3		
Teak	4·5		

A surface or substance may be tested for acidity or alkalinity by wetting the surface with distilled water and applying to it a sheet of neutral litmus paper. Should the litmus turn red this indicates the surface is acidic. If the surface is alkaline the litmus paper will turn blue in colour.

Stains in timber

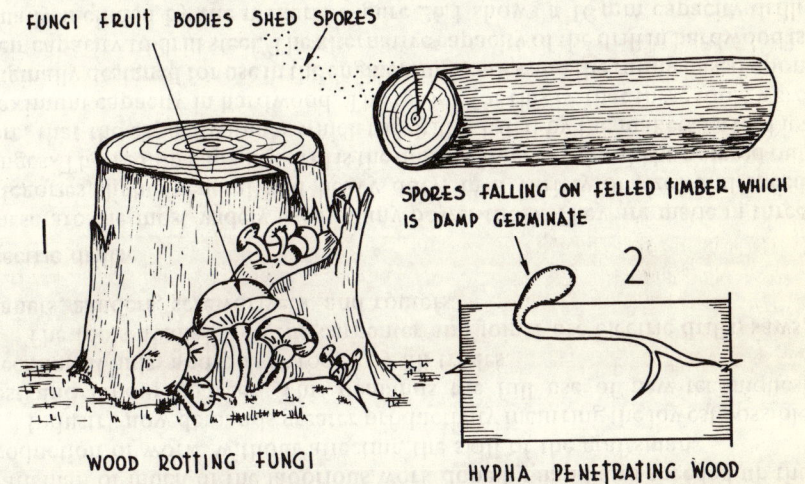

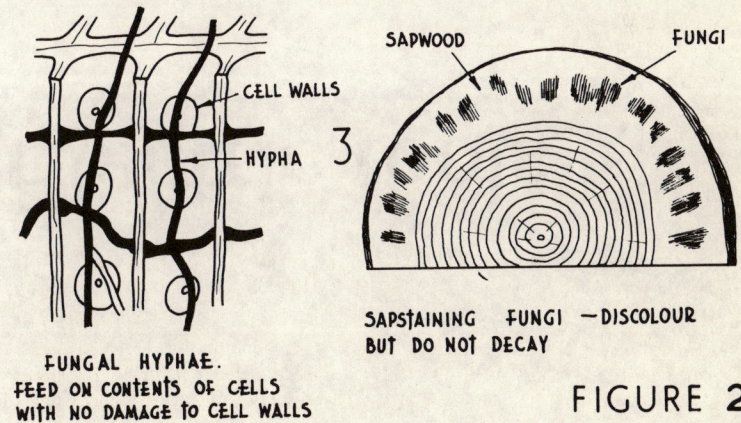

Apart from the disfiguration of certain timbers when ironwork and the timber are used together, other stains can be due to various causes, and discolorations are usually detrimental only when the timber is to be finished in a natural colour. Although discoloration may be apparent, no deterioration is present in the wood in most cases. It is important, however, to differentiate between those stains which have no effect on strength properties and those which indicate the early stages of decay.

A close similarity is to be found between the grey streaks of mineral stain and those of wood-rotting fungi. A satisfactory test may be made by using a knife to raise a splinter of wood; if this breaks off short, giving a brash fracture, the weakening is probably due to the presence of dote.

The most common types of stain in timber may be grouped as follows:

1. Those caused by harmless fungi, including mould fungi and sap-staining fungi.
2. Those caused by wood-rotting fungi, dote, indicating areas in the early stages of attack.

STAINS IN TIMBER

FUNGI FRUIT BODIES SHED SPORES

SPORES FALLING ON FELLED TIMBER WHICH IS DAMP GERMINATE

WOOD ROTTING FUNGI

HYPHA PENETRATING WOOD

CELL WALLS

HYPHA

FUNGAL HYPHAE. FEED ON CONTENTS OF CELLS WITH NO DAMAGE TO CELL WALLS

SAPWOOD

FUNGI

SAPSTAINING FUNGI —DISCOLOUR BUT DO NOT DECAY

FIGURE 25

3. Those arising in the wood during growth, 'mineral stain' and 'stone' caused by frost and soil conditions.
4. Those caused by chemical contamination – 'iron stain'.
5. Those arising due to oxidation in the wood cell contents.

Natural weathering does produce quite marked colouring changes in many timbers. Discoloration due to seasoning and bleaching or darkening by sunlight should also be considered.

Mould fungi, shown in Fig. 25.1, is found on timber when sufficient surface moisture is present to encourage growth, that is, when the moisture content exceeds 20 per cent. These organisms feed on the carbohydrates in the cell structure and are capable of breaking down the wood substance (Fig. 25.2). Any staining to the immediate surface of the wood by mould fungi can be removed by brushing, planing, or sanding.

The most common shades of staining are green, grey-blue, pinkish, and occasionally white. Small black dots in the staining denote fruit-bodies of the growth.

The spores of these fungi (Fig. 25.2) can remain dormant for a considerable time to break out at a later date should conditions be favourable, resulting in speedy destruction.

The blue-staining of timber has already been dealt with under seasoning.

Dote is an infection by wood-rotting fungi in the early stages of attack. Such fungi often enter the standing tree or the log before conversion, resulting in particular colour effects with each infection. Heart rots cause the wood to darken, becoming streaky. Brown, grey, and reddish are the most common colour effects.

Certain discolorations of timber can occur in seasoning. After a log is converted and the surfaces of the timber are left exposed to the atmosphere, oxidation takes place, producing chemical compounds which are sometimes coloured; usually grey and brown markings result. This discoloration only occurs in unseasoned timber where the surface is wet. Correct sticking and protection of timber in drying is therefore most important in preventing staining defects.

Planing and sanding will usually remove sticker marks on the surface of the timber. Stains on finished stock may have to be treated by bleaching, several coats may be required.

Chapter 6

Electric tools

Portable electric tools

The building industry has at its disposal a wide range of portable electric tools for use both in workshops and on the site. These tools have relieved the craftsman of much of the laborious work done by hand and speeded up the production of work, without affecting the skill of the craftsman.

Industry now demands greater productivity incurring the lowest possible cost and labour charges. This demands the full use of new techniques, mechanical aids, and power tools, by all trades.

The tools available to the carpenter and joiner are electric drills, saws, planers, sanders, screwdrivers, and routers.

Electric drills

These are the most widely used of any power tools. They are made in three categories, light, general and heavy duty, in a variety of sizes and speed ranges. The 12 mm capacity drill is the most popular. It should be pointed out here, that the capacity under which the drill is listed by the makers, is half its maximum capacity in hardwood. The reason for this is, that these tools were originally designed for use in the engineering trades and the rating based upon their capacity to drill steel. The alternative capacity of the drill in hardwood is usually included by the makers. Figure 26.1 shows a 16 mm capacity drill.

A wide range of accessories and attachments is available for use with the power drill. An ingenious attachment to a drill is the holesaw, which will cut clean round holes up to 100 mm diameter in any material that a hacksaw will cut. A chisel mortising attachment is also available, suitable for small mortises.

A drill stand is shown in Fig. 26.2. The stand supports the full weight of the drill, freeing the hands for drilling accuracy and speed. The feed arm leverage allows tremendous pressure to be exerted for high-speed heavy-duty drilling.

POWER TOOLS

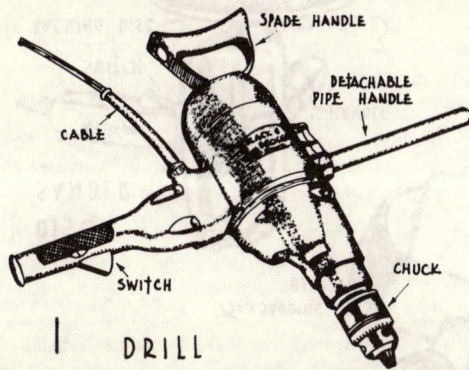

DRILL — 1

SPADE HANDLE
DETACHABLE PIPE HANDLE
CABLE
SWITCH
CHUCK

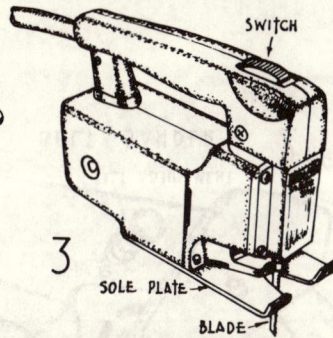

RECIPROCATING SAW — 3

SWITCH
SOLE PLATE
BLADE

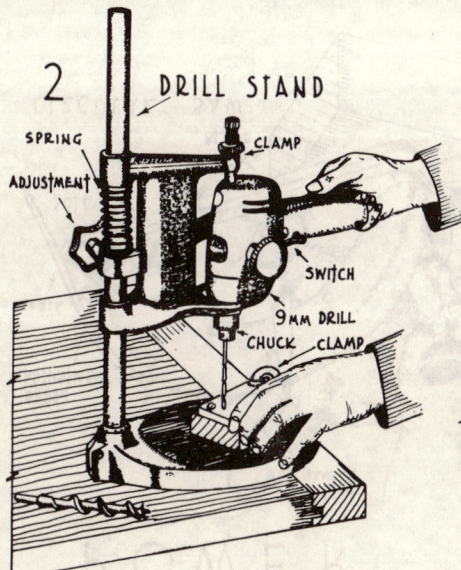

DRILL STAND — 2

SPRING
ADJUSTMENT
CLAMP
SWITCH
9MM DRILL
CHUCK CLAMP

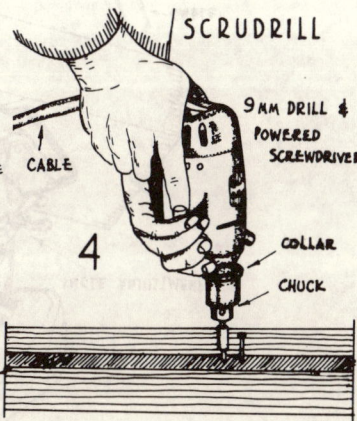

SCRUDRILL — 4

9 MM DRILL & POWERED SCREWDRIVER
CABLE
COLLAR
CHUCK

FIGURE 26

Electric saws (Fig. 27.1).

These are available in 150 mm to 300 mm diameter blade sizes. The most popular size for general use by the joiner is the 175 mm models with a cutting depth of 60 mm. They can be adapted by changing the blade to cut brick, tile, cement, asbestos, cast-iron guttering, stone, plastics, and a wide variety of composite materials.

Saws are adjustable for angle cuts up to 45 degrees, such as occur at rafter ends, and for depth. They are fitted with a telescopic safety guard so designed as to cover the saw blade completely. As the guard is spring-loaded, it uncovers the blade as it enters the work and snaps back over the teeth on the completion of the cut. An outside fence works in the normal way and the saws may be used for ripping or crosscutting as desired. Most of these saws are provided with a pointer to allow a marked cutting line to be easily followed.

Sabre saws

An illustration of this versatile saw is shown in Fig. 26.3. The chief use of the saw is in making internal cuts and intricate shapes in wood or metal. A variety of blades for different materials makes them adaptable for use by various trades.

When using the saw the work should be cramped down, as the cutting action tends to cause considerable vibration. This is due to the saw cutting on its up stroke which tends to pull the machine into the wood. Generally, boring a hole through the wood is the most convenient way of starting the saw on internal panel cuts, although on thin material the saw can cut its own starting place.

Hardwood up 50 mm thick can be cut, also sheet metal, 12 mm diameter steel bars, aluminium, brass and asbestos sheeting, or wallboards, can all be cut with the special blades.

Figure 28.3 shows a two-speed reciprocating saw. At high speeds timber up to 75 mm thick can be cut, either crosscutting or ripping. With a variety of blades available, composition materials, nails, and screws can be cut; a wide range of metals, plastics, pipes, etc. can be cut at low speeds. The saw can be converted quickly into a jig saw for shaped work, as well as for straight cutting, and will cut flush up or down, left or right without additional attachments. The blades can be mounted in six different ways to operate in any cutting position.

Scrudrill (Fig. 26.4).

This machine can be used as a drill and a screwdriver by the adjustment of a collar. Where large numbers of screws have to be driven this tool is essential by reason of the speed with which the screws can be driven home. Pilot holes should always be made in the timber to take the shank of the screw. This avoids increasing the torque on the screwdriver and splitting of the timber.

Electric planer (Fig. 28.4).

This machine combines in one handy portable tool, the double advantages of machine shop power and speed with the handiness of a jack plane.

The planer has ample power for hardwoods and is fitted with 100 mm wide cutters. The depth of cut may be varied up to 3 mm by instant finger-screw

P O W E R   T O O L S

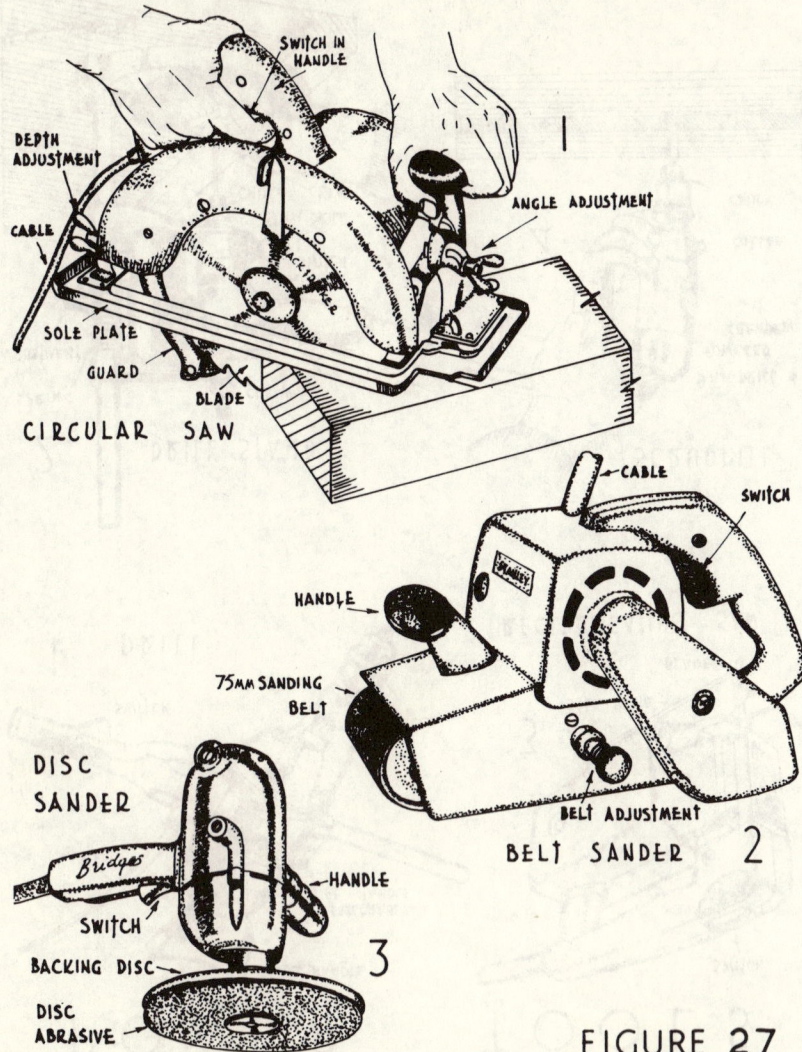

DEPTH ADJUSTMENT

SWITCH IN HANDLE

ANGLE ADJUSTMENT

CABLE

SOLE PLATE

GUARD

BLADE

CIRCULAR SAW 1

CABLE

SWITCH

HANDLE

75mm SANDING BELT

BELT ADJUSTMENT

BELT SANDER 2

DISC SANDER

Bridges

HANDLE

SWITCH

BACKING DISC

DISC ABRASIVE

3

FIGURE 27

adjustment and there is no limit to the width of surface it can plane. A hand grip with an automatic trigger switch ensures safety in operation. The high carbon-chrome steel knives are easily removed for sharpening.

Whether planing with the grain, across the grain, or on end grain, the high-speed cutters of the plane cut without tearing or break-away.

An adjustable fence is available for use when squaring edges of doors, window frames, and similar work.

Portable sanders

These are available in three different types: the disc sander, the belt sander, and the orbital sander.

Disc sander   These sanders range in size from 125 mm to 225 mm diameter (Fig. 27.3). The head has a flexible rubber backing disc and the sanding papers are fixed to this. The disc should not be used flat on the work but at a slight angle with light pressure, to prevent scoring of the surface of the work.

Various accessories are available for use with this machine which include cup grinders, cup wire brushes, and polishing pads.

Belt sander (Fig. 27.2).   This machine is suitable for sanding large flat surfaces and consists of a continuous abrasive band fitted over rollers at each end of the machine. Sanding of framed joinery work with the grain can be accurately carried out and scratching at joints avoided. A dust-collecting bag is fitted to the back end of the machine.

Orbital sander (Fig. 28.1).   This machine is suitable for sanding flat or curved surfaces. The sanding paper is fitted to a flexible base and is ideal for finishing work as it reproduces at high speed the action of hand sanding in either the horizontal or vertical positions.

Electric router   This machine is shown in Fig. 28.2 and works on the same principle as the overhead-router. It is a most versatile machine and is used for moulding, rebating, grooving, etc. An adjustable fence is provided and the machine itself is adjustable for depth of cut.

Attachments, jigs, and templets, and a wide range of cutters are available for use with the router for shaped work, dovetailing, and stair trenching. Plastic cutting and trimming, using carbide-tipped blades, can also be done.

Care and use of electric power tools

1. Check the name-plate voltage range of the tool with the voltage range of the supply line. Serious damage can result if the supply line voltage does not fall within the ranges given.
2. An electric tool should always be earthed while in use. This protection will protect the operator against shock, should the tool develop an earth. Proper earthing is especially important where dampness is present or where abrasive dust is apt to create a short circuit.
3. Never drag the tool around by the cable and avoid any kinks or sharp bends.
4. Wear goggles in all abrasive tool operations.

POWER TOOLS

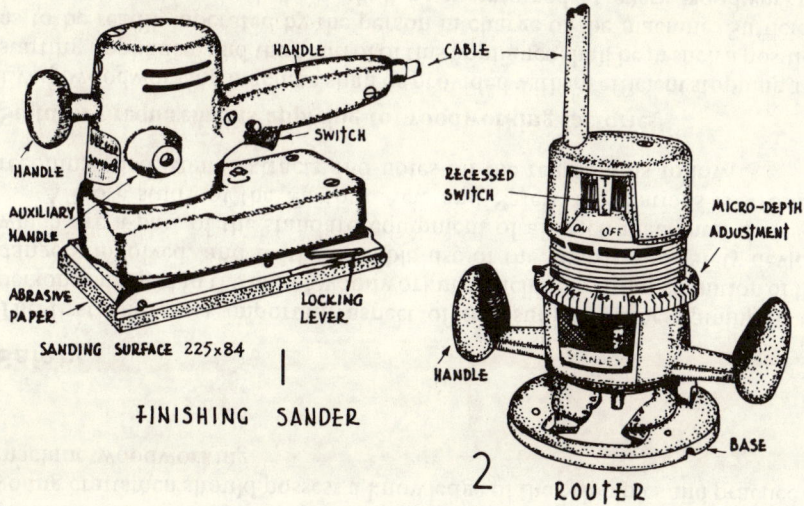

FINISHING SANDER

Labels: HANDLE, CABLE, HANDLE, SWITCH, AUXILIARY HANDLE, ABRASIVE PAPER, LOCKING LEVER, SANDING SURFACE 225×84

1

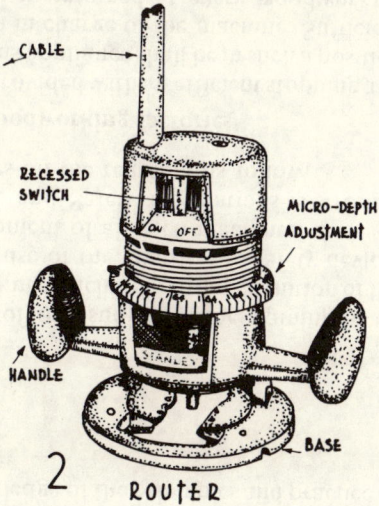

ROUTER

Labels: RECESSED SWITCH, ON OFF, MICRO-DEPTH ADJUSTMENT, HANDLE, BASE

2

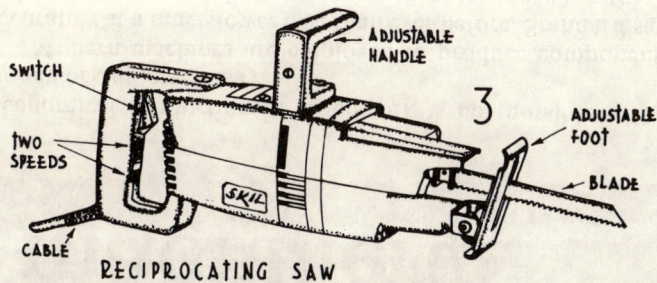

RECIPROCATING SAW

Labels: SWITCH, TWO SPEEDS, CABLE, ADJUSTABLE HANDLE, ADJUSTABLE FOOT, BLADE

3

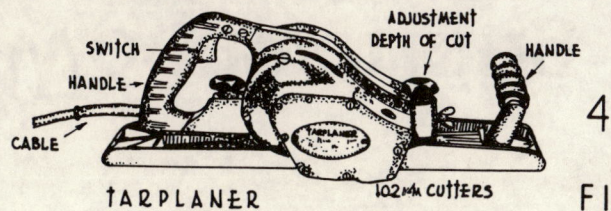

TARPLANER

Labels: SWITCH, HANDLE, CABLE, ADJUSTMENT DEPTH OF CUT, HANDLE, 102 mm CUTTERS

4

FIGURE 28

5. Disconnect all electric tools from the supply point before making any adjustments.

   Correct and adequate lubrication is the most important single factor in determining the life and service of any electric tool. All tools are properly lubricated at the factory, and under normal regular use this lubrication will last for 60 days. Tools used constantly on production will need relubricating more often.

   Bearings and gears contain sufficient grease for many months. Permanent lubricated bearings have sufficient lubrication packed in them at the factory to last the life of the bearings.

6. Keep saws, blades, drills, and cutters sharp and check each for tightness in the machine before starting.

Reduced voltage tools

Safety in the use of portable electric tools always depends upon everything within the earthing circuit being perfect. This requires that there should be no damaged plugs, loose screws, or misconnected cables. To cut down these risks, particularly on building sites, 110 V equipment is recommended. This involves step-down transformers, either fixed or portable, to which lower voltage machines are connected. This lower voltage is regarded as being safe but the transformer is so designed that, in the event of a fault, the operator cannot receive more than a 55 V shock.

Double insulation

Portable tools are now available which are completely safe in that they have an indestructible shockproof plastic body. This gives an independent extra barrier of insulation on top of the ordinary insulation all electric tools have. No earth wire is therefore necessary because the casing cannot become live. In the event of a fault there is no risk to the operator as the current cannot penetrate the protective barrier of double insulation to get to earth.

Chapter 7

Woodworking machinery

Economical production of woodwork is not possible without the use of woodworking machinery.

Modern machines are employed to produce components speedily and accurately at a much lower cost than ever before. Standardisation of wooden components for buildings and furniture, etc. provides the opportunity to manufacture on semi-mass production lines. It is, therefore, important that young craftsmen should possess a knowledge of the principles and practice of machine woodworking.

Safety

The first and most important aspect of this subject is the training of all persons engaged in the use of woodworking machinery to a recognition of the dangers involved, and to the sensible use of the guards and safety devices which are a part of the standard equipment of all such machines.

A close study of the Factory Acts and Safety Regulations is strongly recommended; some extracts and notes on the regulations follow.

Statutory requirements applying to woodworking factories

Every woodworking machine shall be provided with an efficient stopping and starting appliance, and the control of this appliance shall be in such a position as to be readily operated by the person in charge of the machine. Sufficient clear and unobstructed space shall be maintained at every woodworking machine while in motion to enable the work to be carried on without unnecessary risk, and the floor surrounding every woodworking machine shall be maintained in good and level condition, and as far as practicable free from chips or other loose material and shall not be allowed to become slippery.

Where the natural light at a woodworking machine is inadequate and can be improved by the provision of additional or better windows not involving serious structural alteration, or by whitening the walls, the occupier shall take steps to improve the natural light at the particular machine. The means of artificial lighting for every woodworking machine shall be adequate, and shall be so placed or shaded as to prevent direct rays of light from impinging on the eyes of the operator while he is operating the machine.

The Woodworking Machines Regulations 1974

These Regulations impose requirements as to guards and certain other safety devices for woodworking machines used in factories and certain other places to which the Factories Act applies.

The Regulations also impose requirements as to working space, condition of floors, noise, lighting, and temperature in those factories and places and as to the training of persons operating woodworking machines.

These revised Regulations are issued by the Home Office and all persons involved in the use of woodworking machines should make themselves familiar with the current Regulations.

As many firms will be affected by the revised 1974 Regulations, listed below are a number of points which require attention.

1. Extension of bench surface at machine table height on the circular saw, from the back of the saw blade to the taking-off position.
2. The limitation of operations on the surface planing machine.
3. Extraction and collection of waste.
4. Reference to noise limits.
5. The training of operatives.

Regulations – circular saws

The statutory regulations lay down that, that part of the saw blade of every circular sawing machine which is below the machine table be guarded to the greatest extent that is practicable. The machine must be provided with a riving knife which shall be securely fixed below the machine table and be in a direct line with the saw blade, have a smooth surface, be strong, rigid, and easily adjustable and fulfil the following conditions:

1. The edge of the knife nearest the saw blade shall form an arc of circle having a radius not exceeding the radius of the largest saw blade with which the saw bench is designed to be used.
2. The knife shall be capable of being so adjusted that it is as close as practicable to the saw blade so that at the machine table the distance between the edge of the knife and the teeth of the saw does not exceed 12 mm.
3. For saw blades less than 600 mm diameter the knife should project above the table to a height of not more than 25 mm below the top of the saw blade. For blades over 600 mm diameter the knife should project to a height of at least 225 mm above the table.
4. In the case of a parallel plate saw blade the knife shall be thicker than the plate of the saw.

The part of the saw blade of every circular saw which is above the machine table must be guarded with a strong and easily adjustable guard, and must be kept adjusted so that it extends from the top of the riving knife to a point above the top surface of the timber being cut a distance not more than 12 mm. The guard must have a flange of adequate depth on each side of the saw blade and must be kept adjusted so that the flanges extend beyond the roots of the teeth of the saw blade. Where the guard is fitted with an adjustable front extension piece, this must have a flange along its length and so adjusted that the flange extends beyond the roots of the teeth of the saw blade.

Limitations on use. No circular saw shall be used for cutting any rebate, tenon, mould, or groove, unless that part of the saw blade or other cutter which is above the table is effectively guarded.

No circular saw shall be used for a ripping operation (other than any operation involved in cutting a rebate, tenon, mould, or groove) unless the teeth of the saw blade project throughout the operation through the top surface of the timber being cut.

Push-sticks. A suitable push-stick shall be provided and kept available for use at every circular sawing machine which is fed by hand.

Removal of material cut by circular sawing machines where any person (other than the operator) is employed at a circular sawing machine in removing, while the blade is in motion, material which has been cut, the machine table shall be constructed or extended over its whole width by at least 1·200 m.

Narrow bandsawing machine

The saw wheels of every narrow bandsawing machine and the whole of the blade, except that part of the blade which runs downwards between the top wheel and the machine table, shall be enclosed by a guard or guards of substantial construction.

That part of the blade of every bandsawing machine which is above the friction disc or rollers and below the top wheel shall be guarded by a frontal plate which is as close as is practicable to the saw blade and has at least one flange at right angles to the plate and extending behind the saw blade.

The friction disc or rollers of every bandsawing machine shall be kept so adjusted that they are as close to the surface of the machine table as is practicable having regard to the nature of the work being done.

Planing machines

Limitations on the use of planing machines. No planing machine shall be used for cutting any rebate, recess, tenon, or mould unless the cutter is effectively guarded.

Cutter blocks for planing machines for surfacing. Every planing machine for surfacing which is not mechanically fed shall be fitted with a cylindrical cutter block.

Table gap. Every planing machine for surfacing shall be so designed and constructed as to be capable of adjustment so that the clearance between the cutters and the front edge of the delivery table does not exceed 6 mm and the gap between the feed table and the delivery table is as small as practicable.

Bridge guards. Every planing machine for surfacing shall be provided with a bridge guard which shall be strong and rigid, have a length not less than the full length of the cutter block and a width not less than the diameter of the cutter block, and be so constructed as to be capable of easy adjustment both in a vertical and horizontal direction.

Every bridge guard shall be mounted on the machine in a position which is approximately central over the axis of the cutter block and shall be so constructed as to prevent its being accidentally displaced from that position.

Adjustment of bridge guards. While a planing machine is being used for surfacing, the bridge guard shall be so adjusted as to enable, so far as is practicable, the work being done at the machine to be done without risk of injury to persons employed.

When a wider surface of squared stock is being planed, the bridge guard shall be adjusted so that the distance between the end of the guard and the fence does not exceed 10 mm, and the underside of the guard is not more than 10 mm above the upper surface of the material.

Cutter-block guards. In addition to being provided with a bridge guard every planing machine for surfacing shall be provided with a strong, effective, and easily adjusted guard for that part of the cutter block which is on the side of the fence remote from the bridge guard.

Provision and use of push-blocks. When a wider surface of squared stock is being planed and by reasons of the shortness of the material the work cannot be done with the bridge guard adjusted as required by the Regulations, a suitable push-block having suitable handholds which afford the operator a firm grip shall be provided and used.

Combined machines. That part of the cutter block of a combined machine which is exposed in the table gap, shall, when the machine is used for thicknessing, be effectively guarded.

Protection against ejected material. Every machine used for thicknessing shall be provided on the operator's side of the feed roller with sectional feed rollers, or other suitable devices which shall be of such a design and so constructed as to restrain any workpiece ejected by the machine.

Vertical spindle moulding machine

Every detachable cutter for any vertical spindle moulding machine shall be of the correct thickness for the cutter block or spindle on which it is to be mounted and shall be so mounted as to prevent it from becoming accidentally detached.

False fences. Where straight fences are being used for the purposes of the work being done on a vertical spindle moulding machine, the gap between the fences shall be reduced as far as is practicable by a false fence.

Provision of jigs. Where by reason of the nature of the work being done at a vertical spindle moulding machine it is impracticable to provide, as per the Regulations, a guard enclosing the cutters of the machine so that they are effectively guarded, but it is practical to provide, in addition to the guard required to be provided by the Regulations, a jig or holder of such a design and so constructed as to hold firmly the material being machined and having suitable handholds which afford the operator a firm grip, the machine shall not be used unless such a jig or holder is provided.

Guards for protection against ejected parts. Every guard provided for the cutters of any vertical spindle moulding machine shall be of such a design and so constructed as to contain any part of the cutters or their fixing appliances or any part thereof in the event of their ejection.

Provision and use of back stops. Where the work being done on a vertical spindle moulding machine is work in which the cutting of the material being machined commences otherwise than at the end of a surface of the material and it is impracticable to provide a jig or holder, the trailing end of the material shall, if practicable, be supported by a suitable back stop where this would prevent the material being thrown back when the cutters first make contact with it.

Provision of spikes or push sticks. Where the nature of the work being performed at a vertical spindle moulding machine is such that the use of a suitable spike or push stick would enable the work to be carried on without unnecessary risk, such a spike or push stick shall be provided and kept available for use.

Machines driven by two-speed motors. Where the motor driving a vertical spindle moulding machine (other than a high-speed routing machine) is designed to operate at two working speeds the device controlling the speed of the motor shall be arranged so that the motor cannot run at the higher of those speeds, without first running at the lower of those speeds.

Extraction equipment and maintenance

The blade of a sawing machine shall not be cleaned by hand while the blade is in motion.

Extraction of chips

Effective exhaust appliances shall be provided and maintained at every planing machine used for thicknessing other than a combined machine for surfacing and thicknessing, every vertical spindle moulding machine, every multi-cutter moulding machine, every tenoning machine, every automatic lathe, for collecting from a position as close to the cutters as practicable and to the extent that is practicable, the chips and other particles of material removed by the action of the cutters and for discharging them into a suitable receptacle or place.

Maintenance and fixing

Every woodworking machine and every part thereof, including cutters and cutter blocks, shall be of good construction, sound material, and properly maintained.

Every woodworking machine, other than a machine which is held in the hand, shall be securely fixed to a foundation, floor, or to a substantial part of the structure of the premises, save that where this is impracticable, other arrangements shall be made to ensure its stability.

Lighting

In addition to the requirements of the principal Act and the Factories (Standards of Lighting) Regulations the following provisions shall have effect in respect of any work done with any woodworking machine:

1. The lighting, whether natural or artificial, for every woodworking machine shall be sufficient and suitable for the purpose for which the machine is used.
2. The means of artificial lighting for every woodworking machine shall be so placed or shaded as to prevent glare and so that direct rays of light do not impinge on the eyes of the operator while he is operating such machine.

Noise

Where any factory, or any part, is mainly used for work carried out on woodworking machines, the following provisions shall apply:

1. Where on any day any person employed is likely to be exposed continuously for 8 hours to a sound level of 90 dB(A) or is likely to be subject to an equivalent or greater exposure to sound –
   (a) such measures as are reasonably practicable shall be taken to reduce noise to the greatest extent; and
   (b) suitable ear protectors shall be provided and made readily available for the use of every such person.
2. All ear protectors provided shall be maintained and shall be used by the person for whom they are provided.

Training

No person shall be employed on any kind of work at a woodworking machine unless he has been sufficiently trained at machines of a class to which that machine belongs in the kind of work on which he is to be employed.

Every person, while being trained to work at a woodworking machine, shall be fully and carefully instructed as to the dangers arising in connection with such machine, the precautions to be observed, the requirements of the Regulations which apply, and the method of using the guards, devices, and appliances required by the Regulations.

Selection of machines

The type of articles to be made is the guide to the type of machinery which will be required, and after considering the various operations involved a selection of suitable machines is made. The choice is also dependent on the volume of work to be done, because the range of machines now available includes designs which may be used for the production of small miscellaneous work, or high-speed and large-capacity machines used for repetition work in large quantities. When it is necessary to duplicate machines to maintain balanced production it is important to keep to one maker, so that spare parts, etc. are standard.

CIRCULAR SAW

MITRE FENCE
SLIDES ON EACH SIDE
OF SAW IN SLOTS

ADJUSTABLE
GUARD

6 SPRING SET TEETH

7 SWAGE SET TEETH

RIVING KNIFE

RIPPING
FENCE

GRADUATED SCALE

HANDWHEEL
FOR RISE & FALL
ADJUSTMENT

CANTING SAW
ADJUSTMENT

8 PUSH STICK

SWITCHES

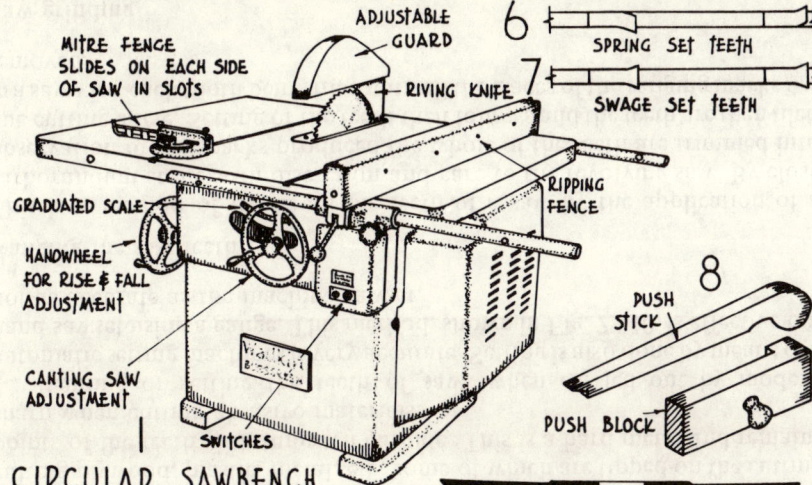

1 CIRCULAR SAWBENCH

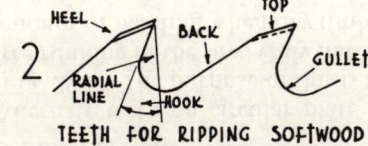

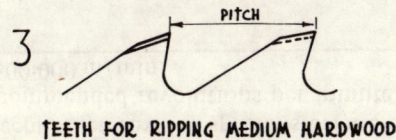

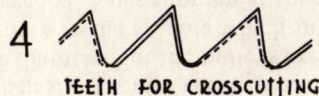

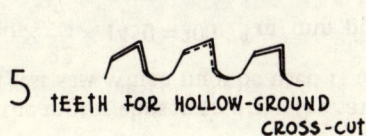

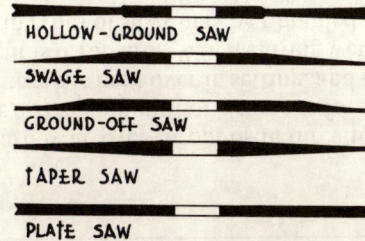

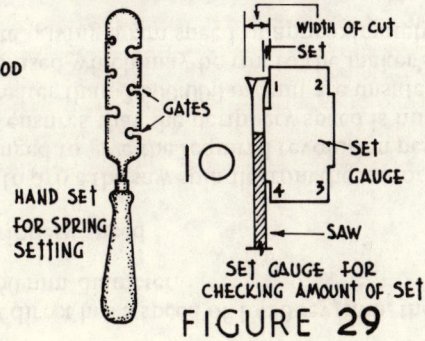

PUSH BLOCK

2 HEEL top BACK GULLET RADIAL LINE HOOK
TEETH FOR RIPPING SOFTWOOD

3 PITCH
TEETH FOR RIPPING MEDIUM HARDWOOD

4 TEETH FOR CROSSCUTTING

5 TEETH FOR HOLLOW-GROUND CROSS-CUT

11 HOLLOW-GROUND SAW
SWAGE SAW
GROUND-OFF SAW
TAPER SAW
PLATE SAW

GATES

HAND SET
FOR SPRING
SETTING

10 WIDTH OF CUT
SET
SET GAUGE
SAW
SET GAUGE FOR
CHECKING AMOUNT OF SET

FIGURE 29

The following are a number of the machines which are to be found in most joinery or woodworking shops:

1. The handfed circular sawbench.
2. The dimension sawbench.
3. The crosscutting machine.
4. The narrow bandsawing machine.
5. The surface-planing machine.
6. The combined surface-planing and thicknessing machine.
7. The chain and chisel mortising machine.
8. The single-end tenoning machine.
9. The spindle moulding machine.
10. The routing machine.

All the above machines are discussed in more detail in the following section.

The handfed circular sawbench

This machine has a sturdy base, which may be cast iron or fabricated steel sheet, with a machine-planed cast-iron table (Fig. 29.1). The saw is mounted on a spindle which runs on ball-bearings and is driven by an electric motor housed under the table. The whole saw assembly may be quickly raised or lowered by means of a handwheel when operations require the raising and lowering of the saw within the table. Access to the saw is by removal of a finger plate. The securing nut is turned loose in a clockwise direction when removing the saw from the spindle.

Uses

This type of sawbench is used mostly for ripping small sizes and crosscutting up to 75 mm deep. It is also useful for square and bevel cutting of small panels to net sizes. When larger work is to be cut, the table may be extended at each end. The saw assembly may be canted when bevel sawing to 45 degrees if required. A machined groove is placed in each side of the table-top parallel to the line of the saw in which may be inserted a crosscutting fence for square or bevelled cuts.

Saws

Ripping and crosscutting operations require different saws for each. These are illustrated in Figs. 29.2 and 29.3, showing teeth for ripping saws, and Figs. 29.4 and 29.5 for crosscutting saws. It will be observed that the principal difference is in the hook angle of the teeth. For ripping, the fronts and tops of the teeth require very little bevel, while for crosscutting the fronts and tops of the teeth are filed to a distinct bevel to produced the knife-edge essential for severing the fibres of the wood. Saws are made from spring steel and the outer rim has tension imparted to it by spreading the steel outwardly from the centre. The hole for the saw spindle should be a close fit, a saw which is a loose fit is more difficult to keep accurately circular with all the teeth cutting.

Centrifugal force

All objects, when revolving, generate centrifugal force and because of this force, the rim speed or periphery speed of the saw is limited to 3 000·000 m/min. This means that the driving spindle speed must be related to the diameter of the saw.

Example

If the motor chosen to drive the saw direct has a speed of 1 450 rev/min, the largest saw which may be used is 660 mm diameter.

$$660 \times \frac{22}{7} \times 1450 = 300 \cdot 73 \text{ m/min periphery speed}$$

When short centre vee belts are used to drive the saw spindle from the motor, the diameters of the pulleys are arranged to give the required revolution per minute to the spindle which in turn ensures that the periphery speed is not exceeded. Saws running at speeds greater than 3 000·000 m/min are unsafe, except when saws of special steel are used which may be run to the maker's recommended revolutions per minute, giving a rim speed of approximately 3 600·000 m/min.

Setting the teeth

Alternate teeth on circular plate saws are bent over in opposite directions (Fig. 29.6). The purpose of this is to give a small amount of clearance for the free running of the saw. Only half of the tooth is bent over in setting, and the amount of bending when dry timber is cut is 0·100 mm. When cutting wet or green timber more set is required. Certain types of saws may be obtained for cutting plywood, plastic, metal, etc, some of which are tipped on the cutting points of the teeth with tungsten carbide. This is a hard metal and remains sharp when cutting abrasive materials.

Bending or setting the teeth of saws when carried out by modern automatic setting machines is very accurate. Setting is also done by means of a hand sawset using a gauge. This method, shown in Fig. 29.10, is effective but not as accurate as the machine sawset.

Ranging the saw teeth

This is a method of truing all the teeth of a saw by the application of a carborundum stone with discretion and care to the revolving saw. By close observation of the sparks produced the whole of the teeth are trimmed into one cutting circle. Setting of the teeth then follows and the teeth are then filed in a saw vice, each tooth being filed until all the traces of the ranging marks are removed.

Saw grinding

A sawmill, having a large number of saws to maintain, will possess an automatic saw grinder of which there are many excellent types. Large- and small-diameter saws may be ground truly circular, and all the teeth perfectly formed for the purpose for which the saw is to be used.

Operating the saw

A skilful sawyer considers his own safety first. The crown guard is a metal cover firmly held on a stout bracket, poised in position to cover the top of the saw. It may be raised or lowered to suit the depth of the cut. To prevent the rising portion of the saw from lifting the wood, a riving knife is fixed in line with the saw cut and 6 mm away from the teeth. The knife is curved to the shape of the saw. Risk of contact is also avoided by the regular use of push sticks and push blocks instead of the fingers when completing the last 150 mm of a cut. These are shown in Fig. 29.8.

Packing the saw

Most saws are provided with a recess on each side of the saw where it enters the table, these recesses are to receive packings and a mouthpiece. The packings are made from thick felt and assist in keeping the saw cutting in a straight line. The mouthpiece is a hardwood insert which prevents spelching on the underside of the timber at the point of exit of the teeth.

Heating, lighting, and cleanliness of the floor are important for safety and comfort of the operator. An efficient extractor system for the speedy removal of the sawdust is also required.

Types of saw blades

These are illustrated in Fig. 29.11.

Hollow-ground saw

These saws are hollow-ground from the collar area to the tooth, giving clearance in working without the need for setting the teeth. At the collar area in the centre of the plate it is parallel, and ground with a hollow taper from this point to the rim of the saw as shown. They are used mainly for accurate dimensioning and often termed 'dimension saws'.

Swage saw

This type of saw is ground off from the collar area in a straight taper to the rim. It is used for cutting thin boards. As the saw is finer at the rims, it has more teeth with shallower gullets than the plate saw.

Ground-off saw

This type has a thin edge and is ground as shown on one side only. As the saw kerf is small with this saw, it is used for cutting thin boards and veneers.

Taper saw

This form of saw is tapered, as the name suggests, on both sides from the collar area to the rim. They are used for splitting boards with very little waste in the cut.

Plate saw

This is the normal parallel saw used for all kinds of sawing both ripping and crosscutting. They may be spring or swage set.

Swage set

This is a method of saw-setting and must not be confused with swage saws. In swage setting the steel at the point of the teeth is pressed out to an equal distance on each side of the saw so that each tooth clears both sides of the saw.

SAW OPERATIONS

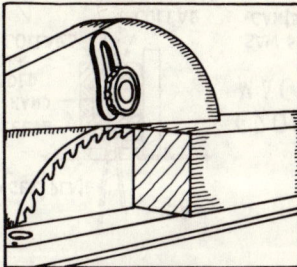

STRAIGHT RIPPING

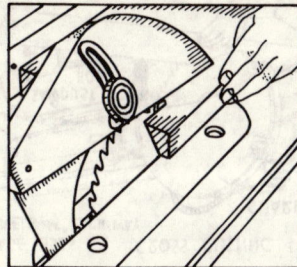

BEVEL RIPPING

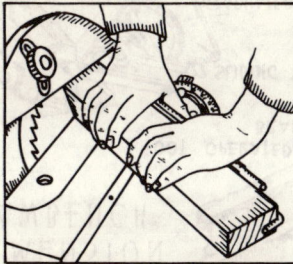

STRAIGHT OR COMPOUND MITRING

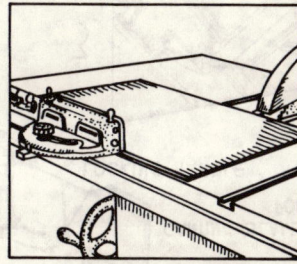

CROSSCUTTING

MOULDING

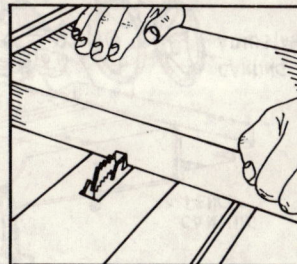

GROOVING OR TRENCHING

FIGURE 30

Swage setting is shown in Fig. 29.7.

Dimension saw

This machine is illustrated in Fig. 31.1. It is used, as its name suggests, for cutting to accurate dimensions off the saw, either ripping or crosscutting. Bevelled work, oblique work, bevel and mitre cuts, grooving, housing, and rebating are other operations which may be carried out on this versatile machine (illustrated in Fig. 30).

The sawbench is fitted with a normal ripping fence and a swivel crosscutting or mitre fence (Fig. 31.3). This is graduated in degrees and allows crosscutting and mitring to any angle, both simple and compound, to be achieved with remarkable accuracy. The saw, which is carried in a tilting arbor, cants to an angle of 45 degrees and is fitted with a graduated scale to denote the saw's position.

The double-mitre fence (Fig. 31.2), is in two parts, both made to pivot from a central pin fixed in the sliding table. Degree angles are marked on the table for accurately setting the fences, the principle angles being positively located by a spring plunger. A gauge bar may be fitted and provided with a turnover stop determining the exact length of material cut off.

The sliding table, which rolls on rollers, has an opening movement to facilitate saw changing and to allow the use of various types of saws, trenching, or grooving heads to be used.

Saws of 406 mm and 457 mm diameter are used and, being of stouter gauge than the normal saw of the same diameter, they are run without packings. The saw should be adjusted high enough to cut the material and no higher.

Drunken saw

Figure 31.4 shows this adjustable grooving saw, often termed a 'wobble saw'. It consists of a stout gauge saw with wedge-shaped collars screwed to it. A sleeve, with locating notches, passes through both the saw and collars. Two further shaped collars, identical to those screwed to the saw, are positioned on the sleeve so that the taper is opposite that of the saw collars. The unit is held in the desired adjustment by tightening on the spindle of the machine. The tool is used mostly on the spindle moulder but may also be used on a number of other machines.

Adjustment is made after slackening the tightening nut by holding the outer collars and moving the saw. This movement makes the eccentric collars tilt or level the saw due to turning. When the collars are paired the saw is level, and by turning the saw one way or the other an increase in the width of the groove is obtained. This width of groove or cut is, therefore, dependent on the amount of wobble.

DIMENSION SAW

ADJUSTABLE GUARD

RIVING KNIFE

CANTING FENCE

SINGLE MITRE FENCE

SLIDING TABLE

Wadkin

CANTING SAW ADJUSTMENT

1

DIMENSION SAWBENCH

SAW RISE & FALL ADJUSTMENT

CANTING SCALE & POINTER

LOCKING HANDLE

FOOT OPERATED BRAKE

SLIDING TABLE

GRADUATED SCALE

DOUBLE MITRE FENCE

2

3

ADJUSTABLE GUARD

STOP

SAW CARRIAGE RISES FALLS & SWIVELS 45° EACH WAY

CROSS·CUTTING FENCE

GUARD

SWITCH

BRAKE

FINGER PLATE

SAW

SLEEVE

LEFT HAND THREADED SPINDLE

COLLARS

COLLAR

4

DRUNKEN OR WOBBLE SAW

5

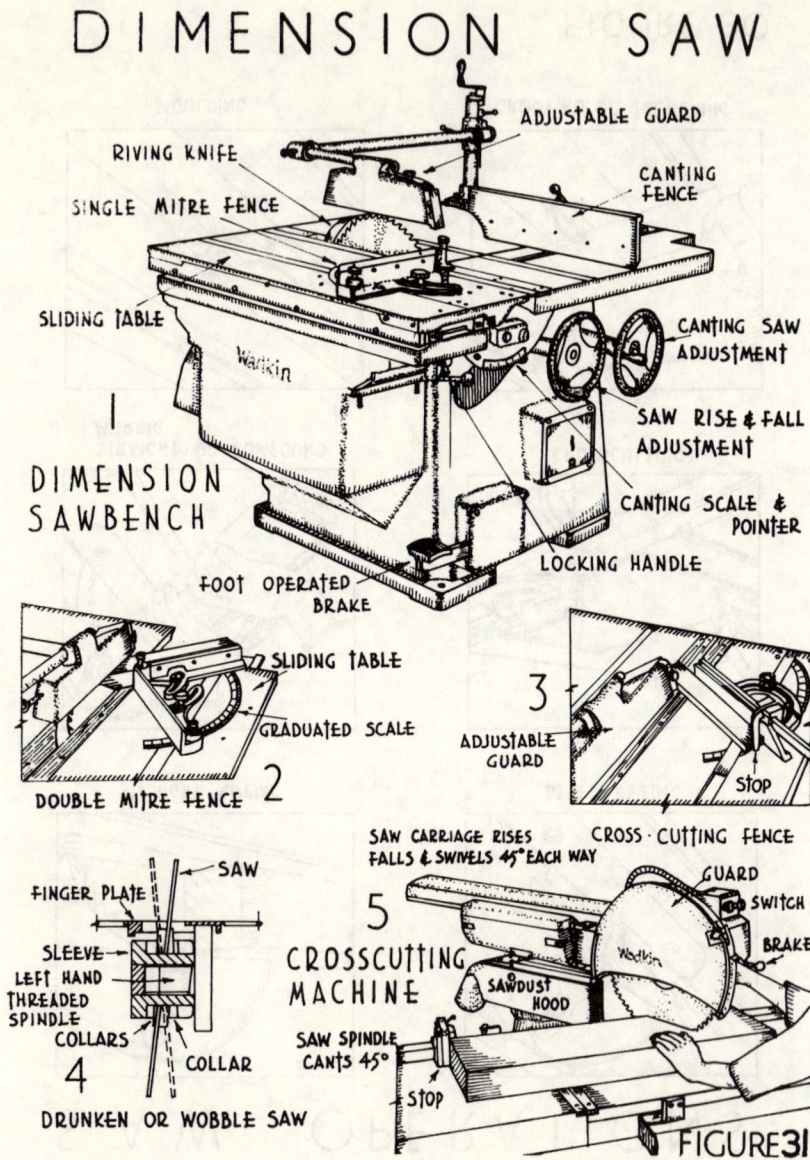

CROSSCUTTING MACHINE

Wadkin

SAWDUST HOOD

SAW SPINDLE CANTS 45°

STOP

FIGURE 31

Crosscutting machine

There are a number of forms and sizes of this machine; which has increased in popularity over recent years as a result of its versatility.

The machine shown in Fig. 31.5 has a 355 mm diameter saw with direct drive. The saw unit rotates horizontally through 360 degrees and fits to any angle from horizontal to vertical. It can be locked in any position along the arm which swings 45 degrees either way. With the saw canting and the carriage to swivel crosscutting, bevel crosscutting, mitring, compound angle-cutting, ripping, and bevel ripping can be done. In addition, by fitting dado or trenching heads, cutterblocks, and moulding blocks, many other operations are possible.

A table on each side of the saw will be found useful. To these, gauge bars, rules, and stops are often used. Length gauges set to the different lengths required enable timber to be cut quickly and accurately, without any measuring and the minimum of waste.

Planing machines

Following the sawing of timber to required sizes for joinery, the next operation consists of planing all the sawn surfaces, straight, flat, smooth, without twist, parallel in width and thickness. It is a trade practice to allow 1·500 mm for each planed surface, and this reduction in finished (planed) sizes is allowed for in setting out framing to sizes. Some joinery manufacturers find that 1·500 mm allowance for each planed surface is sufficient and 4·500 mm is allowed for two surfaces.

Design

Machines for planing are of various designs and sizes, from the small bench type with a capacity 152 mm wide up to the largest which will plane 915 mm wide. Many cutterheads of different design are used, but it should be noted that only the circular safety cutterhead is used on surface planing machines, to comply with safety regulations. Mostly, these are of the rotary planing type, the cutterheads having two, three, or four knives revolving at 4 500 rev/min, as shown in Figs. 32.2 and 33.2.

In workshops where space is available, separate machines are used. One for surface planing and square-planing edges, and another machine for planing to width and thickness. When space is limited, a combined surface planer and thicknesser may be preferred (illustrated in Fig. 34.1). This machine, however, has only half the production capacity compared with separate machines.

Surface planing (Fig. 33.1).

This operation consists of passing the wood smoothly over an infeed and outfeed table, between which the cutterblock is mounted on high quality ball-bearings. Each table is adjustable in height, to regulate the depth of the cutting. The tables are machined and ground to a perfectly flat surface, and providing the wood is kept firmly held to the surface of the outfeed table during planing, a perfectly flat surface is formed, straight and without twist.

PLANING MACHINES

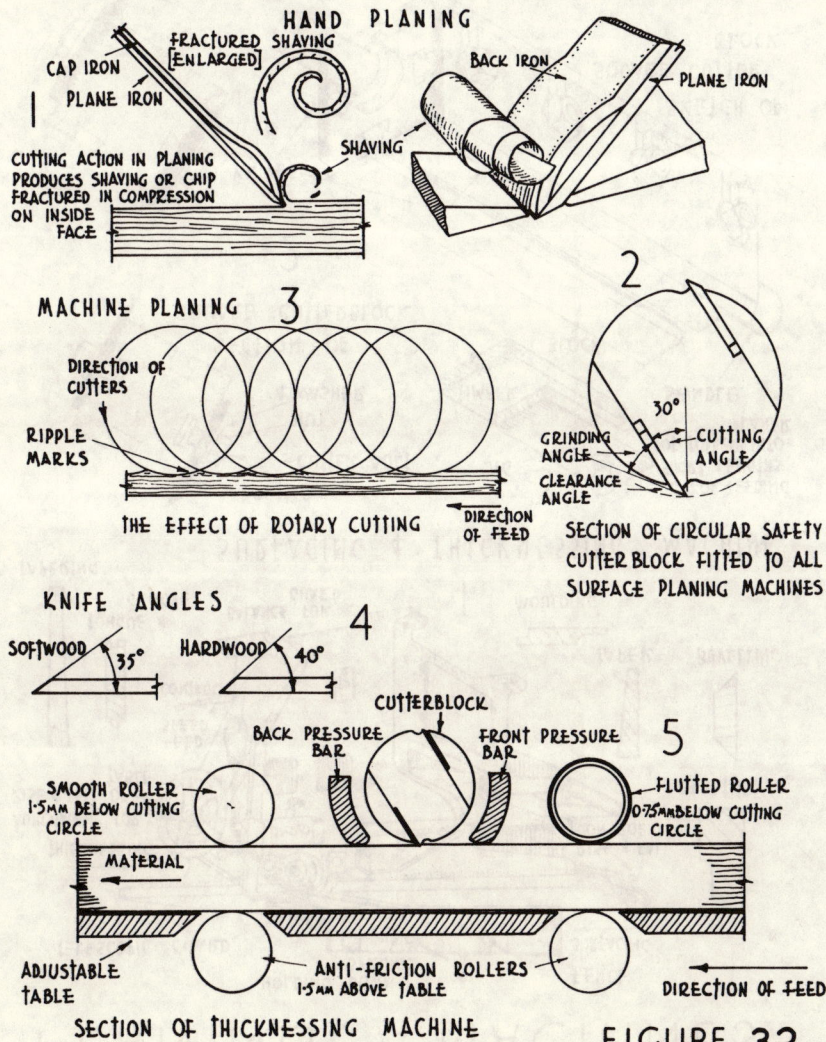

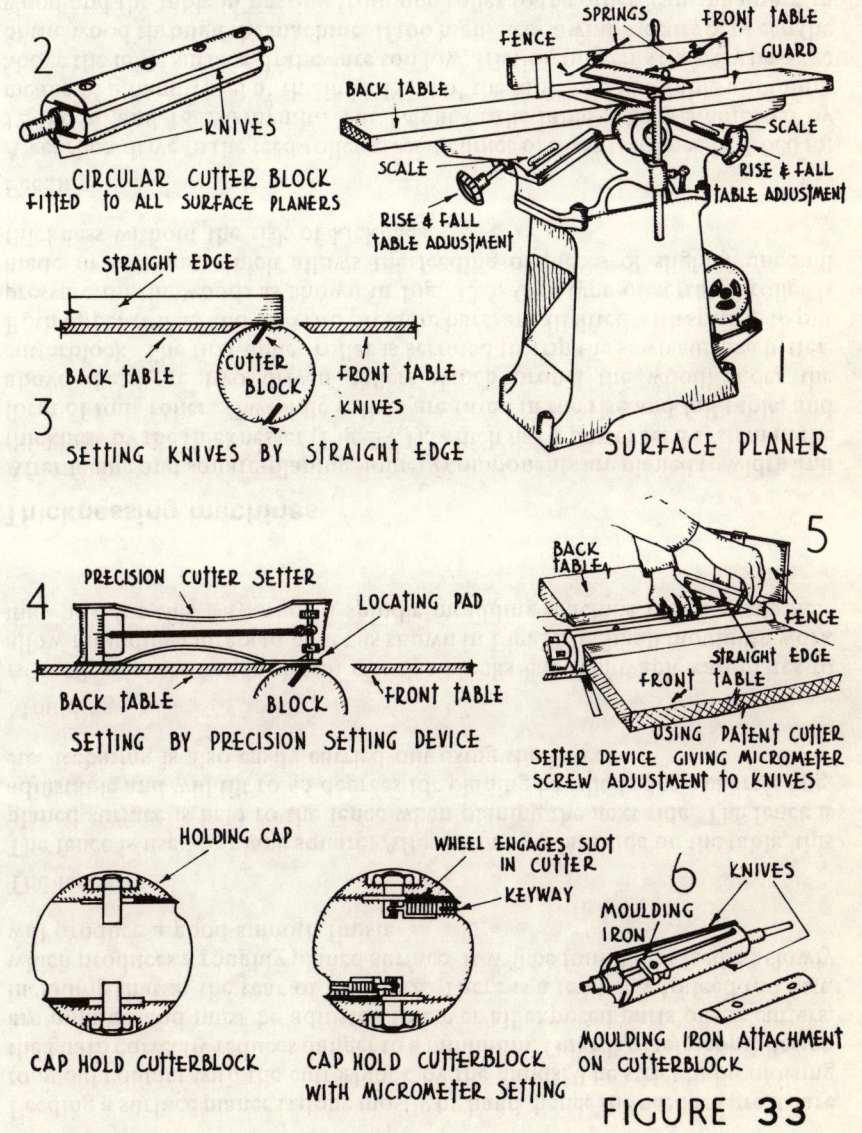

HAND PLANING

CAP IRON
PLANE IRON
FRACTURED (ENLARGED)
SHAVING

1

CUTTING ACTION IN PLANING
PRODUCES SHAVING OR CHIP
FRACTURED IN COMPRESSION
ON INSIDE FACE

SHAVING

BACK IRON
PLANE IRON

MACHINE PLANING

3

DIRECTION OF CUTTERS
RIPPLE MARKS

DIRECTION OF FEED

THE EFFECT OF ROTARY CUTTING

2

30°
GRINDING ANGLE
CUTTING ANGLE
CLEARANCE ANGLE

SECTION OF CIRCULAR SAFETY
CUTTER BLOCK FITTED TO ALL
SURFACE PLANING MACHINES

KNIFE ANGLES

4

SOFTWOOD 35°   HARDWOOD 40°

CUTTERBLOCK

BACK PRESSURE BAR   FRONT PRESSURE BAR

5

SMOOTH ROLLER 1·5mm BELOW CUTTING CIRCLE

FLUTTED ROLLER 0·75mm BELOW CUTTING CIRCLE

MATERIAL

ADJUSTABLE TABLE

ANTI-FRICTION ROLLERS 1·5mm ABOVE TABLE

DIRECTION OF FEED

SECTION OF THICKNESSING MACHINE

FIGURE 32

PLANING MACHINES

2

KNIVES

CIRCULAR CUTTER BLOCK
FITTED TO ALL SURFACE PLANERS

STRAIGHT EDGE

BACK TABLE   CUTTER BLOCK   FRONT TABLE

KNIVES

3

SETTING KNIVES BY STRAIGHT EDGE

SPRINGS   FRONT TABLE
FENCE   GUARD
BACK TABLE
SCALE
SCALE
RISE & FALL TABLE ADJUSTMENT
RISE & FALL TABLE ADJUSTMENT

1

SURFACE PLANER

PRECISION CUTTER SETTER

LOCATING PAD

4

BACK TABLE   BLOCK   FRONT TABLE

SETTING BY PRECISION SETTING DEVICE

5

BACK TABLE
FENCE
STRAIGHT EDGE
FRONT TABLE

USING PATENT CUTTER
SETTER DEVICE GIVING MICROMETER
SCREW ADJUSTMENT TO KNIVES

HOLDING CAP

WHEEL ENGAGES SLOT IN CUTTER
KEYWAY

6

MOULDING IRON

KNIVES

CAP HOLD CUTTERBLOCK

CAP HOLD CUTTERBLOCK
WITH MICROMETER SETTING

MOULDING IRON ATTACHMENT
ON CUTTERBLOCK

FIGURE 33

PLANING MACHINES

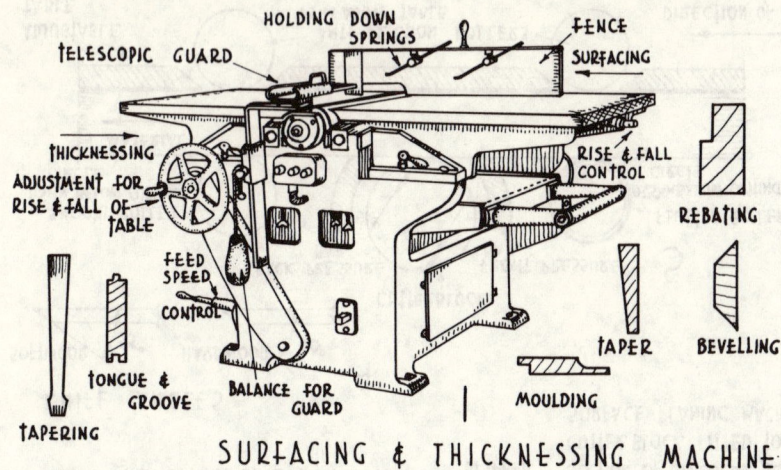

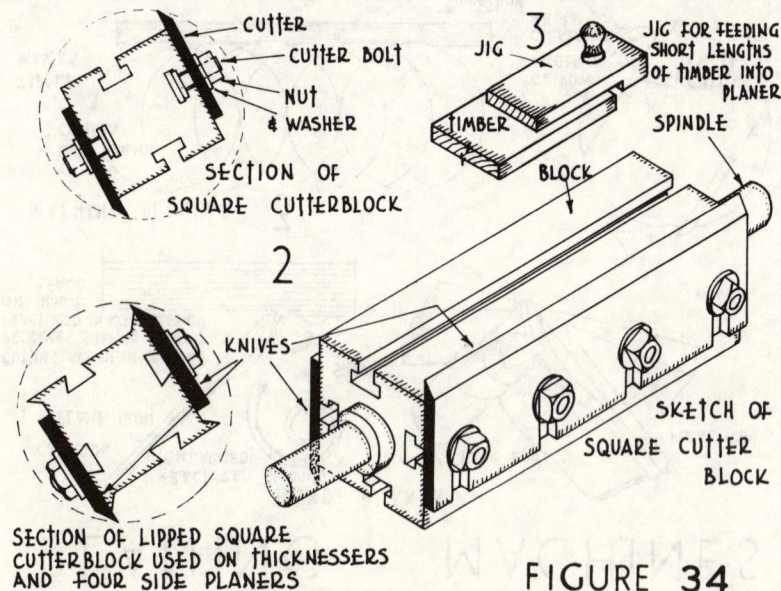

HOLDING DOWN SPRINGS

FENCE

SURFACING

TELESCOPIC GUARD

THICKNESSING ADJUSTMENT FOR RISE & FALL OF TABLE

RISE & FALL CONTROL

REBATING

FEED SPEED CONTROL

BALANCE FOR GUARD

TAPER

BEVELLING

TONGUE & GROOVE

MOULDING

TAPERING

SURFACING & THICKNESSING MACHINE

CUTTER
CUTTER BOLT
NUT & WASHER

SECTION OF SQUARE CUTTERBLOCK

JIG FOR FEEDING SHORT LENGTHS OF TIMBER INTO PLANER

JIG 3

TIMBER

SPINDLE

BLOCK

KNIVES

SKETCH OF SQUARE CUTTER BLOCK

2

SECTION OF LIPPED SQUARE CUTTERBLOCK USED ON THICKNESSERS AND FOUR SIDE PLANERS

FIGURE 34

When the cutters become dulled with constant use, the tables may be drawn away from the cutters and a carborundum oilstone used to restore a sharp cutting edge to each cutter.

Feeding

Feeding a surface planer is done mostly by hand, hence the need for great care to avoid contact with the cutterblock by the hands. The strict habit of using the guard correctly reduces danger to a minimum. Guards of telescopic design are popular and must be adjusted to cover all exposed parts of the cutters, including that at the rear of the fence. There is a tendency to feed too fast, which produces a roughly planed surface. It will be found that feeding slowly will produce a good smooth finish.

The fence

The fence is used to plane square. After forming a face side on the table, this planed surface is held to the fence when planing the next side. The fence is adjustable and will tilt to 45 degrees for planing bevelled edges, chamfering, etc. Rebating is also easily carried out using the fence.

Moulding

A portion of the outer end of circular blocks has removable cap-plates to allow moulding cutters to be set as shown in Fig. 33.6. Small moulding work may be done this way when a spindle moulding machine is not available.

Thicknessing machines

After facing and square-planing, joinery components are planed to width and thickness by the thicknesser (Fig. 34.1), which has a power feed system in the form of four rollers. Two idle rollers are fitted in the rise and fall table, and above these are two driven rollers which propel the wood under the cutterblock. The first upper roller is serrated to grip the sawn surface better. Both upper rollers and also two pressure bars, are all fitted with springs to put pressure on the wood, as shown in Fig. 32.5. One type of serrated roller is made in sections, which allows the feeding of pieces of slightly unequal thickness without the risk of kickback.

Feeding

A gearbox drive to the feed roller gives a choice of three feed speeds, 7·620 m, 12·190 m, and 18·290 m/min. The height of the table rollers is adjusted by means of a handwheel at the infeed end of the table to a workable minimum above the table surface. If they are too low, friction prevents the easy passage of the wood through the machine. If too high, 'see-sawing' occurs between the wood and the table in passing from one roller to the other, causing jumps in the planed surfaces.

Cutterblocks

Although circular cutterblocks are compulsory on surface planners, (Fig. 33.2), they are sometimes fitted to thicknessers, also triangular and square blocks are used, as shown in Fig. 34.2. Setting the knives or cutters accurately, so that all are cutting equally, is very difficult and a truing or jointing device is used to trim all the cutting edges into one cutting circle. This is done by winding a small carborundum stick over the cutterblock while it is in motion. It is carefully adjusted to touch all the cutting edges.

Surface finish

The planed surface of timber is governed by the size of the cuttermarks (Fig. 32.3) and is used as a guide to the quality of finish required for different types of work.

For outside joinery 3·175 mm per cut or 32 mm for ten cuts.
For inside joinery 1·250 mm per cut or 13 mm for ten cuts.
For cabinet work 0·800 mm per cut or 8 mm for ten cuts.

Example

$$\frac{\text{No. of metres/min of feed} \times 100}{\text{No. of cutters cutting} \times \text{rev/min of cutterblock}}$$

If there are two cutters cutting and the cutterblock revolves at 4 500 rev/min on a feed speed of 18·000 m/min it is simply found by dividing the number of millimetres per minute of the feed by the number of cuts per minute:

$$\frac{18 \times 1\,000}{2 \times 4\,500} = 2 \text{ mm per cut, or 20 mm per ten cuts}$$

It will be obvious that, if four cutters were used instead of two, the feed speed may be doubled and still maintain the same surface finish.

Peripheral speed

In addition to the machine cutters being in good condition, they must have a good constant speed on the periphery. This is the circle described by the cutting edge as shown in Fig. 32.2. The speed can be checked when the diameter and the speed in revolutions per minute of the cutterblock are known.

The following formula may be used to calculate the peripheral speed of the block;

$$\frac{\text{Diameter of cutting circle (mm)} \times 3\frac{1}{2} \times \text{rev/min}}{1\,000} = \text{m/min}$$

Example

Diameter of cutting circle of block 125 mm; speed 4 200 rev/min.

$$\frac{125}{1000} \times \frac{22}{7} \times \frac{4\,200}{1} = 1\,650\cdot000 \text{ m/min}$$

Peripheral speed = 1 650·000 m/min

Cutter setting device

This is shown in Fig. 33.5, and is a patent cutter setting device giving very fine micrometer screw adjustment to the knives, which must be set parallel with the back surfacing table and with 1·500 mm projection from the cutterblock. An ordinary straight-edge is placed on the back table overhanging the cutterblock and the cutter is adjusted in relation to it by means of a key operating two micrometer screws. The screws are movable in either direction to adjust the cutter in or out of the cutterblock.

Jointing

Cutter jointing has already been mentioned. Jointing is the operation of bringing two or more cutting edges into a true cutting circle so ensuring that each is cutting to the exact depth of the other. This operation can also be termed tracking, and must be done with special care.

Jointing is done by an abrasive stone or stick carried vertically in a special jig over the cutterblock. It is first adjusted until the stone just touches the tip of the stationary cutters. With the machine in motion the stone is travelled across the full length of the block and returned. The cutterblock is then stopped and examined for even contact with the abrasive. If this is not so the operation is repeated. Each contact increases a second heel on the cutters. This second heel should not exceed 0·800 mm.

Cutter care

Cutters should be kept in balanced sets by ensuring that the cutters have equal dimensions after grinding and that the cutter edge is straight and parallel to the back edge.

For general work, knife angles for softwoods and hardwoods are recommended, as shown in Fig. 32.4.

When a very fine finish is required in dry softwoods and hardwoods a slight front bevel is given. For wet or green timber the cutting bevel may be decreased 5 degrees, but the front bevel should not be given.

Cutters should be kept sharp when in position by using a fine-grade oilstone dipped in paraffin. Allow the stone to rest lightly and flat on the bevel and pass over the cutter with a rotating action a few times. Give about two strokes on the full length of each knife on the face side to remove all burrs from the cutting edge.

A heel, not greater than 0·800 mm wide on the bevel, should not be allowed before removing and regrinding. When the heel becomes too wide the knives may heat up or have a hammering effect on the wood and more than normal power will be required to run the cutterblock.

The narrow bandsaw

This machine may be found in most workshops where wood is worked, such as joinery, coachbuilding, and pattern-making. It is also largely used in cutting plastics, metal, and cloth.

Design

The design consists of a vertical main frame, on which two wheels are

THE NARROW BANDSAW

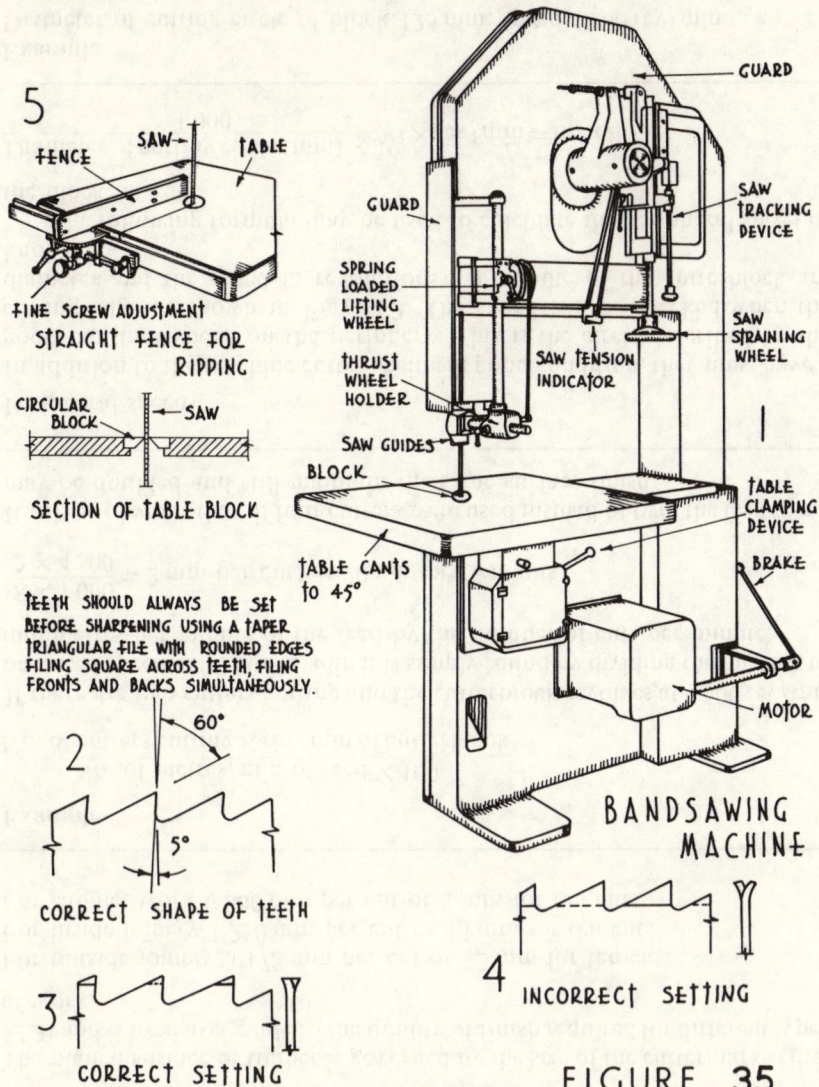

5

FENCE   SAW   TABLE

FINE SCREW ADJUSTMENT

STRAIGHT FENCE FOR RIPPING

CIRCULAR BLOCK   SAW

SECTION OF TABLE BLOCK

TEETH SHOULD ALWAYS BE SET BEFORE SHARPENING USING A TAPER TRIANGULAR FILE WITH ROUNDED EDGES FILING SQUARE ACROSS TEETH, FILING FRONTS AND BACKS SIMULTANEOUSLY.

2   60°   5°

CORRECT SHAPE OF TEETH

3

CORRECT SETTING

GUARD

GUARD

SPRING LOADED LIFTING WHEEL

THRUST WHEEL HOLDER

SAW TENSION INDICATOR

SAW GUIDES

BLOCK

TABLE CANTS TO 45°

SAW TRACKING DEVICE

SAW STRAINING WHEEL

TABLE CLAMPING DEVICE

BRAKE

MOTOR

BANDSAWING MACHINE

4 INCORRECT SETTING

FIGURE 35

mounted, one above the other, which carry an endless spring steel sawblade (Fig. 35.1). The saw runs through a slot in the centre of a cast-iron worktable at a convenient height, and may be used for straight or curvilinear cutting. The lower wheel may be driven directly from an electric motor or by means of short centre vee belts. The upper wheel, which is adjustable vertically, is driven by the saw blade and is fitted with a variable tensioning device to regulate the tension of the sawblade, at the same time acting as a shock absorber preventing breakage of the sawblade when heavy loads occur during cutting.

The wheels

Machines in general use have wheels of 762 mm and 915 mm diameter, to use sawblades 13 mm to 38 mm wide, 5·100 m and 6·000 m long. Small bandsaws used for general purposes in joinery workshops may have wheels of 304 mm diameter. Disc-type wheels on modern machines are an improvement on the older spoked wheels, which caused cold draughts of air which affected the hands of the operator. Rubber tyres are secured to the rims of the wheels to eliminate slip, and also to prevent damage to the saw teeth. The accumulation of resin on the rubber tyres is partly avoided by having fixed brushes which clean the tyres as they revolve. Guards, in the form of hinged doors made from sheet metal or fibreglass, completely enclose all moving parts, except that part of the saw in use.

The upper wheel is mounted on a pivot plate to tilt it forwards or backwards as required to control the running position of the saw on the wheels. The saw will tend to run on the highest circumference.

Saws

Very narrow saws from 3 mm wide are used for cutting curves of small radius. As the radius increases, wider saws may be used, depending on how the work may be turned on the saw. It will be understood that the pitch of the teeth must be less for narrow saws to retain the tensile strength of the blade. To obtain the length of saw required, add the circumference of one wheel to twice the distance between the wheel centres. Saws may be obtained ready jointed to length with the jointing either brazed or butt welded. When automatic setting machines are used an even number of teeth is required for alternate setting. The actual cost of setting and sharpening a saw is also the price of a new saw if done by hand. Hence, it is essential to have automatic setting and sharpening equipment to maintain a large number of saw blades. The shape of the teeth, having 60-degree gullets, shown in Fig. 35.2, is maintained by using triangular files which sharpen by filing the fronts and backs of the teeth in the same stroke. Correct and incorrect setting of saws is shown in Figs. 35.3 and 35.4.

Guides

To prevent the saw from being forced off the wheels during cutting, the back edge is supported by hardened steel discs, revolving on ball-bearings, one fixed beneath the table, the other being adjustable to the depth of the cutting and carried on a bracket above the table. Hardwood or soft-metal guides are also provided to hold the saw in line sideways.

Brakes

Modern bandsaws have brakes fitted to one or both wheels, to stop the saw in an emergency or after use. Braking after use is important for safety as the saw will continue to run idly for a long time after switching off the motor.

Jigs for repetition sawing

Many ingenious devices are used on the bandsaw for repetition work, such as the cutting of the curved rails for chairs, and cutting circular discs, etc. by rotating the work on a centre-pin. Any such methods, which do not require laborious marking out, save time and cost.

Using the bandsaw for conversion

A ripping fence (Fig. 35.5), may be fitted to the table when using the bandsaw for flat or deep cutting. When fitted with a wide saw it becomes a substitute for a band re-saw, within limits, having a capacity 355 mm deep. The table will tilt to an angle of 45 degrees when bevelled sawing is required on straight or circular work. A metal chute is fitted under the table to deflect the sawdust into the extractor system.

Tensioning

Incorrect tension or tightness of the blade over the saw pulleys will end in saw breakage. When placing a saw on the pulleys it is tightened until the correct tension is reached according to the tension indicator which gives the correct tension for the width of blade in use.

Tracking

Every saw has slightly different running characteristics on a bandsaw machine due to the condition of the steel ribbon from which it is made, the brazed joint, and the tension in the blade ribbon. This is compensated by using a crowned or slightly curved rubber on the wheels and providing the top wheel with a slight tilting movement. By slackening the small locking handwheel and adjusting the tracking handwheel the top wheel can be tilted slightly until the saw blade runs or tracks centrally.

This is important because the blade then passes in a straight line between the top and bottom wheel, and does not snake. When the latter occurs the back of the saw keeps hitting the guide runner and damaged guides result.

The chain and chisel mortising machine

The most frequently used joint in woodworking is the mortise and tenon; it is important, therefore, that fast and accurate cutting of mortise holes and tenons may be carried out using highly efficient machines.

Design

Various designs of machines include:

The single chisel.
The single chain.
The combined chain and chisel (illustrated in Fig. 36.1).

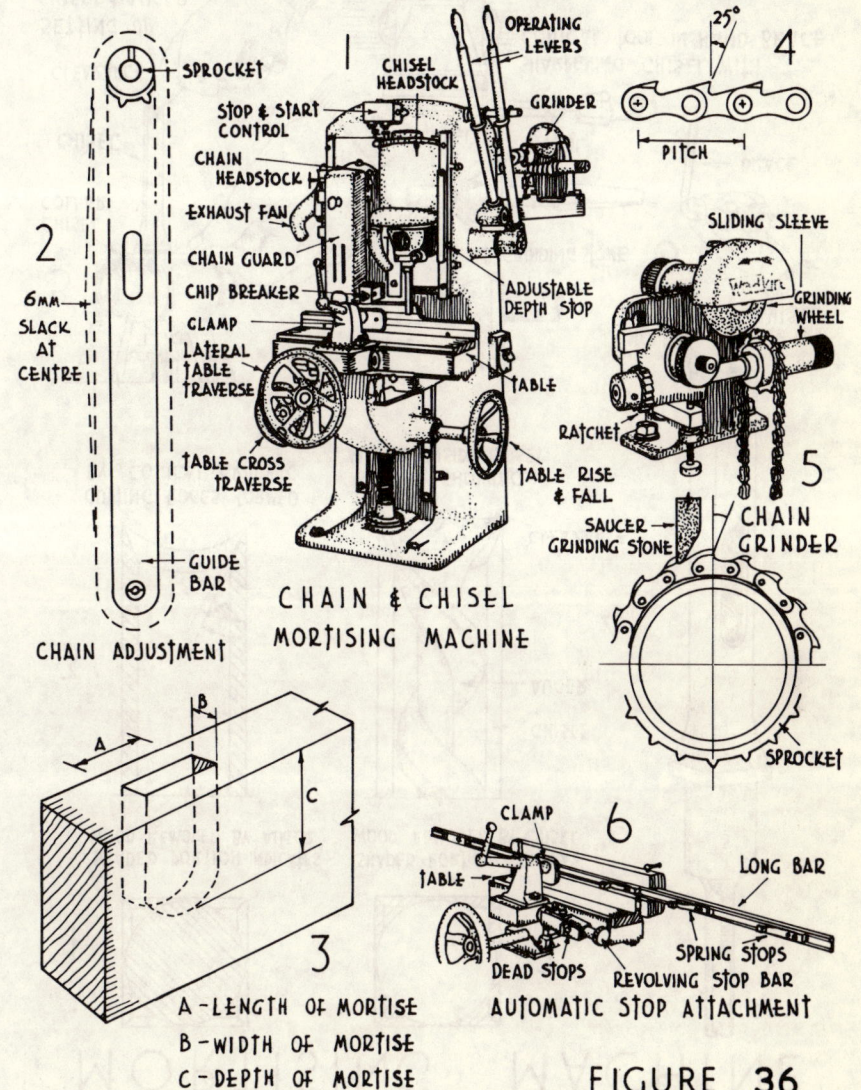

MORTISING MACHINE

1 — CHAIN & CHISEL MORTISING MACHINE

Labels: OPERATING LEVERS, CHISEL HEADSTOCK, GRINDER, SPROCKET, STOP & START CONTROL, CHAIN HEADSTOCK, EXHAUST FAN, CHAIN GUARD, CHIP BREAKER, CLAMP, LATERAL TABLE TRAVERSE, ADJUSTABLE DEPTH STOP, TABLE, TABLE CROSS TRAVERSE, TABLE RISE & FALL, GUIDE BAR

2 — CHAIN ADJUSTMENT — 6MM SLACK AT CENTRE

3 — A-LENGTH OF MORTISE, B-WIDTH OF MORTISE, C-DEPTH OF MORTISE

4 — PITCH — 25°

5 — CHAIN GRINDER — SLIDING SLEEVE, GRINDING WHEEL, RATCHET, SAUCER GRINDING STONE, SPROCKET

6 — AUTOMATIC STOP ATTACHMENT — CLAMP, TABLE, LONG BAR, SPRING STOPS, REVOLVING STOP BAR, DEAD STOPS

FIGURE 36

60

Each of these machines has a heavy cast-iron pedestal with vertical slides, on which the motorised cutter heads(s) move up and down to limit stops by means of hand levers or hydraulic power. Attempts have been made to introduce new methods of mortising, using oscillating cutters and slot boring on the router. Both these methods are very good in certain kinds of production, but the hollow square chisel and the chain cutterheads have firmly established these methods of mortising in most workshops.

The hollow square chisel

Although this form of mortising cutter is not as speedy in production as the chain mortiser, it is useful on jobs calling for very neat and accurate work. The cutterhead consists of a motor mounted vertically on slides. The lower end of the rotor spindle of the motor is fitted with a chuck to receive boring augers of various sizes. The auger in use runs in a hollow square chisel sharpened at the cutting end on all four sides, as shown in Fig. 37.4. The lower end of the auger is set so as to enter the work first cutting a circular hole (Fig. 37.1), leaving only the four corners to be cut by the square chisel, shown in Fig. 37.2. The chisel (Fig. 37.5), is held firmly in position by a collar which is fitted in the lower end of the motor casing.

Fitting the chisel and auger

The correct size of chisel and auger, together with the corresponding sleeve for the auger chuck and collar for the chisel to suit the mortise holes required, are inserted as a complete set.

When tightening the grub screw which secures the auger shank, the chisel is left 3 mm below the collar. When securing the chisel in the collar, it is pushed up to its proper working position and tightened. This allows 3 mm clearance between the cutting edges of the auger and the chisel, as shown in Fig. 37.4. If this clearance is omitted severe damage may be caused due to friction and overheating. Many good chisels have been ruined by this oversight. Apertures in the sides of the chisels allow the wood borings to escape, thus preventing internal clogging.

Maintenance of chisels and augers

Special sets of files may be obtained for sharpening the cutting edges of chisels and augers.

Filing must be carefully carried out, especially on the cutting wings of the auger. A conical carborundum stone which may be fitted to the grinding machine is useful for resharpening the cutting edges of chisels. Also made for this purpose is a set of hard steel reamers of various sizes which, when used in a joiner's brace, act like a countersink and ream the inside edges, as shown in Fig. 37.6. A sectional view showing the grinding angle of the cutting edges of the chisel is shown in Fig. 37.3.

Mortising techniques

When using the square chisel it is the usual practice to mark the wood on both edges and to mortise just past halfway in depth from each edge. This avoids the damage to the underside of the wood which would result from mortising through in one stroke. If a sufficient number of pieces to one pattern is required, jigs may be made to register the position of the holes, to save the

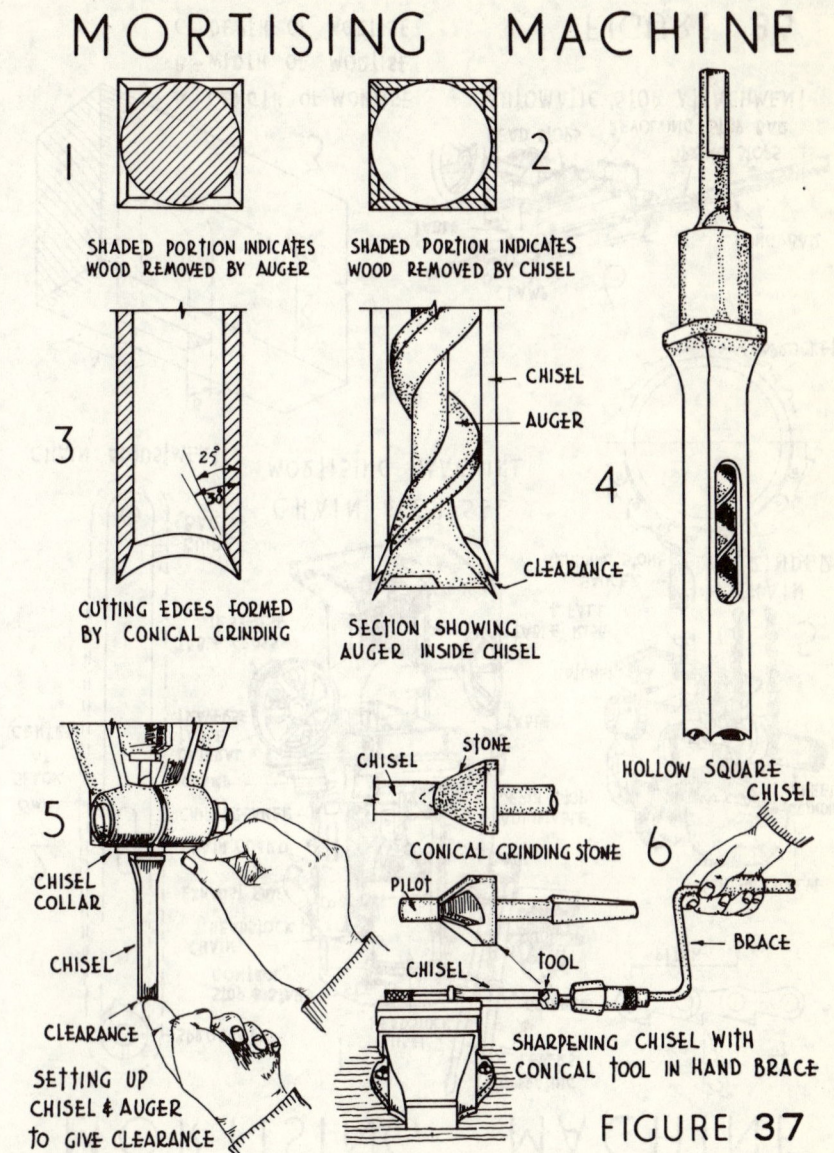

MORTISING MACHINE

1 — SHADED PORTION INDICATES WOOD REMOVED BY AUGER

2 — SHADED PORTION INDICATES WOOD REMOVED BY CHISEL

3 — CUTTING EDGES FORMED BY CONICAL GRINDING

4 — SECTION SHOWING AUGER INSIDE CHISEL / CHISEL / AUGER / CLEARANCE

HOLLOW SQUARE CHISEL

5 — CHISEL COLLAR / CHISEL / CLEARANCE / SETTING UP CHISEL & AUGER TO GIVE CLEARANCE

CHISEL / STONE / CONICAL GRINDING STONE / PILOT / CHISEL / TOOL

6 — BRACE / SHARPENING CHISEL WITH CONICAL TOOL IN HAND BRACE

FIGURE 37

labour of marking. A good practice to ensure good results from the tenoning and mortising machines is to use trying pieces when setting up. A mortise hole is cut in the trying piece working to the lines of a mortise gauge. The operator on the tenoning machine uses the trying piece to ensure that the tenoning is a fit in the mortise hole, and that the face sides are flush when the joints are assembled.

Chisels for hand-operated mortising machines range from 6 mm to 25 mm. Mortise holes above 25 mm should be made on a power-operated machine.

The square chisel works better when cutting its full size of hole. The practice of cutting half the width of the chisel causes undue side thrust. By using special cutters it is possible to utilise the mortising machine for cutting the pocket pieces of pulley stiles, or mitring glass beads, etc.

The chain cutterhead

A horizontal motor fitted with a spindle, which projects to the front and rear, carries a chain drive sprocket on the front and a small belt pulley on the rear end, and revolves at 3 000 rev/min. A bracket attached to the motor casing carries the chain guide bar which is fitted with a roller-bearing wheel. This directs the chain into the cut, each chain link having a cutting tooth. A chain having thirty-five links will cut 105 000 chips/min out of the wood being mortised. This type of chain cutter has been developed to fell trees and crosscut large logs driven by small petrol engines.

Selection of chains

For joinery work three pitches of chains (see Fig. 36.4), are made for small, medium, and large mortise holes.

Pitch 13 mm for small holes 5 mm to 8 mm wide; 19 mm to 35 mm long.
Pitch 16 mm for medium holes 5 mm to 8 mm wide; 22 mm to 38 mm long.
Pitch 22 mm for large holes 6 mm to 32 mm wide; 38 mm to 75 mm long.

Chain gear (Fig. 36.2), is manufactured and supplied in sets comprising chain sprocket, guide bar, and chain. The width of the chain bar governs the length of the mortise hole, as shown in Fig. 36.3. Longer mortise holes are cut by traversing the table and using two or more strokes. Traversing should never be attempted without removing the chain from the work, as this practice puts excessive strain on the chain and bar. The depth of the holes required governs the length of the guide bar. If very deep mortise holes are required it may be necessary to work from both edges.

Chains sometimes break in use and new links and rivets may be fitted to repair them in a manner similar to repairing a cycle chain. Special punches and a small anvil may be obtained to do this.

An attachment to the machine, shown in Fig. 36.5, consists of a small saucer grinding wheel, driven from the pulley at the rear of the chain motor spindle, and a chain holder for grinding the cutting edges of the chain-teeth. The small grinding wheel must be trimmed, using a star wheel dresser, to keep the gullets constant in size. The chain holder is adjustable so as to maintain the correct hook bevel of 25 degrees to the chain-teeth.

Chip breaker

When the chain is in use, the rising side is provided with a piece of hardwood fixed to a weighted bar to prevent spelching by the chain-teeth as they leave the work.

Chain tension

At the top of the chain bar an adjusting screw is provided to ensure that the chain runs freely but not too slack. When the chain may be pulled away from the bar, 6 mm at the middle, as shown in Fig. 36.2, it is at the correct tension. A chain running too slack will not cut a neat mortise.

Safety

A guard is provided to cover the chain when in use and the operator should take every precaution to avoid contact with the chain which would result in very serious injury. The clamp which secures the timber in the machine must always be properly tightened, and should not be released until the chain stops, otherwise there is a danger that accidental contact of the wood with the moving chain will cause the wood to be flung violently out of the right-hand side of the machine, endangering anyone in this vicinity.

An automatic stop attachment is shown in Fig. 36.6. This is an advantage when a fairly large number of pieces have to be worked exactly the same. It ensures that all the pieces are identically mortised without their having to be marked out.

The device is fitted to the fence as shown, to which may be fitted any number of spring stops as required for the job in hand. The long stop bar locates the position of the mortises in the material, working with the end of the material in contact with the appropriate stop.

Any one of four stops can be selected using the lower attachment shown, which works in conjunction with the stop bar. This restricts the longitudinal movement of the table to a set amount corresponding to the length of mortise required.

The single-end tenoning machine

There are many types and sizes of this important machine and the selection is governed by the class of work for which it is to be used.

In a large joinery works where a number of tenoning machines are required, it is an advantage to have various sizes as well as a double-end tenoning machine for quantity production of assorted components. The design of the single-end machine, shown in Fig. 38.1, comprises top and bottom tenoning and scribing cutterheads, and a saw mounted at the rear for cutting tenons to the exact length required. All the heads, which run at 3 000 rev/min on an a.c. supply of 50 cycles, have vertical and horizontal adjustment.

Various items of equipment and attachments may be used which widen the range of operations which may be done. For instance, square cutterblocks may be fitted instead of the tenoning blocks for square turning moulded parts, and saws are often fitted to the scribing heads for slotting and forking

TENONING MACHINE

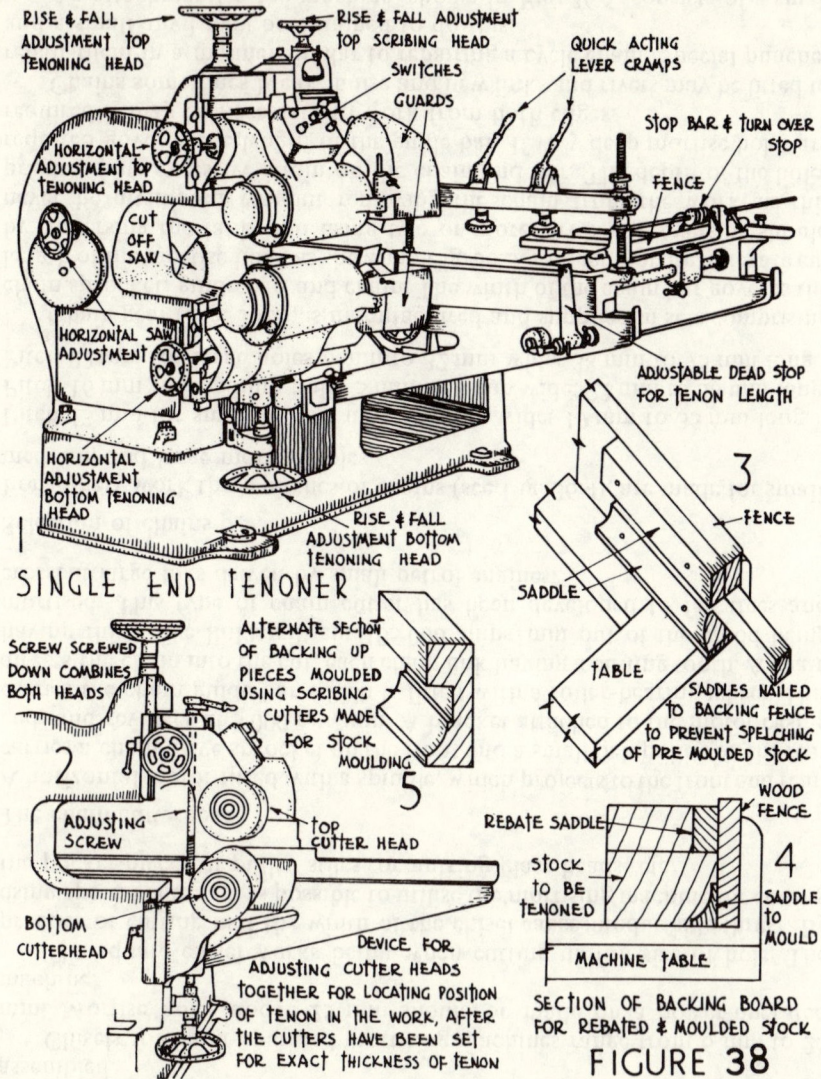

RISE & FALL ADJUSTMENT TOP TENONING HEAD

RISE & FALL ADJUSTMENT TOP SCRIBING HEAD

QUICK ACTING LEVER CRAMPS

SWITCHES

GUARDS

HORIZONTAL ADJUSTMENT TOP TENONING HEAD

STOP BAR & TURN OVER STOP

FENCE

CUT OFF SAW

HORIZONTAL SAW ADJUSTMENT

ADJUSTABLE DEAD STOP FOR TENON LENGTH

HORIZONTAL ADJUSTMENT BOTTOM TENONING HEAD

RISE & FALL ADJUSTMENT BOTTOM TENONING HEAD

1

SINGLE END TENONER

SCREW SCREWED DOWN COMBINES BOTH HEADS

ALTERNATE SECTION OF BACKING UP PIECES MOULDED USING SCRIBING CUTTERS MADE FOR STOCK MOULDING

3

FENCE

SADDLE

TABLE

SADDLES NAILED TO BACKING FENCE TO PREVENT SPELCHING OF PRE MOULDED STOCK

2

ADJUSTING SCREW

5

TOP CUTTER HEAD

BOTTOM CUTTERHEAD

DEVICE FOR ADJUSTING CUTTER HEADS TOGETHER FOR LOCATING POSITION OF TENON IN THE WORK, AFTER THE CUTTERS HAVE BEEN SET FOR EXACT THICKNESS OF TENON.

REBATE SADDLE

STOCK TO BE TENONED

WOOD FENCE

4

SADDLE TO MOULD

MACHINE TABLE

SECTION OF BACKING BOARD FOR REBATED & MOULDED STOCK

FIGURE 38

operations. Motors of greater horsepower are often fitted for certain special work when required.

The pieces to be tenoned are held by clamps against a fence on a table fitted with anti-friction rollers which runs along horizontal rails set at right angles to the cutterheads. The fence is adjustable to an angle of 45 degrees for cutting bevelled tenons in staircase work, etc.

The importance of close cooperation between the mortising machine and the tenoning machine operatives is described in the previous work on the mortising machine. Trying pieces, cut from the actual stock to be used and having a sample mortise hole in one piece and a sample of the tenoning on the other, are kept at hand to provide an occasional check, and as a record of the settings of the machines. Most machines are provided with locking devices on the slides of cutterheads and fences to prevent any movement after setting which may occur due to vibration.

Setting the machine

Having determined the position of the tenon in relation to the face side of the timber, a mortise gauge is set and a mortise hole cut in the trying piece, by either chain or hollow square chisel, exactly to the gauge lines. The end of the part to be tenoned is then marked with the gauge and clamped against the fence, face side down to the table, and in a position so that the cutterheads may be rotated by hand. It may be seen at a glance in which direction the cutterblocks must be moved to conform to the gauge lines. Having done this, a trial cut is made and tried in the mortise hole, adjustments may then be made as required to ensure that (a) the tenon fits the mortise (not too thick or thin) and (b) the face side registers flush with the face side of the mortised piece. The design of the cutterheads has always favoured an adze type which has a skew cut for clean cutting across the grain. Recent designs have tended towards a square-cut type which operates quite as well and is less costly. These cutters are more easily ground and reset compared with the adze type which must be ground to a curved templet supplied with the machine. To set adze-type cutters a planed-up board is clamped on the table and the cutterblock is wound until the sweep of the adze cutter makes contact with the surface of the board. The cutters may then be removed, ground, and reset to the surface of the board in the exact position as before.

Trenching across the grain is carried out by using trenching cutterheads of various designs. Some types found to be useful may be expanded to suit the exact width of the trench required, these are shown in Figs. 39.4 and 39.5. Another type in regular use is the dado head, shown in Fig. 39.3; this cutterhead comprises two outer saws of special tooth design to cut the sides of the trenches, and inside cutters to remove the wood between the saws. Inside cutters are of various thicknesses and may be combined to cut the exact width of trench required up to 50 mm wide. Where an air line is available, air-operated pressure clamps may be used to hold the work securely in position on the table, saving a lot of time and effort of the operator.

When scribing is required the cutters to be used are ground to the required profile, having first set out geometrically the true shape of this profile which is proportionally enlarged to suit the particular approach angle. This approach angle varies with the size of the cutterblock, and to keep the shape of the cutter constant, a sheet metal templet of the profile may be made

TENONING MACHINE

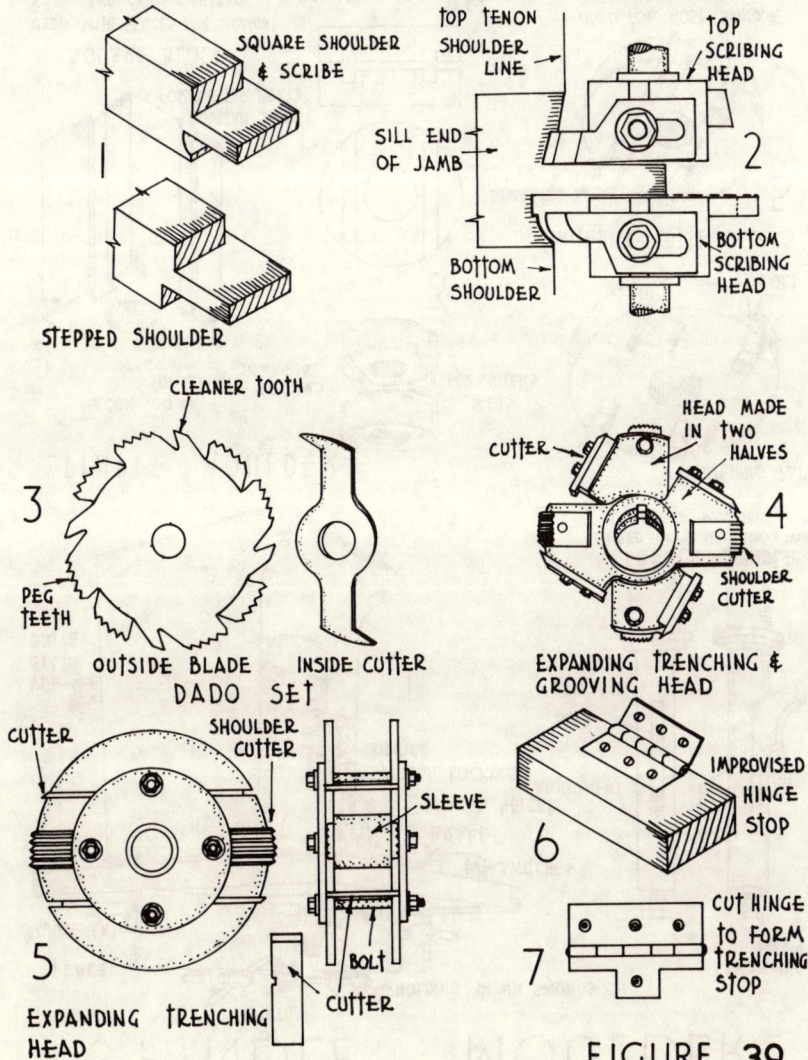

SQUARE SHOULDER & SCRIBE

STEPPED SHOULDER

1

TOP TENON SHOULDER LINE

SILL END OF JAMB

BOTTOM SHOULDER

TOP SCRIBING HEAD

BOTTOM SCRIBING HEAD

2

CLEANER TOOTH

PEG TEETH

OUTSIDE BLADE   INSIDE CUTTER

DADO SET

3

CUTTER

HEAD MADE IN TWO HALVES

SHOULDER CUTTER

EXPANDING TRENCHING & GROOVING HEAD

4

CUTTER

SHOULDER CUTTER

SLEEVE

BOLT

CUTTER

EXPANDING TRENCHING HEAD

5

IMPROVISED HINGE STOP

6

CUT HINGE TO FORM TRENCHING STOP

7

FIGURE 39

---

to work to when regrinding is required on long runs. The cutters are then attached to the blocks, care being taken to allow the top of the cutter to be 1·500 mm above the top of the scribing cutterblock as shown in Fig. 39.2; this ensures that the block will run freely without 'rubbing' on the tenon. A setting templet is required when setting the cutters in position on the cutterblock. This may be made from a small piece of plywood on which the location of the cutter has been set out. A careful machinist will keep his setting boards which saves considerable time when a repetition of the setting is required. When the cutters are set correctly on the block it is then wound into its correct position in relation to the tenon. It is necessary to use the top scribing head for certain types of work, and in setting this the method is the same as that for the bottom head. When all cutterheads are correctly set, the motorised saw at the rear of the machine may be set to cut off the tenons to exact length.

It is the practice in some mills to run stock on the moulding machines and mortising and tenoning operations follow on the pre-moulded stock. This calls for special preparation of the fence of the tenoning machine to prevent damage by spelching where the cutters emerge.

In such cases a negative section of the moulding is attached to the table fence, as shown in Fig. 38.3, 38.4, and 38.5, to act as a chip breaker. It has been found that a clean cut is produced on certain settings by pulling the work on to the cutters instead of pushing it forward also reducing the traverse of the table. When it is practical, several components are clamped in the machine instead of cutting each separately, thus increasing production.

A long threaded screw may be inserted on some tenoning machines, to connect the top and bottom heads when set up. This device, shown in Fig. 38.2, allows the adjustment of both heads together when the tenon is required in an alternative position. This is done by using the upper handwheel only after releasing the set screws which secure the bottom head.

The vertical spindle moulding machine

For the greatest variety of cutting operations the vertical spindle moulder has for many years been considered to be the most useful.

The introduction of the high-speed router has taken over some of its light cutting and moulding work, but the vertical spindle moulder may be used for both heavy and light cutting operations.

The operator must be keenly conscious of his personal safety since it is generally accepted that this machine may cause serious mutilation of the fingers and hands when used without the recommended safety devices.

Figure 40.1 shows a type of spindle which is in general use and consists of a heavy cast-iron base surmounted by a cast-iron table which is machined to a perfectly flat surface. The base houses the spindle on vertical slides, which permit raising and lowering a distance of 152 mm and the driving motor is carried on a horizontal slide at the rear of the base. The motor and spindle are fitted at the lower end with two-speed vee belt pulleys, giving spindle speeds of 3 000 and 4 500 rev/min. A smaller vee pulley may be quickly fitted to the spindle to give a speed of 6 000 rev/min. These are convenient speeds for general use and cutter equipment is used which is recommended by the makers as safe to run at these speeds, relative to the diameter of the cutting-circle.

SPINDLE MOULDER

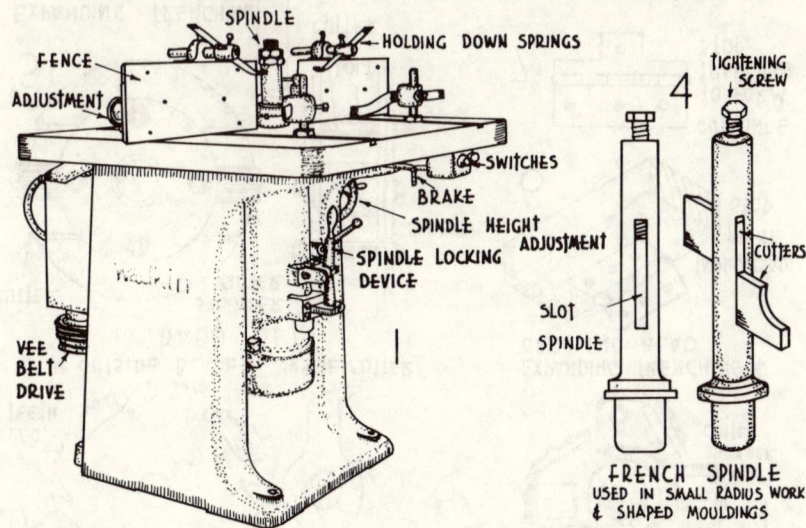

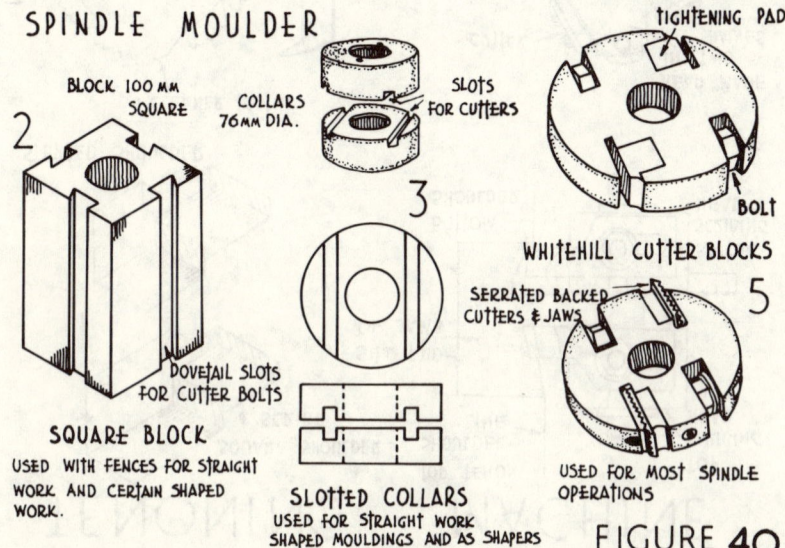

SPINDLE

FENCE ADJUSTMENT

HOLDING DOWN SPRINGS

SWITCHES

BRAKE

SPINDLE HEIGHT ADJUSTMENT

SPINDLE LOCKING DEVICE

SLOT

SPINDLE

VEE BELT DRIVE

SPINDLE MOULDER

1

TIGHTENING SCREW

4

CUTTERS

FRENCH SPINDLE
USED IN SMALL RADIUS WORK & SHAPED MOULDINGS

BLOCK 100 mm SQUARE

2

DOVETAIL SLOTS FOR CUTTER BOLTS

SQUARE BLOCK
USED WITH FENCES FOR STRAIGHT WORK AND CERTAIN SHAPED WORK.

COLLARS 76 mm DIA.

SLOTS FOR CUTTERS

3

SLOTTED COLLARS
USED FOR STRAIGHT WORK SHAPED MOULDINGS AND AS SHAPERS

TIGHTENING PAD

BOLT

WHITEHILL CUTTER BLOCKS

SERRATED BACKED CUTTERS & JAWS

5

USED FOR MOST SPINDLE OPERATIONS

FIGURE 40

Some operations require higher speeds and special designs run at 9 000 and 15 000 rev/min, and it must be remembered that special cutting equipment is required for such speeds. Various types of cutterhead spindles may be used which are quickly interchangeable, being secured by a large phosphor-bronze cap-nut.

The standard loose top piece has a maximum length of 203 mm and is used for most standard cutterblocks. For speeds of 9 000 rev/min a maximum length of 124 mm is used. Longer spindles at very high speeds must be fitted with a top steady bearing to prevent any tendency to whip. The square cutterblocks, shown in Fig. 40.2, are 95 mm wide on each face and of various lengths up to 152 mm. Each face has a dovetailed slot to receive cutter bolts of 16 mm diameter. The cutters used are 9:5 mm thick and have a body of mild steel with a hard high-speed steel face welded on which provides the cutting edge after repeated grinding. Cutters may be obtained ready ground to profile when a section of the required moulding is given to the cutter-makers.

It is also necessary to keep a supply of blanks on hand for miscellaneous jobs.

Cutter bolts are made from high-tensile steel to withstand the continuous strain of tightening, and any yielding to this is readily observed by the nut becoming tight on the threads. Bolts in this condition should be discarded as unsafe since they may break while in use.

Because of its great weight, the square block may be used for very heavy cuts.

Cutters should, whenever possible, be used in balanced pairs. When setting pairs of cutters on the block every effort should be made to have both cutting. Modern setting devices enable this to be done.

Grinding cutters demands care. The metal of the cutter must not become overheated, causing an alteration of the hardness which is essential for a lasting cutting edge. A cupboard, with specially designed recesses, should be used for storing the cutters; this makes selection easier and prevents damage to the cutting edges by careless storage.

When cutting a rebate 12 mm deep, through a 9 mm-thick face board, the cutting circle of a square block is approximately 203 mm diameter, and when a smaller cutting circle is required, slotted collars may be used, shown in Fig. 40.3. Usually they are of 76 mm diameter, 25 mm thick, having parallel slots to receive 6 mm thick cutters. This type of cutterhead is used for cleaning up and moulding curved components and may be fitted with an anti-friction ball-bearing guide to run against the edges of the patterns used.

The projection of the cutters must be kept to a minimum to ensure solid cutting action. Serrated slots and cutters are used to prevent cutters from slipping in or out of the collars.

The French spindle, shown in Fig. 40.4, is also used extensively for shaping operations and sweeps of small diameter. The cutters are also 6 mm thick and made from steel which may be filed to the profile required. A vertical slot through the spindle houses the cutters which are held fast by a securing screw at the top of the spindle.

The cutting action of the French spindle, due to the slightly negative approach angle, is a scraping rather than a cleaving cut and it is due to this that it produces a fairly clean moulding, with or against the grain of the wood. This type of spindle is used a great deal in furniture manufacture.

Solid circular cutterblocks are preferred for many spindle operations.

SPINDLE MOULDER

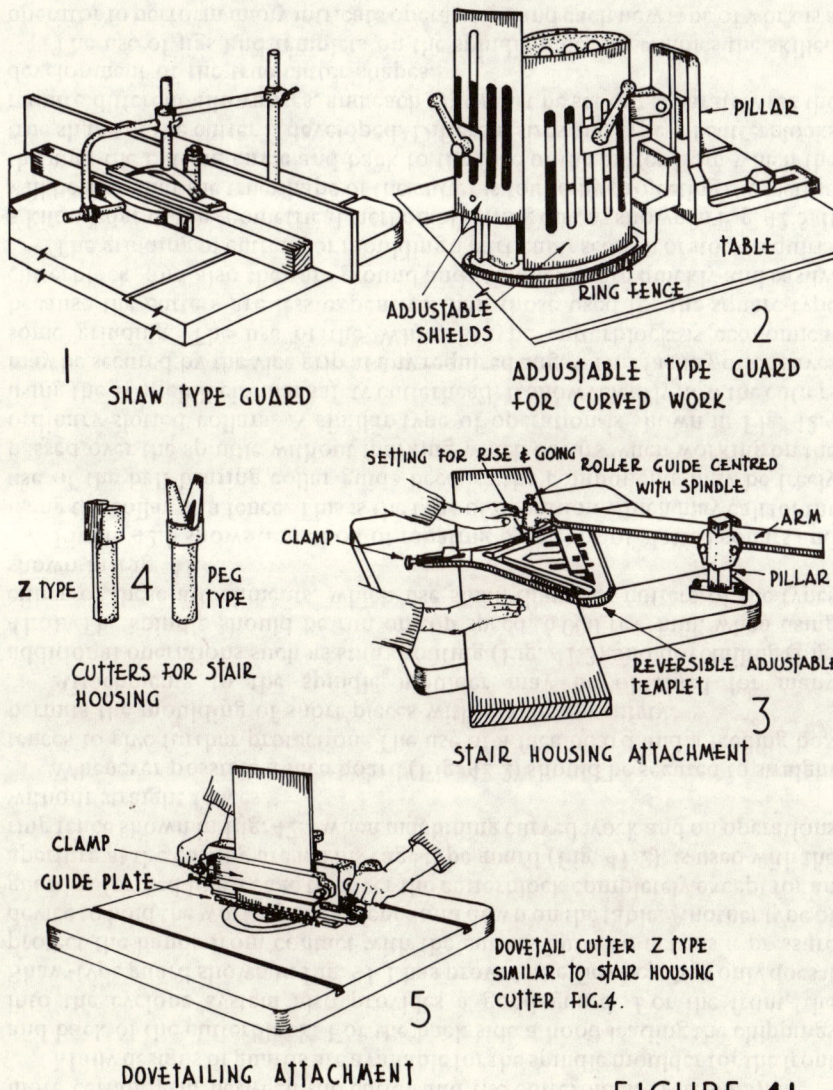

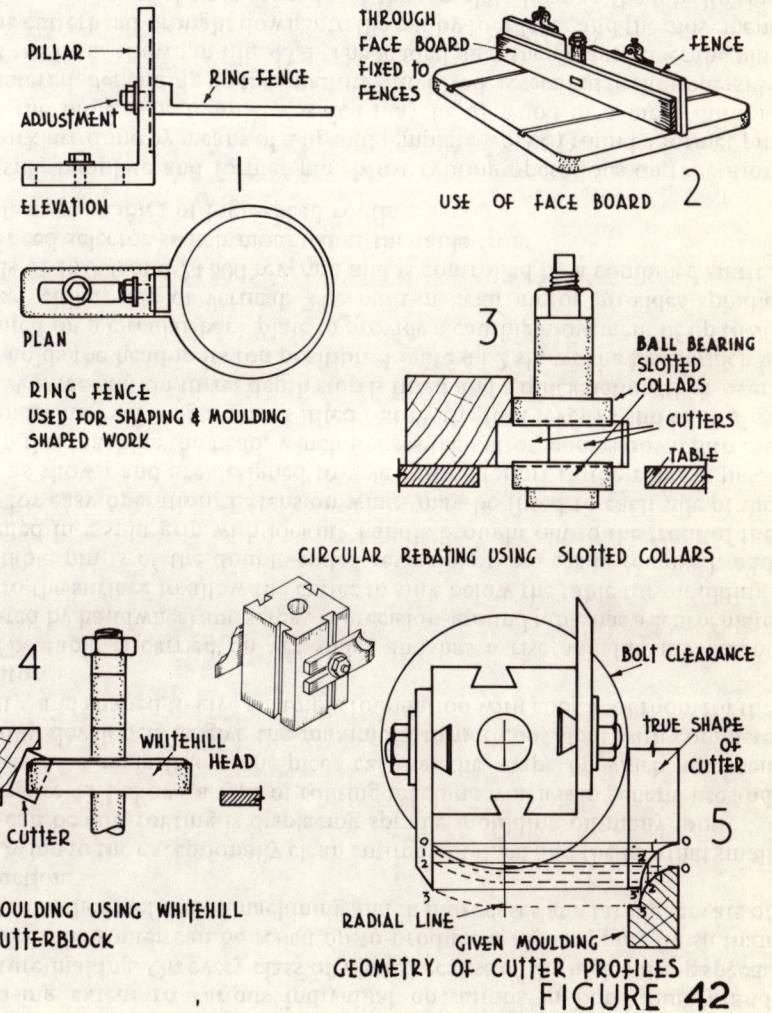

1 SHAW TYPE GUARD

PILLAR

TABLE

ADJUSTABLE SHIELDS

RING FENCE

2 ADJUSTABLE TYPE GUARD FOR CURVED WORK

Z TYPE **4** PEG TYPE

CUTTERS FOR STAIR HOUSING

SETTING FOR RISE & GOING

ROLLER GUIDE CENTRED WITH SPINDLE

CLAMP

ARM

PILLAR

REVERSIBLE ADJUSTABLE TEMPLET

3 STAIR HOUSING ATTACHMENT

CLAMP

GUIDE PLATE

DOVETAIL CUTTER Z TYPE SIMILAR TO STAIR HOUSING CUTTER FIG.4.

5 DOVETAILING ATTACHMENT

FIGURE 41

SPINDLE MOULDER

PILLAR

ADJUSTMENT

RING FENCE

ELEVATION

THROUGH FACE BOARD FIXED TO FENCES

FENCE

2 USE OF FACE BOARD

PLAN

1 RING FENCE USED FOR SHAPING & MOULDING SHAPED WORK

BALL BEARING SLOTTED COLLARS

CUTTERS

TABLE

3 CIRCULAR REBATING USING SLOTTED COLLARS

WHITEHILL HEAD

CUTTER

4 MOULDING USING WHITEHILL CUTTERBLOCK

BOLT CLEARANCE

TRUE SHAPE OF CUTTER

RADIAL LINE

GIVEN MOULDING

5 GEOMETRY OF CUTTER PROFILES

FIGURE 42

65

They are of many types and may be made specially to cut any section of moulding required or they may have vice grip fixing for separate cutters pre-ground to any required profile, shown in Fig. 40.5. There is much less noise from circular cutterblocks than from square blocks because there is less air disturbance. Some circular blocks make use of serrated surfaces to obtain more certain grip between the cutter and the cutterblock (Fig. 40.5).

Many designs of guards are available for the spindle moulder for the front and back of the cutterblock. For the back side a hood leading the chippings into the cyclone system also provides a good guard. For the front, the Shaw-type guard shown in Fig. 41.1 has proved its efficiency. Not only does it protect the hands from contact with the cutters but also acts as a pressure device to hold the work up to the fence and down on the table. Another type of guard is formed like a cage to cover the cutterblock completely except for an aperture at the cutting area. This cage type guard (Fig. 41.2), is used with the ring fence shown in Fig. 42.1 when machining curved work and on operations without straight fences.

Whenever possible a face board (Fig. 42.2) should be secured to straight fences to give further protection. The use of a face board and a feeding box permits the moulding of short pieces with maximum safety.

Attachments to the spindle moulder may be obtained for many additional operations such as stairs routing (Fig. 41.3) and dovetailing (Fig. 41.5). The spindle should be run on top speed, 6 000 rev/min, when using either of these attachments, which use small-diameter cutters of the types shown in Fig. 41.4.

Figure 42.3 shows a method of rebating by means of slotted collars and using the collar as a fence. This is the type of operation which may call for the use of the ball-bearing collar guide because the components may be freely passed over the spindle without marking which occurs when working on the ordinary slotted collars. A similar type of operation is shown in Fig. 42.4 using the Whitehill circular safety cutterhead. It shows clearly how the cutters may be secured by the vice grip at any required angle. This facility often saves some grinding. The use of the Whitehill-type cutterblock is economical because the cutters are less expensive than those used for the square-type cutterblock, and also they are ground and set much more quickly and easily.

The grinding of cutters for moulding a particular section of stock requires a knowledge of the geometrical method of setting out, as shown in Fig. 42.5. It will be seen that the true shape of the cutter is found by projecting the section through the cutting circle and back to the face of the cutter from which the true shape of the cutter is developed. Different sizes and types of cutterblocks require different allowances, and each type must be set out separately for the development of the true cutter shapes.

The use of jigs and templets on the spindle moulder enables the skilled operator to perform many intricate operations, and each new type of work is a challenge to the inventive resources of the craftsman who must always set up and operate the machine with maximum safety.

Electric feeding appliances are used on the spindle moulder. These may be run at various feed speeds giving a constant surface finish and providing maximum safety and minimum fatigue for the operator.

The router

The high-speed router has become an accepted production tool in every branch of woodworking. A tremendous variety of jobs and operations can be done on this machine and are no longer confined to the production on a repetition basis of speciality wooden articles, but is being applied to an increasing extent to various individual operations in both joinery and furniture-making. On every class of work, because of the high cutting speeds involved, the router can be relied on to produce a higher grade finish than from any other method of machining and in most cases at a far higher rate of production.

Owing to the exceptionally clean cutting obtained and the fact that small radii can be cut, routing is displacing spindle moulding on many jobs.

Figure 43.1 shows a type of routing machine which is in general use and consists of a main-frame one-piece casting, the shape of which has been carefully developed to give the maximum throat consistent with complete rigidity, and shaped to give an ample foundation with good foot room for the operator.

The table is carried on vee slides and has a rise and fall movement actuated by handwheel and screw. A precision-ground table has a centre plate let into the surface to allow the cutter to sink below the table for moulding. The table pin is of the double-ended reversible type, easily removed, and mounted in a split grip with locking handle brought out to the front of the table for easy operation. Extension wings may be fitted to each side of the table as shown and are designed to give extra support to the routing jigs.

In this machine the head, which houses the cutter, comes down into the job under its own weight and is lifted out by the foot lever as shown in Fig. 43.3. A three-station turret depth stop is fitted and a quick acting 'flick-over' lever holds the head in its top position. Figure 43.2 shows the head which is mounted on a circular back plate to provide a canting movement of up to 90 degrees either side of vertical. The built-in head motor provides spindle speeds of 18 000 and 24 000 rev/min and is controlled by a combined starter and speed selector switch mounted at the table front.

Basic principles of high-speed routing:

1. Using template and former pin. Most routing operations on repetition work are done by means of a jig and template worked round a former pin in the table. The template, which may be of wood or a more durable material, depending on the quantities involved, is secured to the underside of the jig as shown in Fig. 43.4. The loaded jig is then placed over the pin, the cutterhead brought down into the cut by foot lever and the movement of the jig round the pin reproduces the template form in the job. By this method any shape with external or internal profiles or recesses can be handled.

2. Using straight fences. Increasing use is now being made of the router in conjunction with a straight fence shown in Fig. 43.6. A finer quality of machining is thus being obtained on many common operations in joinery, or cabinet work, such as half lapping, grooving, moulding, shaping, haunching and relishing, corner locking, etc.

3. Using the former pin as a guide fence. Owing to the exceptionally clean

ROUTER

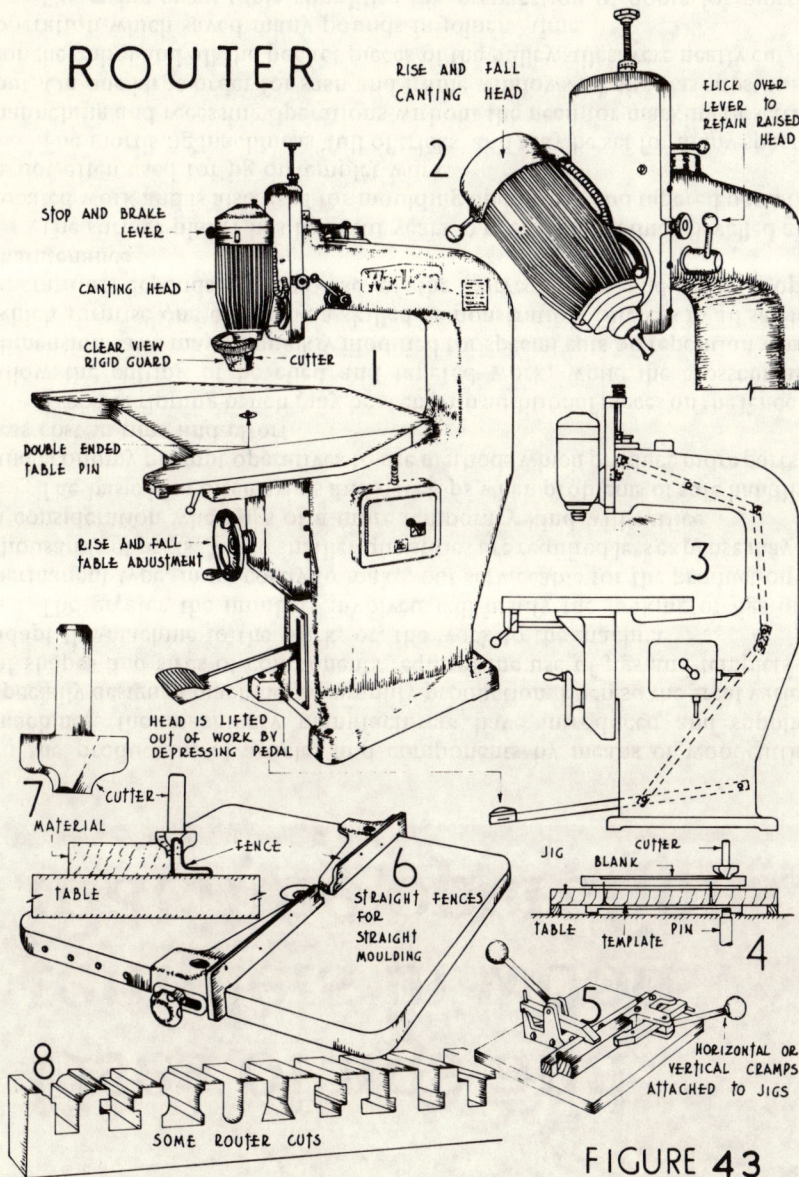

RISE AND FALL CANTING HEAD

FLICK OVER LEVER TO RETAIN RAISED HEAD

STOP AND BRAKE LEVER

CANTING HEAD

CLEAR VIEW RIGID GUARD

CUTTER

DOUBLE ENDED TABLE PIN

RISE AND FALL TABLE ADJUSTMENT

HEAD IS LIFTED OUT OF WORK BY DEPRESSING PEDAL

CUTTER

MATERIAL

FENCE

TABLE

STRAIGHT FENCES FOR STRAIGHT MOULDING

JIG

CUTTER BLANK

TABLE

TEMPLATE

PIN

HORIZONTAL OR VERTICAL CRAMPS ATTACHED TO JIGS

SOME ROUTER CUTS

FIGURE 43

cutting obtained and the fact that small radii can be cut, routing is displacing moulding by spindle on many jobs.

The table-pin is used as a guide against which the template runs. With this method of working both pin and cutter must be of the same diameter and a range of pins in stepped sizes up to 50 mm are available to suit a wide range of cutters. Figure 43.5 shows a sketch of the universal cramp for attaching to router jigs to hold the work.

Examples of some router-cuts are shown in Fig. 43.8 with a typical moulding cutter shown in Fig. 43.7.

These are solid shaped cutters obtainable to order from the makers of the machine and ground to any required shape. Other cutters available are flat knife chucks, tongue and groove cutter arbours, expanding cutter heads and eccentric chucks. These can be divided into two flutes and single flute cutters and the single flute cutter, again, into two groups: (a) concentric and (b) eccentric. Concentric types are always single flute and are for use in eccentric chucks. The cutting edge is not relieved, relief is provided by the eccentricity of the chuck or spindle. The cutting size can be varied slightly by adjusting the chuck. Eccentric types are used in concentric chucks, the cutters are relieved on the cutting edge and the maintenance of this relief is vital throughout the life of the cutter. Eccentric cutters can have single or double flutes.

The cutters are made from High Speed Steel, but those for cutting man-made boards like chipboard need Tungsten Carbide tips.

Sharpening is best done on the corner of an oilstone or rectangular section slipstone by placing the flute over the corner of the stone rubbing the cutter forward and back along it. The flute angle permits stoning in this way and a perfect finish is given to the edges. Remove any burr on the outside of cutting edge by a light stroke with the slipstone. After a period of use and repeated sharpening, a router cutter may need regrinding. This is best done by trained personnel using special equipment.

All tungsten carbide tipped tools will give prolonged service on man-made timbers. Specialist equipment however is needed to grind these and users without the equipment and expertise should consult a specialist or the manufacturer.

Chapter 8

Uses of jigs and templets in wood machine operations

In the production of articles and components by means of woodcutting machines, the machinery manufacturers have introduced and supplied specially designed machines to simplify production. Even so the great variety of shapes and sizes of components requires the use of jigs and templets to adapt the machine to the work, or, the work to the machine.

The greater the numbers involved will justify the making of jigs of a permanent type, more costly to make, but serviceable for the production of thousands of parts. When smaller quantities are required less expense may be a consideration when jigs of a more temporary kind will suffice.

The basic idea often arises and develops when problems of safe handling and economy prompt operatives to use methods which produce more parts at less cost in time and effort.

Even the ripping bench may be seen with additional pieces on the fence to allow the cutting of bevelled and tapered work, while the crosscut and dimension saws may be quickly modified for special cuts on repetition work, which surprise one on seeing a skilled demonstration. Success in all sawing operations depends on the use of the correct saws and their proper maintenance.

The surface planer has been for years expected to produce bevelled and rebated work and is also used for moulding operations and tapered parts but is not often used for jig or templet work.

The mortising machine is 'full of tricks' and may be set for many spacing haunching and recessing operations without the need for marking or setting out. On one large order for sash and frame windows a knife was substituted for the chisel and all the pocket pieces of the pulley stiles were neatly cut, an operation which saved many pounds in joiners' time.

The swing-away table simplifies the preparation of doors for mortise locks and bevelled mortising for spandrel framing.

The high-speed router is a very valuable tool in the furniture, joinery, and soft metals industries and is almost entirely dependent on jigs and templets for its tremendously growing uses in fast and accurate cutting operations.

The design and making of the jigs has become a highly skilled part of routing production on shaping large quantities of components at great speed. When a small number off are required it is, of course, quite easy to make jigs quickly and cheaply which will last for the duration of the work.

The designer of router jigs should be acquainted with the following:

1. Method of loading and unloading jigs and secure holding of the pieces.
2. Manipulation with complete safety.
3. The cutting action and cutter selection.
4. Motion analysis and fatigue of operator.

The most popular method of control of the cutting is by means of a guide pin which is located in the table at the same centre as the cutter which engages in a groove formed in the underside of the jig, and to the exact size and shape of the required article.

1. The loading and unloading of jigs should be by an assistant who will load one jig while the operator is performing the cutting operation.
2. All jigs must be provided with handles at a safe distance from the cutter and give clearance to the perspex guard. Owing to the high working speed of the cutter (18 000–24 000 rev./min.) the slightest contact with the hands may result in serious mutilation.
3. Examination of the cutters indicates the direction of the cutting, and care must be taken to avoid kickback which may occur through engaging the work with the cutter in the wrong direction.

Components are usually cut net size to fit snugly in the jig and secured by some type of clamp, of which there are many kinds, the best being the quick-release type – often a simple turn button will suffice. Thought must be given to the cutting regions when fitting the clamps and it is an advantage to allow them to swivel clear of the cutter, resetting again when past the cutter. A pressure device may be fitted which surrounds the cutter, and when this device is used clamps are not required. Impaling the pieces on small spikes is sometimes enough to hold the parts to be cut, care being taken to blow out the dust after unloading.

The spindle moulder may be used for a multitude of varying operations and offers great scope in the use of templets and jigs. The first consideration must be the safety of the operator's hands. With this in mind it is better that the man making the jig is fully conversant with spindle operating techniques.

When provided with well-designed jigs the most difficult cutting operations may be carried out with safety and ease, for example the shaping of circular and shaped pieces and even such small items as heels for ladies' shoes. Fresh adaptations of this versatile machine are now appearing in the form of high-speed shaping machines, some with a revolving table, some with a static table. A very modern version has a gang of six spindles on one machine, any one of which may be put into operation giving a choice of six profiles without dismantling any cutterheads.

Another high-speed spindle shaper, with a static table, is used for external shaping and dimensioning components with a moulded edge such as table-tops, casement sashes or chair parts, etc. A sprocket with variable speeds is fitted in the table, and by engaging with a chain fitted to the jig feeds the work past the cutting head without the need for any hand feeding.

However, such machines are expensive and a guarantee of full-time running is the only justification for having them installed in the machine shop.

Production sequence

The floor area of a woodworking factory imposes limits in the quantities and dimensions of the type of articles which may be manufactured.

Large and small factories sometimes find a need for more space to carry out projects to a required rate of production and it follows that the area available must be used wisely and economically.

The intake, storage, and treatment of the raw materials, the processing by machinery, the assembly finishing and the final storage prior to despatch, all require adequate space in order to avoid congestion, which hampers production in the manner of a traffic hold-up.

Timber and plywood, etc. is obtainable in many stock sizes and buying should be confined to current needs in order to avoid 'dead stock', keeping in mind that huge stacks of timber are not making profits, whereas their value would, if in the bank. By careful planning aided by modern drying facilities stocks may be kept to a minimum.

Handling of large quantities calls for the use of forklift trucks which speedily and economically unload and reload in a ratio of twenty men per machine when compared to manual handling. Provision must always be made to allow the forks to enter and lift by placing suitable spacing timbers in the load. By this method a load of 6–10 tonnes may be handled in 15–20 minutes.

A good practice is to crosscut the timber to required lengths before kiln drying, the capacity of the kiln is then used to maximum purpose and the useless drying of waste offcuts is avoided.

This practice needs the careful placing of one or two covered crosscut saws between the timber stacks and the kilns. The dried timber is issued to the machine department and when a costing system is in operation a 'check in' is passed on to the cost office.

A straight-line edger is often required to convert stock sizes to framing, usually arranged by one flat cut, e.g. 2/75 × 50 mm EX. 150 × 50 mm. A machine of this type is thus well placed nearest to the entrance.

Note: Entrance doors should be easily opened and closed to avoid loss of heat.

A 900 mm sawbench (power feed) or a small re-saw is placed adjacent to the straight-line edger for any deeper cutting.

The four-side planing and moulding machines may be placed at the outfeed end of the saws, in some instances the feed to the moulders is immediately continuous from the saw by feed track.

Depending on the volume of production the foregoing arrangement may be duplicated when varieties of section are required simultaneously.

The foregoing plant is fairly popular in many works where large quantities of joinery and cabinet-framing are produced.

Additional machine equipment will vary greatly as the articles to be made require varying machine operations, and where joinery is produced we may find a number of tenoning and mortising machines forming the next group dealing with this important and ever popular method of jointing framing. Of this group the most productive are the double-end tenoning and gang or multiple-mortising machines. Single-end tenoning and single-head mortising machines are very necessary to cope with a variety of work and small quantities, and may be interspersed in the layout to give output in the general flow production. This applies also to many other machines essential to production such as in the following:

Spindle moulding
Stair trenching (double)
Dovetailing
Routing
Bandsawing
Tenon haunching
Dimension sawing
Boring, etc.

An ever-increasing use of the many sandpapering machines available proves to be very rewarding in reducing cost and giving high-quality surfaces to receive paint or even polish finish.

The fully machined components may be placed in orderly array in a 'finished part store' from which any quantities may be issued to the assembly department as required.

In the up-to-date assembly department as much as possible is achieved by pneumatic pressure devices on which jigs may be arranged to receive the components which are assembled at the touch of a floor pedal-valve. Joints are secured by wood or metal pins and may be pre-treated with paint or glue by dipping or brushing. The assembled items may be 'dressed' at the joints on an overhead belt sander and passed on to a painting and loading department.

Paint spraying in properly constructed booths with extractor fans or water curtains to absorb the surplus paint fumes is an economical process and, still better where it is employed, 'curtain spraying' of paints and polishes, etc. has revolutionised these finishing operations. Inspection for faults is advisable before delivery.

Most of the small woodcutting machines are designed to be so self-contained that, apart from the electrical supply and waste extraction, it is quite normal to reinstate machines where a fresh position would prove to be more suitable when changing over to new types of production. An efficient management will take advantage of this reorganisation when it arises in order to improve the flow-line production methods in up-to-date establishments.

Workshop planning and production

Joinery manufacture requires that management give due consideration to all aspects of planning, the layout of the building, work areas, and available machines. No management which does not mechanise its main production can exist for long in today's highly competitive market.

The planning of the workshop will depend on the type of work being produced and the quantity, with an awareness of having the fundamental machines sited to allow a good and even flow of work, along with adequate working areas. While it is not easy to decide what is a typical layout for joinery production, generally a simple plan will meet most requirements.

The distance between machines and benches will be governed by the maximum lengths of timber being machined and how materials are to be handled.

In deciding the direction of the flow of work the work may pass forward in a direct line from one machine to another with a minimum of passing backwards and forwards until it reaches the despatch area. Therefore, the flow line should be as short as possible to minimise costs in time and handling.

Floor areas require to be more generous at the input end of the shop where the timber handled from the store is in the most bulky state. The first operation is usually that of crosscutting to length, so that the crosscut saw should be sited close to the timber store. Dimensioning or cutting to size follows and, where available, is carried out on the band saw or on the main ripsaw.

Plywoods, blockboards, and the like will usually be cut on the dimension saw or alternatively a panel wall saw positioned close to the plywoods rack.

Moving the material forward in the work flow, small quantities are planed on the various planing machines. Large quantities may be directed to the four-cutter where the processing of all faces of the sawn stock can be carried out at one passing.

Marking out the timber for jointing and the sections for moulding are next carried out by the marker-out to suit framing requirements, the positioning of mortises and tenons are marked on the material and pass in the work flow to the mortise and tenoning machines. Finally the material passes to the spindle moulding machine and the router when necessary.

Cleaning up of members is carried out before assembly at the benches in the joiners' shop. This is done using sanding machines where available and may be drum, belt, or disc sanding machines. Alternatively, portable hand-power sanders are used.

Assembly at the finishing end of the flow line requires an area for joiners' benches and finishing processes, storage, and despatch areas.

Sheets 1 and 2 show two examples of buildings providing machine layouts.

Sheet 1.2 illustrates an almost ideal layout from input to finish, which is possible in a new building or an existing building of similar floor area.

So far, the planning of the workshop has been concerned with the layout of the various areas and the flow line of material in joinery production.

The building itself, whether it is newly constructed, an adaptation of an existing building, or one already established, must conform to the current Building Regulations and to other statutory legislation concerning health, safety, and welfare. Reference to these regulations should be made (see pages 74–75).

The Woodworking Machines Regulations are dealt with on page 48.

Adequate lighting, both natural and artificial, must be provided along with sources of heating. Extraction plants for the removal from machines of wood waste, chips, sawdust, and dust from sanding machines also require consideration.

Extraction systems

Extraction systems vary according to the size of the plant and are an essential part of planning in any works, whether it is a small workshop or a large factory, as a production aid and a safety consideration. The main ducting usually connects through a power-driven fan directly to a cyclone with branch points connected to the main duct at machine locations.

Sweep-up points are connected and sited at intervals throughout the shops and usually fitted with spring-operated covers so that, when in use, floor areas can be quickly cleared.

Where machines have more than one cutterhead, as in the case of the outside planing and moulding machine, smaller branches are arranged to each cutterhead.

The cyclone has a cowling through which air from the plant is extracted, allowing the waste material to fall into a collecting chamber. This requires the cyclone to be sited close to the collecting chamber.

Dust from sanders must be extracted by separate arrangements since the mixture of the dust with the correct proportion of air can form an explosive danger and fire hazard when assisted by grits from sanding belts and papers.

Sheet 1.3 shows a sketch of part extraction arrangements to machines, with the plan in Sheet 1.4.

It will be realised that the removal of waste and the cleaning of machines and workshops, along with the disposal of the waste, is a costly operation and emphasises the value of an efficient extraction system.

Apart from complete dust-extraction plants to precise machine locations, smaller systems of units for confined spaces, suitable for spindle moulders, small planers, and crosscut or circular sawbenches are available.

Noise limits

Reference to noise limits is made in the Woodworking Machines Regulations and needs to be referred to here in connection with the planning of workshops. Managers must ensure that the requirements of those regulations are complied with where noise limits are imposed by the Health and Safety at Work Act.

The regulations state that no worker should be exposed to an equivalent continuous A-weighted sound level greater than 90 dB(A) during an 8-hour working day. Where this level is exceeded, noise-control measures must be adopted and hearing protectors must be provided to avoid impairment of workers' hearing.

Research into noise abatement with regard to woodworking machinery is continuing and methods of controlling or reducing noise at source investigated.

Circular saws have a high-pitched whistle when idling, rotating cutterblocks projections generate noise by the rapid compression and release of air. Helically bladed cutterblocks have been developed which have the cutting edge continuously in contact with the work.

Noise emitted by existing machines may be reduced by enclosing the machine with a form of heavy barrier around the machine with the inside lined with sound-absorbent material. Openings are necessary in an enclosure for infeed and outfeed requirements, with provision for windows and chip extraction incorporated in the enclosure design.

MACHINE SHOP LAYOUT

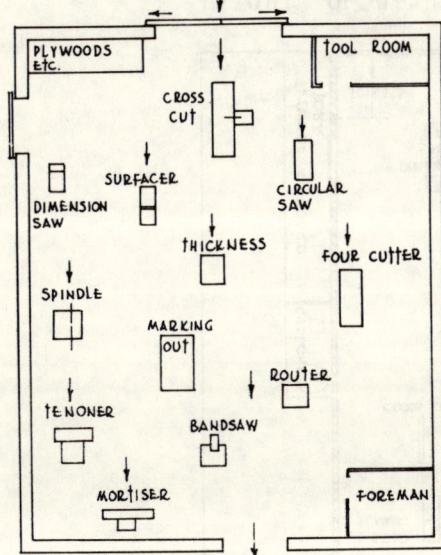

PLYWOODS ETC.

TOOL ROOM

CROSS CUT

SURFACER

DIMENSION SAW

CIRCULAR SAW

THICKNESS

FOUR CUTTER

SPINDLE

MARKING OUT

ROUTER

TENONER

BANDSAW

MORTISER

FOREMAN

WORKFLOW

MACHINE SHOP LAYOUT FOR PROGRESSIVE PRODUCTION.

1

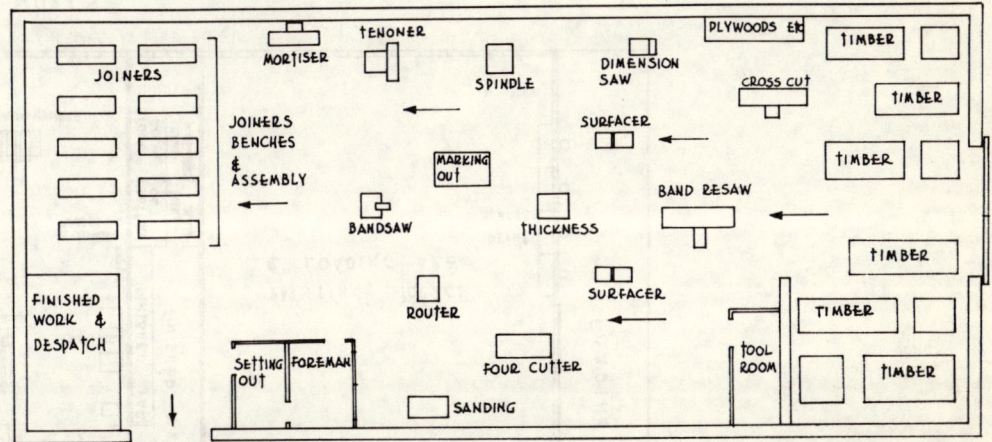

PLYWOODS ETC

TIMBER

JOINERS

MORTISER

TENONER

SPINDLE

DIMENSION SAW

CROSS CUT

TIMBER

JOINERS BENCHES & ASSEMBLY

SURFACER

TIMBER

MARKING OUT

BANDSAW

THICKNESS

BAND RESAW

FINISHED WORK & DESPATCH

SETTING OUT

FOREMAN

ROUTER

FOUR CUTTER

SANDING

SURFACER

TIMBER

TOOL ROOM

TIMBER

MACHINE SHOP LAYOUT, TIMBER STORE & JOINERS SHOP FOR PROGRESSIVE PRODUCTION.   2

WORK FLOW

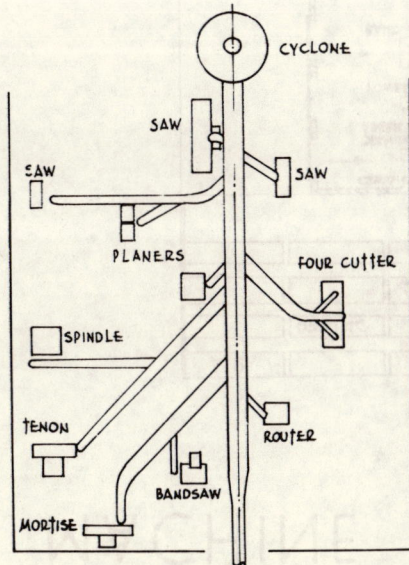

CYCLONE

SAW

SAW

SAW

PLANERS

FOUR CUTTER

SPINDLE

TENON

ROUTER

BANDSAW

MORTISE

4

WASTE EXTRACTION TO ABOVE MACHINE SHOP LAYOUT.

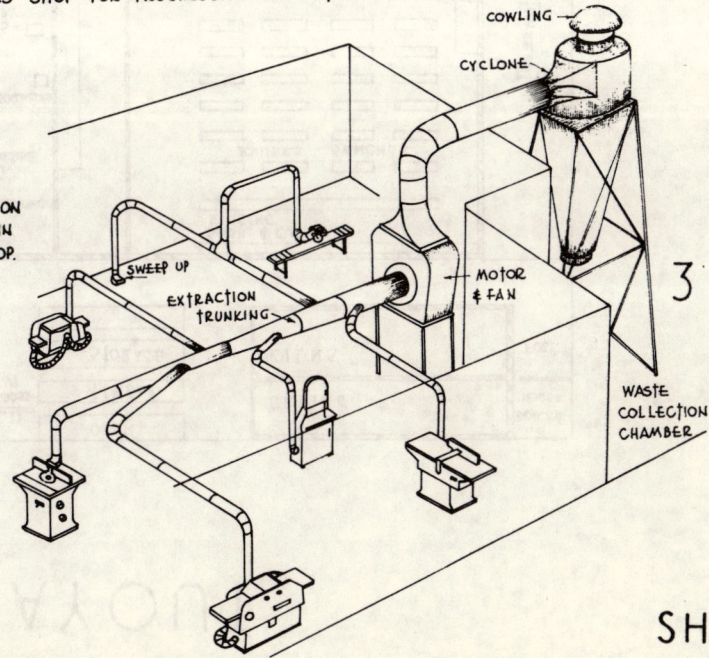

COWLING

CYCLONE

WASTE EXTRACTION TO MACHINES IN SMALL WORKSHOP.

SWEEP UP

EXTRACTION TRUNKING

MOTOR & FAN

3

WASTE COLLECTION CHAMBER

SHEET 1

MACHINE SHOP LAYOUT

AIR

DRYING

SHED

CROSS-CUT

OPEN
TIMBER
STORAGE

CROSS-CUT

OPEN
TIMBER
STORAGE

DRYING

KILNS

BOILER HOUSE

FUEL

CROSS CUT

CROSS CUT

SASH & CUPBOARD HINGING

STRAIGHT LINE EDGER

4 SIDED PLANERS

STAIR TRENCHING

POWER SAW

& MOULDERS

GANG MORTISE

JOINERS BENCHES

BAND RE-SAW

HAUNCHER

4 SIDED PLANER

MACHINE AREA

DOUBLE END TENONER

S P I N D L E S

DESPATCH OFFICE

SINGLE END TENONERS

ASSEMBLY AREA

CROSS CUT

FRAME CRAMP

CYCLONE

DOUBLE END TENONERS

BELT SANDER

ROUTERS

FINISHED PART RACKS

FINISHED PART STORE

HINGE RECESSING

SASH CRAMPS

PAINTING, STORAGE & LOADING AREA

DRUM SANDERS

HINGE RECESSING

SETTING OUT

FORE- MAN

PAINT SPRAYS

BAND SAW

DIMENSION SAWS

TOOL ROOM

IRON- MONGERY

PLYWOOD & BOARD STORE

TOILETS

FRAME CRAMP

DRUM SANDER

DOVETAILING

LOADING BAY [COVERED]

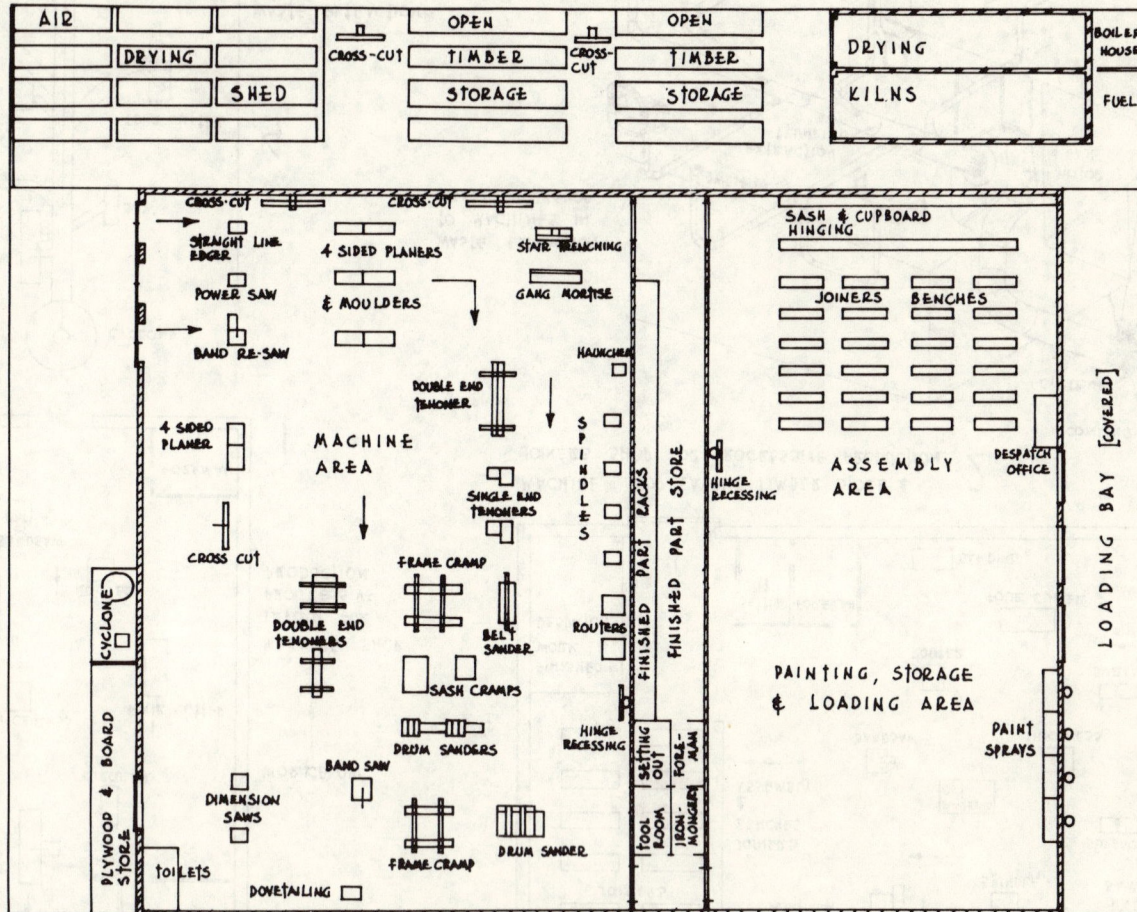

LAYOUT OF MEDIUM SIZE JOINERY WORKS

SHEET 2

Disposal and use of wood waste

Wood waste may be disposed of by contracting to have it taken away, or by selling privately. Large works may have in-built systems using the wood waste for warm-air heating of the workshops, drying kilns, and hot presses.

Individual extraction is arranged from each machine and the waste conveyed to the collecting chamber where the wood shavings are separated. They are then fed into a burner, complete with heat exchanger which produces hot air for warming the shop in cold weather. The heated air is conveyed along ducting, usually in the roof structure, and into the workshop via numerous ventilating outlets.

Chapter 9

British Standards Institution

The British Standards Institution is the recognised body in this country responsible for the preparation and publishing of national standards.

This body is composed of various technical and industry committees assisted by advisory committees, covering a wide range of British Standards.

The scope of standards work includes: glossaries of terms, definitions, and symbols; methods of test; specifications for quality, safety, performance, or dimensions; preferred sizes or types; codes of practice.

Most raw materials, components and fittings used in building are covered by a British Standard Specification and as with any standard, whoever is concerned with the standard should possess the standard, and have detailed knowledge of its contents. This is most difficult for those engaged in the construction industry where some 1 100 standards are involved, and presents a problem to architects, building and quantity surveyors, consulting engineers, clerks of works, merchants, contractors, estimators, and site staffs.

A British Standard handbook, *Summaries of British Standards for Building, 1974*, which includes Codes of Practice, is available to meet the needs of those mentioned above. Detailed summaries are included and are intended to give sufficient information to enable detailed drawings to be prepared, specifications and bills of quantities to be drawn up, and for materials to be ordered and checked by staffs. British Standards are usually abbreviated to BS.

Codes of Practice

These, as has been indicated, are issued by the British Standards Institute and set out the best current practice in the light of up-to-date knowledge and experience in particular fields of activity. Codes of Practice are usually abbreviated to CP.

Generally, British Standard Specification and Codes of Practice will satisfy the requirements of the Building Regulations.

Building Regulations

These regulations are designed to set a minimum standard for all building work and to safeguard public safety and health.

The Building Regulations are administered by local authorities through the building surveyors department. Building inspectors within the department are responsible for the checking of plans submitted for approval, to ensure that they conform to the Building Regulations. Following approval and work on site commencing, site inspections are carried out by the building inspectors as set out in the regulations.

Government publications

There are many publications issued by government departments providing technical information on special aspects of building.

The Building Research Establishment is the main government organisation concerned with research and development for the construction industry. It comprises the Building Research Station, the Fire Research Station, the Princes Risborough Laboratory, and the Scottish Laboratory. Building research publications are usually abbreviated BRE.

The National Building Agency references by the RIBA Services Ltd publish bulletins, reports, and research papers on matters of current technical interest.

Organisations

Numerous organisations covering all aspects of construction, including all the building centres in areas throughout the country, are other sources where information is readily available.

Trade literature

Manufacturers and trade associations publish technical information and descriptive literature on the products and techniques concerned. This information is now classified and a system known as SfB has been adopted to this end. All literature so classified is to the metric A4 size, 297 mm × 210 mm with the SfB classification reference at the top right-hand corner of the sheet.

The SfB system uses letters and numbers as symbols of the main processes used in building. Capital letters indicate the main divisions with functional items such as doors, windows, and stairs indicated by numbers within brackets. Building materials are indicated by lowercase letters.

Journals

A further source of reference and means of investigating new trends in materials and techniques for the student is through trade journals. These may include: *Building Trades Journal*; *Building Materials and Technology*; *Woodworking Industry*, and the *Timber Trades Journal*.

BSI certification trade mark

This is a registered trade mark owned by the British Standards Institution. Its presence on or in relation to a product is an assurance that the goods have been produced under a system of supervision, control, and testing, operated during manufacture and including periodical inspection of the manufacturer's works. The marking may only be used by those licensed under the certification mark scheme operated by BSI. The marking is termed the BSI Kitemark.

British Standards relating to joiners' work

BS	459	Doors, Pts. 1, 2, 3, and 4
BS	565	Glossary of terms relating to woodwork
BS	585	Wood stairs
BS	644	Wood windows, Pts. 1, 2, and 3
BS	648	Schedule of weights of building materials
BS	745	Animal glue for wood
BS	913	Wood preservation by means of pressure creosoting.
BS	1000	Timber and woodworking industry
BS	1186	Quality of timber and workmanship in joinery
		Pt. 1 Quality of timber; Pt. II Quality of workmanship
BS	1192	Building drawing practice
BS	1195	Kitchen fitments
BS	1202	Nails
BS	1203	Synthetic resin adhesives for plywood
BS	1204	Synthetic resin adhesives for wood
BS	1210	Wood screws
BS	1215	Oil stains
BS	1227	Hinges
BS	1285	Wood surrounds for steel windows and doors
BS	1297	Grading and sizing of softwood flooring
BS	1444	Cold setting casein adhesive powders for wood
BS	1445	Plywood manufactured from tropical hardwoods
BS	1567	Wood door-frames and linings
BS	1579	Connectors for timber
BS	2572	Phenolic laminated sheet
BS	3444	Blockboard and laminboard
BS	3544	Methods of test for polyvinyl acetate and adhesives for wood
BS	3583	Information about blockboard and laminboard
BS	3827	Glossary of terms relating to builders' hardware
		Pt. 1 Locks; Pt. 2 Latches; Pt. 3 Catches; Pt. 4 Door, drawer, cupboard, and gate furniture
BS	3842	Treatment of plywood with preservatives

BS 4047 Grading rules for sawn home-grown timber
BS 4072 Wood preservation
BS 4092 Domestic front entrance gates
BS 4169 Glued laminated structural members
BS 4471 Dimensions for softwood
BS 4512 Methods of tests for plywood
BS 4787 Internal and external wood door sets, door leaves, and frames
BS 4978 Timber grades for structural use

Codes of Practice

CP 112 The structural use of timber
CP 151 Doors and windows including frames and hinges
CP 153 Windows and rooflights
CP 201 Wood flooring

Health and Safety at Work Act

The Health and Safety at Work Act introduces new principles to deal with the problems created by people at work. It covers their health and safety and also related problems such as danger to the public and damage to the neighbourhood.

The purpose of the Act is to provide the legislation required to promote, stimulate, and encourage high standards of health and safety at work.

The aim is to promote safety awareness and effective safety organisation and performance.

There are four parts to the Act with Part 3 directed to amending the law relating to Building Regulations.

It should be pointed out that the statutory requirements in the previous legislation covering many different Acts of Parliament and occupations will continue in force. But in the future the technical provisions and new regulations and codes of practice will be integrated into this new Act. For example, integration of the law applies to the guarding of woodworking machinery and other forms of machinery.

The Act places a duty and responsibility on the employer, employee, the self-employed, and the designer, manufacturer, or supplier of articles and substances for use at work. Duties are also placed on persons having control of premises.

Employers must safeguard the health, safety, and welfare of their employees. This applies in particular to the provision and maintenance of safe plant and systems of work, and covers all machinery, equipment and appliances used.

The employer has a duty to provide any necessary information and training in safe practices.

Designers, manufacturers, and suppliers of articles must ensure they are safe when properly used and must test articles for safety in use. Information about the use of the article and conditions regarding its safety must be supplied.

Employees have a duty to take reasonable care to avoid injury to themselves or to others by their work activities, and to cooperate with employers and others in meeting statutory requirements.

Enforcement of the existing Acts or regulations and the new Act will be by the Inspectorate, with powers to issue prohibition notices, improvement notices, or prosecute any contravention of the Act.

Chapter 10

Specifications

The specification in the construction industry is one of the documents prepared by the architect enabling his ideas and thoughts on the type and standard of construction to be communicated to the other members concerned with actual construction work.

The principal aim of the specification is to state in words the standard of work and the quality of finish required in the construction work. It will define matters that are not sufficiently clear from the drawings and should not therefore repeat that information, but be complementary to the drawings.

It will do this by describing the building, its construction, the type of materials, and the standard of workmanship in every aspect of the work.

The specification will cover all the trades employed in the work following the sequence of the Standard Method of Measurement. This is a document published by the Royal Institute of Chartered Surveyors and the National Federation of Building Trade Employers and provides a uniform basis for measuring building works and embodies the essentials of good practice.

Tendering for contract work is carried out by contractors in competition and is based on drawings and specifications prepared by the architect, along with a bill of quantities prepared by a quantity surveyor in accordance with these drawings, specifications, and instructions.

A specification is not intended to tell the contractor how to do his job but must describe the final result required by the architect.

The following specification headings covering the work of the carpenter and joiner, set out the information needed by the quantity surveyor: information which he in turn will pass on to the builder in the bill of quantities.

Carpentry

1. Type and/or stress grading of timber.
2. Preservatives.
3. Strutting for floor joists.
4. Battens to suspended ceilings.
5. Noggings for plasterboard and other linings.
6. Firrings.
7. Insulation in floors and roofs.
8. Cistern casings.
9. Tie rods, fabricated shoes, straps.
10. Bolts and other connectors.

Joinery

1. Types of timber
   (a) softwoods;
   (b) hardwoods;
   (c) ply, blockboard, hardboard etc.;
   (d) plastic-faced boards;
2. For each of the following:
   (a) boarded flooring;
   (b) strip flooring; state: (i) timber, (ii) jointing, (iii) margins, (iv) finish.
3. Plain and matchboarded linings.
4. Panelled linings.
5. Particulars of the following where not shown on detail drawings:
   (a) doors and frames;
   (b) window and frames;
   (c) borrowed lights;
   (d) lantern lights;
   (e) cornices, friezes, and dados;
   (f) sub-frames;
   (g) cupboard units;
   (h) staircases;
   (i) shelves;
   (j) counters;
   (k) bar fittings;
   (l) special furniture;
   (m) special fittings.
6. Information on the following where not shown on detail drawings or schedules:
   (a) ironmongery for doors;
   (b) ironmongery for windows;
   (c) ironmongery for fittings;
   (d) fixing cramps for joinery;
   (e) dowels for joinery;
   (f) water bars;
   (g) handrail brackets;
   (h) special ironmongery.

Specification clauses for joiners' work

Scope of work

The work shall consist of the manufacture, delivery to the site, and fixing in the building of all joinery described in the specification and shown on the drawings, including the supply and fixing of the following:

1. Metal, straps, lugs, and dowels.
2. Priming, preservatives, and polishing.
3. All ironmongery specified or shown on the drawings.

Materials

Hardwoods and softwoods shall comply with BS 4978 and CP 112, AMD 1265. Reference should be made to the stress grading of timber on page 81.

Plywood for exterior use shall be British made to comply with BS 1455: 1963, bonding W.B.P.

Grade 1, where varnished.
Grade 2, where painted.
Grade 3, where hidden.

Blockboard, veneers, plastic veneers, glues, screws, and nails to be specified.

Preservative treatment

Where this is specified, the timber is to be of the correct moisture content specified and free from surface moisture and dirt. Treatment is to be carried out after all cutting and shaping is completed and care is to be taken to avoid damage to surfaces of treated timber in subsequent handling.

Note: Methods of treatment: see pages 27–28.

1. Pressure impregnation.
2. Hot and cold open tank treatment.
3. Dipping or steeping.
4. Brushing or spraying.

Priming

Where priming is specified, the timber shall be coated with red or white lead.

Aluminium base primers may be used when the timber is particularly resinous.

Moisture content

The moisture content of the timber when the joinery is manufactured and delivered to the site shall come within the limits given in the chart on page 22 for each use, and shall be maintained until the building is completed.

OR

The moisture content of the timber used for internal joinery is to be 10 per cent and that used for external doors and frames is to be 16 per cent when the joinery is delivered to the site, and these moisture contents are to be maintained until the building is finished.

Workmanship and manufacture

All 'wrought' timber is to be sawn, planed, drilled, or otherwise machined or worked to the correct sizes and shapes shown on the drawings or specified.

Where the 'nominal' dimensions are stated for 'wrought' timber an allowance of 3 mm shall be permitted for each wrought surface. The full-size detail drawings shall be held to show the 'actual' dimensions.

Finish

When natural finish or finish for staining, clear polishing, lacquer, or varnishing is specified, the timber in adjacent pieces shall be matched or uniform or symmetrical in colour and grain. Surface finish to be specified.

Shrinkage

The arrangement, jointing, and fixing of all joinery works shall be such that shrinkage in any part and in any direction shall not impair the strength and appearance of the finished work, and shall not cause damage to contiguous materials or structures.

Fabrication

The joiner shall perform all necessary mortising, tenoning, grooving, matching, tonguing, housing, rebating, and all other works necessary for correct jointing. He shall also provide all metal plates, screws, nails, and other fixings that may be ordered by the architect or that may be necessary for the proper execution of the joinery works specified. The joiner shall also carry out all works necessary for the proper construction of all framings, linings, etc. and for their support and fixing in the building.

Joints

The joinery shall be constructed exactly as shown on the architect's details. Where joints are not specifically indicated they shall be the recognised forms of joints for each position. The joints shall be made so as to comply with BS 1186.

Loose joints are to be used where provision must be made for shrinkage or other movements acting other than in the direction of the stresses of fixing or loading.

Glued joints are to be used where provision need not be made for shrinkage or other movements in the connections, and where sealed joints are required. All glued joints shall be tongued or otherwise reinforced.

Members in construction to be joined by gluing are to be of similar conversion. Adequate pressure should be applied to glued joints to ensure intimate contact and maintained while the glue is setting.

Mixing, application, and setting conditions should be in accordance with the maker's instructions. Organic glues or casein may be used as adhesives for joints in non-load-bearing internal work and for joints in work where the moisture content is always less than 16 per cent.

For work under damp conditions (moisture content normally 20 per cent or more or conditions liable to fungal attack) resin-type adhesives are to be used. See pages 33–40.

Mouldings

All moulded work shall be accurately worked to the full-size details supplied by the architect. All mouldings shall be worked on the solid, except where otherwise stated.

Bent work

Where 'bending' is specified, the work is to be performed by saw-kerfing, keying, backing-a-veneer, laminating, or steaming and shall be carried out to the satisfaction of the architect.

Circular work

When circular work is specified it shall be built up with an appropriate number of pieces cut to the required shapes. The pieces shall be put together in two or three thicknesses so that they break joint, and shall be secured with oak keys and wedges or with oak pins.

Measurements for joinery

The contractor is to take all measurements for joinery works at the building, and not from the architect's drawings, except where the work is specified to be 'built in'.

Built-in joinery

Where joinery works are specified to be built in or erected in position before the surrounding or enclosing works of the main building carcass have been carried out, it shall be the responsibility of the contractor to ensure that the joinery works are set plumb and true, and shall not be damaged or displaced by subsequent operations. Where necessary the joinery works shall be temporarily encased and braced. The contractor shall also provide and secure suitable anchors or other fixings so that these may be built in to the carcass while it is being constructed. The anchorage connections shall be constructed so that they will permit settlements in the building carcass without stressing or otherwise loading the joinery works.

 The contractor shall also provide and fix temporary strips of other suitable packings to all edges or surfaces of the joinery works that are to be enclosed by the materials of the building carcass.

Fixed-in joinery

Where joinery works are specified to be fixed in or inserted in the positions they are to occupy after the surrounding or enclosing carcass has been constructed it shall be the responsibility of the contractor to ensure that the necessary fixings are incorporated in the carcass; alternatively the contractor shall construct such groundworks as are required to provide a suitable base and fixing for the joinery works.

 The contractor is to secure fixed in joinery works to that they are plumb and true to the shapes and dimensions shown on the working drawings and details. Joinery works shall not be fixed in position until after all floor, wall, and ceiling surfaces have been formed or constructed, unless otherwise stated.

Joinery assembled *in situ*

Where joinery works are specified to be assembled *in situ*, and all stresses of support and fixing are to be engaged in the building carcass, it shall be the responsibility of the contractor to ensure that the necessary fixings are incorporated in the carcass; alternatively the contractor shall construct such groundworks as are required to provide a suitable base and fixing for the joinery works.

 In situ joinery works shall not be executed until after all floor, wall, and ceiling surfaces have been formed or constructed, unless otherwise specified.

Drawings

Work is not to commence until the architect has approved the manufacturer's full-size setting-out drawings. Suggestions which the manufacturer may wish to make for modifying the construction and joints shown on the architect's drawings will be considered when the shop drawings are examined.

Inspection

Facilities are to be given for the architect to inspect all work in progress in shops and on the site.

Preservative treatment

External joinery

1. The backs of all softwood window frames and external door-frames and any other bedded surfaces built into or in contact with external walls shall be given two coats of a suitable sealer before delivery to the site, and before, and in addition to, the first coat of primer.
   OR
2. All surfaces of all external joinery including softwood window-frames and casements, external door-frames and doors, fascias, sills, and other exposed joinery to be pressure impregnated.

Priming

Joinery which is prepared for painting is to be knotted and primed before the work is despatched to the site. Where adjustments are made on site the priming is to be made good.

Transparent finishes

The joinery is to be given the first coat of finish before being despatched to the site.

Delivery

None of the joinery is to be delivered until it is required for fixing in the building. Joinery which does not require to be built in as the work proceeds is not to be brought to the site and fixed until the building is enclosed, and the heating is in operation.

Transport and protection

The joinery is to be kept under a waterproof cover during transit and it is to be similarly covered and kept clear of the ground on the site. It is to be handled and stacked carefully to avoid damage.

Defective work

Should any shrinkage or warping occur or any other defects appear in the joiner's work before the end of the defects liability period, such defective work is to be taken down and renewed to the architect's satisfaction, and any work disturbed in consequence must be made good at the contractor's expense.

Quality of timber and workmanship in purpose-made joinery

There are certain fundamental principles of the joiner's craft which are vital to all good work:

1. An understanding of timber.
2. A complete mastery of his tools.
3. A sound knowledge of woodworking machines.
4. Appreciation of design and construction.

In understanding his material the joiner must have a sense of what can be made with it, and a recognition of its limitations and shortcomings. The finished item of joinery is controlled by the nature of the material. Swelling and shrinking, which may take place according to changes of humidity in the surrounding atmosphere, must be considered and methods employed to combat this tendency of the timber to move, both in the design and construction of the work in hand.

The choice of timber in particular conditions where decay may result is a further important condition, and reference to the lists of timbers and their uses in later chapters should be studied.

The development of present-day woodworking machinery has brought into being the skilled machinist, the joiner setter out, and the specialist firms of joinery manufacture. The smallest joiners' shop invariably has a number of machines: these require the guidance of the craftsman in control, and maintenance, and it is the quality of the work which is important, whether it is hand or machine made.

The purpose and design, along with the material, will decide the form of construction of the joinery work. This is divided into two classes:

1. Preparation and assembly in the workshop.
2. Fitting and fixing of the joinery on site.

Work in the shop is further divided into sections, depending on the layout and methods employed in preparing the joinery.

Where standard units of joinery are manufactured all the processes are carried out on the various machines with final assembly of the parts done by hand.

Work required for a particular purpose may be carried out more economically by hand methods rather than the more costly preparation of grinding cutters and the setting up of specialised machines.

High-class joinery usually refers to individual work. The items dealt with in this book can be said to fall in this category – in work other than domestic.

Where different species of timber are used in a unit or sub-assembly of joinery, care should be taken that such mixing does not lead to distortion or unequal reaction to paint or polish finish which would render the finished work unsatisfactory. Timber should be clean with a uniform texture and colour and should work with a bright silky lustre, be properly seasoned with a moisture content reduced to suit the particular job for which it is used.

The moisture content of the timber when the joinery is manufactured and despatched should be within ± 2 of the percentage moisture content that it is expected to attain in service. Reference should be made to the moisture content chart on page 22.

Timber

Specifications do not usually differentiate between timber for carpentry and timber for joinery and may be judged to refer to softwoods rather than hardwoods.

It may be assumed that work in which hardwood is specified will be of a higher standard than that using softwoods, and in certain respects will be free from some of the defects associated with the latter. Because there are very many species of hardwoods in use, against relatively few species of softwoods, the species of hardwood to be used in any joinery item should be named in the specification.

This will depend on appearance and the timber's suitability for the purpose required. Reference to the various species of timbers on pages 5–8 should be made.

In practice, it is not usual to specify a particular grade of hardwood but is assumed that, for joinery, the timber will be of the best quality or chosen as most suitable for the job.

Grades of hardwood vary from country to country as do the names, but certain terms are used in the trade for square-edged boards:

FAS, firsts and seconds or equivalent are the best quality.
SELECTS and No. 1, Common or equivalent are the next lower quality.
Hardwoods imported in log form are usually graded A, B, and C. After conversion the planks are graded as below.
FAS should provide clear timber free from the defects described elsewhere under that chapter heading.
Selects should contain defects of a minor nature only.

In the softwood species used in joinery, the grades have previously been 'unsorted' for Scandinavian timbers – 'unsorted' being a mixture of seconds, thirds, and fourths, along with some 'firsts'. Firsts are the best quality available and are referred to as 'clears'. The next grade down is 'fifths' and is not considered good enough for high-class joinery.

American timbers are graded differently with 'clears' and 'door stock' the first quality and 'select merchantable' the second quality. This is an equivalent grade to the 'unsorted' from Scandinavia.

As different countries have different methods and grades of timber, there is a need for standardisation in the trade. The changes now coming into effect in stress grading of timber should help to meet this need.

Finishes for joinery

Items of joinery used externally in building work are freely exposed to sun and rain. This means that under such conditions the timber is affected and that it swells in wet weather and shrinks in dry weather, this movement causing checking or splitting.

The colour of the timber is also affected, resulting in the colour substances being washed out and the occurrence of bleaching, developing into a grey colour. This usually affects boards more at the bottom than the top, giving a streaky and uneven appearance and surface.

Suitable treatment to finished surfaces may considerably reduce these effects on the timber. This treatment can be classified under four headings: preservatives, paints, stains, and varnishes.

The preservation of timber and the types of preservatives in use are fully covered on pages 25–28. Reference should also be made to the work on treatment of plywoods for exterior use on page 36 and also the work on acid effects on timber on page 42.

Paint

The painting of exterior joinery is long established and well known. The essential requirements are that a good-quality paint is used and that the manufacturer's recommendations are followed. Most important is the preparation of the timber surface before painting when sanding, knotting, and filling are followed by priming, undercoating, and the final finish coat.

Stains

There is a wide range of colours of stain for exterior use available from clear to black. Most common stains are those which allow the grain to show through. Some stains soak into the surface layers of the wood while others provide a definite film on the surface.

Stains are simple to apply, an even coat being applied by brush, and where the finish may be matt, semi-matt, semi-gloss, or gloss.

Maintenance is by washing down and cleaning the surface followed by further costs when required. This may be from 2 years onwards, depending on exposure and pigmentation levels.

Varnishes

A varnish provides a protective transparent coating to the surface of the timber giving a matt, semi-gloss, or gloss finish.

The life of a varnish finish is usually no more than 2 years before cracking of the surface occurs in exposed situations. Where maintenance is neglected at this stage restoration can be extremely costly – particularly where much moulded timber work is involved – to achieve the required appearance.

Cracking of the surface requires the area to be sanded to a firm edge, the staining out of bleached parts followed by two coats of varnish, and finally two coats to the whole surface.

It should be pointed out about all these materials that they do not retain the natural colour of the wood unless renewed frequently, and that their main function is to retard moisture movement by soaking into the wood.

Boiled or raw linseed oil, used alone or with other materials, is perhaps one of the most common treatments of surfaces. Although checking is reduced using linseed oil, it is unsuitable for colour retention. When the oil is brushed in very little surface film remains and soon disappears with weathering. Where a thick film is applied and drying is delayed, becoming tacky in direct sunlight, dirt is collected and the surface finish affected accordingly.

Interior finishes

Purpose-made joinery and indeed any joinery product for internal use requires some form of finish.

The reasons for applying a finish to timber, apart from preventing decay from fungi and insects (see page 25), are:

1. To prevent its absorbing moisture.
2. To prevent damage to the surface in situations where acids, alcohols, and the like are used.
3. To enhance the appearance of the timber and enable it to be kept clean.

The type of finish applied to the work will be determined largely by cost. Materials after planing by the planing machines will contain machine marks caused by the circular motion of the machine's cutterblock and its cutters. Such ripple marks will show through a painted surface and in the case of general joinery the smoothing of the surfaces and the flushing of joints will usually be done by sanding with glasspaper.

For a painted finish papering is better done before assembly and may be crosspapered. This is sanding across the grain at an angle to form a key, along with the normal grain structure of the timber, for the paint finish.

Most joiners' shops are now equipped with portable powered tools, and the range of powered hand sanders now available are used to reduce hand sanding by the joiner at the bench.

Hardwoods used in special fittings and furniture require different preparation and finish to softwood items. Surfaces require preparing with the grain when sanding and not across it. In polished work, the preparation of the surfaces is most important and costly. Various grades of glasspaper are used, the surface dampened to raise the grain, followed by further sanding and applications of staining, filling, and polishing necessary to produce a high gloss finish enhancing the natural characteristics of the timber.

With the introduction of cellulose and other synthetic polishes a full and glossy finish can be obtained which is quicker and less costly than using french polish. These finishes may be applied either by brushing or spraying to give a glossy or matt finish. The larger joinery firms include a spray department at the end of the work flow line which requires special booths and equipment operating under controlled conditions.

The simplest finish for hardwoods, which is reasonably cheap and effective, is wax polish. The grain of the timber is filled with the wax in paste form (beeswax and turpentine) using a soft cloth, and burnished to provide a sheen finish.

Stress grading of timber

Timber is the oldest and probably the most versatile material in the history of construction, and being a natural product is perhaps the only indispensable material used in building today. It has found a favoured place as a structural material in competition with steel, concrete, and plastics in modern building as a result of technological advances made in obtaining and formulating data on solid and built-up structural components, combined with new techniques in jointing.

The engineer requires to be provided with data on timber in a form that he can use in accordance with his theory of structures. He must be provided with a material on a sound structural basis with high mechanical properties to establish timber as one of the most economic materials for use in building.

The applications of timber fall into three main classes: structural, semi-structural, and non-structural. Structural uses include walls, floors, and roof work. When using timber in non-structural work, the main considerations are appearance and workability. Depending on the purpose for which it is to be used, a visible examination of the surface of the sawn timber is usually all that is necessary with no regard to its classification for strength or durability. For joinery work the timber should be clean with a uniform texture and colour. It should also work with a bright silky lustre and be properly seasoned with a moisture content reduced to suit the particular job for which it is used.

Timber used in structural work has been handicapped in the past by the problem of jointing the solid members together efficiently. Where traditional methods of jointing are used, the junction between the members under load weakens them seriously. Previously, components were limited in size, shape, and quality available, and the techniques which have been developed to overcome these limitations are in the jointing together of members using timber connectors, brackets, and hangers, and in laminating using waterproof adhesives. The latter has resulted in the fabrication of members which have no limitations in size, shape, or condition of use. It may be said that the size is limited from a fabrication point of view only by the capacity of the laminating equipment used at the workshop.

A structual member in a building or structure is referred to as one which carries a load. Rafters, ceiling joists, floor joists, struts, and purlins are structural members used in house-building. Larger span buildings such as churches, schools, and factories employ laminated beams, arches, and trusses in their construction.

The characteristics which distinguish timber from other materials are due to its structure and form of growth. No kind of tree produces timber which is free from defects and blemishes, and for these reasons it is not easy to class timber into exact grades for structural purposes. It can, however, be graded on the number and form of such defects, known as grading for strength or stress grading.

The strongest timber is likely to be that which is free from all defects, knots, shakes, sap, waney edges, and other defects, and to be straight grained. If there were enough of this 'clear' grade of timber available in large quantities to meet all requirements the need for grading would not arise. This is not the case, hence the need for stress grading.

A number of important changes have recently been introduced within the framework of earlier British Standards on timber grading and now include both visual and machine stress grading.

For visual stress grading, the principles applied in Canada and the United States of America of knot area ratio has been adopted as a means of determining the maximum possible knots for a given grade.

Two standard grades have been established for visually graded timber, namely, General Structural grade (GS), and a higher grade, Special Structural grade (SS).

Stress grading for timber required to carry loads does not allow strength-reducing characteristics such as large knots in critical positions or inclined grain. Where knots occur it is necessary to estimate whether the knot, if it is on the edge of the material, occupies more or less than half the thickness of the timber. One-quarter of the thickness qualifies for 75 grade and one-half the thickness for 50 grade. Thus the proposed stress grades for the revised Code of Practice are those which have 75 and 50 per cent of the strength of perfect timber.

There are also two machine stress grades which may be substituted for the two visual grades (GS) and (SS) they are, MGS and MSS, or two grades designated M75 and M50.

Visual grades for laminating timber are specified in three grades LA, LB, and LC.

Mechanical or machine grading

This is based on the knowledge that there is a high correlation between the strength of the timber and its stiffness, or the amount of deflection under load. This deflection is measured as the timber is passed through the machine and a load applied. The original shape of the timber is measured and fed to the machine's computer along with the recorded deflection under load. Comparisons of the two recordings are made and the output is the effective modulus of elasticity from which the maximum allowable stress can be estimated.

The machine also marks the timber with a coloured spray denoting the grade of each member at intervals of 150 mm in its length. The final grading given for the member is that of the lowest grade appearing.

Machine-graded timber must have the following information clearly marked on at least one face:

1. The number of the stress grading machine.
2. The grade of the timber.
3. The BSI Kitemark and the number of the British Standard (BS 4978:1973) or, as an alternative to these, any mark approved by the approving authority (see page 74, BSI Kitemark).

Knots Knots are assessed by their knot area ratio, this is the ratio of the sum of the projected cross-sectional areas to the cross-sectional area of the timber (Sheet 3.1). Knots of less than 5 mm diameter can be ignored.

Fissures The longitudinal separation of fibres including checks, shakes, and splits fall in this category of defects and are measured by the distance between lines enclosing the fissure and parallel to a pair of opposite faces as shown in Sheet 3.2.

STRESS GRADING

KNOT AREA RATIOS = KAR   1

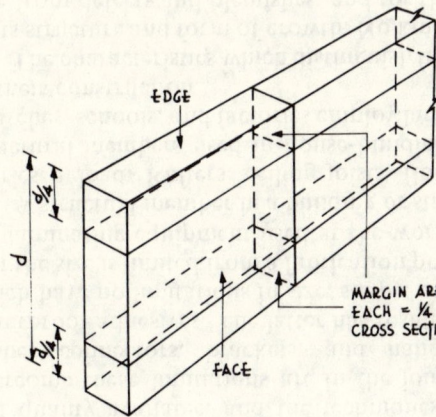

EDGE

d

MARGIN AREAS:
EACH = ¼ TOTAL
CROSS SECTION AREA

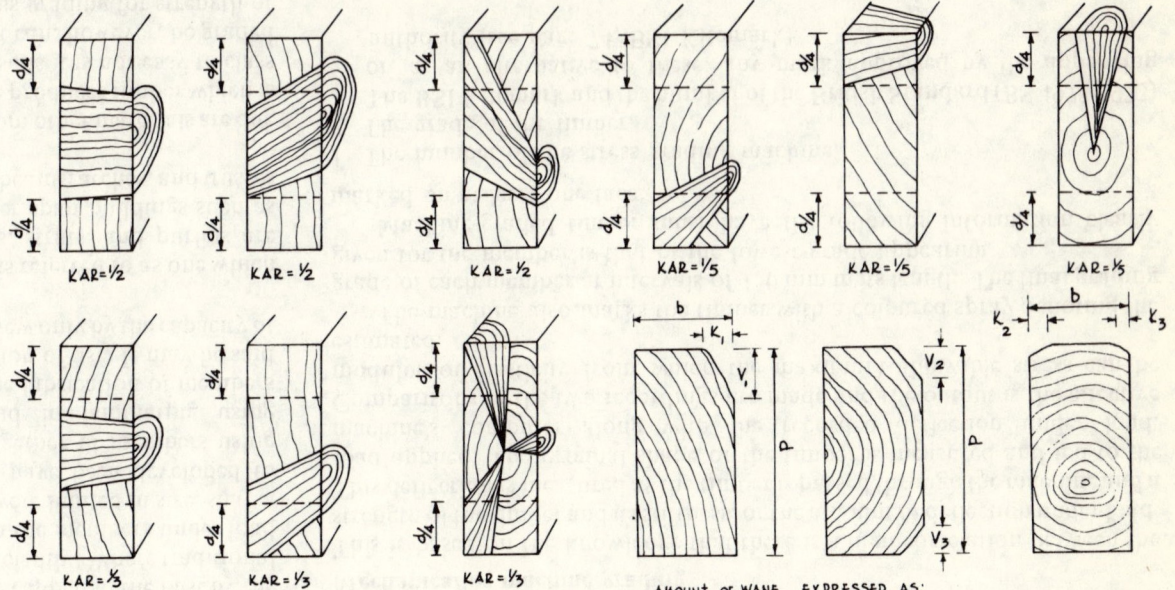

FACE

EDGE, FACE AND MARGIN AREAS

KAR = ½   KAR = ½   KAR = ½   KAR = ⅕   KAR = ⅕   KAR = ⅓

KAR = ⅓   KAR = ⅓   KAR = ⅓

AMOUNTS OF WANE   3

AMOUNT OF WANE EXPRESSED AS:
RATIO $\dfrac{V_1}{d}$ OR $\dfrac{V_2 + V_3}{d}$

RATIO $\dfrac{K_1}{b}$ OR $\dfrac{K_2 + K_3}{b}$

EDGE

d

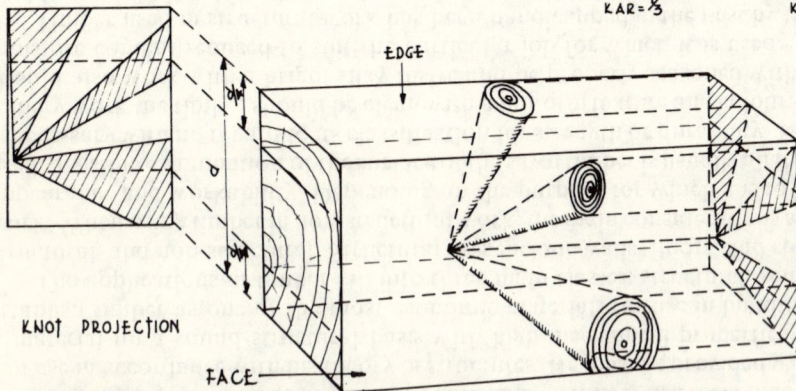

KNOT PROJECTION

FACE

SKETCH SHOWING THREE DIMENSION
VIEW OF A GROUP OF KNOTS AND
THEIR PROJECTION IN CROSS SECTION

MEASUREMENT OF SLOPE OF GRAIN   5

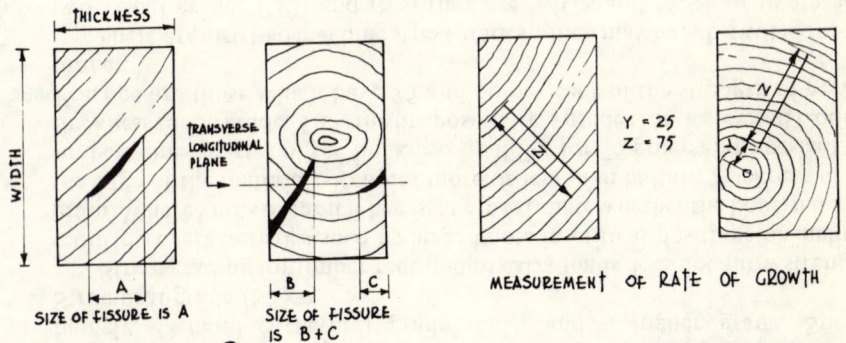

THICKNESS

WIDTH

TRANSVERSE
LONGITUDINAL
PLANE

SIZE OF FISSURE IS A

SIZE OF FISSURE
IS B + C

FISSURES   2

Y = 25
Z = 75

MEASUREMENT OF RATE OF GROWTH

SHEET 3

Slope of grain   Sheet 3.5 shows how this is to be measured, taking a distance sufficiently great to determine the general slope.

Wane   The amount of wane is taken as the sum of the wane at the two arises expressed as a fraction of the surface where it occurs as shown in Sheet 3.3.

Rate of growth   This is shown in Sheet 3.4 and is measured on the end of the timber as the number of growth rings per 25 mm on a radical line 75 mm long.

Resin pockets   Measured as for fissures.

Distortion   Bowing, springing, and twist in timber is measured over a 3 m length.

GS grade   Examples for knot area ratios for GS and SS grades of timber are shown in Sheet 3.1.

Table 10.1

Defect	Grade of timber		Laminated		
	GS	SS	LA	LB	LC
Knot area ratio:			1/10	1/4	1/2
1. In margin	1/3	1/5			
2. Not in margin	1/2	1/3			
Fissures					
1. Defects less than half thickness	Number unlimited	Number unlimited	Unlimited in depth but shall not form		
2. Defects greater than half thickness	Length not to exceed 900 mm	Length not to exceed 600 mm	an angle less than 45°		
Slope of grain	1 in 6	1 in 10	1/18	1/14	1/8
Wane	Not to exceed one-third of surface where it occurs	Not to exceed one-quarter of surface where it occurs	Not permitted		
Rate of growth	4 rings/25 mm	4 rings/25 mm	No limit		

Grade stresses

The grade stresses in sawn timber and machine stress-graded timber are given in the appropriate tables in BS Code of Practice CP 112, Part 2 to which reference should be made. The grades of timber with allowable defects are listed in Table 10.1.

Giving of notice and deposit of plans

Building Regulations 1973.

A 10–1. Subject to the provision of paragraphs (2) and (3), any person who intends to –
(a)   erect any building; or
(b)   make any structural alteration of or extension to a building; or

(c)   execute any works or install any fitting in connection with a building; or
(d)   make any material change of use of a building shall, if any provisions of these regulations apply to such operation or such change of use, give notices and deposit plans, sections, specifications and written particulars in accordance with the relevant rules of Schedule 2.

Schedule 2 – general

Rule A. The following provision shall be observed in relation to the giving of notices and the deposit of plans, sections, specifications and particulars referred to in the other rules of this schedule:
(1)   Notices and other particulars shall be in writing.
(2)   Drawings shall be executed or reproduced in a clear and intelligible manner with suitable and durable materials. Plans and sections shall be to a scale of not less than 1 : 100 or if the building is so extensive as to render a smaller scale necessary, not less that 1:200; block plans shall be to a scale of not less than 1:1250; and key plans shall be to a scale of not less than 1 : 2500. The scale shall be indicated on all plans, sections and other drawings and the north point on all block plans and key plans.
(3)   Every notice, drawing or other document shall be signed by the person required to furnish it to the local authority or by his duly authorised agent, and if it is signed by such agent it shall state the name and address of the person on whose behalf it has been furnished.
(4)   Every such document, together with a duplicate thereof, shall be sent or delivered to the offices of the local authority.
   This rule implies that drawings should be of a high standard of draughtsmanship and presentation on linen or reinforced paper, with at least one copy of all the plans submitted drawn using a drawing ink.

Erection of buildings

Rule B. The following are the notices to be given and the plans, specifications and particulars to be deposited by a person intending to erect a building, which is neither wholly or partially exempted within the meaning of regulation A5:
(1)   Notice of intention to erect a building not wholly or partially exempted from the operation of these regulations.
(2)   Particulars, so far as necessary to show whether the building complies with all such requirements of these regulations as apply to it, of –
(a)   the intended use of the building; and
(b)   the materials of which the building will be constructed; and;
(c)   the mode of drainage; and
(d)   the means of water supply.
(3)   A block plan showing–
(a)   the size and position of the building and its relationship to adjoining buildings; and
(b)   the width and position of every street adjoining the premises; and
(c)   the boundaries of the premises and the size and position of every other building and of every garden, yard and other open space within such boundaries.
(4)   A key plan showing the position of the site when it is not sufficiently

identifiable from the block plan.

(5) A plan of every floor and roof of the building and a section of every storey of the building, upon which shall be shown, so far as necessary to enable the local authority to determine whether the building complies with these regulations –

(a) the levels of the site of the building, of the lowest floor of the building and of any street adjoining the premises, in relation to one another and above some known datum; and

(b) the position of the damp-courses and any other barriers to moisture; and

(c) the position, form and dimensions of the foundations, walls, windows, floors, roofs, chimneys and several parts of the building; and

(d) the intended use of every room in the building; and

(e) the provision made in the structure for the protection against fire and for insulation against transmission of heat and sound.

Alterations and extensions

Rule D. The following are notices to be given and the plans, sections, specifications and particulars to be deposited by a person intending to make any alteration of or extension to a building:

(1) Notice of intention to alter or extend a building.

(2) In the case of alterations not involving any extension of a building –

(a) the plans and sections required by item 5 of Rule B or Rule C of the alterations and of the building so far as affected by the alterations, so far as necessary to establish whether the proposals comply with these regulations; and

(b) a key plan showing the position of the site when it is not sufficiently identifiable from such plans.

(3) In the case of an extension to a building –

(a) the plans, sections, specifications and particulars referred to in items 2, 3, 4 and 5 of either Rule B or Rule C in relation to the extension as if the extension were the building therein referred to; and

(b) the plans and sections as required by Item 5 of Rule B or Rule C of the building so far as affected by the extensions, so far as necessary to enable the local authority to determine whether the proposals comply with the requirements of these regulations.

Providing plans comply with the building regulations and the various statutes, the local authority will give its approval and notification of the passing of the plans within five weeks of receipt.

The rejection of plans by the local authority indicates they do not meet the requirements of the building regulations. In certain circumstances an applicant may appeal to the Magistrates Court where a refusal has been given by the Local authority, or both parties may together request the secretary of State to mediate.

Working drawings

No building can be designed or built without drawings, the same applies to special items of joinery which is to be purpose made. They should show all the necessary information and the greatest care should be used to ensure that drawings are as accurate as possible.

For all drawings other than full size details, figured dimensions are essential. The following scales are recommended;

Location drawings	1 : 2500
Block plans	1 : 1250
Site plans	1 : 500
	1 : 200
General drawings	
(a) Ranges	1 : 100
	1 : 50
	1 : 20
(b) Details	1 : 10
	1 : 5
	1 : 1
Assembly drawings	1 : 20
	1 : 10
	1 : 5

Large-scale details

These comprise enlargements of component parts of assemblies to show the full details.

(1) Cills, heads and jambs of windows and doors.

(2) Mullions and transomes.

(3) Timber sections such as handrails, window sections, joinery details.

(4) Jointing details.

(5) Staircases.

(6) Special fittings and fixtures.

Building drawing practice

BS 1192 : 1969, Metric Units, contains the recommendations for drawing office practice and gives guidance on the production of building drawings so that information is communicated accurately, clearly, without repetition, and with economy of means. The selection of a drawing system for this purpose is most important.

Information concerning a building project is given normally on drawings by the architect, together with written or printed specifications. To facilitate the presentation and to enable information to be found quickly, both drawings and other information should be classified according to the particular type of information to be communicated. Drawings should not give information which could be better included in schedules, specification, or information sheets. The following classifications of the types if drawings and information are recommended:

1. Design stage

Sketch drawings. Preliminary drawings, sketches or diagrams to show the designer's general intentions.

2. Production stage

Location drawings.
(a) Block plans: to identify site and locate outline of building in relation to town plan or other wider context.
(b) Site plans: to locate the position of buildings in relation to setting-out point, means of access, and general layout of site. Site plans also contain information on services, drainage work, etc.
(c) General location drawings: to show the position occupied by the various spaces in building, the general construction and location of principal elements, components, and assembly details.

3. Component drawings

(a) Ranges: to show the basic sizes, system of reference and performance data on a set of standard components of a given type.
(b) Details: to show all the information necessary for the manufacture and application of components.

4. Assembly drawings

To show in detail the construction of buildings; junctions in and between elements and components, and between components.

Conditions of contract

'Conditions of contract' is a legal document the purpose of which is to define the responsibilities of employer, contractor, and architect/engineer, and other non-technical matters.

Specification

A specification is a precise description of materials and workmanship of a project or its parts which are not shown on drawings or in schedules.

Schedules

Schedules give tabulated information on a range of similar items differing in detail, such as doors, windows, manholes, etc. and which are more easily described by words than by drawings, i.e. doors including frames and ironmongery; fixed joinery fittings other than doors and windows.

Bill of quantities

A bill of quantities provides a complete measure of the quantities of materials, labour and any other items required to carry out a project based on the specification, drawings and schedules. The value of a project is obtained by pricing the individual items.

Contract drawings

Contract drawings comprise the following:

1. Conditions of contract.
2. Specification.
3. Schedules.
4. Bills of quantities.
5. Standard details as appropriate.
6. Drawings.

Drawing paper

For work in the earlier stages of construction cartridge paper is generally used. This is in A-size sheets and also in rolls. It is made in three grades. The medium grade is most suitable for students and the smoother of the sides is the 'right side'. Figure 46. shows 'A' sheet sizes drawn to a comparative scale.

In preparing a drawing sheet for student use a border is drawn to 10 mm from the edge of the sheet. Panels to receive the student's name, course, title, date, etc. may be placed at the top of the drawing sheet or, which is more usual, along the bottom.

A typical layout of a drawing sheet used in architects' offices and the like is shown in Fig. 46. This is a more detailed panel layout which may be used horizontally as shown or vertically. Types of lines used on drawings are shown in Fig. 44.

Construction lines are made faint with finished lines firm and even throughout their length.
Hidden details are shown with broken lines of dashes and gaps.
Projection lines are light and broken with gaps.
Centre lines have short continuous lines broken by gaps as shown.
Broken lines have short 'Z' markings at intervals showing an incomplete drawing.
Section lines are used to indicate a 'cut' is made.
Dimension lines are lighter and thinner than construction lines. They are continuous lines with arrowheads touching the projection lines. The dimensions should be placed horizontally or vertically as shown in Fig. 44.

Modular coordination drawing

Figure 44 shows a **modular grid**. This is a planning grid used to define the principle spaces in a building and the basic position of the main external and internal wall elements.

Controlling lines are indicated on a drawing using a broken line of dashes and gaps with a circle at the termination or a full line with a circle.

The use of reference grids, to which sizes and positions of building components may be related, is helpful in the preparation of all types of drawing and particularly when modular coordination is applied to design and construction. The distance between adjacent parallel lines can be 300 mm or 100 mm.

DRAWING OFFICE PRACTICE

Construction lines

Finished lines

Hidden details - dashed

Projection lines - ₵

Centre lines

Broken lines

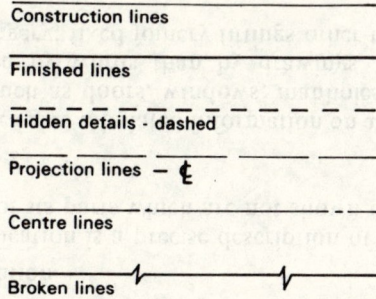

A ▲————▲ A   B •————————• B

Section lines

|← 75 mm →| 28 mm

Dimension lines

types of lines used on drawings

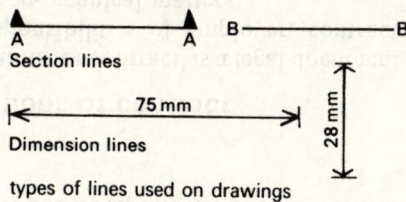

841 mm | 841 mm

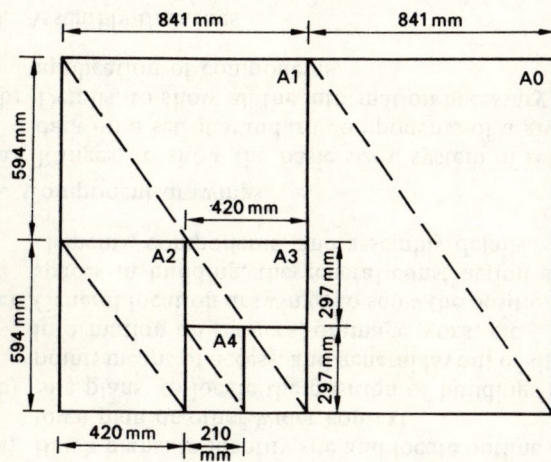

594 mm

594 mm | 420 mm | 297 mm

420 mm | 210 mm

AO 1189 × 841
A1   841 × 594
A2   594 × 420
A3   420 × 297
A4   297 × 210

Drawing paper 'A' sizes

Grid lines 300 X 300

Controlling lines

Modular space

Key dimensions

General dimensions

Component dimensions

Modular co-ordination drawing
Fig 44

Brick

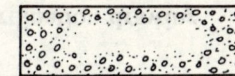
Concrete

Earth

Fibre board

Glass

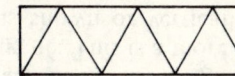
Hardcore

Loose insulation

**Building materials
represented in section**

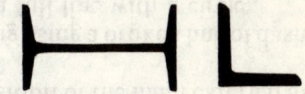
Metal

Partition blocks

Plywood

Screed

Sheet membrane

Stone

Wood
Unwrot   Wrot

Fig 45

PANEL
LAYOUT

Fig 46

architect / planner		drawing title		
job architect				
job title		job no		drawing no revision
		CI/SfB		
scale	date	drawn		checked

180

60

Representation of materials

This is indicated on drawings by some form of shading or hatching on the sections as shown in Fig. 45. They should be accompanied by a descriptive note stating the type of material, its thickness, etc.

Graphical symbols used on drawings

The principal types of symbols used in building drawing practice are graphical. Many of these, as well as other kinds of symbols such as letters, numbers, and signs, are covered by British Standards to which reference should be made. A selection of symbols most commonly used and allied to the details covered by this book are as follows:

Centre to centre	C/C
Centre line	₵
Direction of view	⟶
External	ext
Internal	int
Modular space	⬡
Rise of stair	R
Finished floor level	FFL
Ground level	GL
Diameter	dia
Diameter, inside	I/D
Diameter, outside	O/D
Kilometre	km
Litre	l
Metre	m
Millimetre	mm

Abbreviations for components and materials:

Boarding	bdg
Brickwork	bwk
Building	bldg
BS Universal beam	BSUB
BS Channel	BSC
Cupboard	cpd
Damp-proof course	DPC
Damp-proof membrane	DPM
Drawing	dwg

Hardboard	hdb
Hardwood	hwd
Insulation	insul
Joist	jst
Mild steel	MS
Plasterboard	pbd
Softwood	swd
Tongue and groove	T & G
Wrought iron	WI

Daywork

Daywork means work which under the terms of the contract is to be paid for by time and materials, and not by measurement by the quantity surveyor.

The RIBA form of contract states where work cannot properly be measured and valued, the contractor shall be allowed daywork rates on the prices prevailing when such work is carried out. Such decisions as to whether work shall be carried out under daywork charges will usually be made by the quantity surveyor having considered that the work, or the circumstances under which it has to be carried out, justify a daywork claim.

Vouchers specifying the time daily spent upon the work and the materials employed shall be delivered to the architect or his representative for checking and verification not later than the end of the week following that in which the work was done.

Examples of work in building which the quantity surveyor cannot estimate accurately and therefore justify daywork rates, include underpinning, damage to timber structure by fungi and insect attack, pumping water, doors and windows having to be moved, altered, or additional ones installed as the job goes on, etc.

Recording of daywork

The records are to be kept on a weekly basis showing the number of hours worked, the materials used, and the use of any plant for the duration of the job. Such information will be recorded by the foreman with copies for the architect and the clerk of works, in order that a check can be kept on the contractor's daywork claims.

Time sheets

The basis of determining labour costs is accurate time-keeping, and production can only be fully operational if each operative's time is accounted for and recorded.

There are two types of time sheets normally used, daily where daywork or jobbing is being done, and weekly suitable for site work on contract, or repetition work.

Estimating

1 cubic metre (m^3) = 35·315 cubic feet (ft^3).
Timber priced at £7·00 per ft^3 = £247·20 per m^3.
A standard of timber (165 ft^3) = 4·675 m^3.
1 ft^3 of timber = 0·0283 m^3.
1 m^3 of timber = 0·214 of a standard.

Softwood: unsorted redwood £140·00 per m^3.

Hardwood: West African mahogany	£7·75 per ft^3
50 mm utile	£8·60 per ft^3
50 mm sapele	£8·60 per ft^3
50 mm European beech	£6·00 per ft^3
50 mm Japanese oak	£12·70 per ft^3
50 mm teak	£16·56 per ft^3

Plywood:	
6 mm ext. birch plywood	£2·28 per m^2
9 mm ext. birch plywood	£3·00 per m^2
12 mm ext. birch plywood	£3·95 per m^2
18 mm ext. birch plywood	£5·86 per m^2

Blockboard:	
16 mm birch faced	£3·37 per m^2
18 mm birch faced	£4·00 per m^2
22 mm birch faced	£4·22 per m^2
25 mm birch faced	£4·93 per m^2

Estimating for joinery

Work carried out by the estimator may be listed under two headings:

1. Pre-tender procedure,
2. Acceptance of tender procedure.

In part 1, the estimator is required to examine and assess all the work involved in the drawings and specifications and form of tender for the job under consideration.

From the drawings and specifications, quantities of materials and items for manufacture are taken off and estimates prepared for each, including the labour involved in the construction and fixing of the items. These items are costed out and allowances made for overheads in completing the estimate.

In part 2 on acceptance of the tender, the estimator prepares materials and progress schedules and allocates time for the departments involved in the work – machine shop, joiners' shop, and finishing department.

Costing procedure, it can be said, can be considered under three headings, namely, materials, labour, and overheads.

Fluctuating costs of materials and labour must be constantly watched and adjusted as necessary. These frequent changes in values at the present time, of course, do not make the work of the estimator more easy but the reverse, yet provision for such changes needs to be made.

In considering joinery for both private and public clients, ranging from special individual items to mass-produced units, care and attention to detail and to methods of production is necessary.

Further considerations to be borne in mind will include the quality of the work, its cost, construction, finish, and the time factor or delivery date.

1. Materials

Timber is bought from the timber merchant in one of two forms: either sawn through and through, or square edged into scantling sizes, having sawn or planed finish. Softwood prices vary according to size; wide boards, planks, and the like being more expensive than battens, etc. Timber prices are quoted or sold by the cubic metre, but particular scantling sizes are commonly quoted per 100 linear metres. The price varies according to the finish of the timber, whether the merchant is involved in more machining of scantling sizes, incurring more waste than sawn material, will be shown in the price.

Along with other building materials, purchasing timber in bulk is usually more advantageous and a reduction in price for minimum lots of 200 m of particular scantlings are often quoted.

Since the introduction of the metric system in this country, the so-peculiar figures now in use require to be fully understood:

$$1 \text{ sq. metre } (m^2) = 1\,000 \text{ mm} \times 1\,000 \text{ mm}$$
$$= 1\,000\,000 \text{ mm}^2$$
$$1 \text{ cub. metre } (m^3) = 1\,000 \text{ mm} \times 1\,000 \text{ mm} \times 1\,000 \text{ mm}$$
$$= 1\,000\,000\,000 \text{ mm}^3$$

Here 1 000 000 may be written and used as 10^6 which is 10 raised to the power of six in calculations. Sawn softwood, carcassing quality quoted at £5·00 per 10 lineal metres of 100 mm × 50 mm equals £50·00 per 100 m. To change this price for the timber into the price per cubic metre requires the following calculation:

$$\frac{£50 \times 10^6}{100 \times 100 \times 50} = £100·00 \text{ m}^3$$

Further adjustments to this figure need to be made, depending on the job for which the timber is to be used, but first the following allowances or material constants must be considered.

Waste

Softwood for joinery PSE (planed square-edged):

1.	For general joinery	10–15%
2.	For specific purpose from stock	20%

Hardwood for joinery PSE:

1.	Mahogany	33%
2.	European oak	20%
3.	English oak	50%+
4.	Teak/Iroko/Japanese oak	35%
5.	Utile/Sepele	33%
6.	Walnut	33%

Plywoods, blockboards, hardboards:
1. From stock — 20%
2. Purchased for specific purpose — 10%

Then to the price of £100 per metre cube must be added the waste constant of 10 per cent for general joinery:

1 m³ of carcassing timber 100 mm × 50 mm	£100·00
Waste 10%	£10·00
Price per cubic metre	£110·00

The number of lineal metres of any scantling size in 1 m³ is found by dividing 1 m³ by the cross-sectional area of the timber, i.e. how many lineal metres of 100 mm × 50 mm in 1 m³?

$$\frac{1\,000\,000}{100 \times 50} = 200 \text{ m}$$

Price per lineal metre = $\dfrac{\text{price per metre cube} \times \text{cross-section}}{10^6}$

i.e.

$$\frac{£110·00 \times 100 \times 50}{10^6} = £0·55 \text{ per lineal metre}$$

This formula may be written as follows: multiply the cross-sectional area of the timber in square millimetres by the price in pounds and move the decimal point six places to the left, to give the price of the timber in decimals of a pound per lineal metre for the scantling size.

The reader will now understand that the pricing unit of timber is millimetres × millimetres × lineal and that the use of a calculating machine is essential. An alternative, of course, is the slide rule.

Timber costs

Market costs delivered plus an unloading and stacking weighting. Allow ¾ hour for labourer offloading and stacking at £0·79 per hour = £0·59. Therefore

1 m³ of unsorted redwood costs	£140·00 delivered.
Offloading and stacking	£0·59
Waste 10%	£14·00
Cost per m³	= £154·59

Converted timber:
100 mm × 50 mm at £154·59 = £0·77 per lineal metre
150 mm × 50 mm at £154·59 = £1·15 per lineal metre
150 mm × 25 mm at £154·59 = £0·57 per lineal metre

2. Consideration of labour costs

The first consideration of labour costs is to determine as accurately as possible the cost per hour of employing a tradesman or labourer. These rates will, of course, change from time to time and adjustments will therefore require to be made as necessary.

Analysis of cost to employ a tradesman and a labourer based on a 40-hour week plus 9 productive hours of overtime at December 1976:

	£	£
40-hour basic week	37·00	31·40
Joint Board supplement	11·00	10·20
Guaranteed bonus	4·00	3·60
Incentive bonus	3·00	2·50
9 hours overtime (1¼ × 92)	11·33	9·26
Tool money	0·20	
Seven public holidays $\frac{56 \times 1·30}{49}$	1·49	1·29
2% guaranteed wage	1·04	0·95
National Insurance	5·80	4·98
Holiday credit	4·05	4·05
Proportion of National Insurance to cover holidays	0·17	
Sick pay scheme	0·30	0·14
CITB and training	0·41	0·20
Redundancy provision	2·14	1·90
	£81·93	£70·77
	÷ 49	÷ 49
Per hour	£1·67	£1·44

It will be seen from the above analysis that the inclusive hourly cost of tradesman and labourer is £1·67 and £1·44, respectively, exclusive of any paid travelling time or expenses and is termed an 'all-in' wage rate.

3. Overheads and profit

There is no rigid rule for fixing the percentage addition in respect of overheads and profit. Overheads are a variable factor and depend largely on the individual firm, its size and therefore other commitments – cost of maintaining workshops, rates, lighting, heating, capital outlay, loans, number of staff employed, salaries, pensions, insurances, telephone, printing, replacement of machines, etc. Profit is usually based between 10 and 15 per cent and is intended to provide the employer with a satisfactory living, provision for reserve capital, and to meet losses should these occur.

There are two ways of dealing with the items of overheads and profit, these again will vary from firm to firm. In the first method the labour item, when it is costed, is doubled and termed 'oncost'. Therefore, this shows that the oncost item is 100 per cent to cover the overheads. In the summary a profit of 10 per cent is used. In the second method, in the summary 45 per cent is added to the materials and labour to cover overheads and profit items.

So far, material costs, the all-in wage rate, overheads, and profit have been dealt with. Any further plus rates must now be considered, and will include extras paid to the setter-out, the marker-out and any specialist machinists where these are employed. Purpose-made joinery, comprising the majority of items in this book, will take rather more than the average percentage of setting-out and marking-out time than less complicated production items.

Cost factors in operating machines

There are three main cost factors in operating any type of woodworking machine in the production of purpose-made joinery. These are:

1. Setting up the machine.

2. Running the machine.
3. Preparing special cutters.

The setting-up times for machines will vary from machine to machine; the outside planing and moulding machine used for a complicated section to be machined may have up to 2 hours allowed, whereas, a simple planing machine may require only seconds. This means that machining a short run will be fairly costly and a long run economical.

Special sections of material require special cutters to be prepared by the machinist, both for sticking the section and scribing the moulded intersections. Again, extra cost is involved, resulting in short runs becoming very expensive.

Setting-up times   The following times are allowed for setting up and passing inspection for various machines. The allowances for grinding and sharpening cutters will be decided by the quantity of pieces to be worked and on the hardness of the timber.

Bandsaw	Tighten saw, set fence and operate power		15 mins
	Changing saws		30 mins
Sawbench	Set fence	per job	5 mins
	Changing saws	per day	45 mins
Dimension bench	Setting up straight cuts	per job	15 mins
	Setting up bevel and mitre cuts	per job	30 mins
Cross-cut	Setting up	per job	5 mins
Four-cutter	Setting up complete	per job	150 mins
	Change of section	per job	45 mins
	Change of size	per job	30 mins
Band re-saw	Changing saws	per day	45 mins
Gang mortise	Setting up, changing all chains for mortising stiles		210 mins
	Changing over to rails		130 mins
Chain and chisel	Setting up	per job	30 mins
Tenoner single	Setting up	per job	60 mins
Tenoner double	Setting up	per job	90 mins
Spindle	Setting up	per job	45 mins
Router	Setting up	per job	15 mins
Drill	Setting up	per job	10 mins
Trenching	Setting up	per job	15 mins

Machine times per hour:

Cross-cut	Up to 100 mm × 50 mm	60 pieces per hour
	Up to 225 mm × 75 mm	40 pieces per hour
	Over 225 mm × 75 mm	30 pieces per hour
Sawing	Two cuts for framing	45 m per hour
	One cut for framing	150 m per hour
	Deep cut 225 mm	90 m per hour
	Up to 200 mm	120 m per hour
Planing	Framing	165 m per hour
	225 mm × 75 mm	90 m per hour
	Up to 225 mm × 75 mm	120 m per hour
Thicknessing	Framing	210 m per hour
	225 mm × 75 mm	150 m per hour
	Up to 225 mm × 75 mm	180 m per hour
Four-cutter	Framing	195 m per hour
	Upwards of framing	150 m per hour
	Hardwood	90 m per hour
Tenoner single end		60 pieces per hour
Tenoner double end		180 pieces per hour
Boring	Under 12 mm	180 holes per hour
	Over 12 mm	60 holes per hour
Trenching	Up to 50 mm	18 m per hour
Mortise	Gang mortise	60 pieces per hour
	Hand mortise	40 pieces per hour

It must be understood that machine times, and the machine rates which follow, are only approximate and can only be based on individual establishments.

Woodworking machines and labour rates.   Machine rates and labour rates now require to be considered, as applied to each item of joinery when costing as follows:

	Machine rate (£)	Labour (£)	Total/hour (£)
Cross-cut saw	0·53	1·67	2·20
Bandsaw	0·53	1·67	2·20
Dimension saw	0·53	1·67	2·20
Mortiser	0·53	1·67	2·20
Tenoner	0·53	1·67	2·20
Spindle	0·53	1·67	2·20
Router	0·53	1·67	2·20
Dovetailer	0·53	1·67	2·20
Surfacer	0·58	1·67	2·25
Thicknesser	0·58	1·67	2·25
Belt sander	0·60	1·67	2·27
Band re-saw	0·83	1·67	2·50
Power sawbench	0·83	1·67	2·50
Moulding machine	4·00	1·67	5·67
Three-drum sander	1·95	1·67	3·62

Pricing doors

There is a wide variety of types of both interior and exterior doors with three or four types of flush doors and panelled doors.

The construction of doors also vary between those produced by manufacturers and doors purpose-made for special purposes, often referred to as 'one-off'. Such constructions affect both quality and price.

Panelled doors are described by the number of panels and their finish. Reference should now be made by the reader to the panelled door which has been set-out in Sheet 26, along with a cutting list. A typical estimate for a general joinery establishment will now be dealt with based on the work so far. The following item in a bill of quantities may be presented.

Item 1. No. 20, 50 mm (nominal) softwood five-panelled doors ovolo moulded both sides; stiles, top rail, and intermediate rails Ex. 100 mm × 50 mm, bottom rail 200 mm × 50 mm, 9 mm plywood panels; 746 mm wide by 2·040 m high.

Material to include waste, softwood £140·00 per m³: Exterior grade plywood £26·50 per 10 m².

20/2/2·100 stiles	100 × 50 × 70p =	£58·80
20/1/750 top rails	100 × 50 × 70p =	£10·50
20/4/750 inter rails	100 × 50 × 70p =	£42·00
20/1/750 bottom rails	200 × 50 × 140p =	£21·00
100/600 panels	300 × 9 =	£47·70
	Total	£180·00

Labour:
Rate: £2·00 per hour to include insurance, holidays, etc.

Machine rates:
Cutting, mortise, tenon, spindle   at £2·20 per hour
Planing, sanding, edging   at £2·25 per hour
Machining, setting out

20 hours at £2·20	=	£44·00
15 hours at £2·25	=	£33·75
Assembly		
25 hours at £2·00	=	£50·00
	=	£127·75

Summary, twenty, five-panelled ovolo moulded doors:

Materials	£180·00
Labour and machines	£127·75
On cost	£127·75
	£435·50
Profit 10%	£43·55
Delivery	£1·00
Cost of twenty doors =	£480·05
Cost per door =	£24·00

In the above costings it will be seen that each item of timber in the doors has been costed separately, using the price of each member per lineal metre (see page 89).

The same answers will be arrived at if all the members are cubed, the results added together and priced at £140 per m³ as follows:

20/2/2·100 stiles	100 × 50 =	0·420 m³
20/1/750 top rails	100 × 50 =	0·075 m³
20/4/750 inter rails	100 × 50 =	0·300 m³
20/1/750 bottom rails	200 × 50 =	0·150 m³
	Total	0·945 m³

Then

0·945 m³ at £140 per m³ = £132·30		=	£132·30
Plywood panels		=	£47·70
		Total	£180·00

The same twenty, five-panelled doors will be costed using the second method (see page 89) in dealing with overheads. Summary:

Materials	£180·00
Labour and machines	£127·75
	£307·75
Profit and overheads 45%	£138·48
Delivery	£1·00
Cost of 20 doors =	£447·23
Cost per door =	£22·36

Item 2. A 100 mm × 63 mm softwood door-frame for a 2·040 mm × 826 mm × 50 mm nom. door, jambs and head rebated, moulded, and check throated. Materials redwood to include waste £140 m³.

2/2·110 jambs	100 × 63 × 88p =	£3·72
1/1·050 head	100 × 63 × 88p =	£0·92
		£4·64
	Waste 5%	£0·23
		£4·87

Labour:

Machining, mortise, tenon sanding	25 min at £2·20 =	£0·91
Assemble	20 min at £2·00 =	£0·66
		£1·57

Summary:

Materials	£4·87
Labour	£1·57
Overheads 45%	£2·89
Delivery	£1·00
Cost	£10·33

Comparison:
Costing of door frame in hardwood, utile

Material, utile £7·75 per ft³	=		£274 per m³
Waste 33%	=		£90
			£364 per m³
2/2·110 jambs	100 × 63 × £2·29 =	£9·66	
1/1·050 head	100 × 63 × £2·29 =	£2·40	
		£12·06	
	Waste 5%	£0·60	
		£12·66	

Labour:

Machining, mortise, tenon, sanding	25 min at £2·20 =	£0·91
Assemble	20 min at £2·00 =	£0·66
		£1·57

Summary:

Materials	£12·66
Labour	£1·57
Overheads 45%	£6·40
Delivery	£1·00
Cost	£21·63

It is now proposed to look at an item of purpose-made joinery for a particular situation and therefore a 'one-off' job – the church doors and frame detailed in Sheet 37.1 in Japanese oak.

Description in specification:

No. 1   100 mm × 75 mm rebated, chamfered and sunk moulded door frame in Japanese oak with gothic head to suit a pair of 64 mm doors 2·475 mm by 778 mm.

No. 1   Pair of 64 mm oak doors 2·475 m by 778 mm with Gothic headed top rails, each leaf in three 32 mm solid panels, vee jointed and flush on the inside and with rebated and vee-jointed meeting stiles.

Note: Reference to the jointing of all the members must be made on Sheet 37.

Specialist joinery, as in this example, should be priced to the details in Sheet 37.1. As the material quantities are relatively small, the material cost will be high. It should also be remembered by the reader that, due to the nature of the work, as in almost all specialist work, site visits, setting-out, marking out, supervision and machining of members for short periods needs to be considered and taken account of.

Detailed drawings by the architect are not accurate enough to set out such work and can only mean that allowance must be made for site visits and for accurate site dimensions to be taken.

Material: Japanese oak at £12·70 per ft³ = £12·70 × 35·315 = £448·50 per m³.

Frame:

2/1·300 × 100 × 75 × £3·36 =	£8·73
4/450 × 125 × 75 × £4·20 =	£7·56

Doors:

Stiles	2/1·300 × 125 × 64 × £3·58 =	£9·30
	2/2·325 × 125 × 64 × £3·58 =	£9·48
Top rails	2/1·350 × 225 × 64 × £6·54 =	£17·41
Muntins	2/1·900 × 75 × 64 × £2·15 =	£8·17
	2/1·725 × 75 × 64 × £2·15 =	£7·41
Panels	2/1·950 × 175 × 32 × £2·51 =	£9·78
	2/1·800 × 175 × 32 × £2·51 =	£9·03
	2/1·650 × 175 × 32 × £2·51 =	£8·28
Bottom rails	2/0·780 × 225 × 64 × £6·45 =	£9·99
		£105·14
(see page 88)	Waste 35%	£36·79
	Total	£141·83

Labourer's rate	£1·77 per hour
Joiner's rate	£2·00 per hour
Machinist's rate	£2·00 per hour

Machine rates:

Cutting-machinist	1½ hours at £2·00 =	£3·00
Sorting-labourer	1½ hours at £1·77 =	£2·65
Planing	½ hour at £2·25 =	£1·12
Marking out	4 hours at £2·00 =	£8·00
Mortise	½ hour at £2·20 =	£1·10
Tenon	1½ hours at £2·20 =	£3·30
Bandsaw	½ hour at £2·20 =	£1·10
Spindle	7 hours at £2·20 =	£15·40
Assembly	40 hours at £2·00 =	£80·00
Sander	¼ hour at £3·62 =	£0·90
Spraying – polish	1 hour at £2·00 =	£2·00
	Total	£118·57

Ironmongery (antique):

Black mock hinges £3·60 (pair)	£7·20
100 mm brass butts £6·20 (1½ pair)	£12·40
Three-lever mortise lock, two keys	£1·18
Black lock lever set £3·74 (pair)	£3·74
Black studs, dozen sets £1·14	£4·56
Cranked tower bolts £2·00 each	£4·00
	£33·08

Summary:

Materials	£141·83
Labour and machines	£118·57
On cost	£118·57
Ironmongery	£33·08
	£412·05
Profit 10%	£41·20
Total cost	£453·25

The above estimate highlights several interesting factors in specialised work of this nature where, although relatively small quantities of timber are used, the proportional value of material to labour is high, having regard to the amount of bench work and joiner's time involved. This indicates the level to which the prices of prime quality hardwood timbers have risen.

Labour constants: at £1·67 per hour.
Fit and hang doors including fixing of butts.

	Time (hours)	Price (£)
Lightweight flush door on 75 mm butts	1·25	2·08 each
Solid core flush door on 100 mm butts	1·50	2·50 each
Plastic laminate secured with contact adhesive	1·25	2·08 per m²
Fix only ironmongery to softwood rising butts 75 mm (pair)	1·00	1·67 per unit
rising butts 100 mm (pair)	1·15	1·92 per unit
Helical spring hinges and blanks 75 mm (pair)	1·30	2·17 per unit

	Time (hours)	Price (£)
100 mm (pair)	1·40	2·38 per unit
Overhead door springs	2·00	3·34 per unit
Floor spring double action	4·00	6·68 per unit
Check action floor spring	5·00	8·35 per unit
Finger plates	0·20	0·33 per unit
Bales catches	0·50	0·84 per unit
Kicking plate 750 mm	0·33	0·55 per unit
Pull handle 450 mm	0·33	0·55 per unit
Flush sliding door pull	0·30	0·50 per unit
Letter plates	1·50	2·50 per unit
Track for sliding cupboard	0·375	0·62 per m
Drawer pulls	0·10	0·16 each
Centre door knob	0·50	0·83 each
Rim latch	0·75	1·25 each
Rim dead lock	0·50	0·83 each
Yale latch	1·50	2·50 each
Mortise lock and furniture	1·75	2·92 each
Rebated mortise lock	2·50	4·17 each
Rim lock and furniture	1·00	1·67 each
Till lock	0·75	1·25 each
Barrel bolts 150 mm	0·20	0·33 each
Barrel bolts 300 mm	0·40	0·66 each
Flush bolts 300 mm	2·00	3·34 each
Indicating bolts	1·00	1·67 each
Necked bolts	0·25	0·41 each
Monkey tail bolts 450 mm	0·50	0·83 each
Panic bolts single	1·00	1·67 each
Panic bolts double	1·50	2·50 each
Espagnolette bolt	3·00	5·01 each
Quadrant stays	0·50	0·83 each
Casement stays	0·40	0·66 each
Casement fasteners	0·60	1·00 each

Unit rates:
150 mm black japanned barrel bolt and fixing to softwood:

No. 1 barrel bolt including screws 56p	£0·56
Joiner 0·20 hour at £1·67	£0·33
	£0·89
Add profit 20%	£0·18
Price each	£1·07

Unit rate = £1·07 each.

Rebated mortise lock and furniture and fix to hardwood doors:

No. 1 rebated mortise lock and furniture	£4·25
Joiner 2·50 hours at £1·67	£4·17
	£8·42
Add profit 20%	£1·68
Price	£10·10

Unit rate = £10·10 each.

Standard model 'Briton' 2004 series pneumatic overhead door closer and fixing to softwood:

No. 1 2004 series door closer and screws	£23·27
Joiner 2 hours at £1·67	£3·34
	£26·61
Add profit 20%	£5·32
Price each	£31·93

Unit rate = £31·93 each.

Chapter 11

Applied geometry

Mouldings

Mouldings are important architectural features used in construction to improve the appearance of purpose-made items of joiner's work – projections, arches, doors, windows, etc.

Roman mouldings are formed from arcs of circles, whereas Grecian mouldings are often made up of elliptical curves. The following mouldings are used in building work.

Figure 47. The **ovolo** is a quadrant of a circle with a fillet usually at both ends of the moulding.

Figure 48. The **scotia** moulding comprises two quadrants of differing radii giving a curve concave in shape.

Figure 49. The **cavetto** is a moulding with a concave curve, the opposite of the ovolo, and is formed by a quarter of a circle.

Figure 50. The **cavetto** has the same height as in Fig. 49 but less projection, giving a flatter curve.

Figure 51. The **cyma recta** moulding is often termed the ogee mould and set out within a square.

Figure 52. The **cyma reversa** moulding is known as the reverse ogee.

Figure 53. A further method of setting out the **cyma recta** is shown when the height and the projection of the moulding are not the same.

Figure 54. A further method of setting out the **cyma reversa** of unequal height and projection is shown and is similar to the previous example.

Mouldings (Grecian)

Some examples of the classic mouldings used by the Greeks are shown in Figs. 55–60. The profiles of these mouldings are in use today and their construction is based on conic sections other than the circle. The student should notice particularly the proportions which are clearly indicated and the difference between the Grecian and Roman types.

Reference should be made to the section on conic sections (below,

p. 97) especially the method of drawing the curves. Following the setting out by geometrical methods of a large number of Grecian mouldings, the student should practise drawing the same mouldings freehand.

Figure 55. To set out a **cyma recta** in a rectangle:
1. Draw the rectangle and divide the centre lines into a number of equal parts.
2. Set up perpendiculars from the horizontal divisions to intersect with radial lines from the vertical divisions.
3. Through the points of intersection complete the parabolic curve.

Figure 56. To set out a **cyma reversa** in a rectangle:
The construction is carried out in a similar manner to that in the previous example.

Figure 57. To set out a **cavetto** in a rectangle:
It will be seen that this moulding is a quarter ellipse and the geometrical setting out is similar to that on p. 100.

Figure 58. To set out a **scotia** in a parallelogram:
This setting out follows that in the previous example and consists of a semi-ellipse.

Figure 59. To set out a **parabolic ovolo** moulding:
This moulding, it will be seen, is a parabolic curve similar to half the cyma reversa in Fig. 56.

Figure 60. To set out a **hyperbolic ovolo** moulding:
1. Divide the height and width (projection) of the moulding into the same number of divisions.
2. From the points of division in the width radiate lines to point A.
3. Make AB equal to AO and from B radiate lines to the vertical points of division.
4. Through the points of intersection draw the hyperbolic curve.

In a polygon the number of angles equals the number of sides, and the sum of the external angles equals the sum of the internal angles, 360 degrees.

Figure 61. To draw an **octagon** in a given square:
1. Draw the given square ABCD.
2. With the corners of the square as centres and radius AO equal to half the diagonal, draw arcs cutting the sides of the square giving the angular points of the octagon.
3. Join the angular points to complete the figure.

To construct a regular **hexagon** given the diameter:
Figures 62 and 63 are constructed using compasses, 60 degree set-square and tee-square.

On a given diameter, tangents to the circle are drawn using the set-square (as illustrated in hexagonal nuts and bolts, Fig. 64).

Figure 65. To construct any number of regular **polygons** on a given side (hexagon and nonagon):
1. Let AB be the given line and bisect it to give the centre line.
2. Erect a perpendicular from B equal to AB.
3. Join AC, cutting the centre line in point 4.
4. With centre B and radius AB draw the quadrant AC, cutting the centre line in point 6.

APPLIED GEOMETRY

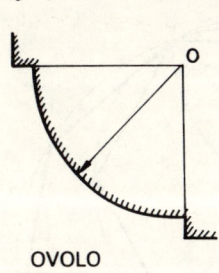

OVOLO

Fig 47

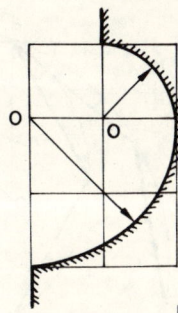

SCOTIA

Fig 48

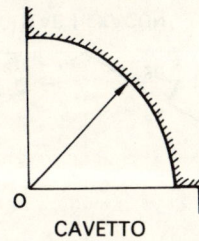

CAVETTO

Fig.49

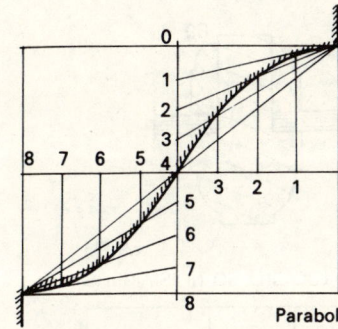

CYMA RECTA

Parabolic mouldings

Fig 55

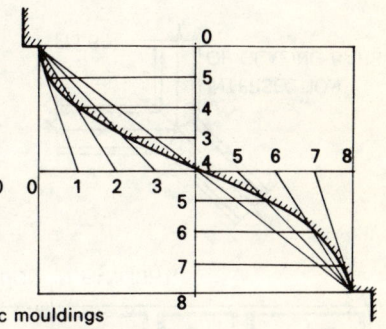

CYMA REVERSA

Fig 56

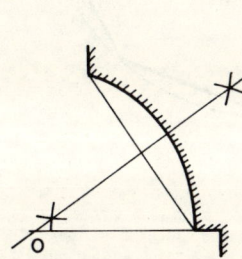

CAVETTO

Fig 50

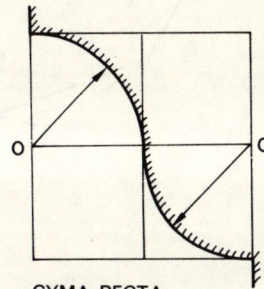

CYMA RECTA

Fig 51

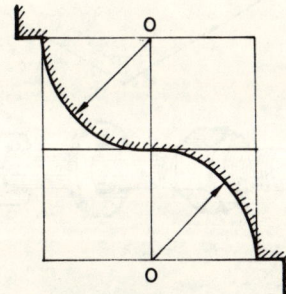

CYMA-REVERSA

Fig 52

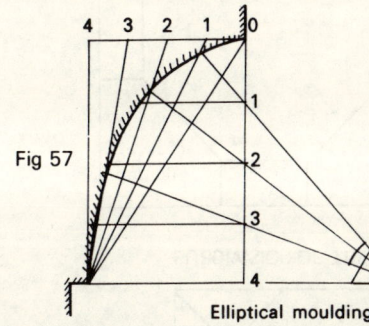

Fig 57

Elliptical mouldings

CAVETTO

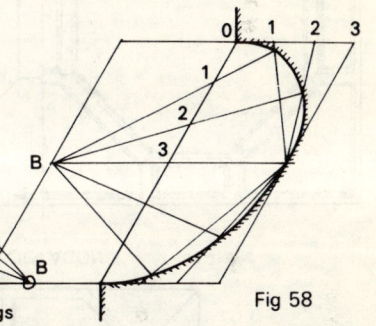

Fig 58

SCOTIA

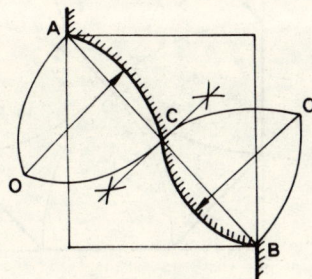

CYMA-RECTA

Fig 53

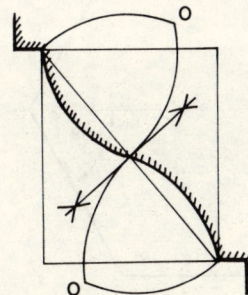

CYMA REVERSA

Fig 54

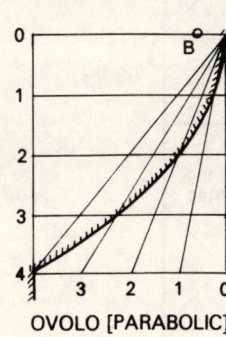

Fig 59

OVOLO [PARABOLIC]

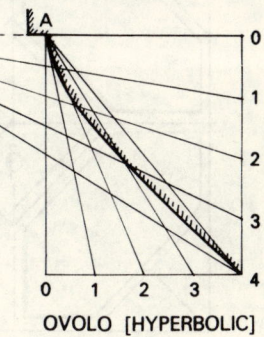

Fig 60

OVOLO [HYPERBOLIC]

APPLIED GEOMETRY

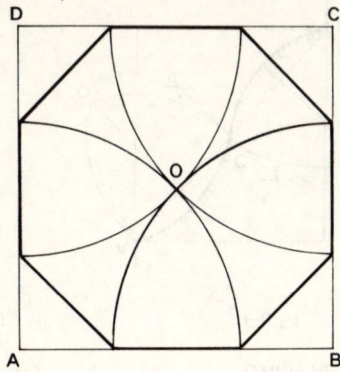

Fig 61

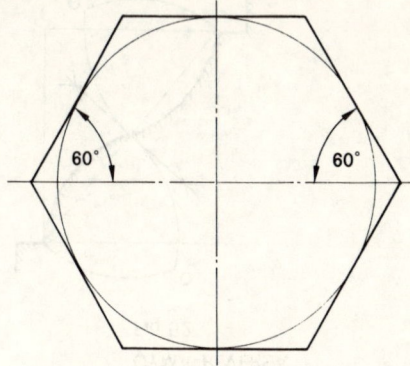

Fig 62

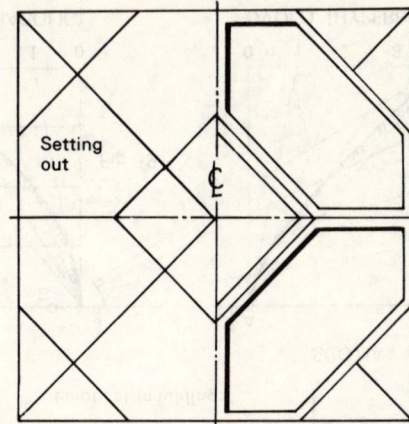

Fig 66

SUBDIVISION OF THE OCTAGON

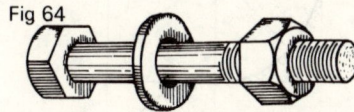

Fig 64

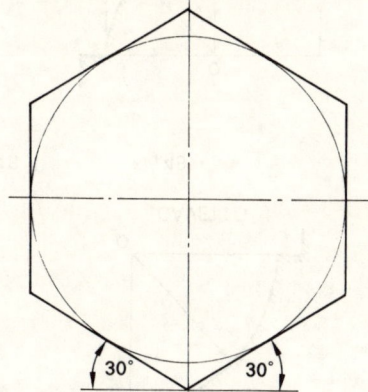

REGULAR HEXAGON

Fig 63

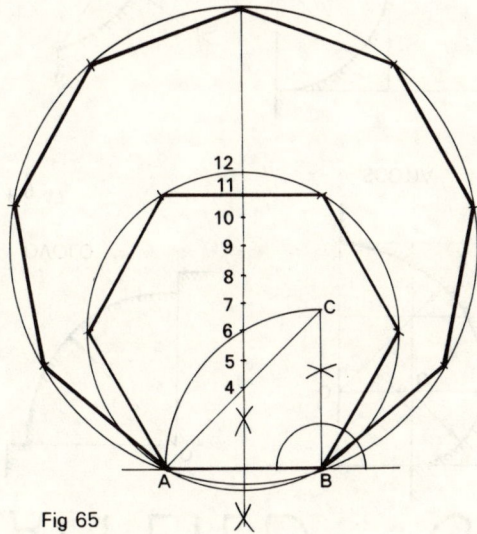

Fig 65

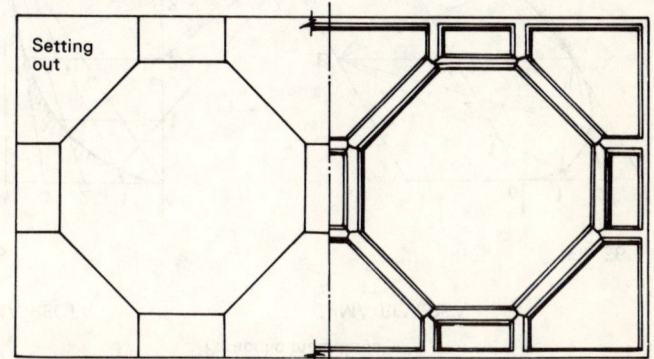

SUBDIVISION OF THE OCTAGON IN FANLIGHT

Fig 67

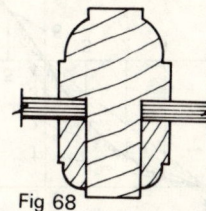

Fig 68

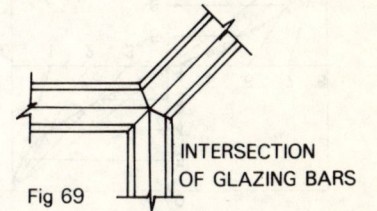

INTERSECTION OF GLAZING BARS

Fig 69

5. Halve the distance between the points 4 and 6 to give point 5 and step this distance along the centre line to give the points 7, 8, 9, 10, etc. These points are the centres of circumscribing circles of polygons having sides equal to AB.

Subdivision of the octagon

Figures 66 and 67. The division of an octagon is shown. The preliminary setting out is shown to the left of the centre line with subsequent filling of an ornamental nature to the right. This type of work is to be found in glazed fanlights to entrances of public buildings, banks and the like, where high-class joinery is employed.

A typical glazing bar section with glazing beads is shown in Fig. 68. The jointing of the glazing bars is shown in Fig. 69. Unless the mitre halves or bisects the over-all angle the members will not coincide. The centre lines of the bars are first set out, all the mitres intersecting on this line.

When any two mouldings intersect in one plane, the intersection is known as a mitre. Where two mouldings are of the same width the line of the mitre will be straight and at 45 degrees.

Figure 70 shows two straight mouldings and the bevel to which each will be cut.

Figure 71 shows two mouldings of the same width meeting at an angle of 120 degrees. The mitre line, as in the previous example, is a line drawn from the intersecting point of the two outer edges to the intersection of the two inside edges.

Figure 72. To reduce a given moulding in width only:
1. Draw the given architrave mould and the required mould in outline to determine the mitre line.
2. On the face of the given mould select a number of points and project these to the mitre line.
3. Draw vertical lines downwards from these points on to the required section. Transfer the various thicknesses from the given mould to determine the outline of the required mould.

Figure 73. When a straight moulding intersects a curved moulding and are both of the same section, the mitre is curved:
1. Draw the straight moulding and divide the outline into suitable divisions 0, 1, 2, 3, 4, 5, 6.
2. Draw the section of the curved moulding using the same points of division as in the straight moulding.
3. From these points of division draw lines parallel to the direction of each moulding to intersect each other in order as at 0', 1', 2', 3', 4', 5', 6'.
4. Trace through the points to give the curved intersection of the two mouldings.
   The curved mitre is an example of the practical application of loci.
   When two curved mouldings of equal radii and width intersect, the joint or mitre line will be a straight line. When two curved mouldings of equal width but of different radii intersect, the mitre line is curved.

A further example of a straight and curved moulding intersecting is shown in Fig. 74. The working is similar to that in Fig. 73.

If for any reason a straight mitre were required, this would mean that the intersection would be straight and consequently the points along the intersection would alter. These new positions would need to be turned into the line AB, Fig. 73. From these new points set up lines at right angles to AB and transfer corresponding heights from the given section, which will result in a new section for the curved moulding. In the majority of cases the distortion to the curved mouldings would be too pronounced and the curved mitres would have to be adopted.

Figure 75. When a head architrave moulding, owing to insufficient height, has to be reduced in width and made to mitre with the wider jamb architrave:
1. Draw the jamb or given architrave.
2. Set out the width of the smaller moulding, in this case 50 mm, and determine the line of the mitre AB.
3. Project ordinates from points 1, 2, 3, 4, 5, 6, 7, 8 to the line of the mitre and at right angles to contact corresponding coordinates projected from the side of the required moulding in points 1', 2', 3', 4', 5', 6', 7', 8'.
4. Complete the required moulding section.
5. The section of the moulding at the mitre AB is obtained by projecting ordinates at 90 degrees to the mitre line to intersect with ordinates as previously.

Conic sections

The ellipse, parabola, and hyperbola are obtained as sections of a cone, and hence are called 'conic' sections (Fig. 76).

The ellipse, section E–E shown in Fig. 76. This is the section made by a plane which cuts the cone obliquely across opposite generators.

The parabola, section C–C is the section made by a plane which is parallel to a generator.

The hyperbola, section D–D is the section made by a vertical cutting plane.

The horizontal section of a cone is a circle section B–B and the vertical section taken on the centre line is a triangle, section A–A.

The parabola

When a cone is cut parallel to its generator or inclination, the section obtained in outline is a curve shown in sketch Fig. 77. The craft student will meet this curve frequently in the course of his studies: the curve of a bending moment diagram in structural calculations; arch and centre construction.

Figure 78. To construct a **parabola** within a rectangle:
1. Set out the rectangle ABCD.
2. Divide AD into a number of equal parts and AE into the same number.
3. From the points of division draw lines to F and perpendiculars from AE.
4. Through the points of intersection draw the curve.

APPLIED GEOMETRY

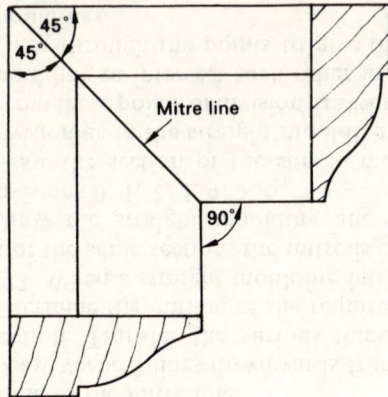

Fig 70

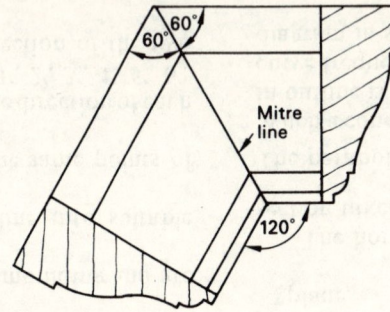

Fig 71

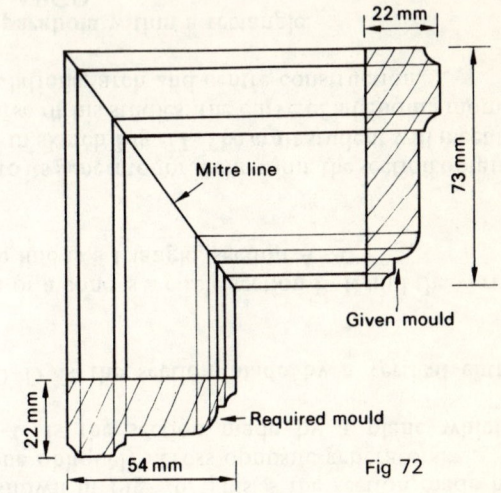

Fig 72

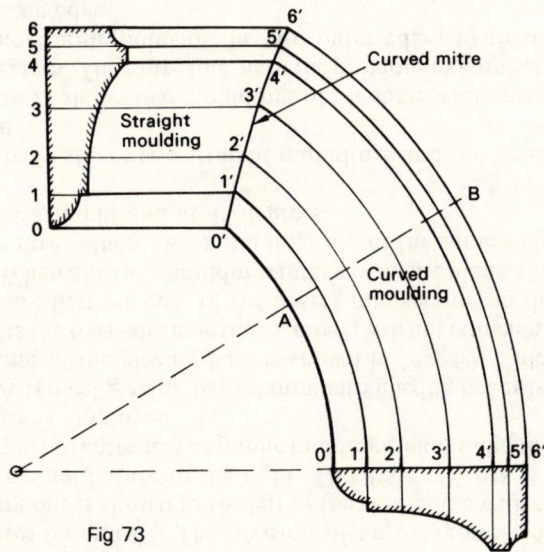

Fig 73

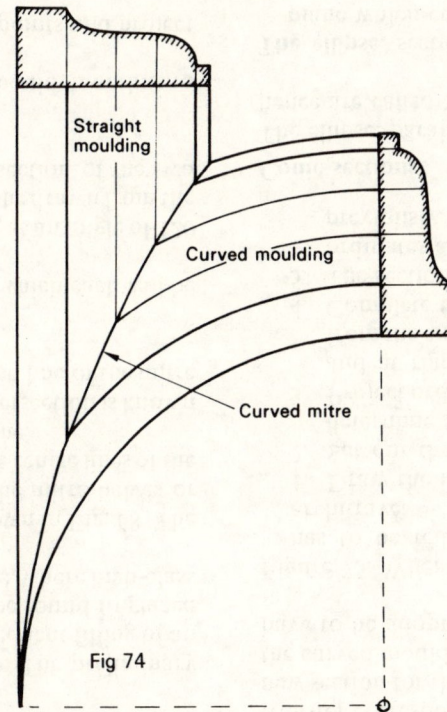

Fig 74

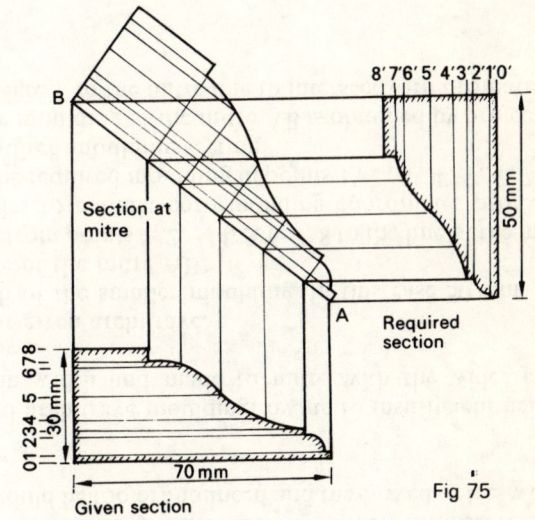

Fig 75

APPLIED GEOMETRY

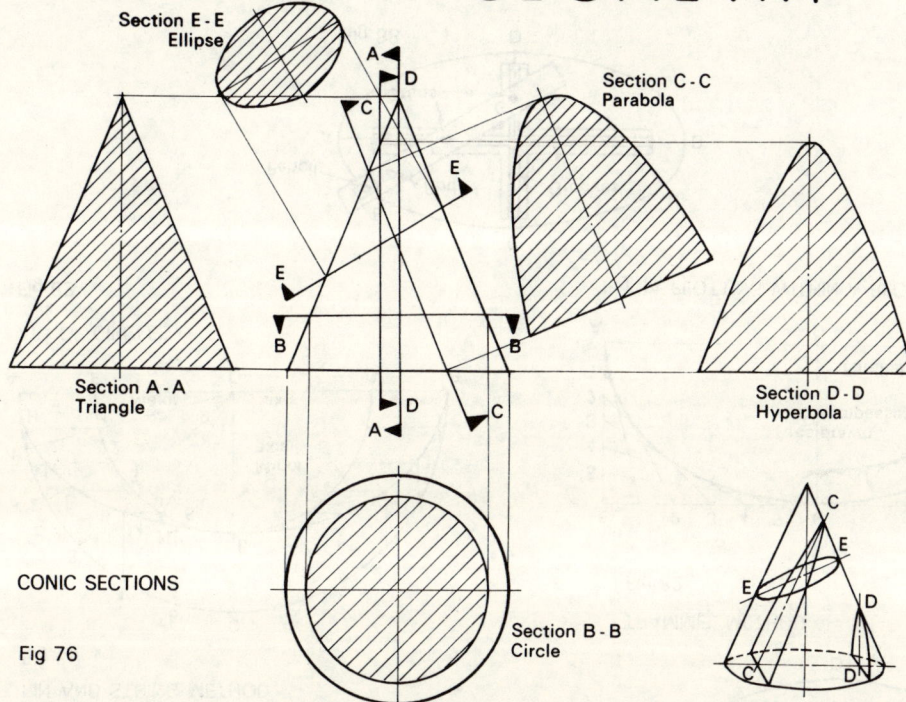

Section E-E
Ellipse

Section A-A
Triangle

Section C-C
Parabola

Section D-D
Hyperbola

CONIC SECTIONS

Fig 76

Section B-B
Circle

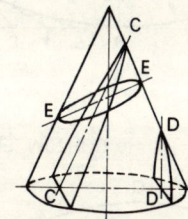

CONIC CURVES

Fig 77

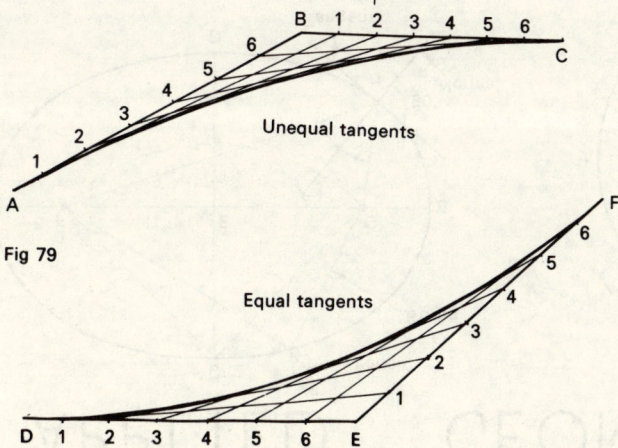

Unequal tangents

Fig 79

Equal tangents

Fig 80   TRUE SHAPE OF CONE SECTIONS

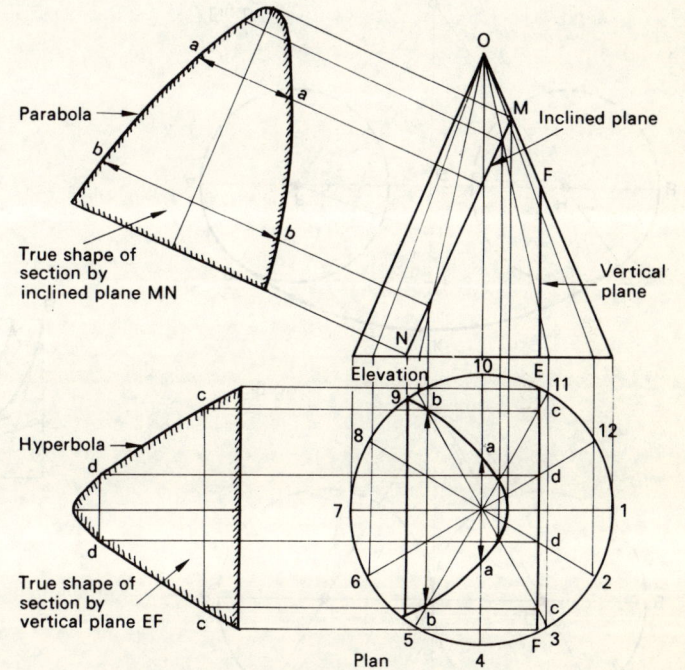

Parabola

True shape of
section by
inclined plane MN

Inclined plane

Vertical
plane

Elevation

Hyperbola

True shape of
section by
vertical plane EF

Plan

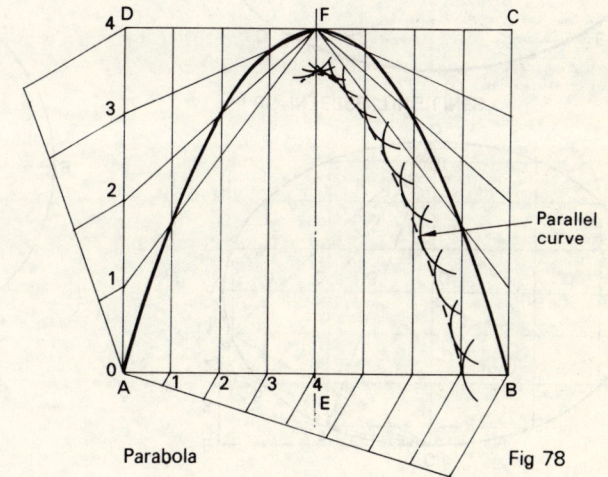

Parallel
curve

Parabola   Fig 78

APPLIED GEOMETRY

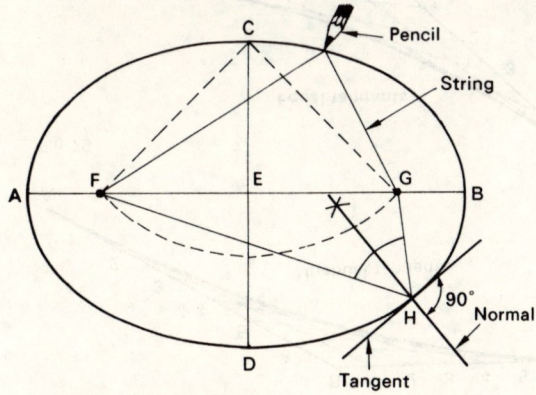

PIN AND STRING METHOD
Fig 81

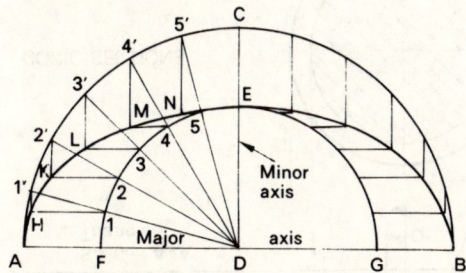

AID OF CIRCLES

Fig 83

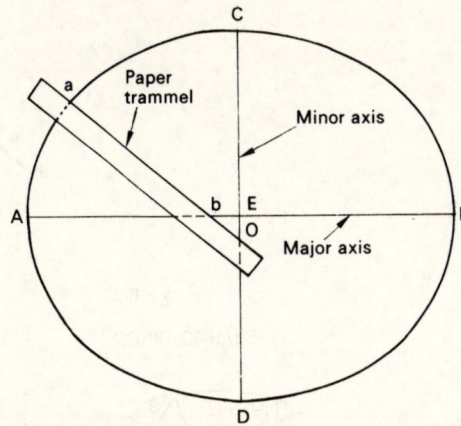

TRAMMEL METHOD

Fig 82

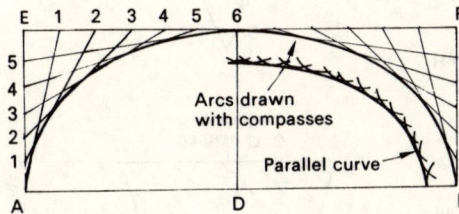

Fig 84 PLOTTING WITHIN A RECTANGLE

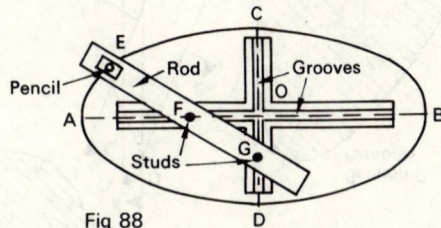

SETTING OUT ELLIPSE

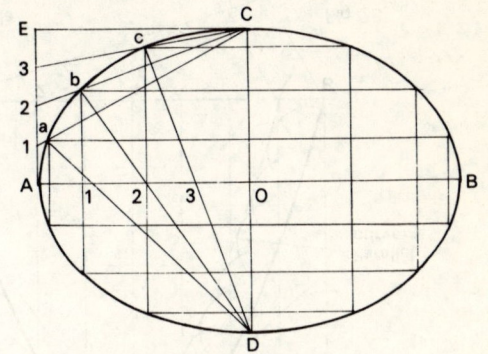

Fig 85 INTERSECTING LINES

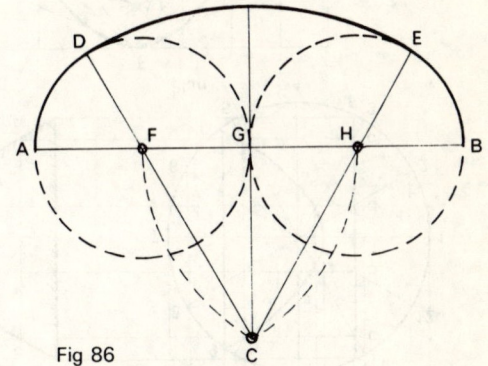

Fig 86

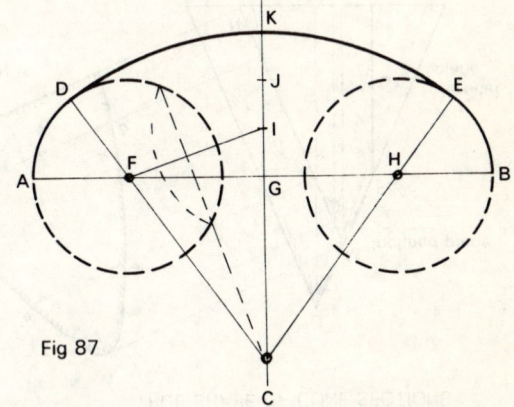

Fig 87

Fig 88

The method of drawing a curve parallel to a given one which is not a circular arc, is shown to the right of the centre in Fig. 78.

Figure 79. To construct an **easing** or parabolic curve other than by compasses using predetermined unequal tangents:
1. Let AB and BC be the unequal tangents.
2. Divide B1 and B6 into a similar number of parts on each tangent line.
3. Join 1–1, 2–2, etc. A fair curve touching the tangents gives the required easing.

Figure 80 shows the plan and elevation of the cone with the projection of the true shape of the cutting sections by two planes. The first plane MN is inclined and parallel to the slant side and the section to be determined is a parabola.

The second plane EF is vertical and the section in this case is a hyperbola. In both cases the widths on each side of the centre lines are taken from the corresponding plan positions.

The **ellipse** is a plane figure bounded by one continuous curve, described about two points called foci, and the sum of the distances from a point in the curve to the foci is constant. The ellipse is derived from the section of a cone or cylinder, made by a cutting plane inclined to the axis.

The **major axis** is the longest distance that can be drawn across the figure. The **minor axis** bisects the major axis and is perpendicular to it. The major axis contains two points called the **foci** of the ellipse.

Figure 81. To draw an **ellipse** by means of a string, the major and minor axes being given:
1. Let AB and CD be the major and minor axes and E the centre.
2. From C with radius AE describe two arcs cutting AB in the foci F and G.
3. Insert pins at the foci F and G and fix a looped string round them; place a pencil in the loop and make the string taut.
4. Move the pencil as shown and trace the elliptical curve.

To draw a **tangent** and a **normal** at a given point in the circumference of an ellipse:
1. Let ACBD be the given ellipse and H the given point.
2. Join HF and HG and bisect the angle FHG; the bisector is the required normal.
3. Draw a line through H at 90 degrees to the normal to give the required tangent.

Figure 82. To draw an **ellipse** by trammel method, the major and minor axes being given:
1. Set out the axes as previously.
2. On a strip of paper with a straight edge mark off O equal to AE and ab equal to CE.
3. Place the paper so that O is on the minor axis and b is on the major axis. a is then a point on the curve which is marked.
4. Move the trammel still keeping O and b on the two axes when other points will be obtained.
5. Through these points complete the curve.

Figure 83. To draw an **ellipse** by means of auxiliary circles, the axes being given:
1. Let AB be the major axis and ED half of the minor axis.
2. On AB as diameter, construct a circle ACB and from the same centre D with radius equal to half the minor axis describe a circle FEG.
3. Draw any number of radii of the circles as D11', D22', etc., and through the points of division on the outer circle draw lines parallel to CD.
4. Through the points of division on the inner circle draw lines parallel to AB. The intersections of the two sets of lines will give points on the curve.

Figure 84. To plot an **ellipse** within a rectangle:
1. Draw the rectangle ABFE.
2. Divide AE and EC into the same number of equal parts.
3. Join A1, 12, 23, etc. and through the points of intersection draw the curve.

A parallel curve drawn through arcs drawn using compasses is shown to the right of the centre.

Figure 85. To draw an **ellipse** by 'intersecting lines' method:
1. Draw the axis and construct the rectangle AOCE.
2. Divide AE and AO into the same number of equal parts and draw C1, C2, C3.
3. From D draw D1 to meet C1, D2 to C2 and D3 to C3.
4. Through the points draw a quarter of the ellipse and complete the figure either by repeating the process for the other quarters, or by symmetry as shown.

Figure 86. To describe using compasses a figure resembling an **ellipse**, the major axis being given:
1. Divide the major axis AB into four equal parts in the points F, G, and H.
2. On FH construct an equilateral triangle and produce the sides of the triangle as to points D and E.
3. With F and H as centres and radius AF describe the circles ADG and BEG.
4. With C as centre and radius CD, describe the arc DE to complete the semi-ellipse. Complete the other half of the ellipse in the same manner.

Figure 87. To describe using compasses a figure resembling an **ellipse**, the major and minor axes being given:
1. Let AB be the major axis and GK the semi-minor axis.
2. Divide GK into three equal parts in J and I and make AF and BH each equal to two of these parts.
3. Join FI.
4. With F and H as centres and radius AF describe the circles AD, BE.
5. Bisect the line FI meeting the axis KG produced in C.
6. From C through F draw CFD and from C with the radius CD describe the arc DKE completing the semi-ellipse.

Figure 88. To set out an **ellipse** by trammel method as in the workshop. The essential features of the arrangement are a framed wooden cross, with uninterrupted grooves as shown, a lath or rod with a pencil holder at E, and wooden dowels at F and G:
1. Set out the major and minor axes AB and CD.

APPLIED GEOMETRY

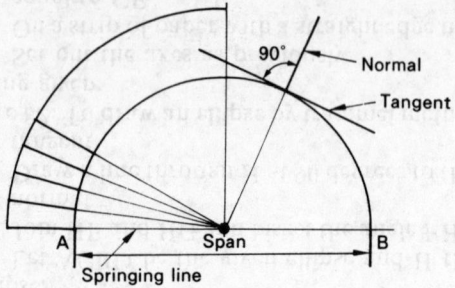

Fig 89   SEMI-CIRCULAR

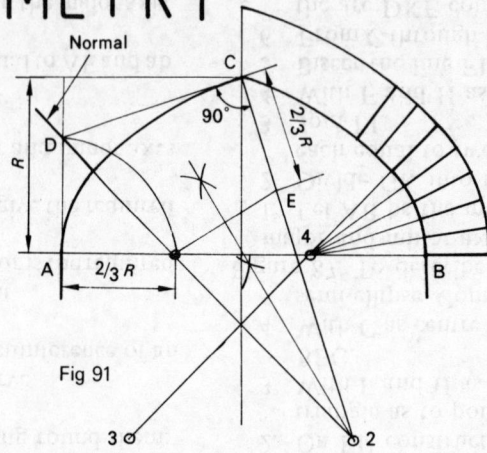

Fig 91

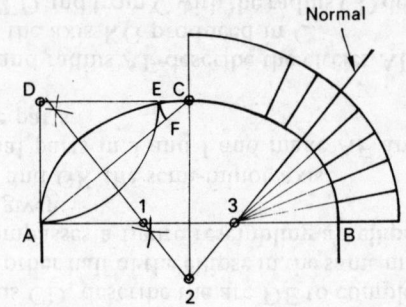

Fig 92 THREE-CENTRE

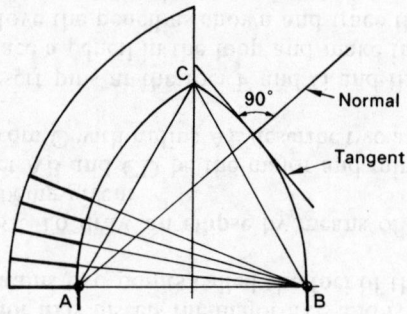

Fig 90   EQUILATERAL

FOUR-CENTRE

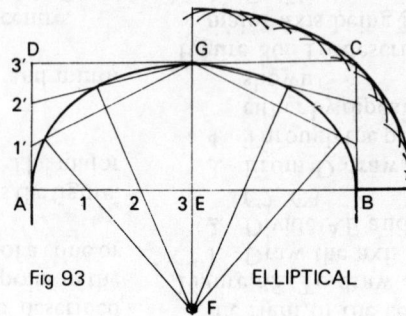

Fig 93   ELLIPTICAL

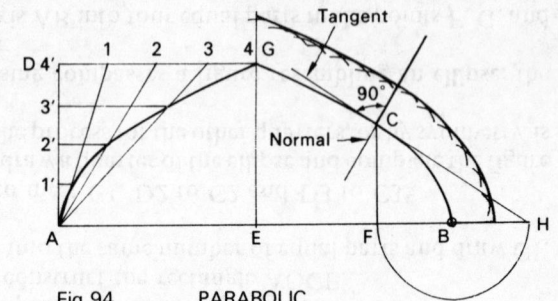

Fig 94   PARABOLIC

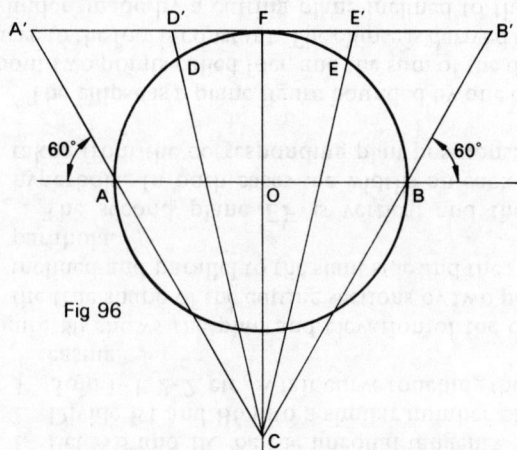

Fig 96

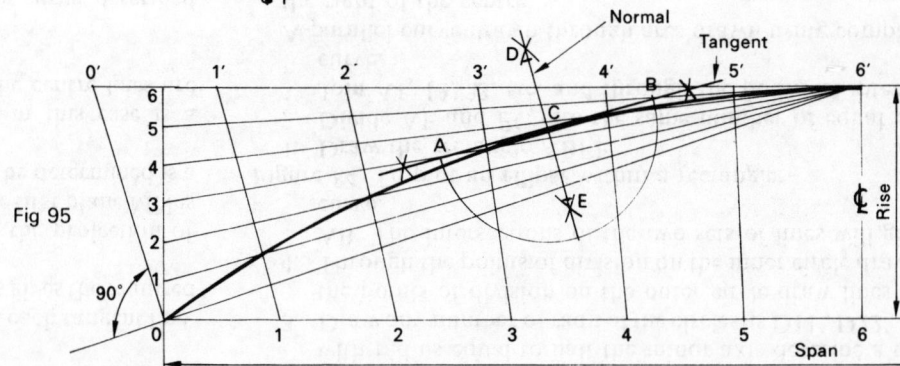

Fig 95

2. Lay the framed cross upon the axes so that the centre lines coincide with them.
3. The dowels are next fixed into the rod at F and G, such that EF, EG are equal to half CD and AB, respectively.
4. Move the pencil to the right or left with F and G always in contact with the grooves, to trace the required ellipse.

Setting out arches

In the construction of brick or stone arches it is necessary to set out the outline of the arch in order that the size and shape of the timber framing used as a support to the arch under construction may be determined, and also the bricks or stones forming the arch. The necessary data to construct an arch outline consist of:

1. The architectural style.
2. The span or width between supports.
3. The rise or height of the arch.

The highest point of the arch is called the crown, and the springing the points where the arch curve begins.

Figure 89. The semicircular arch:
All joint lines radiate to the centre from which the arch outline is struck.

Figure 90. The equilateral arch. In the setting out of this arch the span equals the radius. To set out this arch:
1. Let AB be the span.
2. With points A and B as centres and radii equal to AB draw the arcs AC and BC.

Figure 91. The four-centre arch. This arch often termed a 'tudor arch' is similar in outline to the parabolic arch shown in Fig. 94:
1. Draw the springing line AB and set out the rectangle as shown.
2. Divide the rise into three equal parts and from point D draw DC.
3. With A as centre and radius $\frac{2}{3} R$ obtain the first centre 1.
4. From C measure $\frac{2}{3} R$ to point E at 90 degrees to DC and join E1 which is $\frac{2}{3} R$ from A.
5. Bisect E1 to cut CE produced at 2, the second centre.
6. Set out centres 2 and 4 from 1 and 2 to complete the arch outline.

Figure 92. The three-centre arch. The outline of this arch is approximately elliptical:
1. Set out the span and rise and complete the rectangle.
2. With D as centre and radius DA draw the quadrant AE.
3. With C as centre and radius CE draw the arc to cut AC in F.
4. Bisect AF and produce to cut the springing line in AB to give centre 1 and the centre line to give centre 2.
5. Complete the arch outline using centres 1, 2 and 3.

Figure 93. The elliptical arch:
1. Set out the span and rise as at ABCD.
2. Divide AD and AE into equal parts as 1', 2', 3' and join these points to the point G.
3. Make EF equal to EG and from F radiate lines through points 1 and 2 to intersect G1' and G2'. A free hand curve through these intersections

will give the required curve.
The method of drawing two parallel curves using compasses set to scribe a series of arcs at the required distance is also shown in this example.

Figure 94. The pointed parabolic arch. To set out the arch outline:
1. Construct the rectangle from the span and rise data.
2. Divide AD and DG into the same number of equal parts and draw the construction lines as shown.
3. Through the intersections draw a freehand curve to give the required outline.

To draw a **tangent** and **normal** at any point on the parabolic curve:
1. Select the point C on the curve and drop a perpendicular to cut AB in F.
2. With B as centre and radius BF draw the arc to give point H.
3. From H draw through C, this is the tangent.

The normal is drawn at 90 degrees to the tangent at the point C.

Figure 95. To set out a segmental arch or curve of large span where the centre is inaccessible:
1. Set out half the span and the rise and construct the rectangle as shown.
2. Divide the base line into a number of equal parts and the rise into the same number as 0, 1, 2, 3, 4, 5, 6.
3. Draw the diagonal and erect a perpendicular at point 0 to 0'.
4. Divide 0'–6' into the same number of parts as the base line and join the points 1–1', 2–2', etc.
5. Radiate lines from 1, 2, 3, 4, 5 and 6 to point 6' at the crown and where these intersect with 1–1', 2–2', 3–3', etc., draw a fair curve, the outline of the arch.

A normal or joint line is shown set up at point C.

Figure 96. To find the length of the **circumference** of a circle:
1. Let AB be the diameter of the circle.
2. Using the 60 degree set-square draw lines through points A and B meeting in point C.
3. Draw a tangent to the circle at F produced to cut CA and CB produced in A' and B'.

The line A'B' is equal to half the length of the circumference of the circle.

To find the length of an **arc** of circle, DE:
1. Let DE be the arc.
2. After completing the above setting out draw from C lines through points D and E to cut the tangent A'B' in D'E'.

The line D'E' is the length or development of the arc.

Cone development

Figure 97 shows in sketch form how a cone is developed by unrolling its surface. The apex or point of the cone remains in one place as the circular base unrolls around it for a distance equal to its circumference.

Figure 98. To develop the surface of the cone:
1. Draw the plan and elevation of the cone.
2. Divide the plan into a number of equal parts as 1 to 12.
3. With O as centre and the slant height or generator as radius, draw an arc.
4. Step off the distances 1–2, 2–3, 3–4, 4–5, etc. Form the plan along the

APPLIED GEOMETRY

SKETCH OF CONE
DEVELOPMENT

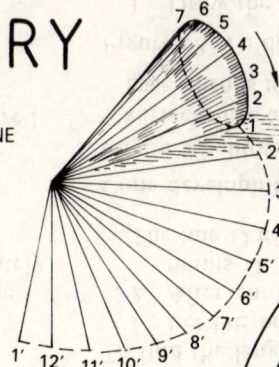

Fig 97

Ellipse

Inclined
plane

True shape of
section

Development
of surface frustrum

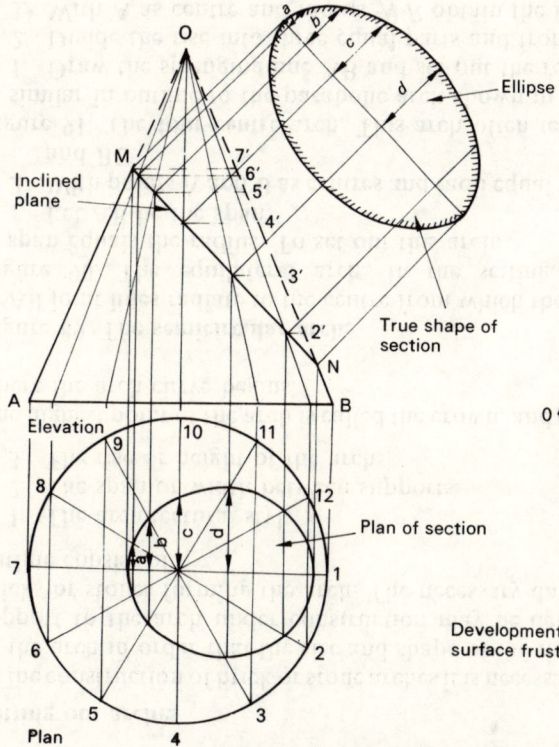

CONE SECTIONS AND DEVELOPMENT

Fig 99

Elevation

Plan of section

Plan

Development
of
whole surface

Development
of frustum

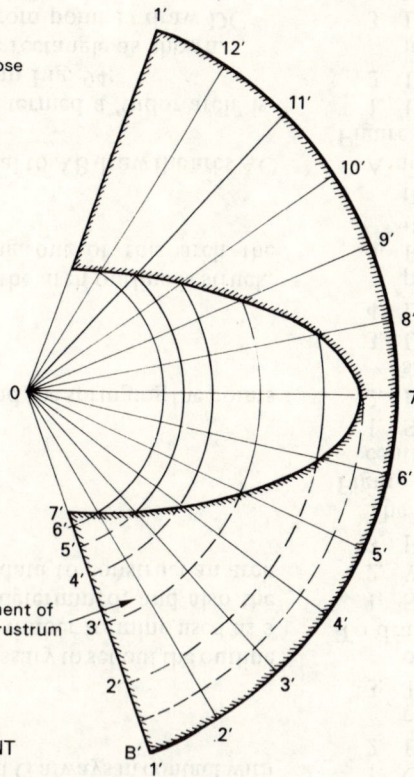

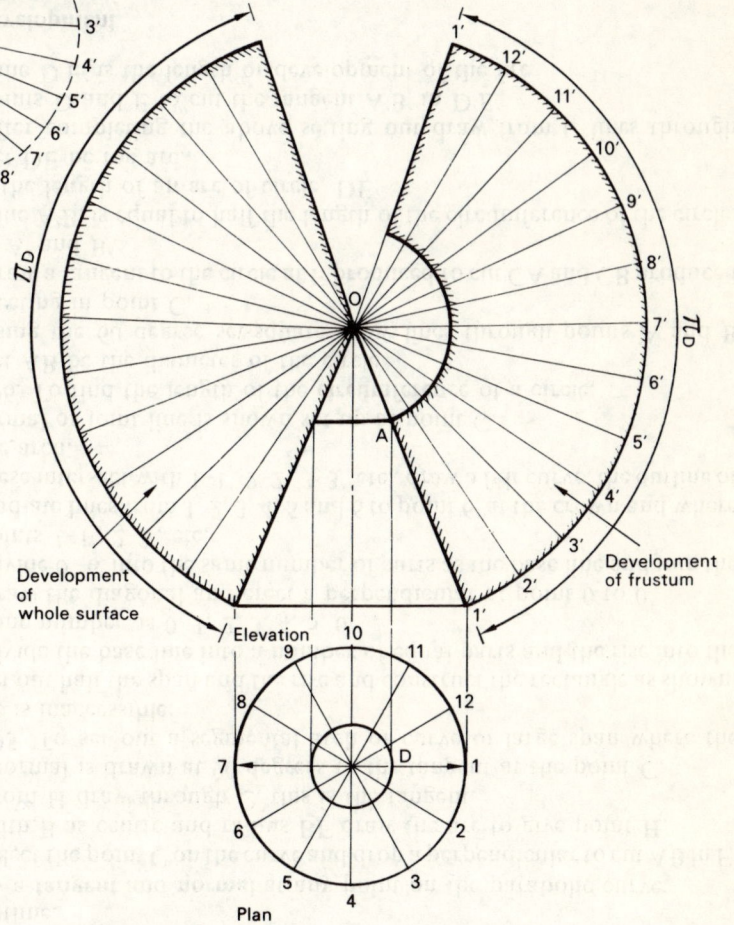

CONE DEVELOPMENT

Fig 98

Elevation

Plan

arc so that its length is equal to the circumference of the cone's base.

5. Join the last point to O at the apex to complete the development.

Frustum of cone

When a cone is cut parallel to its base the portion remaining is called the frustum. The development of this portion of the cone is shown to the right of the centre in Fig. 98.

Figure 99. The plan and elevation along with the plan and true shape of the section of the cone is similar to the example. To develop the frustum of the cone cut by the inclined plane MN:
1. Take out horizontal lines from where the radiating lines cut the inclined plane to touch OB.
2. With O as centre and each of the points on OB as radius draw the arcs.
3. From B' set off the equal divisions 1–2, 2–3, 3–4, 4–5, etc. taken from the plan and draw from these points to O.
4. Through the intersection of the radiating lines and the arcs draw in the curve.

Surface developments

Many objects are constructed by cutting flat sheets to accurate dimensions, and by cutting, rolling, or folding them up into the finished shape after which they may be secured at the joint. The joint may be made by gluing, riveting, welding, brazing, or bolting the edges together.

The basic shapes met with in practice are usually flat, cylindrical or conical, and the following examples will illustrate the constructional methods involved in developing such surfaces.

Examples of the practical application of work on the cylinder in building construction are many. Skew arches. soffit and face moulds, etc.

Figure 100 shows the plan and elevation of a semicircular arch with parallel jambs. The auxiliary elevation viewed in the direction of the arrow marked 'A' follows from previous work. To determine the development of the soffit necessary to cut the laggings in the construction of the arch centre:
1. Divide the elevation into a number of equal parts 0–8 and project these points to the plan.
2. Clear of the plan, step off the distances 0–8 taken from the elevation as at 0', 1', 2', 3', 4', etc. and erect perpendiculars.
3. Horizontals taken from both faces of the arch in plan to intersect with verticals from 0', 1', 2', 3', 4', etc. give points through which both edges of the soffit development are drawn.
(*Note*: The curves of development are not parallel, although when the pattern or development is cut and bent over the arch centre in construction, the surface between the curves appears parallel.) At the crossing of the two semi-elliptic curves the soffit is generated by a level line. This level line moves round the semicircle at the fixed distance of the radius from the centre.)

Figure 101 shows the development of the frustum of a hexagonal pyramid made by an inclined plane.

The true shape of the section is shown projected clear to the right on a new cutting plane M'N', on to which points brought out horizontally from the elevation are projected.

The dimensions a and b are plotted on each side of a centre line, the dimensions being taken from the plan view of the section.

Figure 102 shows the sketch of an octagonal turret base intersecting a main roof centrally at the ridge.

Figure 103 shows the surface development of the base which is carried out in a similar manner to that in the previous example.

Figure 104 shows a sketch of a solid, ovoid in section, with a sloping end.

Figure 105. To set out the ovoid section:
1. Draw two tangential circles of radii 150 and 75 mm with centres A and B.
2. Deduct the radius of the smaller circle from the larger circle to give point C.
3. Join BC and bisect it to cut CA produced in D. This gives the centre for the tangential connecting arc.
4. A line from D through B gives the point of contact between the arcs. The surface development of the ovoid (Fig. 106) is carried out in a similar manner to that described previously.

Loci

Definition

The word **locus** means path, the plural being loci. When a point moves under some definite law or conditions it will generate a line straight or curved, such that any point on the line, curve or path forming the locus will satisfy the conditions. Similarly, a surface may be generated by a line, straight or curved, moving under a fixed law.

Figure 107. The locus of a point B is shown which moves at a fixed distance from a straight line XY and gives a straight line locus.

Figure 108. The locus of a point B which moves at a fixed distance from another fixed point O – like a point on the hand of a watch – gives a circle as the locus.

Figure 109. The locus of a point C moving at equal distances from two other fixed points A and B is shown. The locus $C_1C_2C_3$ generates a straight line, and if repeated on the other side of the line joining AB would bisect the straight line AB. This may be considered a geometrical proof of the method of bisecting any straight line.

Figure 110. To inscribe a circle within a triangle using **loci** to find the centre.
1. Construct the triangle ABC.
2. Draw parallel to AC the lines numbered 1 to 6 at equal distances.
3. Set off a series of points numbered 1 to 6 on AC and draw lines parallel to AB.

APPLIED GEOMETRY

Fig 100

SOFFIT DEVELOPMENT

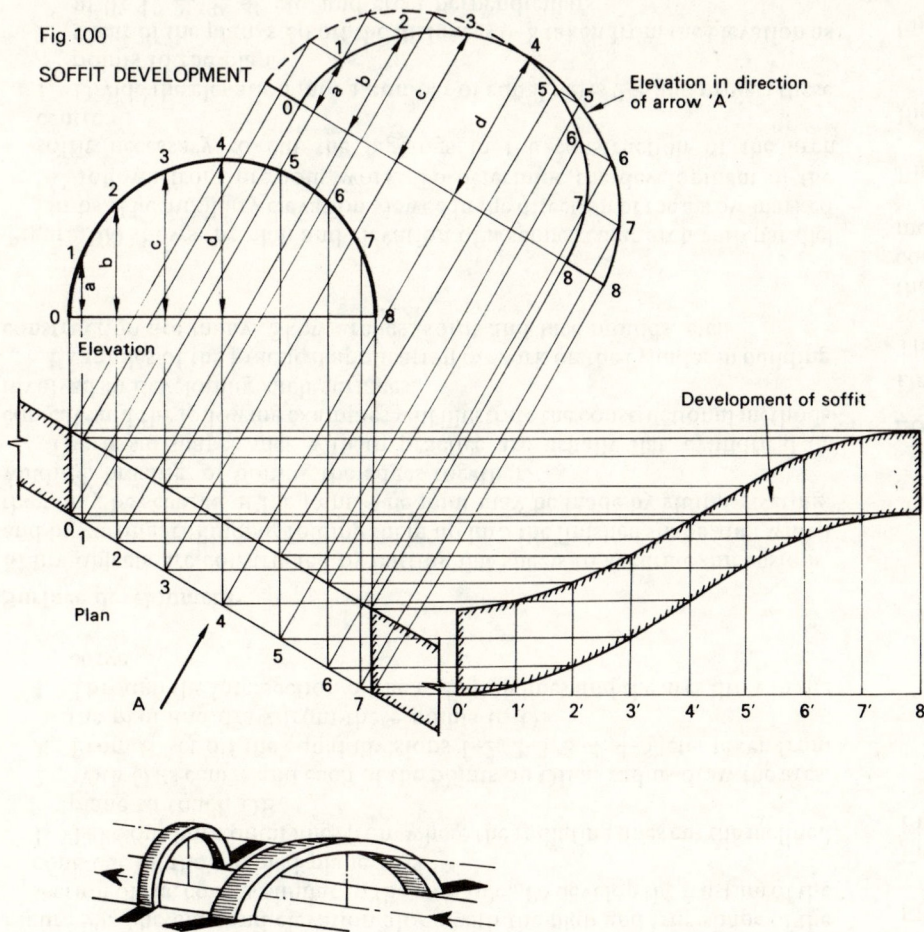

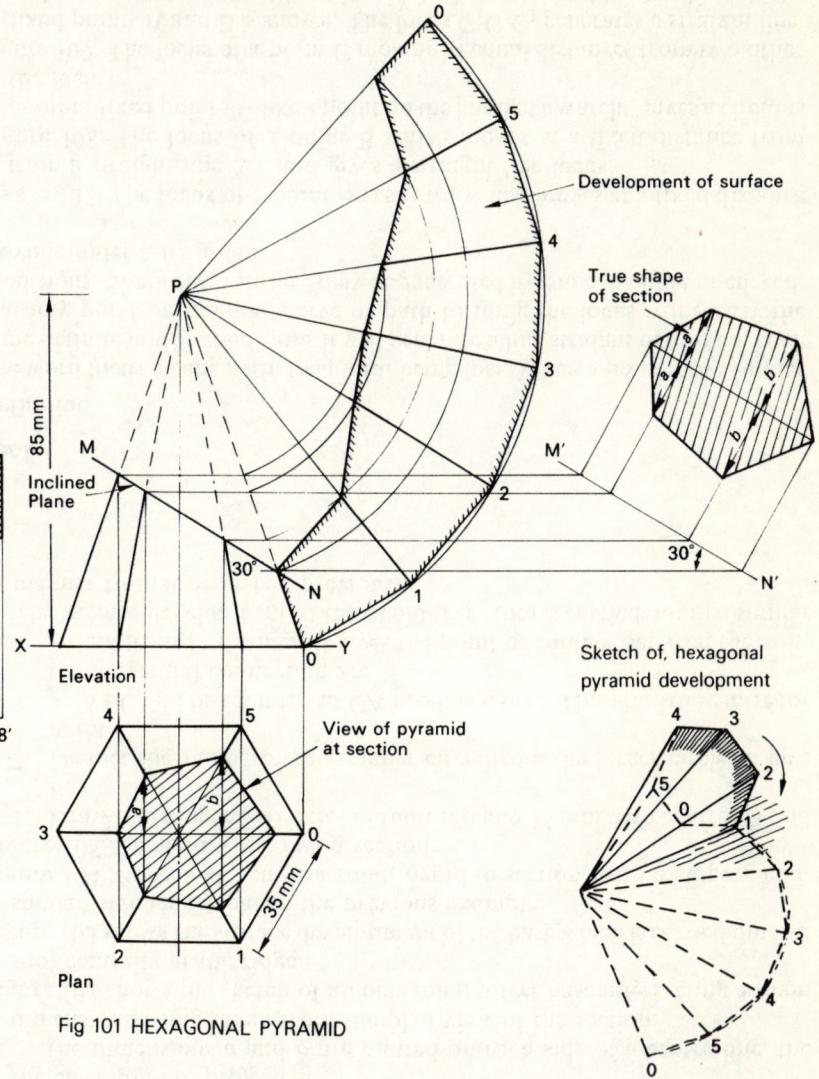

Elevation in direction of arrow 'A'

Elevation

Development of soffit

Plan

A

85 mm

P

Development of surface

True shape of section

Inclined Plane

M   M'

30°   30°

N   N'

X   Y

Elevation

View of pyramid at section

35 mm

Plan

Sketch of, hexagonal pyramid development

Fig 101 HEXAGONAL PYRAMID

APPLIED GEOMETRY

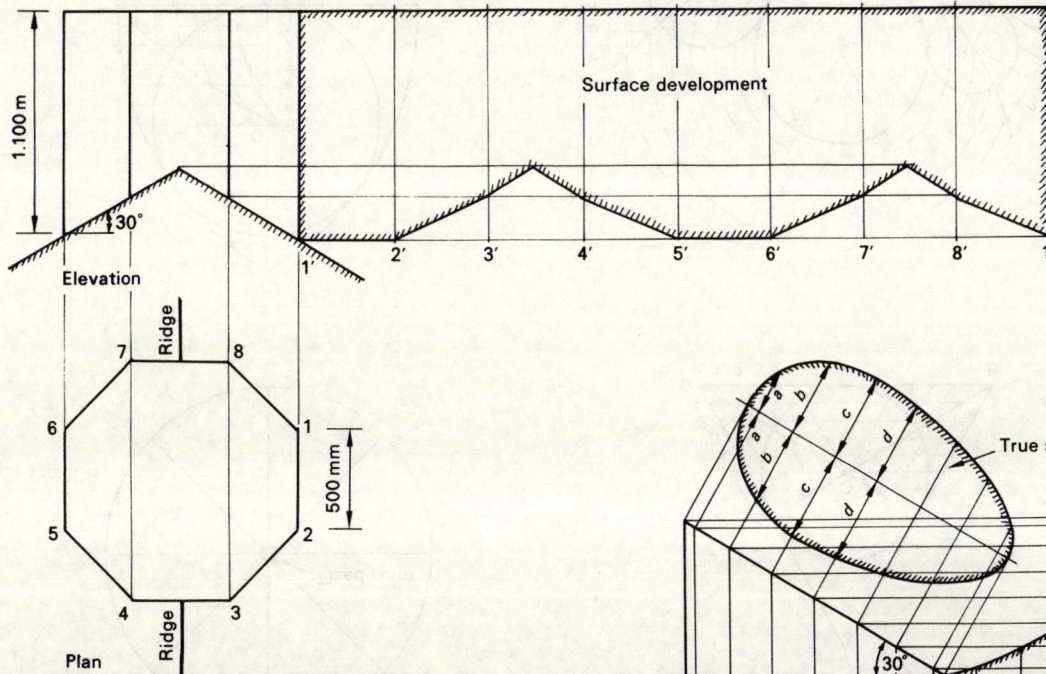

Surface development

1.100 m

30°

Elevation

Plan

Ridge

7   8

6   1

5   2

4   3

Ridge

500 mm

Fig 103 PRISM DEVELOPMENT

2' 3' 4' 5' 6' 7' 8' 1'

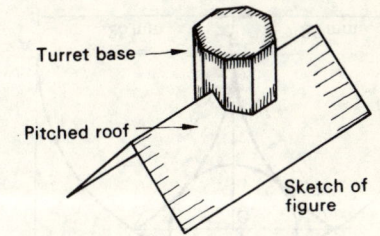

Turret base

Pitched roof

Sketch of figure

Fig 102

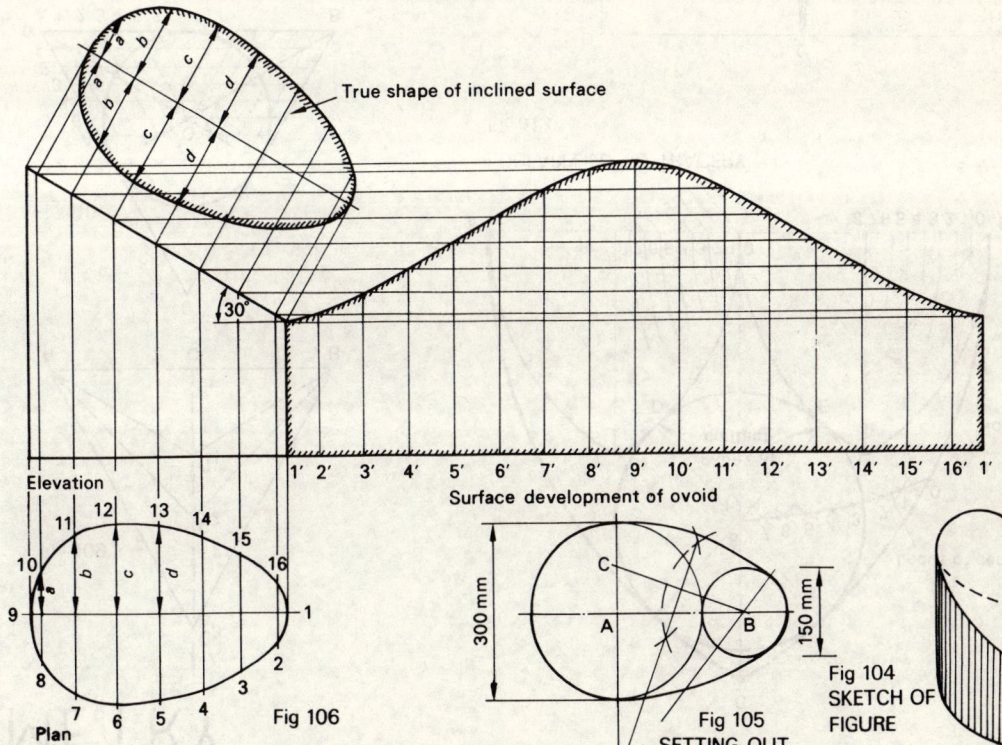

True shape of inclined surface

30°

Surface development of ovoid

1' 2' 3' 4' 5' 6' 7' 8' 9' 10' 11' 12' 13' 14' 15' 16' 1'

Elevation

11 12 13 14

10   15

9   16

           a b c d

8   1

7 6 5 4 3 2

Plan

Fig 106

300 mm

C

A   B

150 mm

**Fig 105
SETTING OUT**

D SECTION

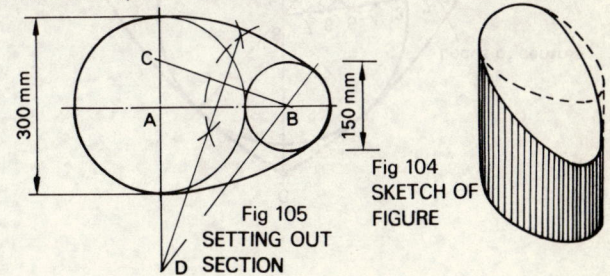

**Fig 104
SKETCH OF
FIGURE**

APPLIED GEOMETRY

Point B moves at a constant distance from XY

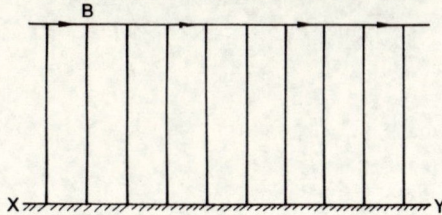

Fig 107

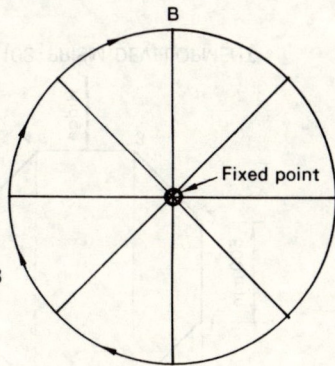

Fixed point

Fig 108

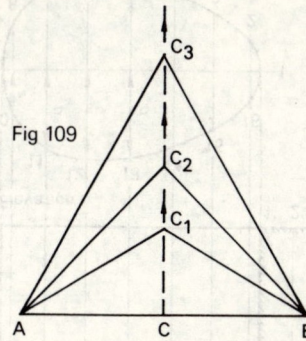

Fig 109

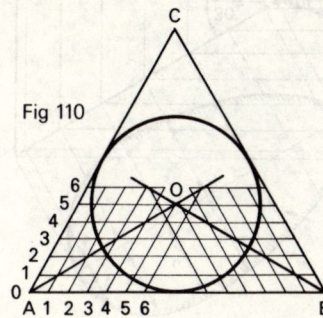

Fig 110

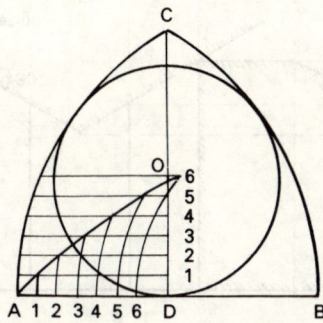

Fig 111

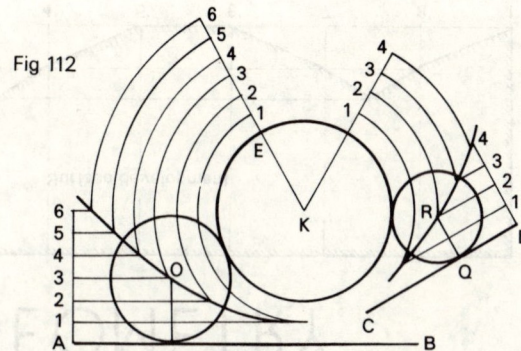

Fig 112

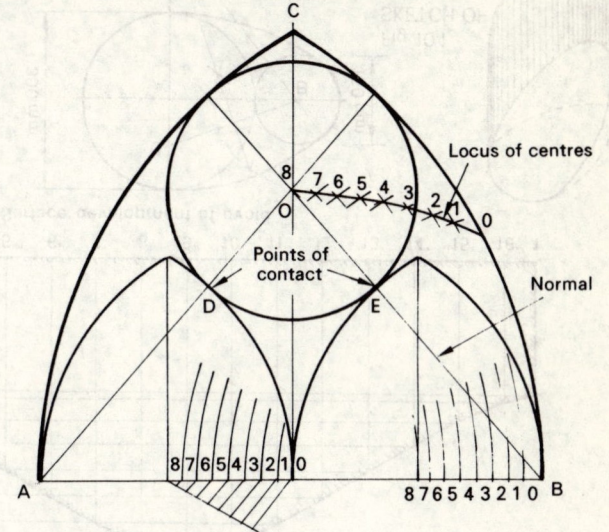

Locus of centres

Points of contact

Normal

LOCI APPLIED TO TRACERY

Fig 113

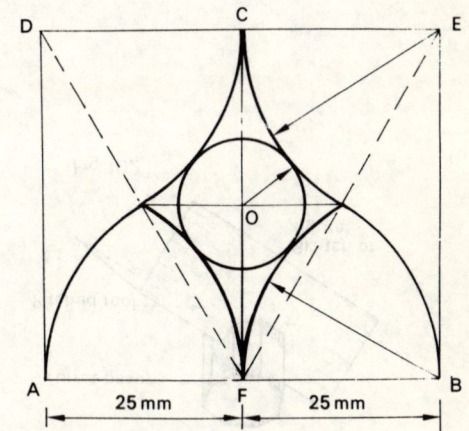

25 mm   25 mm

Fig 114 EQUILATERAL ARCH

APPLIED GEOMETRY

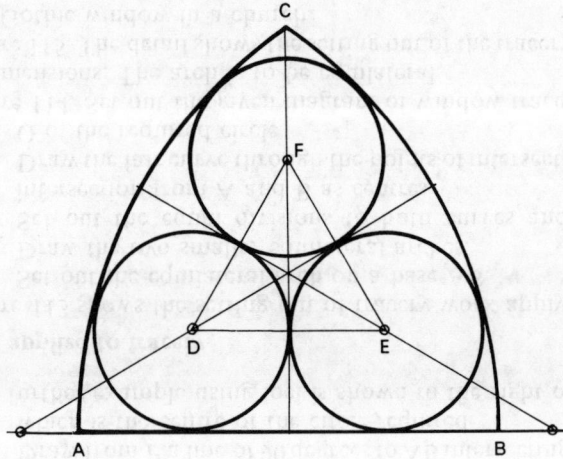

Fig 116 POINTED ARCH

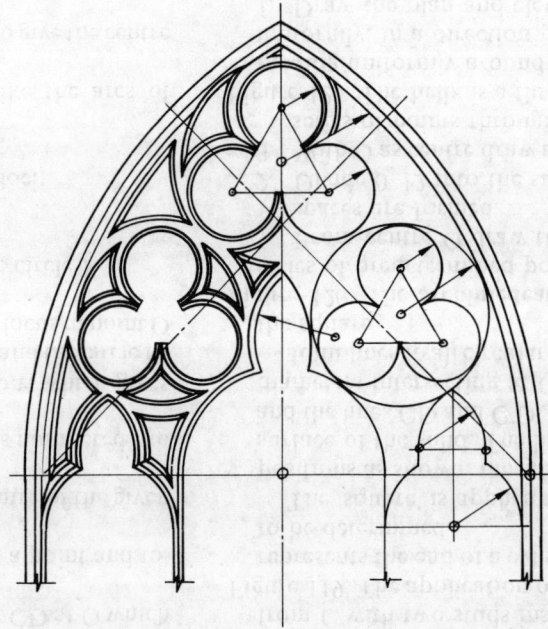

APPLICATION OF TRACERY TO GOTHIC WINDOW

Fig 115

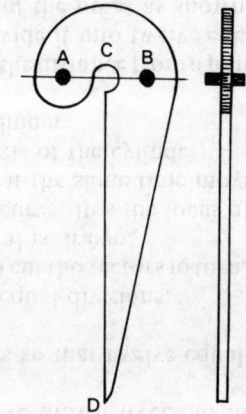

Fig 117

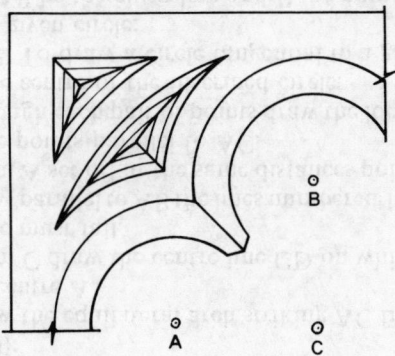

CENTRING 'SQUARE'

Fig 118

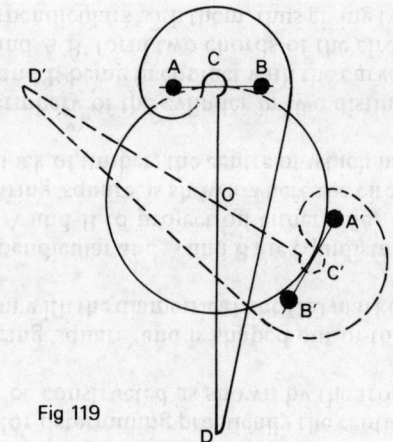

Fig 119

APPLICATION OF SQUARE

4. Through the corresponding points 1–1, 2–2, etc. draw the line which will bisect the angle CAB.
5. Draw the bisector of angle CBA as before.
6. The loci intersect at point O to give the centre of the circle.

Figure 111. To inscribe a circle within an equilateral arch using **loci** to find the centre (this is a similar problem to the previous one illustrated in Fig. 110):

1. Draw the equilateral arch striking AC from the centre B and CB from the centre A.
2. From C draw the centre line CD on which the centre for the inscribed circle must fall.
3. Draw parallel to AB the lines numbered 1 to 6 on CD at equal distances.
4. From A set off at the same distances points 1 to 6 and draw lines from these points parallel to AC.
5. Through each pair of points draw the locus to intersect CD at O which is the centre of the inscribed circle.

Figure 112. To draw a circle tangential to a given line from a point and to touch a given circle:

1. Let AB be the given line and P the point and K the centre of the given circle.
2. From A and E mark a series of corresponding distances numbered 1 to 6.
3. Draw parallel and concentric lines making intersections which gives points on the locus moving at equal distances from AB and the circle K.
4. Draw from P a line of 90 degrees to AB intersecting the locus in point O which is the centre of the circle required.

A further example using loci is shown to the right of the circle K.

Loci applied to tracery

Figure 113 shows the setting out of tracery work applying loci:

1. Set out the equilateral arch on a base AB.
2. Draw the two smaller equilateral arches.
3. Set out the equal divisions to both curves and strike the arcs of intersection from A and B as centres.
4. Draw the fair curve through the points of intersection to give the centre O of the required circle.

Figure 114. Set out the given diagram of window tracery using the figured dimensions. The arch is to be equilateral.

Figure 115. The detail shows the setting out of the tracery work at the head of a Gothic window in a church.

Figure 116. Setting out three tangential circles within two equal arcs as in tracery work:

1. Draw the equilateral triangle DEF.
2. Describe the three circles with centres D, E and F and radius equal to half the sides of the triangle.
3. Draw AB tangential to the two lower circles.
4. Produce the bisectors of the angles of the triangle to give the centres of the arcs on AB.

Figure 117. A spandrel formed in tracery work is shown where it is necessary to find the intersection line between the mouldings. The spandrel is formed by a horizontal, a curved and a vertical line:

1. Set out a series of equal and parallel distances as in the previous work on loci, from each of the three lines.
2. Through the points of intersection draw the required mitre lines. The intersection between the two curved mouldings gives a curved intersection and the intersection between two straight lines at 90 degrees give a 45 degrees mitre.

Figure 118. A useful instrument used for determining practically the centres of circles by chordal bisection may be constructed as shown by the front and edge views in Fig. 118.

The instrument is termed a centring 'square' and is shaped out of thin hardwood, hardboard, or aluminium with the diametrical line CD marked on it.

The chordal line AB is drawn perpendicular and A and B are equidistant from C with two studs inserted at A and B to project on either side.

Figure 119. The application of the centring 'square' is shown where the circle represents the end of a cylindrical block of timber, the centre of which has to be determined.

The 'square' is applied to the periphery of the cylinder in two distinct positions as shown, the studs at A and B being in contact with the curved surface of the solid. The lines AB and A′B′ form two chords of the circle and the lines CD and C′D′ make perpendiculars with them, thus giving two diameters intersecting at O the centre.

Joint lines to all circular and elliptical curves may also be obtained using the square.

Figure 120. The **archimedean spiral** is a curve drawn freehand through a series of predetermined points:

1. From centre O draw the radius vectors so that twelve equal angular spaces are formed.
2. Divide 0–12 into the same number of equal divisions.
3. With O as centre draw a series of arcs to cut the vectors in turn, giving a series of points through which the spiral is drawn.

Figure 121. The **helix** is a three-dimensional curve. It is the locus of a point moving uniformly around a cylinder and at the same time moving, also uniformly, in a direction parallel to the axis of the cylinder:

1. Draw the plan and elevation of the cylinder.
2. Divide the plan into twelve parts.
3. In the elevation, take the pitch, which is the distance from a point on the curve to the next similar point, and divide it into twelve equal parts.
4. Number the points and plot the locus of the point as shown.
5. The development of the cylinder gives the true distance moved by the point as the hypotenuse giving a straight line. Using the division points from the plan, project horizontally from the elevation of the curve to vertical projectors from the division points in development.

Figure 122. The square threaded screw shown is helical in construction. Two cylinders are to be dealt with, the outer or bolt diameter and the other with the inner or core diameter. Four helices have to be drawn, two on the cylinder with the core diameter and two on the cylinder with bolt diameter:

1. Set out the helix as in the previous example.
2. Measure pitch distances from a number of suitable points and repeat

APPLIED GEOMETRY

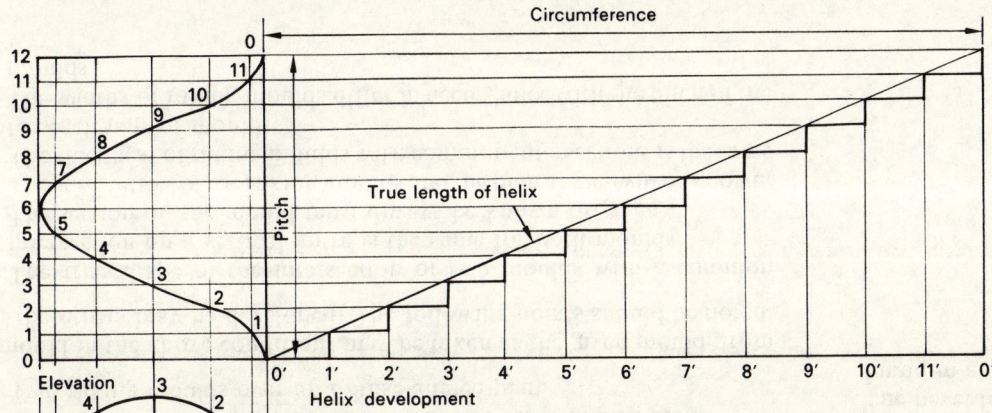

Circumference

True length of helix

Pitch

Elevation

Helix development

Fig 121

Plan

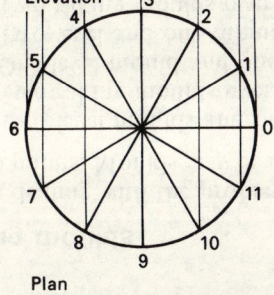

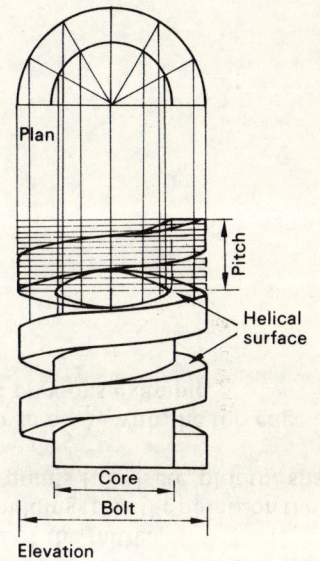

Plan

Pitch

Helical surface

Core

Bolt

Elevation

Fig 122

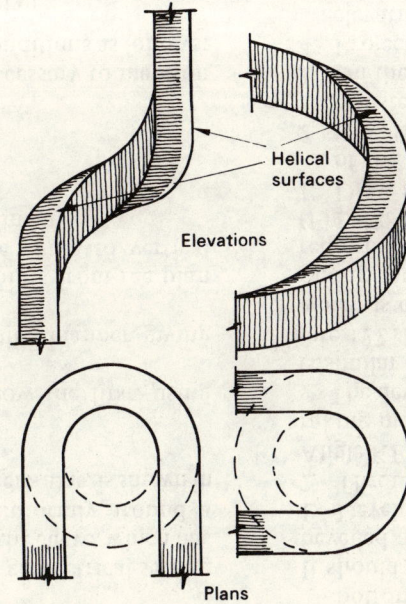

Helical surfaces

Elevations

Plans

Fig 123

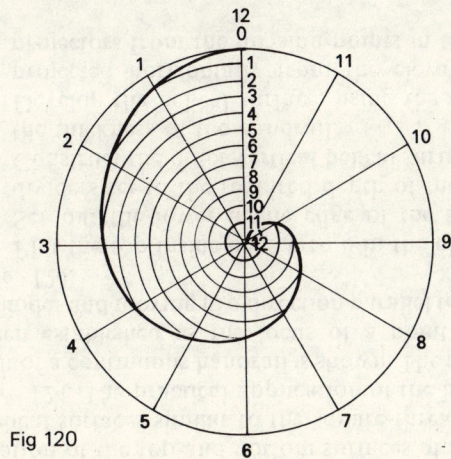

Fig 120

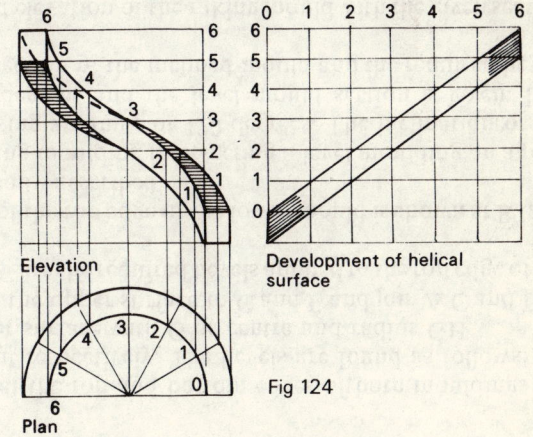

Elevation

Development of helical surface

Plan

Fig 124

the helix construction already set out.

In the example shown, the core has been omitted from the first turn to indicate the appearance of a spring of the same size.

Figure 123. A further example of the helix in handrailing shows the formation of the top and bottom surfaces of the wreath. These surfaces are helical surfaces similar to the square-threaded screw.

Figure 124. The practical application of the helix in the geometrical setting out of a continuous handrail is shown. The plotting of the curve, which has been established as the locus of a point moving uniformly around a cylinder and moving in a direction parallel to the cylinder's axis, is shown in Fig. 124.

1. Plot the top inside edge 0 to 6 in the elevation.
2. Set out the lower inside edge of the surface below the first, using dividers set to the required depth of the rail.
3. Construct the outer vertical helical surface in a similar manner, giving the thickness of the handrail.
4. Develop the helical surface, using the division points from the plan projected horizontally from the elevation of the curve to vertical projectors from the division points in the development.

Raking moulds

Before dealing with the problems on this subject, it is necessary to mention certain points. Moulds may be required to suit varying conditions as follows:

1. Two level moulds and one raking over a square plan.
2. Two raking moulds over a square plan.
3. One level mould and one raking over an obtuse-angled plan.
4. Two level and one raking over obtuse- and acute-angled plans.
5. Two raking moulds over an obtuse-angled plan.

Any mould in the above conditions may be taken as the given mould, from which the others may be developed. The following points should be noted:

1. The true shape of the intersection of two moulds with a common intersection on a vertical mitre is the same for both moulds.
2. The section of one mould must always be known or given.
3. The inclination of the raking moulds and the plan angles will be known.
4. The thickness of raking moulds with common intersections is the same for each pair of moulds.
5. The widths of raking moulds differ in accordance with the pitch of the moulds.

Raking mouldings

Figure 125 shows a raking moulding on a square plan mitred with level mouldings at the top and bottom. To find the required bevels and the mould sections, Fig. 126:

1. Draw the plan and elevation of the raking mould with the given section of the raking mould on the elevation.

2. Set off the distances 1, 2, 3, etc. on the given section at both the top and bottom moulds.
3. From these points, draw perpendiculars to intersect with corresponding elevation lines or arrises of the moulding.
4. Through the intersections plot the shape of the sections of the top and bottom moulds.

It should be clear that the top and bottom edges of both mouldings are bevelled and undercut respectively. The bevels are found as follows:

1. Develop the upper surface with G as centre and radius GH.
2. Through K, draw the upper surface to C and E and join A'C and DE. Angle KED and KCA' are the required bevels applied to the top edge of the raking moulding.

The development of the top edge of the lower mould is shown at B" and is similar to that already described.

Figure 127 shows a raking moulding intersecting a level moulding on a plan which is obtuse, making an angle of 120 degrees. The inclination of the raking mould is 30 degrees and the level mould section is given. It is required to find the section of the inclined mould and the required bevels (Fig. 128):

1. Draw the plan and elevation of the raking mould with the given section of the level mould on the elevation.
2. Set off the distances 1, 2, 3, etc. as before on the given section, and transfer these from a normal to the slope, drawn from G.
3. On the plan of the level mould, draw ordinates onto the mitre line AB.
4. Project these points from A to intersect with the arrises drawn in elevation to complete the elevation of the mitre.
5. Draw the elevation of the raking moulding from the points on the mitre and where the arrises intersect with points 1, 2, 3, etc. plot the shape of the required inclined moulding.

The bevel for the inclined mitre is shown at A, with the top edge bevels found in a similar way to those in the previous example.

APPLIED GEOMETRY

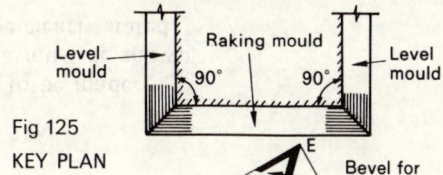

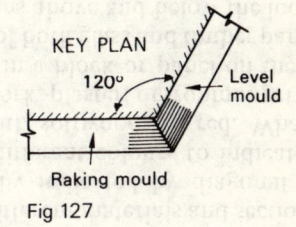

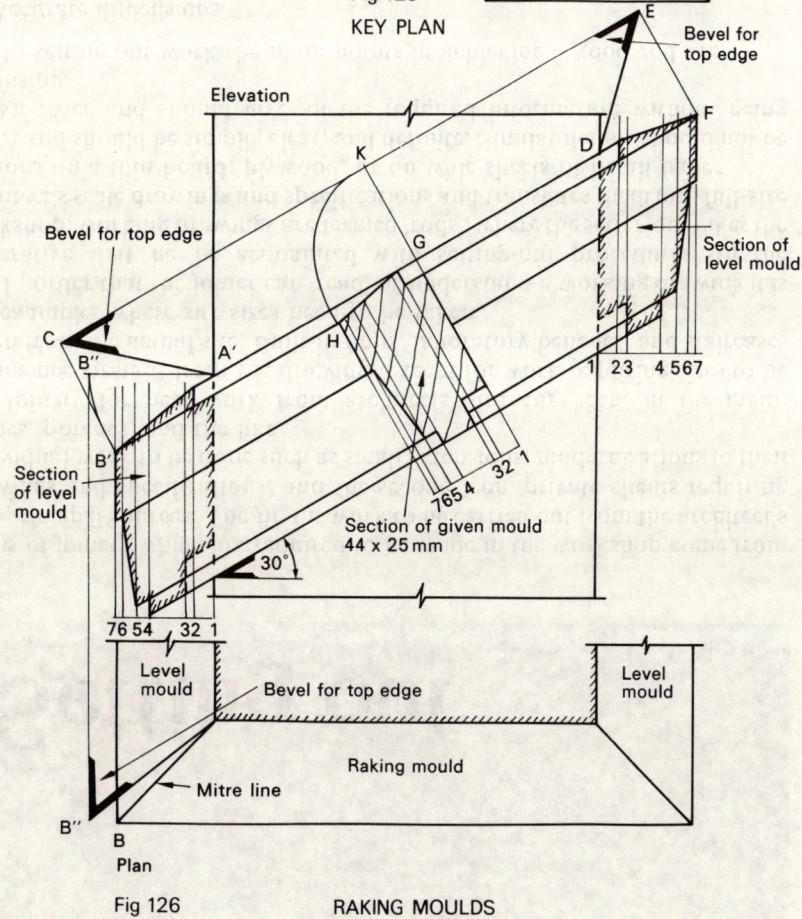

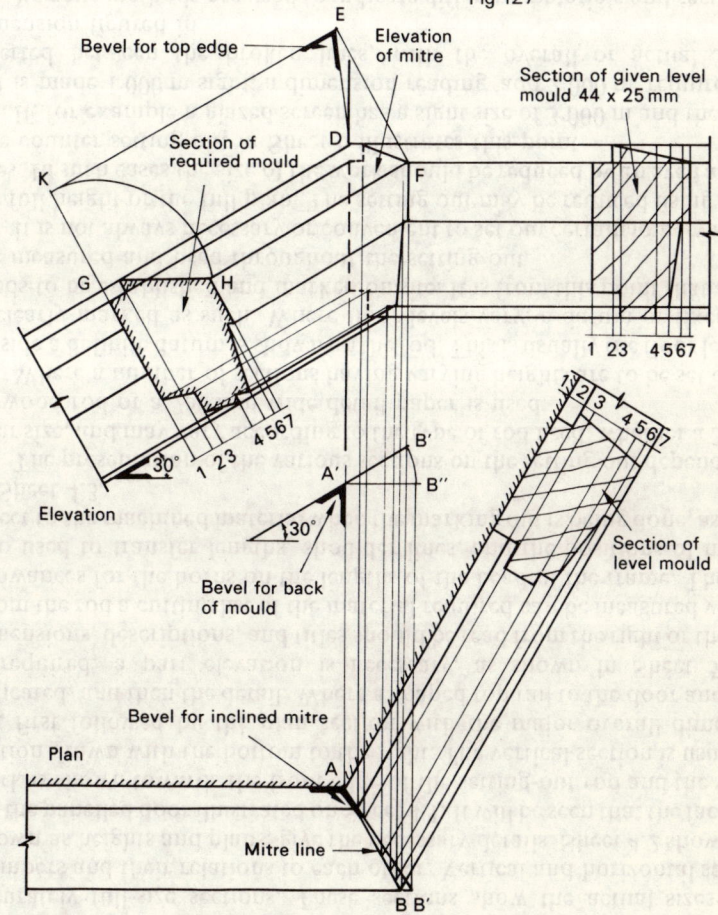

Level mould Raking mould Level mould
90° 90°

Fig 125
KEY PLAN

KEY PLAN

120°

Level mould

Raking mould

Fig 127

Elevation

Bevel for top edge

Bevel for top edge

Section of level mould

Section of given mould 44 x 25mm

Mitre line

Bevel for top edge

Level mould

Level mould

Raking mould

Plan

Fig 126   RAKING MOULDS

Bevel for top edge

Elevation of mitre

Section of given level mould 44 x 25 mm

Section of required mould

Bevel for top edge

Elevation

Bevel for back of mould

Bevel for inclined mitre

Plan

Mitre line

Section of level mould

Fig 128   RAKING MOULDS

113

Chapter 12

Setting out

Items of joinery which are required to be made in the workshop come from two principal sources. The first is work to be carried out from the architect's drawings and specifications, and the second from private clients requiring individual work to be done such as small extensions, modernisations to their homes, porches, and the like.

Joinery for new work from architects' drawings can, in the main, commence straight from the drawings except for work requiring sizes to be taken from the actual site. Built-in units, laboratory benches, and staircases are examples where site sizes need to be taken.

In order that the joiner can 'read' or understand a working drawing it is imperative that he be acquainted with setting-out procedures. In the workshop, working drawings are termed 'rods', where the setter-out takes the architect's scale drawings and specifications and translates them into full-size sections on a thin board, plywood, or on wide sheets of detail paper.

A rod should be simple, clear, and definite. Similar details should not be drawn twice and should give all the required information without being confusing.

In setting out work the main points in achieving a good rod are:

1. Accurate dimensions.
2. Clarity.
3. Allowances for any fixing need to be made.
4. Construction details of the work must be shown.
5. Shopwork and sitework must be clearly stated.

It will be clear from the above that a setter-out must have a clear understanding of orthographic projection. He must also have an understanding of his material and an appreciation of design and construction of joinery items. A sound knowledge of the use of woodworking machinery in respect of its aids to the joiner, its production and limitations is very necessary.

The setter-out must also have an appreciation of the required allowances for fitting and fixing of the joinery on site. Special considerations need to be given to work of large sizes, as in wall panelling, from the point of view of handling, transporting, and getting the framing upstairs and through doors.

Finally the ability to be able to visualise the form the completed work will take, carrying the job through its various stages of construction to completion to the satisfaction of the architect.

It has been said that the setting-out on the rod consists of drawing accurately full-size sections. These sections show the actual sizes of the members and their relations to each other. Vertical and horizontal sections, known as heights and plans, give the necessary details. Sheet 4.2 shows a rod for the panelled door illustrated on Sheet 4.1. It will be seen that the face of the work is shown towards the front edge of the setting-out rod and the vertical section drawn with the bottom to the right. The vertical section is usually set out first followed by the plan section with the major overall dimensions indicated, and then the detail. Where a shaped top rail to the door and frame is required, a part elevation is necessary, as shown in Sheet 5.1. All dimensions, descriptions, and titles should be read from the right or the front. From the rod a cutting list of the material required can be measured with any allowances for the horns on the lengths of the head of the frame. The rod is also used to transfer lengths, shoulder lines, and the positions of mortises direct to the machined material when the marking out is being done, as shown in Sheet 4.3.

The presentation of the various sections on the setting out depends upon their size, and may vary according to the type of rod used, whether a 300 mm plywood rod or a 750 mm wide detail paper is used.

Where a number of sections having varying heights are to be set out side by side a definite datum is shown on the rod. This is usually the floor level and is clearly marked as such. Where floor levels vary, a datum or fixed point needs to be established and marked on site. It is from this point that all sizes are measured and used throughout the setting out.

It is not always necessary or convenient to set out certain joinery items to the full height or the full plan. The setting out may be reduced using broken lines. In such cases the size of the work should be reduced by an even amount. The counter setting out in Sheet 6 illustrates this point.

If, for example a glazed screen has a sight size of 3·000 m and the setting out is made 1·000 m sight, a dimension reading 'add 2·000 m' requires to be inserted between the broken lines, with the overall or actual 3·000 m dimension figured in.

Various methods are used to indicate different materials and sections on rods. Groundwork of softwood are usually indicated by diagonal lines. Othere sections are hatched, often using different colours to indicate end grain, i.e. hardwoods hatched in black with softwoods in red. Whatever colours are used for timber sections, brickwork, plaster, or ironmongery, it is advisable to indicate the colour code used in a block or panel on the rod.

In items of work which involve the use of both glass and timber panels as in screens and semi-glazed doors, the sections above and below the lock rail may vary. This can occur either in the height or the plan section, and in setting out one section may be superimposed on the other as shown in Sheet 5.1. Using this method the change of section can be seen immediately from the rod.

Sheets 29 and 30 deal with swing doors. The architect's drawing will indicate the form and type of hinge or floor spring to be used in the schedule

SETTING OUT

746

2·040

MARKING OUT OF STILE

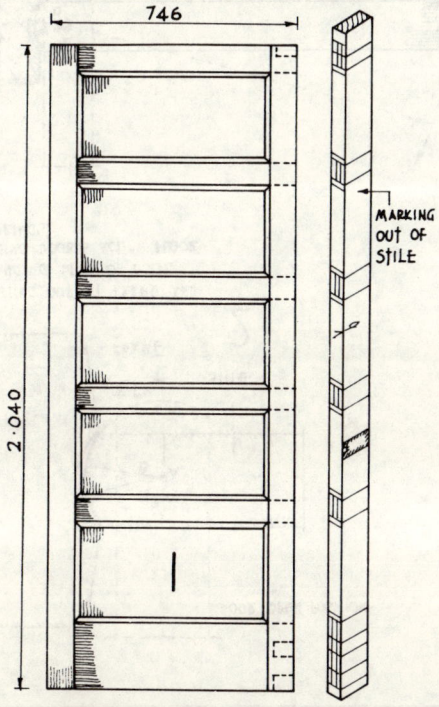

ELEVATION
FIVE PANELLED DOOR

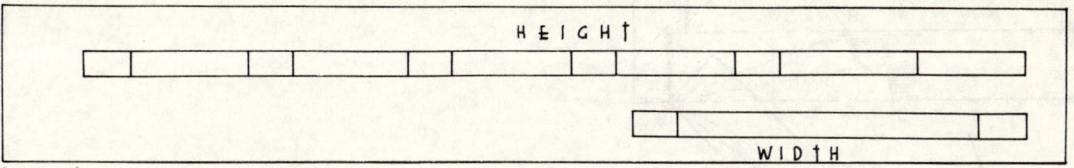

D

D/4

D/4

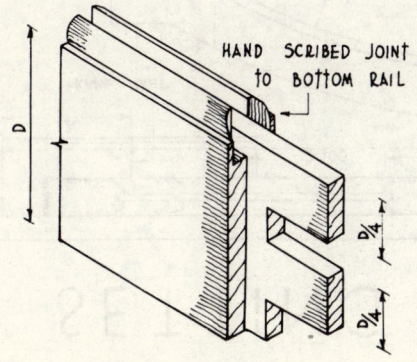

HAND SCRIBED JOINT
TO BOTTOM RAIL

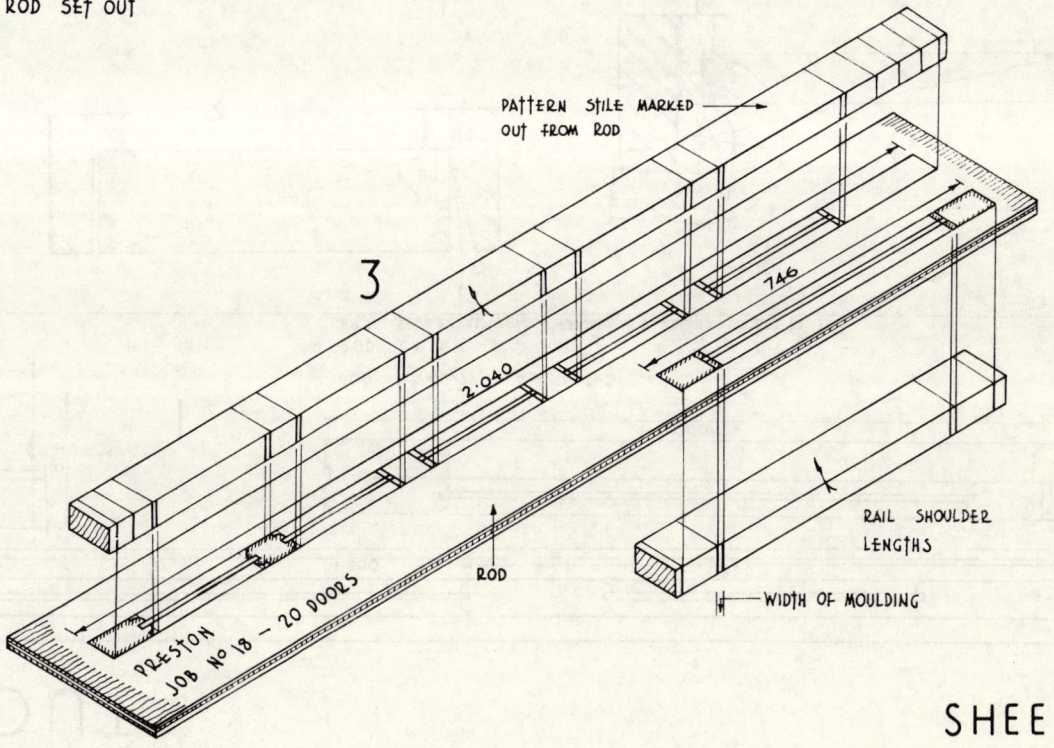

HEIGHT

WIDTH

FIRST STAGE IN SETTING-OUT ROD

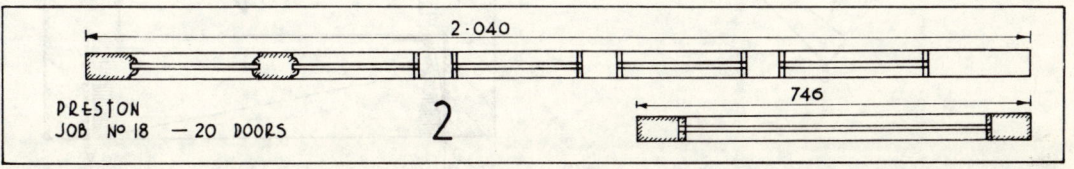

2·040

746

PRESTON
JOB Nº 18 — 20 DOORS

ROD SET OUT

2

3

PATTERN STILE MARKED
OUT FROM ROD

746

2·040

ROD

PRESTON
JOB Nº 18 — 20 DOORS

RAIL SHOULDER
LENGTHS

WIDTH OF MOULDING

SHEET 4

SETTING OUT

1

JOB Nº 42/77   A.E. YATES.
ROD   Nº 18
Nº 5 SEGMENTAL HEADED FRAMES & PANELLED DOORS TO THIS.

FRAME JOINT

SEGMENTAL HEAD
TO FRAME & DOOR

2

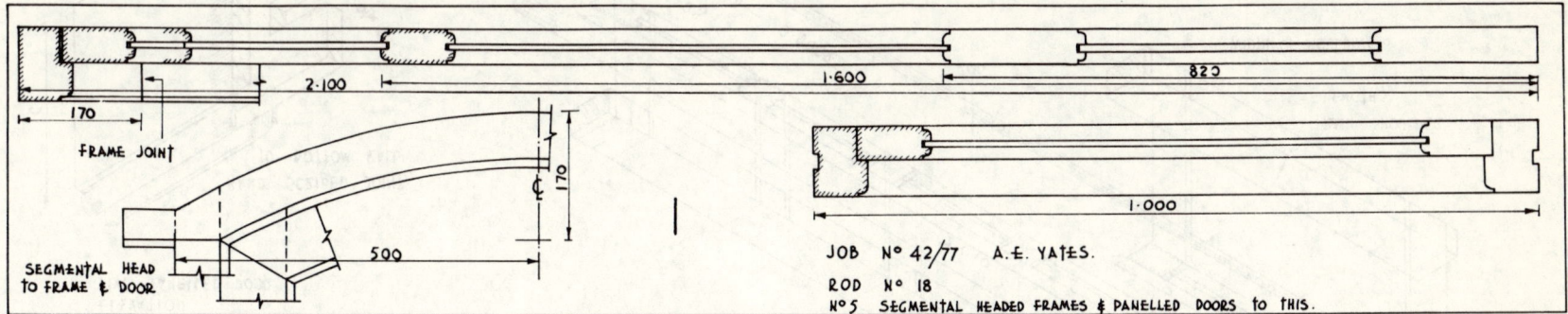

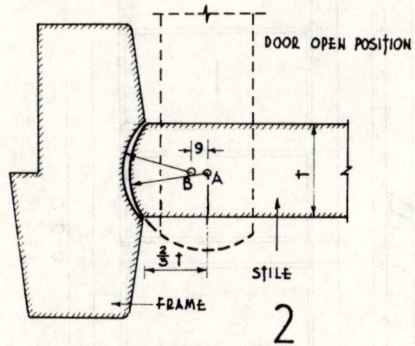

SETTING OUT AT FRAME AND
HANGING STILE OF DOOR
USING DOUBLE ACTION FLOOR
SPRINGS

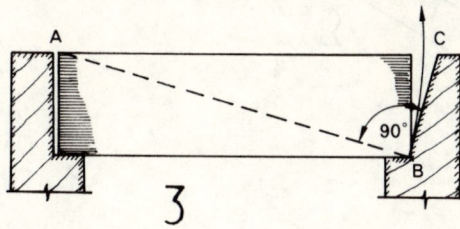

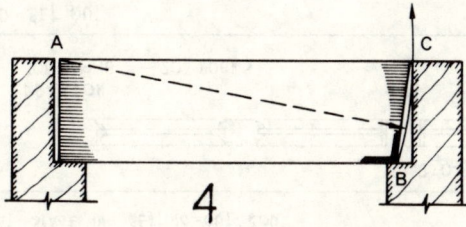

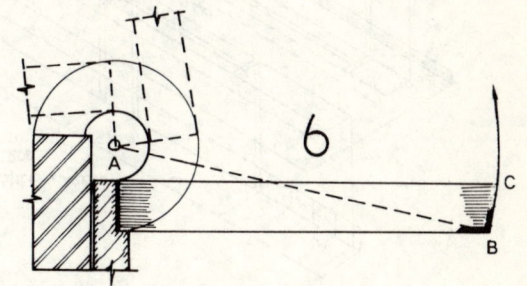

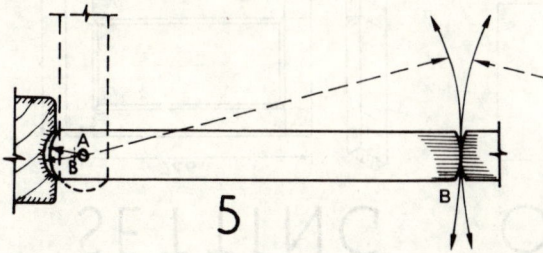

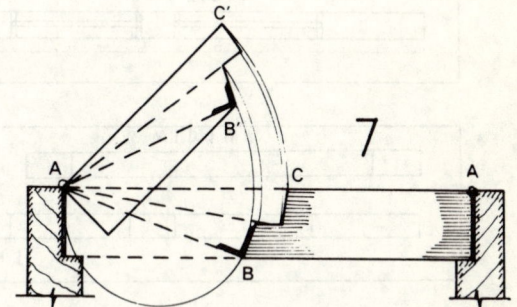

DOOR OPEN POSITION

STILE

FRAME

3

4

5

6

7

SHEET 5

SETTING OUT

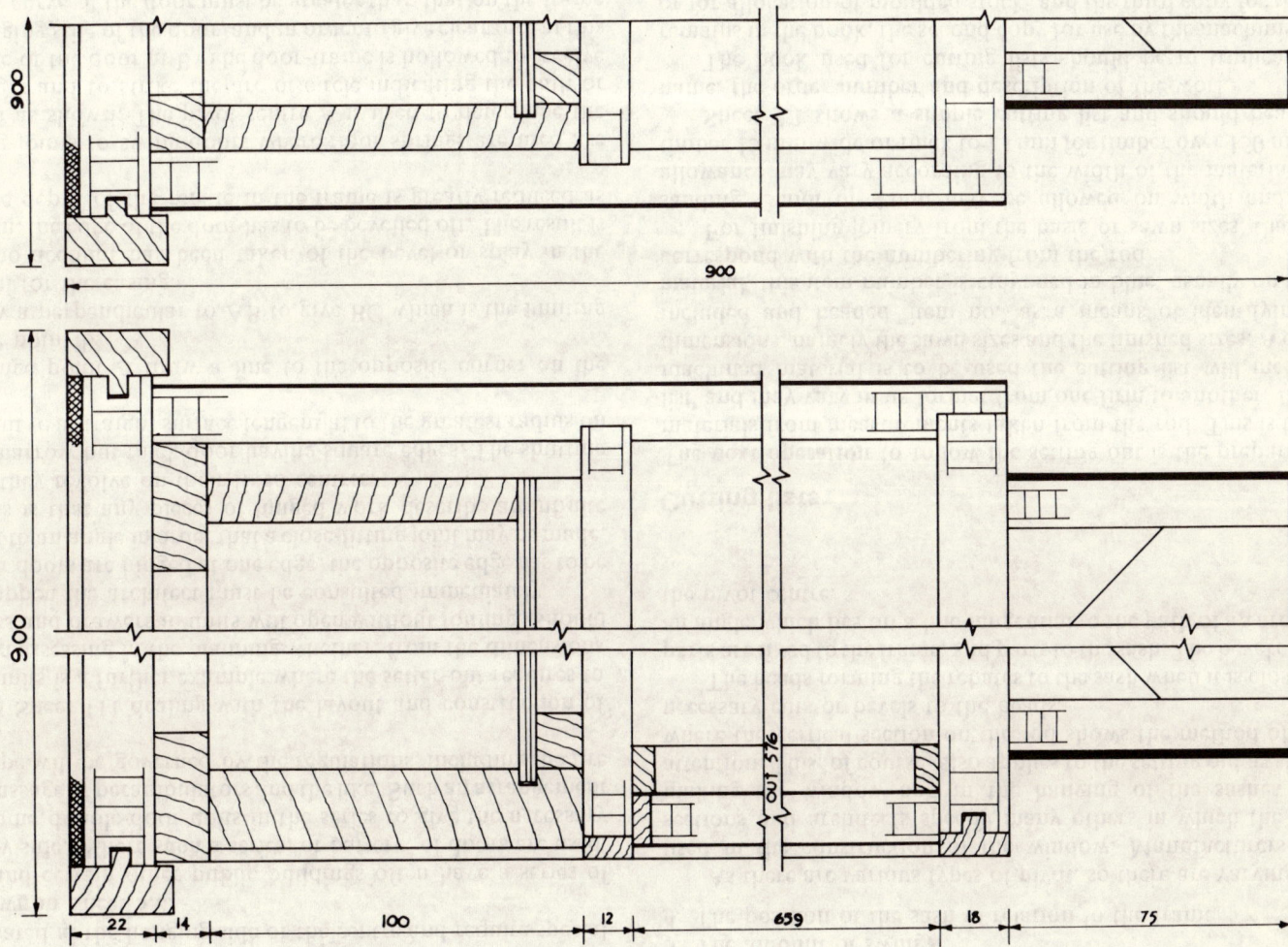

SHEET 6

118

of ironmongery. When setting out the doors and frames special attention needs to be paid where both single-action or double-action swing doors are required. Where single-action springs are used on double doors, a rebated frame may be used with the meeting stiles rebated or left square. 'Pull' or 'push' plates are used on both doors where no centre rebate is used, and on one door only when they are rebated. In the case of doors rotating through 180 degrees, double-action floor springs with top centre pivots are used. The frames are not rebated at the hanging side or the centre and require special setting out, as shown in Sheet 5.2.

Large stores and certain other public buildings often have a series of swing doors side by side. Where such a series, or battery, of doors are used, there should be some double-door units in the series to give the necessary clearance for the passage of perambulators and the like. Such an arrangement of doors of this type will be governed by the regulations, including the fire regulations.

The project on Sheet 111 dealing with the layout and construction of counters and wall units is a further example where the setter-out requires to exercise foresight in assessing in the planning whether, from the dimensions and spacings, doors and drawers in units will open without fouling. Should this be likely to happen the architect must be consulted immediately.

When sashes or doors are hinged at one edge, the opposite edge has to be bevelled or splayed to an angle in order that a close-fitting joint may be made. The reason for this is that any pieces of hinged work describe an infinite number of loci as they revolve on their fixed centres.

Sheet 5.3 shows a narrow but thick door having square edges. The shutting joint BC is set out to a straight surface tengential to the greatest radius on the framing:
1. From the hinge point A draw a line to the opposite corner on the shutting edge point B.
2. From B draw a perpendicular to AB to give BC which is the limiting angle or bevel for the casing.

Sheet 5.4. When no account has been taken of the bevel or splay in the setting out shown, the edge of the door has to be bevelled off. The result is that the effective depth of the rebate in the frame is greatly reduced as shown at B.

Sheet 5.5. Shutting joints to swing doors where floor springs are used are normally curved as shown. The pivot centre A is used to determine the curved door edges and to strike the arc of circle indicating the path or travel of the edge of the door at B. The door-frame is hollowed to receive the rounded hanging stile of the door, and in order to give clearance at this point, the radius curve of the door must be greater than that on the frame to prevent rubbing.

Sheet 5.6. When a sash or door is hinged to open outwards over a reveal, a parliament hinge having its pivot at A outside the framing is used. This type of hinge increases the bevel at the shutting edge due to the pivot position.

Sheet 5.7. Thick, narrow folding doors require to have the joints on the shutting edges set out and made to fit correctly. For reasons given in connection with the previous work, doors having the usual square rebates on the closing stiles would not open. Setting out should follow the previous examples.

Pivot-hung windows, their design and construction are dealt with elsewhere in this book.

The construction of the window is governed by a number of factors:

1. The type of pivot,
2. The method of opening – horizontally or vertically,
3. The amount of swing,
4. The position of the sash in relation to the frame.

As there are various types of pivot, so there are varying timber sections used in the construction of the window. Manufacturers have their own sections and architects specify many others in which the joiner's work in making the window and in the hanging of the sashes requires special attention. This, of course, also applies to the setting out as shown in Sheet 54 where the vertical section on the rod shows the method of determining the necessary cuts or bevels to the beads.

The beads forming the rebates to the sash when it is closed are cut so that parts are fixed to the frame, and parts to the sash. The bevel cut on the beads is an angle which lies on a line tangential to the path of an arc described about the pivot centre.

Cutting lists

The next operation to follow the setting out is the preparation of a list of materials from measurements taken from the rod. This is termed a 'cutting list' and may vary in its format from one firm to another. In all cases where machined material is to be used the cutting list will include two sets of dimensions, namely the sawn sizes and the finished sizes. A column is usually included and headed 'item no.' as a means of identifying each piece of material, this item number is crayoned in blue, usually on the end grain, to correspond with the numbering from the rod.

For finishing joinery from the basic or sawn sizes, that is, planing and sanding, 3 mm or 4 mm may be allowed on width and thickness. This allowance may vary according to the width of the material from 7 mm for timber 15 mm wide or thick to 13 mm for timber over 150 mm wide or thick.

Sheet 7.1 shows a simple cutting list and should bear the customer's name, the order number and description of the work.

The book used for cutting lists should be in triplicate: the top copy remains in the book, the second copy for use by the machinist in the saw-mill or for allocation of moulded stock, and the third copy for the costing clerk's use. This last copy is used in the office for the purpose of costing.

In getting out material from the cutting list, whether it is in the sawn state or manufactured in the form of mouldings, employees involved must be made aware of the importance of working to the sizes and quantities specified. Unless these are strictly adhered to, the final cost analysis made by costing staff will not be accurate. Wasteful cutting, i.e. exceeding the waste allowance, can have serious results in manufacturing costs. Any offcuts from ripping or pieces left by crosscutting should therefore be of such a size that they will easily be used for normal requirements.

CUTTING LISTS

1

CUTTING LIST

DESCRIPTION	JOB Nº	DATE
Nº 6 SINGLE PANEL DOORS	42	2: NOV.
ALL DIMENSIONS METRIC		

NO	DESCRIPTION	L	W	T	FIN. SIZE W	FIN. SIZE T	REMARKS
12	STILES	2·100	100	50	94	44	SOFTWOOD GROOVED & MOULDED
6	BOTTOM RAILS	850	200	50	194	44	Do
6	TOP RAILS	850	100	50	94	44	Do
6	PANELS	1·700			630	6	PLYWOOD

SIMPLE CUTTING LIST FOR
SINGLE PANEL DOORS SHEET

2

CONTRACT:						DATE:			
ORDER NUMBER:			DESCRIPTION :						

No.	SAWN SIZES Length	SAWN SIZES Width	SAWN SIZES thick	FINISH M/C SIZES Length	FINISH M/C SIZES Width	FINISH M/C SIZES Thick	Description.	Section	Materials

ALTERNATE LAYOUT OF CUTTING SHEET

3

CUTTING LIST

DESCRIPTION:	JOB Nº:	CONTRACT:	DATE:
PIVOT & CASEMENT WINDOWS	18	BOLTON CORP⁄	20:11:77
ALL DIMENSIONS METRIC			

INSTRUCTIONS CODE:

be	both ends	tr	trenched
c	cut to dead length	tg	tongued
d	dovetailed	v	vent
h	housed		
k	combed		
m	mortised		
nc	not central		
oe	one end		
s	scribed		
t	tenoned		

PIVOT & CASEMENT WITH VENT LIGHTS

10 OFF 1·578 1·795

ITEM Nº	MEMBER	MAT.	Nº OFF	FINISHED SIZES L	FINISHED SIZES W	FINISHED SIZES T	TOTAL	INSTRUCTIONS
1	SILL	S/W	10	1·920	70	70	19·200	m. be.
2	HEAD	"	10	1·920	70	56	19·200	m. be.
3	JAMBS	"	20	1·388	"	"	27·760	t. be. m RH for tran.
4	MULLION	"	10	1·388	"	"	13·880	t. be. m for tran.
5	TRANSOM	"	10	630	"	"	6·300	t. be.
6	STILES PN.	"	20	1·280	54	46	25·600	k. be. to rails
7	RAILS TOP.	"	10	1·130	"	"	11·300	Do k. be. to stiles
8	" BOT.	"	10	1·130	"	"	11·300	Do Do
9	STILES VENT	"	20	360	46	41	7·200	k. be. to rails
10	RAILS	"	10	540	"	"	5·400	Do k. be. to stiles
11	STILES "	"	10	950	"	"	9·500	Do k. be. to rails
12	RAILS TOP	"	10	540	"	"	5·400	Do k. be. to stiles
13	" BOT.	"	10	540	70	"	5·400	Do Do
14	SUBSILL	"	10	1·920	70	46	19·200	tg.
15	BEADS	"	20	1·280	44	35	25·600	
16	"	"	20	1·130	"	"	22·600	Do
17	WEATHER Mᵈ	"	10	1·920	44	33	19·200	tg.
18	"	"	10	540	50	19	5·400	

COMPLETED CUTTING LIST FOR
WINDOWS SHEET

SHEET 7

An additional allowance is made on members of frames which have to be fitted, door stiles and rails are examples. Members of larger and longer sections like transoms and sills require a larger planing allowance. A finished size of 162 mm × 62 mm would be ordered out of 175 mm × 75 mm on the cutting list.

A minimum of 25 mm is usually allowed on lengths up to 1·800 m rising to 150 mm on lengths of 4·500 m and the like.

Where mouldings are required to be used, these are taken in convenient lengths and not as separate pieces, to allow cutting and mitring to be done from the length of moulding.

Glass is ordered by stating the size in millimetres and giving the height dimension first. A 6 mm cover is usually allowed with a greater allowance on larger sheets. Examples: 3 mm clear sheet glass; 6 mm rough cast; 6 mm Georgian wired cast; 6 mm clear plate glass; 4 mm clear sheet glass.

Further examples of cutting lists are shown on Sheet 7.2 and 7.3. It will be seen that as much information as possible is given, along with a column to provide quantities of the various members for costing purposes.

Site measurements

In considering the responsibilities of the setter-out, reference has been made to joinery requiring sizes to be taken from the actual site and the need for making the required allowances for fitting and fixing of the items within the building.

Before taking actual measurements, a draft plan of the area noting any projections, changes in levels, heating pipes, and the like will be found useful. Measurements can then be taken and dimensions figured in on the plan. Heights of rooms, door, and window openings can be entered on the plan in small circles. Attention should be given to wall angles, these can be checked for accuracy by diagonals along with walls for being plumb to ensure that true sizes are taken.

Levelling needs to be carried out with a level at least 1 m long and a straight edge. Work on the larger sites is often carried out using levelling instruments which give a saving in time.

The equipment necessary for the purpose of taking site measurements and preparing the setting-out rods in the shop later include: a generous sized setting-out bench or table having a flat solid top with a straight bench edge from which to use the tee-square. This may have a parallel blade which can be set square by the insertion of metal pins, or adjusted to any angle for bevelled work by a thumb-screw, having first removed the metal pins. Instruments required include 300 mm, 45-degree, and 60-degree set-squares; trammel pins, french curves, drawing pins, and pencils. For site use, along with a small variety of joiner's tools, a metric folding boxwood rule, 2 m and 3 m steel tapes, a suitable level, straight edge, and laths will be found useful.

Chapter 13

The fire endurance of timber structures

This is generally very considerable. Nevertheless, wherever timber is suggested for use in a structure, the fact which tends to be considered to the exclusion of all others is that wood burns, whereas alternative structural materials do not. Yet timber structures under fire will generally fail less quickly than comparable structures of unprotected metals or pre-stressed concrete structures, both of which are rapidly affected by heat.

In short, fire resistance is a property distinct from combustibility and must be considered separately. These points should be noted in this connection:

1. The strength of timber is not affected by heat: loss of strength is directly proportional to the reduction of cross-section due to the effect of charring which occurs at a steady rate regardless of temperature.
2. The expansion of timber in a fire is very low and side pressure by timber roofs or floors on the external walls does not develop: hence the building does not fracture or collapse in case of fire.
3. The high fire endurance of timber structures allows evacuation of occupants, a long period of salvage and comparatively safe access for fire-fighting.
4. The risk of fire breaking out and spreading is largely determined by the use and contents of a building and not by the materials from which it is constructed. There are other considerations in relation to the fire hazard – for instance, the risk of outbreak of fire and that of the rapid spread of flame in a building.

It is useful to remember these points in this connection:

1. Since the shell of the building is often undamaged, repairs and replacement of burnt-out timber members are easier and less costly than total rebuilding required in the case of collapsed buildings.
2. The amount of burnable materials used in construction is small compared to the fuel content introduced by occupancy.
3. Rapid spread of flame is associated not so much with the structure as with

inflammable contents or with wall and ceiling linings of combustible materials having a high rate of spread of flame. Linings of timber or timber derivatives (such as fibre boards) can, however, easily be treated to bring their spread of flame classification up to the best class.

Glued laminated timber structures

Lamination is the process of building up comparatively thin pieces of timber into larger members which may be either straight or curved. The pieces are bonded together with glue under pressure to produce a structural member of such size, length, shape, and strength as may be required for the purpose. The grain of the laminates is parallel, in contrast to plywood in which the grain alternates with each lamination.

The laminates may be arranged in one of two ways, vertically or horizontally. Vertical lamination is seldom used, and the great majority of structural components are laminated horizontally.

The most important advantage of horizontal lamination is that members may readily be curved. It allows the design of segmental or parabolic arches which are in themselves economical structural forms. It also simplifies the design of fixed joints in such members as portal frames, thereby gaining economy through continuity in the structure. Laminated timber arches and portal frames have an immediate aesthetic appeal which is enhanced by the fine finish which can be given to the material.

Other advantages of horizontal lamination may be summarised briefly. Members may be tapered or moulded to give the greatest strength where it is most needed. Seasoning is relatively easy, due to the small cross-section of individual laminates. Components may be built up to any desired length or cross-section. Material may be chosen which would otherwise be too small to be structurally useful.

Adhesives

One on the basic requirements of lamination is that the bonding material must be as strong and durable as the timber itself. In timber structures the weakness has always been in the joints. With the developments in modern glues, joints, when tested to destruction, break in the timber itself rather than in the glue-line. The types of glue most commonly used are the caseins and synthetic resins. Generally speaking casein glues, which are derived from sour milk, are easy to mix and apply and are not over-sensitive to damp. They make extremely strong joints, but lose much of their strength when wetted and are liable to attack by micro-organisms. They should not be used if the timber is likely to attain a moisture content of 20 per cent. If moisture contents above this figure are expected, a synthetic resin adhesive should be used.

All synthetic resin adhesives are plastics, the best known being those with urea, phenolic, and resorcinol bases. They are manufactured as liquids, powders, or in dry impregnated film, and may be either thermoplastic, which soften with the application of heat, or thermosetting, which do not soften when set, using either hot or cold. Synthetic resins are immune from

attack by moulds and bacteria and are resistent to moisture. There are two principal types: the urea-formaldehyde for interior work, and the phenol-formaldehyde for exterior work with which can be included the resorcinol group.

The urea is supplied in two parts, the resin in the form of either syrup or powder and the hardener. They may be purchased together as a ready-mixed powder to which water is added to make the liquid adhesive. When they are used separately, the resin is applied to one face of the joint and the hardener to the other, the setting takes place when the two are brought together and pressure applied. Both casein and urea glues can be set at room temperature but resorcinol resin glues generally require additional heat, although many resin glues are supplied in forms that set quickly without the application of heat.

To promote setting of the resins within a reasonable time when gluing normally in the workshop, one of two methods or a combination of both is used. Heat greatly accelerates setting, as has already been mentioned. A hardener, often called an 'accelerator' or 'catalyst', is added or applied to the resin. This causes the resin to become thicker, then rubbery and finally promotes setting.

Heat is used in conjunction with hardeners when it can be conveniently applied, and produces setting within a very short time. In the furniture industry heat can be economically and conveniently applied by radio-frequency, low-voltage heating, and heated hydraulic presses to raise the glue-line temperatures, enabling joints to be made in minutes.

Casein glues in powder form have a storage life of about a year if kept in a cool dry place, and resin will remain usable for 3–6 months after manufacture under normal temperature conditions.

A quick-acting hardener shortens the length of time for which resin glues remain usable, this is known as the 'pot life'. This is usually quoted by the makers at various temperatures. As the temperature rises the pot life becomes shorter.

Glues used for building structures must be gap-filling. That is to say they must be capable of bridging gaps up to 1·25 mm in joints where the surfaces may not be in close contact owing to the impossibility of adequate pressure or any inaccuracies. These gap-filling properties are obtained by the addition to the resins of suitable fillers by the manufacturers.

Stress grading of timber is becoming more and more an essential part of timber engineering where the strength of a laminated member is governed by the minimum cross-section of clear timber in it. With solid timber, defects such as knots, shakes, splits, etc. reduce the strength of the particular member, whereas in laminated work these defects may be present in individual laminates, but they do not occur upon one another and therefore only affect the strength of that particular lamina. While lower-grade timber may be used for the centre laminates, the outer laminates should be selected clear grades.

In straight laminated members the thickness of the laminates is governed only by the need to obtain close contact at the glue-lines, and 50 mm nominal material is commonly used. With curved members, however, the timber must be bent without breaking in a dry state, unheated and with glue applied. The thickness of the laminates will depend on the radius of curvature. This should

TIMBER SHELL ROOFS

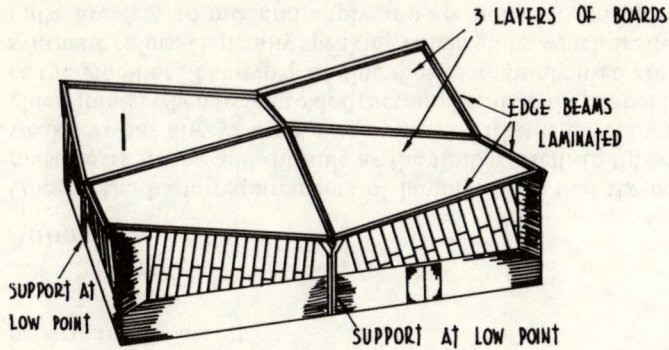

3 LAYERS OF BOARDS

EDGE BEAMS
LAMINATED

1

SUPPORT AT
LOW POINT

SUPPORT AT LOW POINT

SKETCH OF MULTIPLE HYPERBOLIC PARABOLOID SHELL

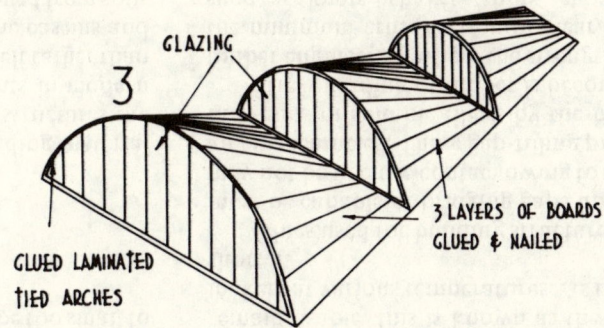

GLAZING

3

3 LAYERS OF BOARDS
GLUED & NAILED

GLUED LAMINATED
TIED ARCHES

SKETCH OF 3 CONOID TIMBER SHELLS

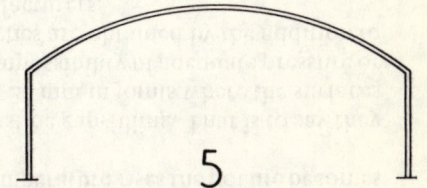

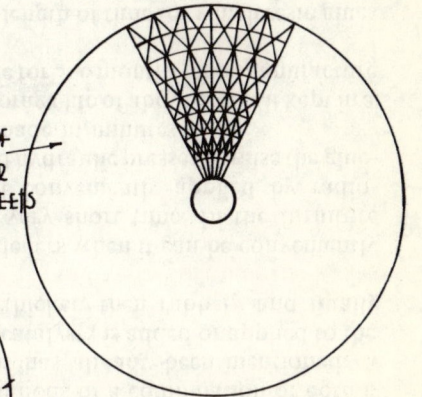

5

SECTIONS PARALLEL TO EDGES
ARE STRAIGHT LINES

SECTIONS PARALLEL TO DIAGONALS
ARE PARABOLAS

D

A'

C'

2

A

C

B

GEOMETRICAL DIAGRAM OF HYPERBOLIC
PARABOLOID SHELL

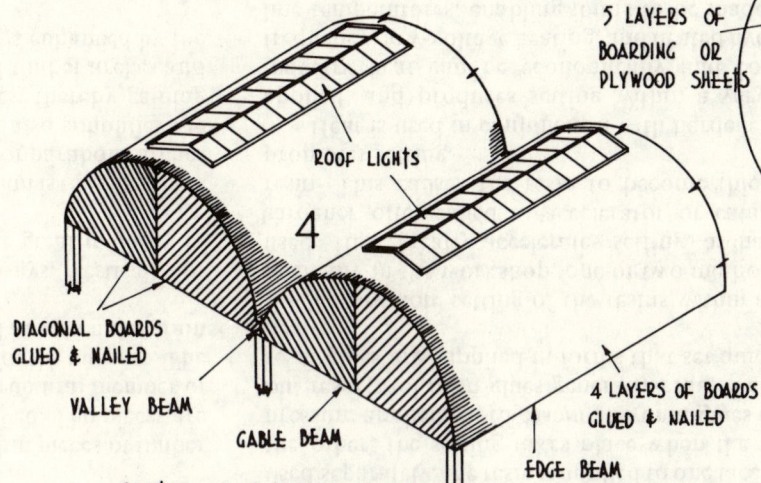

ROOF LIGHTS

5 LAYERS OF
BOARDING OR
PLYWOOD SHEETS

4

SKETCH OF GLUED LAMINATED
DOMED TIMBER SHELL

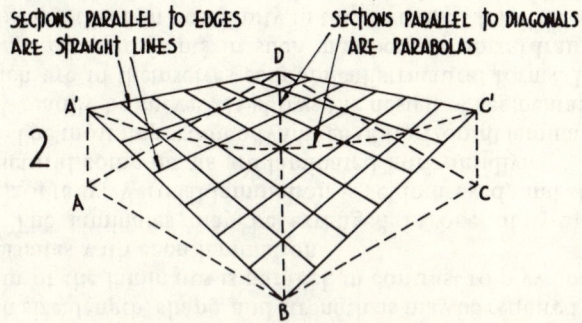

DIAGONAL BOARDS
GLUED & NAILED

VALLEY BEAM

GABLE BEAM

4 LAYERS OF BOARDS
GLUED & NAILED

EDGE BEAM

SKETCH OF 2 BAYS OF BARREL
VAULT TIMBER SHELLS

SHEET 8

be taken as 25 mm thick to 3·750 m radius for practical purposes.

In fabricating glued laminated members the moisture content of the timber is controlled at between 8 and 18 per cent, depending on the type of glue used, and where the member is to be used. The surfaces to be glued are machined to fairly accurate dimensions, and scarf joints are cut for the jointing of laminations in the same layer. Plain scarfs with a slope of not more than 1 in 12 are made on a circular sawbench adapted for the purpose. Some form of jig is necessary for curved members which may be arranged vertically in the form of a centre, or horizontally where wood or metal posts in the floor are used. The jig must be correctly shaped to the profile of the concave side of the member to be fabricated. The first laminate is fastened to the jig and thereafter the remaining laminates are passed through a mechanical glue-spreader and assembled one on top of the other.

The pressure required is normally 600–1 380 kN/m² for softwoods, and the best method of obtaining this is by using closely spaced clamps tightened with powdered nut-runners. A final squeeze-out of glue is obtained with manual ratchet wrenches. Cramps may be spaced as far as 375 mm apart, but on curved surfaces they may be as close as 100 mm. Pressure is maintained from 6 to 12 hours, depending on the type of adhesive used and the shop conditions. When the member is removed from the jig the faces are planed to final dimension and sanded. A preservative seal is applied, and, for protection, components are wrapped in waterproof paper or other suitable covering which is left in place until after erection.

Timber shell roofs

These are now being used quite frequently in modern building as a result of technological advances made in obtaining and formulating data on the new built-up structural components in timber.

A shell roof consists of a thin sheet of material which is light in weight, and rigid as a result of curvature. Where timber is used the shell of the roof will generally consist of three or more layers of boarding placed in different directions and glued and nailed together.

Shell roofs are, by their size, restricted to site construction where centring is used to form the necessary curves. A number of the different constructions of shell roofs are illustrated on Sheet 8.

The hyperbolic paraboloid timber shell roof shown in the sketch in Sheet 8.1 consists of a covering made up of four panels in which each panel is a three-dimensional surface. Sheet 8.2 shows one of the panels constructed in the following manner: four straight lines are arranged to form a square with the two diagonally opposite corners raised. The sides are divided into an equal number of parts and the corresponding points on each pair of opposite sides joined by straight lines.

Although the surface is formed of straight lines parallel to the edges in plan, it is in fact doubly curved. The cross-section through the two raised corners is a parabola (concave upwards) and the cross-section through the low corners is a parabola (concave downwards). All cross-sections parallel to the diagonals are parabolic, and have the same form, only varying in the level of the apex.

The shell need not have all four sides with the same slope. It may have two adjacent sides horizontal and two sides inclined either upwards or downwards. Alternatively it may have two adjacent sides with one slope and the other two sides a different slope. Finally all sides may have different slopes. All these shells are parts of the same geometrical surface (the hyperbolic paraboloid) provided that they are square on plan and have straight sides.

The method of support for the roof will depend on whether single panels are used or whether panels are grouped together as in Sheet 8.1. Although the load of the roof is carried at the low corners it is advisable to give some support at the high corners to limit deflection. Normally curtain walling or panel infilling is used.

Sheet 8.3 shows a second form of shell roof consisting of three conoid timber shells supported on glued laminated tied timber arches.

The Timber Research and Development Association have carried out much research in the design and construction of timber shell roofs in this country and have erected and tested full-scale models at their research laboratories.

The conoids formed in position consist of three layers of boards glued and nailed to each other with vertical glazing used in the spaces as shown, to give the necessary lighting below.

A third form of timber shell roof is shown in the sketch, Sheet 8.4. The construction is of timber barrel-vault shells carried on vertical laminated timber ribs, supported on glued-laminated edge and valley beams. The shells consist of four layers of boards glued and nailed to each other with the gables diagonally boarded. A continuous roof light is used in each shell.

A domed timber shell shown in Sheet 8.5 gives a further example of the wide possible application of this type of roof, the actual shape of the roof determined by a need for an uninterrupted floor space.

As with all timber engineering, design is the concern of the structural engineer while the carpenter's concern is that of fabrication, construction, and erection of the various members.

Sheet 9 details a number of methods used in the fabrication of straight and curved laminated members with various cramping forms for limited production and for quantity production.

Sheet 9.1 shows a method of laminating curved members using timber clamps with metal angle brackets.

A double-ended bolt with nuts and washers to pass through the timber clamps is shown in Sheet 9.2.

A simple timber saddle or jig for the limited production of laminated work is shown in Sheet 9.3.

Sheet 9.4 shows a further type of timber clamp formed from two outer pieces of material with distance pieces used at intervals, allowing the bolts to pass through as shown.

A laminating bed is shown in Sheet 9.5 and is a form of staging suitable for quantity production. Fabricated steel brackets bolted through the staging are used with alternate types of metal cramps detailed in Sheet 9.6.

Sheet 9.7 shows an alternate open-tread laminated stair with metal balusters screwed to hardwood treads and finished with a hardwood plain handrail. It will be seen that the laminated stringers are formed to support the half-space landing and are carried by the end wall.

LAMINATED WORK

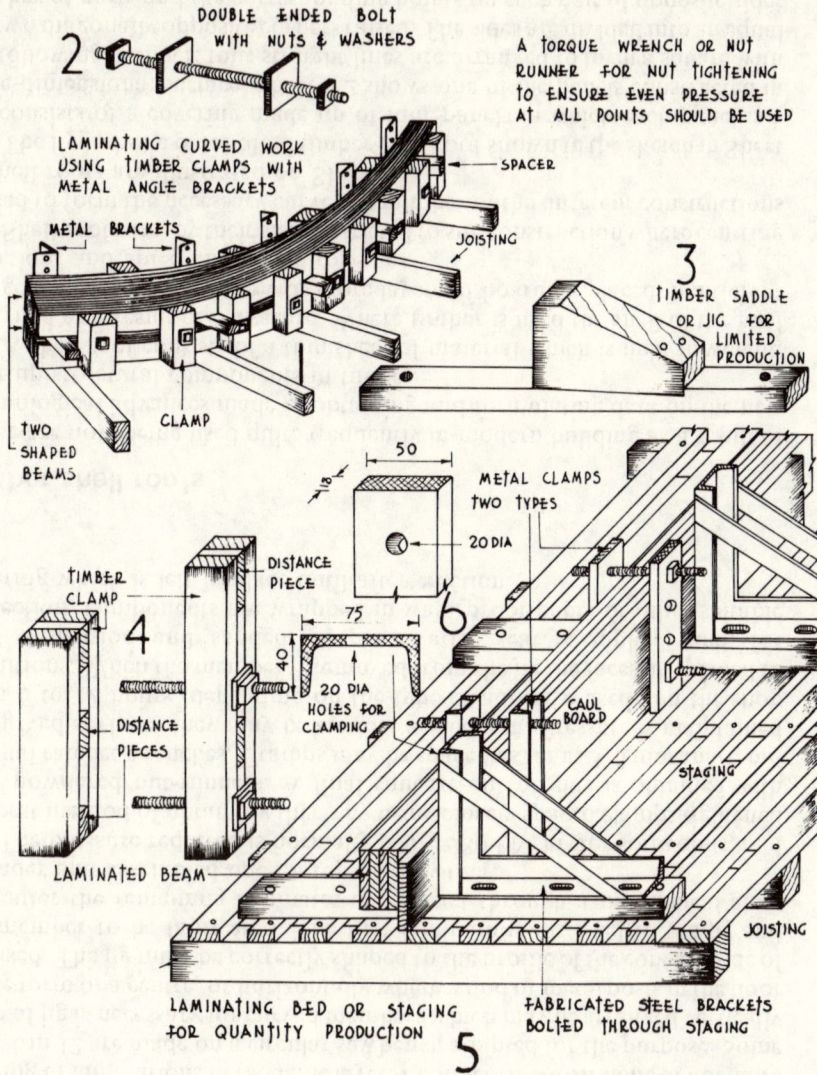

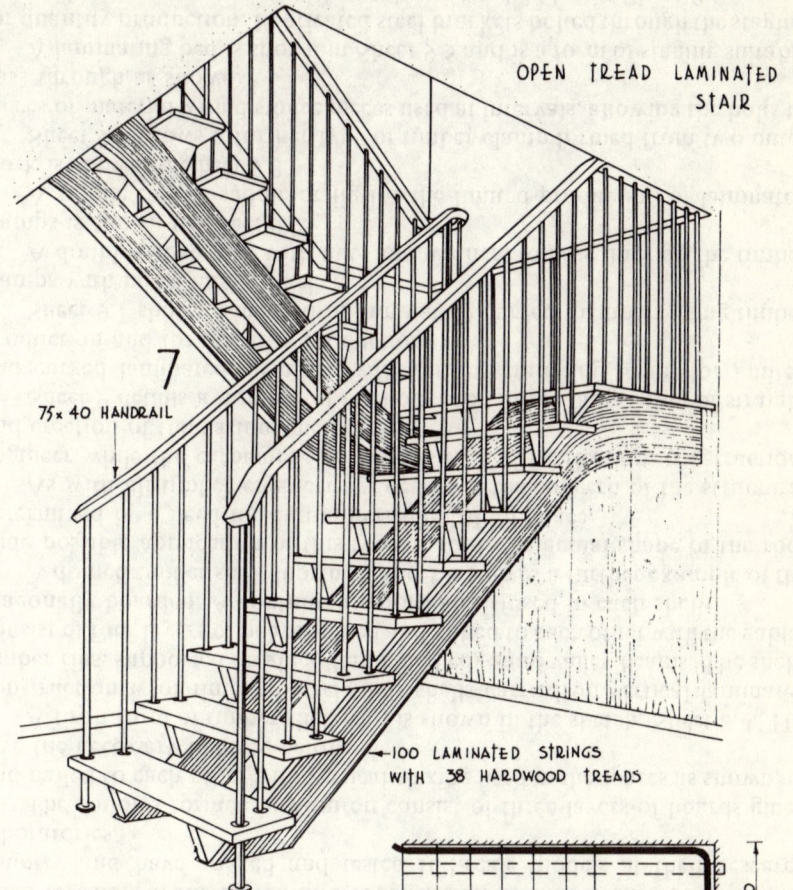

2 DOUBLE ENDED BOLT WITH NUTS & WASHERS

A TORQUE WRENCH OR NUT RUNNER FOR NUT TIGHTENING TO ENSURE EVEN PRESSURE AT ALL POINTS SHOULD BE USED

LAMINATING CURVED WORK USING TIMBER CLAMPS WITH METAL ANGLE BRACKETS

SPACER

METAL BRACKETS

JOISTING

3 TIMBER SADDLE OR JIG FOR LIMITED PRODUCTION

TWO SHAPED BEAMS

CLAMP

TIMBER CLAMP

DISTANCE PIECE

4

DISTANCE PIECES

LAMINATED BEAM

50

13

METAL CLAMPS TWO TYPES

20 DIA

75

40

20 DIA HOLES FOR CLAMPING

6

CAUL BOARD

STAGING

JOISTING

LAMINATING BED OR STAGING FOR QUANTITY PRODUCTION

FABRICATED STEEL BRACKETS BOLTED THROUGH STAGING

5

OPEN TREAD LAMINATED STAIR

75 x 40 HANDRAIL

7

100 LAMINATED STRINGS WITH 38 HARDWOOD TREADS

GOING OF STEPS 250
RISE 162
HEIGHT FL. TO FL. 2·600

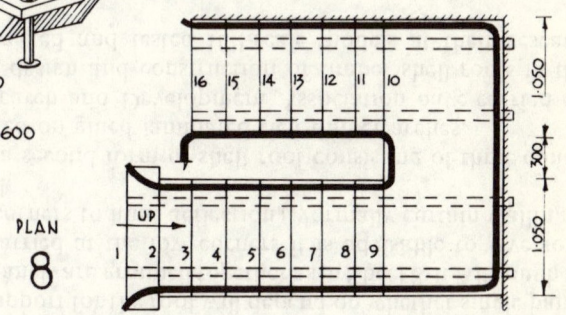

16 15 14 13 12 11 10

1·050

300

1·050

PLAN

8

UP

1 2 3 4 5 6 7 8 9

SHEET 9

Details for setting out the stair and thus the laminated stringers are given in the plan Sheet 9.8 and the accompanying specification.

Hyperbolic paraboloid roof construction

The plan and elevation of a hyperbolic shell roof consisting of four panels are shown in Sheet 10, figs. 1 and 2. In this example the spans are not great for this form of roof, being 4·800 m square with a rise of 1·500 m. Support for the roof is by timber columns at the low points with 22 mm diameter tie rods to resist the outward thrust as shown. To carry the forces around the edges of the shell, edge beams are required, these may either be cut from solid timber or laminated. Solid timber is only used for small shells up to 6·000 m square. Most edge beams are fabricated in the workshop from glued laminated timber.

The slope of the shell varies along the edge beam, this requires the beams to be laminated with a twist in them during construction.

For erection on the site, tubular steel scaffolding is provided to the underside of the shell. The bottom half of the edge beam is put in position followed by three layers of boards glued and nailed and finally the top half of the edge beam. The purpose of gluing the boards in timber shell construction is mainly to increase the stiffness of the membrane.

Sections taken at the high, mid, and low points in the edge beams are shown in Sheet 10, figs. 3, 4, and 5.

Sheet 10.6 shows the tie rod and mild steel anchor fixing at the low points to the timber columns. In order to throw the rain-water clear over the edge of the beam the area shown in the plan at the low points is boarded as shown in Sheet 10.6.

Now that timber shell roofs have become established, new improved nails have been introduced for securing the boarded laminations in the roofing. They are of two types – square twisted shank nails and ringed shank nails.

Square twisted shank nails are used between the laminated membrane and edge beams and are available up to 200 mm long. They are made from high-carbon-content steel and can be driven without bending.

Ringed shank nails are wire nails up to 56 mm in length and are used for board nailing. They do not withdraw easily and as the wood fibres enclose the ringed serrations of the nail shank, the final pressure imparted to the layers of boards is retained. This pressure between the layers when gluing is desirable.

Sheet 11.1 shows a pictorial view of part of a school building depicting further joinery items which may be considered in the project by the student.

The hyperbolic roof is similar to that featured in the previous example and its construction and erection identical. Curtain walling is used below the line of the sloping roof in the form of glazed spandrels which require setting-out to the dimensions given.

Sheet 11.2 deals with the detail of the glazed spandrels and the sizes of the members; the sections are shown in Sheet 11.3 along with the design for the entrance doors and window.

Finger joints

For many years it was realised that by end jointing lengths of timber using modern synthetic resin glues, large economics could be made in the utilisation of timber. The adoption of these techniques was dependent on the development of machinery for cutting and jointing.

Plants are now installed in many parts of the world and in this country specialising in the finger jointing of timber to take account of these economies.

Since the joint strength is comparable with that of unjointed timber, finger jointing is a most valuable addition in the making of laminated structures.

Sheet 12.2 illustrates the scarf joint previously used in laminated work for the jointing of laminates. There is considerable waste in cutting this form of joint as will be appreciated, although the jointing of the laminates may be cut on a circular sawbench adapted for the purpose, thus not requiring a specialised machine. A further disadvantage has been locating the scarf joint where pre-cutting is carried out, holding up the production process.

Sheet 12.8 shows the perfected finger joint produced on a finger-jointing machine. The machine completes the jointing by shaping, gluing, and joining the timber in one operation.

The timber is locked in position under tons of pressure. This enables the fingers to be cut precisely parallel, the glue applied with equal precision, and the timber pressed together with the fingers in perfect mesh. This produces a precision 10 mm joint of a joint quality that meets the requirements of the highest grade of timber strength specification.

Laminated roof beams

The illustrations on Sheet 12 show examples of glued laminated beams for varying spans with alternative fixings to timber stanchions and brickwork.

In Sheet 12.1 a 6 mm mild steel plate is housed into the beam and stanchion on each side and fixed by screwing.

Sheet 12.2 shows a scarf for the jointing of laminates.

A tapered laminated beam suitable for a span of 8·400 m is shown in Sheet 12.4. This has a camber of 19 mm at the centre and is fixed at the eaves to glued laminated legs. The fixing is by two 19 mm bolts bored as shown in Sheet 12.5. Sheet 12.3 shows the fixing of the laminated ties between the beams where mild steel angles are coach screwed to the ties and the beam on each face.

A cranked glued laminated beam is shown in Sheet 12.6. This is carried on padstones where mild steel angles are used with coach screws into the beam and rag-bolts into the padstone. Purlins, 175 mm × 75 mm in section, are spaced at 600 mm centres and notched over the truss and cleated. The roof is boarded, with an asbestos eaves gutter as shown in Sheet 12.7. Matchboarding is used to finish the ceiling with the exposed trusses varnished.

HYPERBOLIC PARABOLOID SHELL ROOF

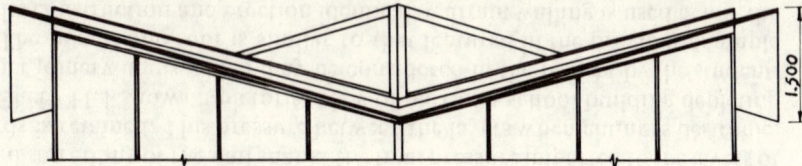

ELEVATION

1.500

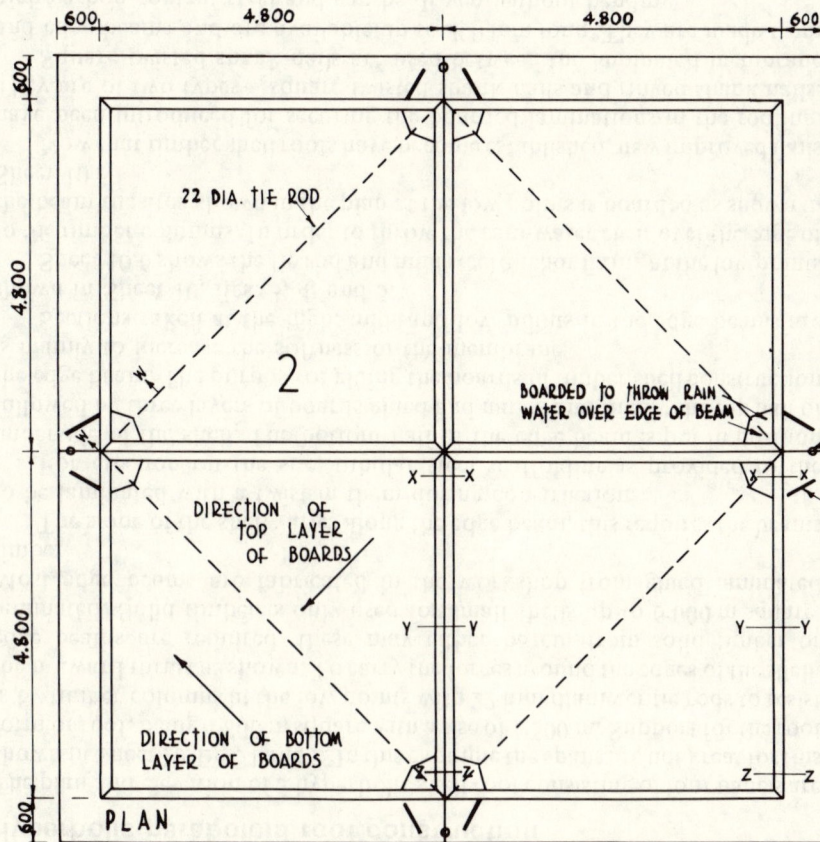

600 | 4.800 | 4.800 | 600

600

4.800

4.800

600

22 DIA. TIE ROD

2

BOARDED TO THROW RAIN
WATER OVER EDGE OF BEAM

RAIN WATER
OUTLET

50×50 CHECK

DIRECTION OF
TOP LAYER
OF BOARDS

DIRECTION OF BOTTOM
LAYER OF BOARDS

X — X

Y — Y

Y — Y

X — X

Z — Z

Z — Z

PLAN

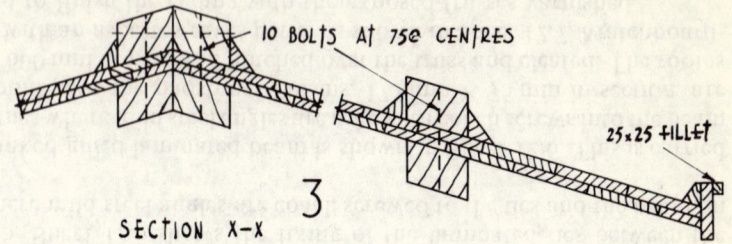

10 BOLTS AT 750 CENTRES

25×25 FILLET

3

SECTION X-X

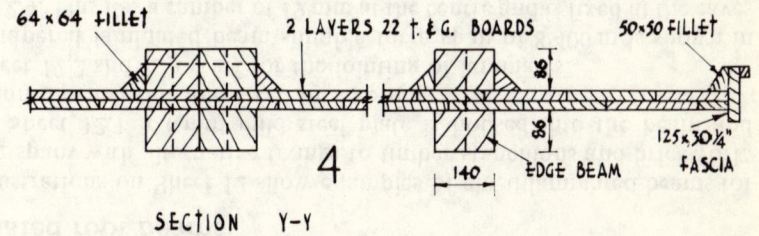

64×64 FILLET

2 LAYERS 22 t. & G. BOARDS

50×50 FILLET

86

86

140

EDGE BEAM

125×30½"
FASCIA

4

SECTION Y-Y

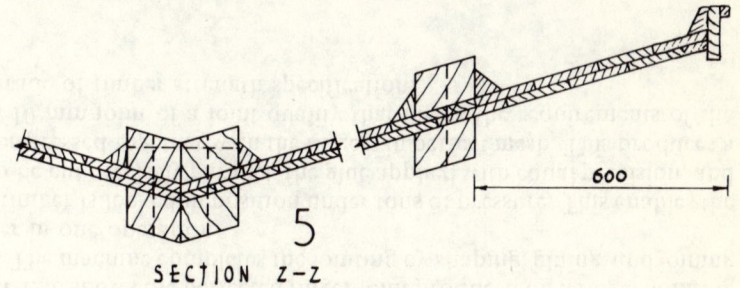

600

5

SECTION Z-Z

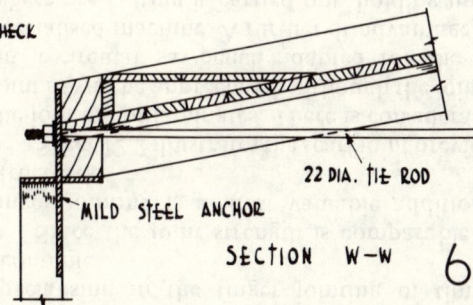

22 DIA. TIE ROD

MILD STEEL ANCHOR

SECTION W-W

6

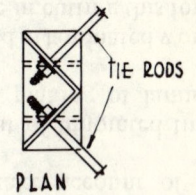

TIE RODS

PLAN

SHEET 10

PARABOLIC ROOF PROJECT

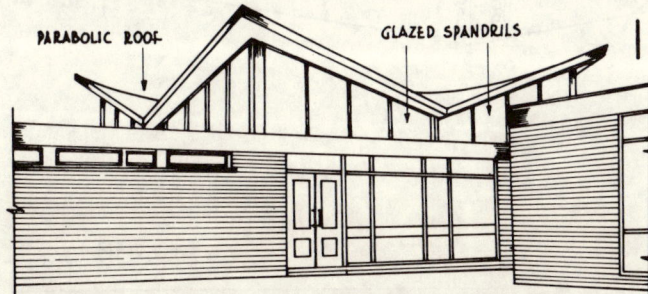

PARABOLIC ROOF

GLAZED SPANDRILS

1

HYPERBOLIC PARABOLOID ROOF SPANDRIL AND WINDOW PROJECT DESIGN

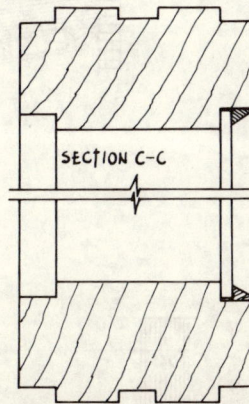

SECTION C-C

SECTION B-B

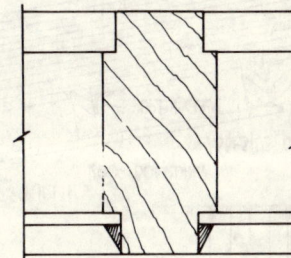

SECTION A-A

3

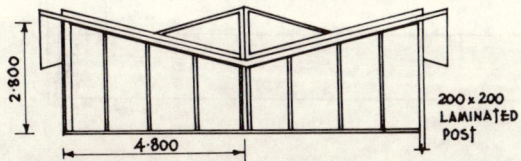

2·800

4·800

200 x 200 LAMINATED POST

ELEVATION OF GLAZED SPANDRIL

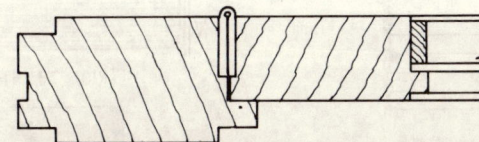

SECTION D-D

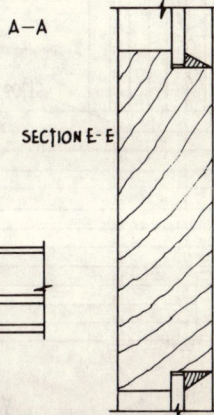

SECTION E-E

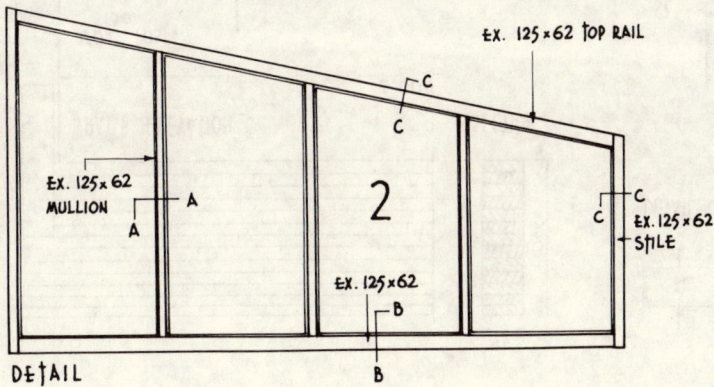

EX. 125 x 62 TOP RAIL

C
C

EX. 125 x 62 MULLION
A
A

2

EX. 125 x 62
B
B

C
C
EX. 125 x 62 STILE

DETAIL

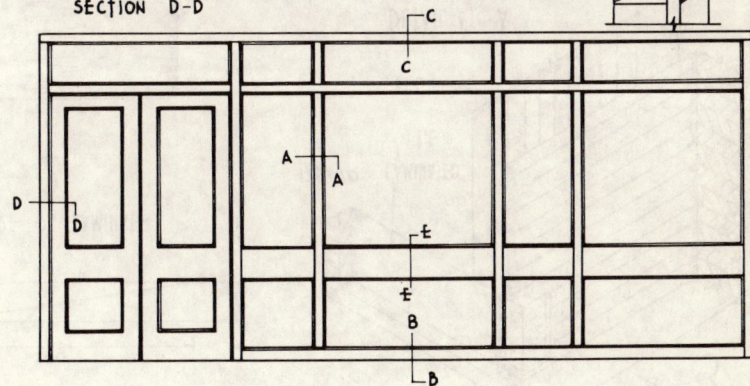

C
C

A
A

D
D

E
E

B

B

SHEET II

127

LAMINATED BEAMS

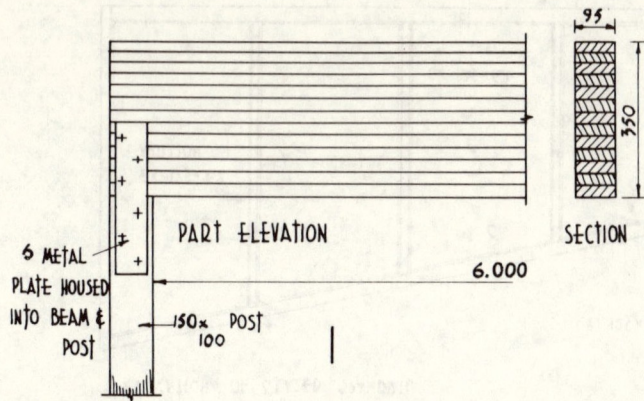

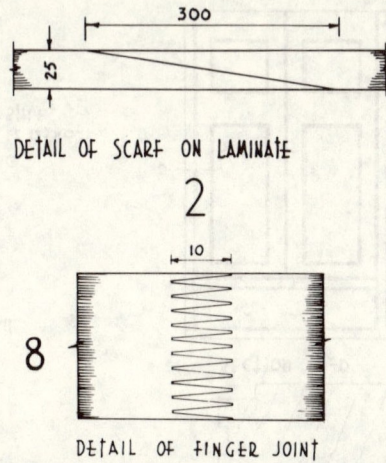

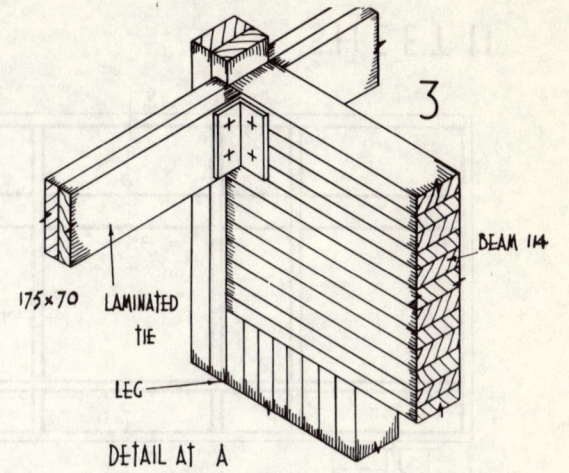

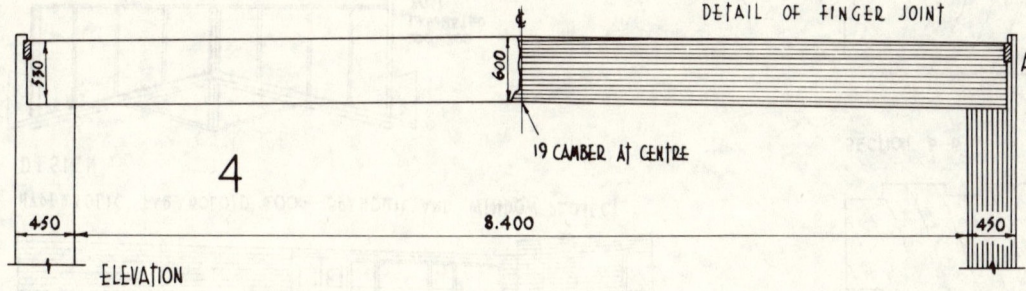

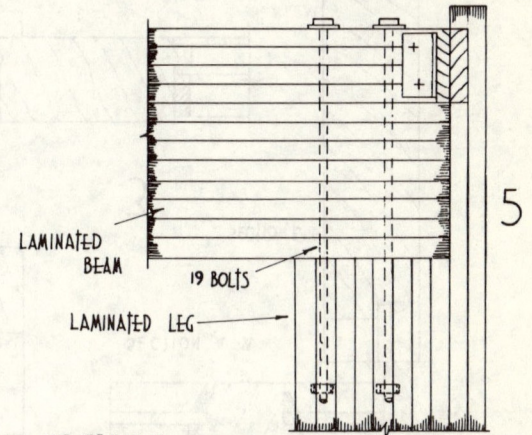

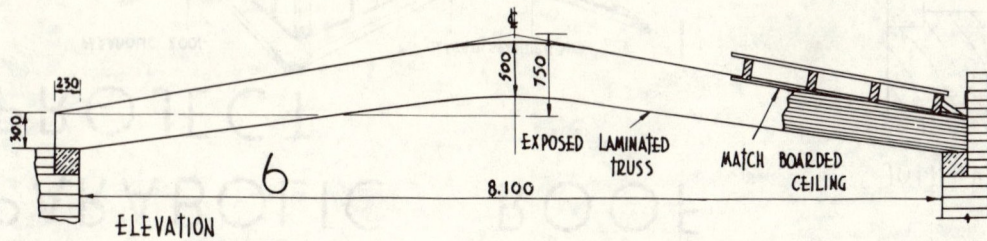

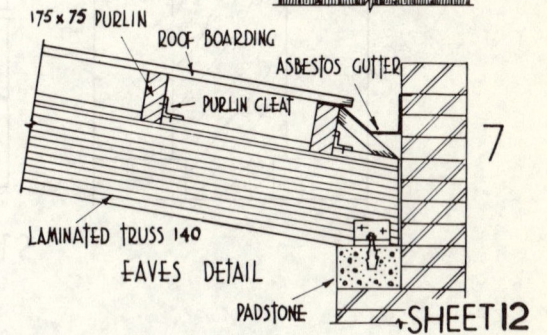

1

PART ELEVATION

SECTION

95

350

6 METAL PLATE HOUSED INTO BEAM & POST

150 x 100 POST

6.000

2

300

25

DETAIL OF SCARF ON LAMINATE

8

10

DETAIL OF FINGER JOINT

3

BEAM 114

175 x 70 LAMINATED TIE

LEG

DETAIL AT A

4

550

600

19 CAMBER AT CENTRE

450

8.400

450

ELEVATION

A

5

LAMINATED BEAM

19 BOLTS

LAMINATED LEG

6

230

500

750

500

EXPOSED LAMINATED TRUSS

MATCH BOARDED CEILING

8.100

ELEVATION

7

175 x 75 PURLIN

ROOF BOARDING

PURLIN CLEAT

ASBESTOS GUTTER

LAMINATED TRUSS 140

EAVES DETAIL

PADSTONE

SHEET 12

B O X B E A M S

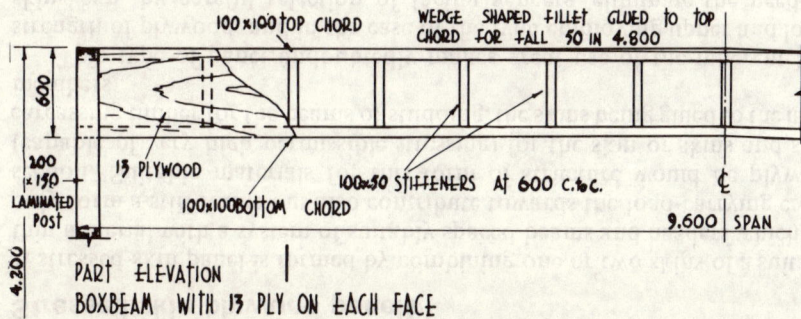

100 x 100 TOP CHORD

WEDGE SHAPED FILLET GLUED TO TOP
CHORD FOR FALL 50 in 4.800

600

200
x 150
LAMINATED
POST

4.200

13 PLYWOOD

100 x 100 BOTTOM CHORD

100 x 50 STIFFENERS AT 600 c.ㅕc.

9.600 SPAN

PART ELEVATION ‖ 1
BOXBEAM WITH 13 PLY ON EACH FACE

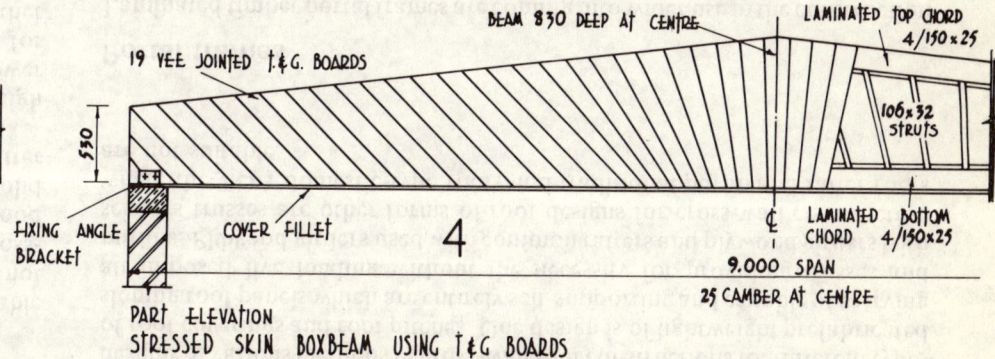

BEAM 830 DEEP AT CENTRE

19 VEE JOINTED T.&G. BOARDS

LAMINATED TOP CHORD
4/150 x 25

530

FIXING ANGLE
BRACKET

COVER FILLET

4

LAMINATED BOTTOM
CHORD 4/150 x 25

9.000 SPAN
25 CAMBER AT CENTRE

PART ELEVATION
STRESSED SKIN BOXBEAM USING T&G BOARDS

106 x 32
STRUTS

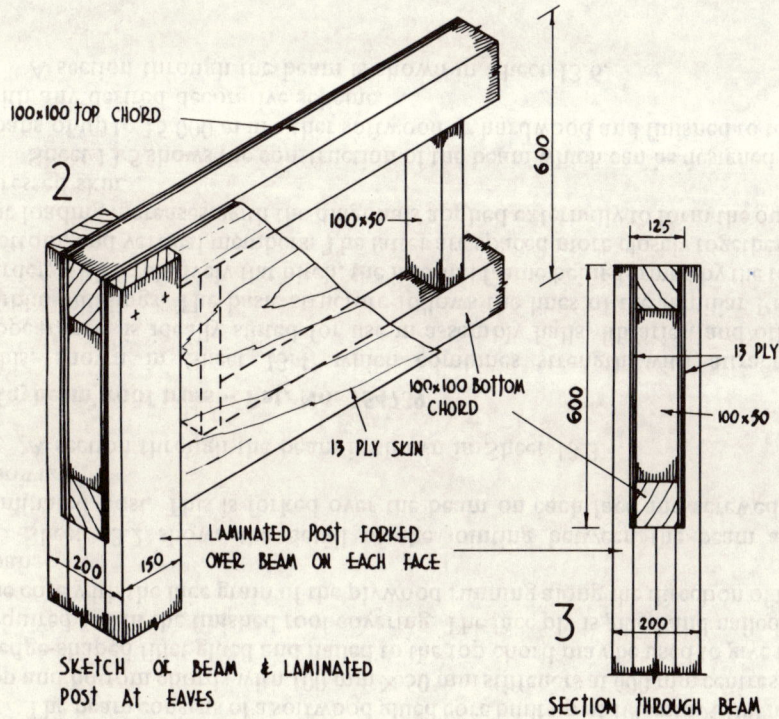

100 x 100 TOP CHORD

100 x 50

600

100 x 100 BOTTOM
CHORD

13 PLY SKIN

2

LAMINATED POST FORKED
OVER BEAM ON EACH FACE

200   150

SKETCH OF BEAM & LAMINATED
POST AT EAVES

125

600

13 PLY

100 x 50

200

3

SECTION THROUGH BEAM

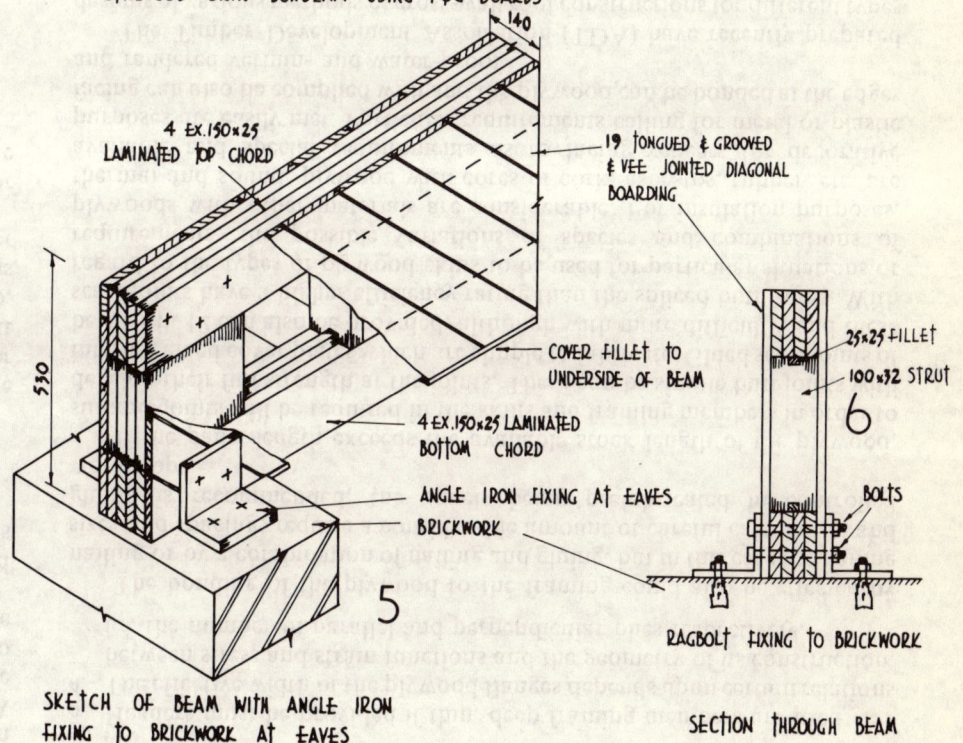

140

4 EX.150 x 25
LAMINATED TOP CHORD

19 Tongued & grooved
& VEE JOINTED DIAGONAL
BOARDING

COVER FILLET TO
UNDERSIDE OF BEAM

4 EX.150 x 25 LAMINATED
BOTTOM CHORD

ANGLE IRON FIXING AT EAVES
BRICKWORK

5

SKETCH OF BEAM WITH ANGLE IRON
FIXING TO BRICKWORK AT EAVES

530

25 x 25 FILLET

100 x 32 STRUT

6

BOLTS

RAGBOLT FIXING TO BRICKWORK

SECTION THROUGH BEAM

SHEET 13

Box beams

Sheet 13.1 shows the part elevation of a box beam faced with 13 mm Douglas fir plywood on each face. This type of beam has a high degree of stiffness combined with a low weight factor and is cheaper than the equivalent laminated beam.

The beam consists of a softwood glued core built up of 100 mm $\times$ 100 mm top and bottom chords with 100 mm $\times$ 50 mm stiffeners at 600 mm centres. A wedge-shaped fillet glued and nailed to the top chord may be used to give the required fall in the finished roof covering. The face ply is glued and nailed to the core with the face grain of the plywood running along the direction of the span.

Sheet 13.2 shows the detail of the jointing between the beam and laminated post. This is forked over the beam on each face and screwed as shown.

A section through the beam is shown in Sheet 13.3.

'Kaybeam' roof truss – Pat. No. 754739

This, shown in Sheet 13.4, which combines strength with attractive appearance, is ideally suited for use in assembly halls, libraries, and other public buildings. The basic structure follows the lines of the familiar Pratt girder, with a relatively flat pitch, the internal frame being formed by the top, bottom, and vertical members. The latter are spaced more closely together as the loading increases, with the diagonals applied externally to form the outer stressed skin.

Sheet 13.5 shows the construction of the beam which can be designed for spans of up to 15·000 m in either softwood or hardwood and finished to tone with any desired decorative scheme.

A section through the beam is shown in Sheet 13.6.

Stressed-skin plywood panels

A stressed-skin panel is formed by combining one or two skins of a suitable thin material with a system of suitably spaced beams and headers which not only form a stiff frame but also contribute towards the load-carrying cross-section. Suitable materials for this form of structure would be plywood (capable of very high permissible stressing) for the skin or skins and solid carcassing timber for the beams or studding, the skins being glued to the latter members.

This type of panel undoubtedly makes great use of the inherent high strength of plywood, and in the case of the type employing upper and lower skins can, by careful selection of facing veneers, eliminate the need for separate materials for the ceiling or internal cladding, depending on whether the unit is used for flooring or for walling.

Being extremely light in weight and thin in cross-section, the panels are ideal for use in prefabricated housing for walls, floors, and roofs, being simple to erect. They also present possibilities for use as bridge decks, for shuttering for concrete, advertisement hoardings, etc. and are capable of acting as slabs spanning simultaneously in two directions with consequent shallower depth of thickness.

For a panel to qualify as a stressed-skin panel, the following criterions must be satisfied:

1. Skins must be glued to the framing.
2. Skins and frames must be continuous or adequately spliced longitudinally
3. Headers must be provided if thin, deep framing members are used.
4. The effective width of the plywood flanges depends upon certain relations between stress and strain functions and the geometry of its construction, i.e. the number of parallel and perpendicular plies respectively.

The bonding of the plywood to the framing could also be effected by nailing or by a combination of nailing and gluing, but in this case the nailing sizes and spacings require a considerable amount of careful calculation and gluing is recommended, the panels being prefabricated in controlled workshops.

If the panel length exceeds the available stock length of the plywood, suitable joints will be required in the skins and framing members in order to develop their full strength at the joints. These can be simple butt joints with internal glued cover plates which are simple to fabricate. Glued scarf joints of bevel 1 in 12 can also be provided, although with more difficulty, and these scarf joints have a higher efficiency rating than the spliced butt joints. With regard to the types of plywood skins to be used for particular situations or requirements, the possible variations of species and combinations of plywoods with other materials are considerable. For insulation purposes, thermal and sound, plywood with cores of cork, asbestos, rubber, etc. are available and special requirements as to facing veneers for decorative purposes are easily met. Particular requirements calling for metal or plastic facing can also be complied with and the plywood can be bonded at the edges and rendered vermin- and water-proof.

The Timber Development Association (TDA) have recently prepared designs of various methods of cross-wall roof constructions for different types of roof coverings and roof pitches. One design is of lightweight prefabricated sloping roof panels which are entirely self-supporting and capable of carrying all imposed live loading without the necessity for providing trusses and purlins. Plywood girders used with common rafters and plywood girders with scissors trusses are other forms of roof designs for cross-wall construction where the TDA domestic-type roofs and traditional purlin and rafter roofs are not suitable.

Portal frames

Laminated timber portal frames are coming into wider use in the construction of churches, halls, gymnasiums, and similar buildings. They may be designed as plywood-faced portals using a softwood core, or as glued laminated arches where a certain dignified architectural appeal is realised and timber's versatility is exploited to the full.

Sheet 14.1 shows the elevation of a plywood-faced portal frame where it will be noted that alternative designs are used in the construction. To the left,

PORTAL FRAMES

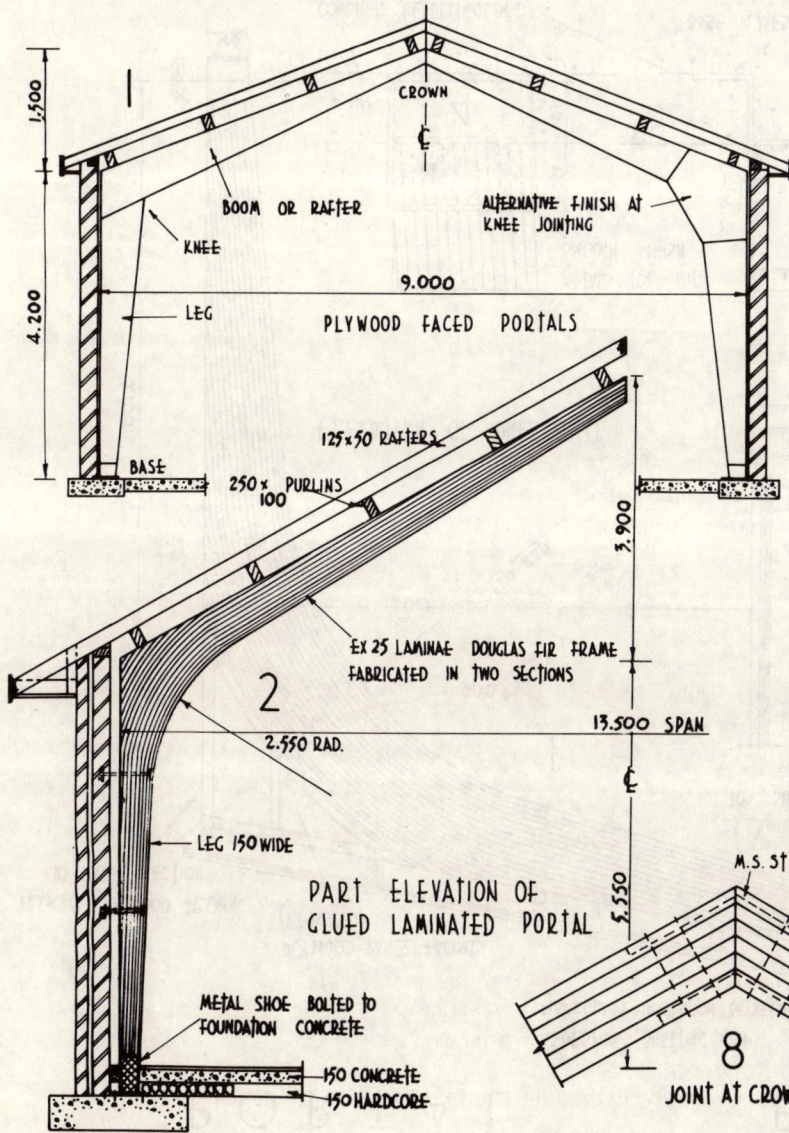

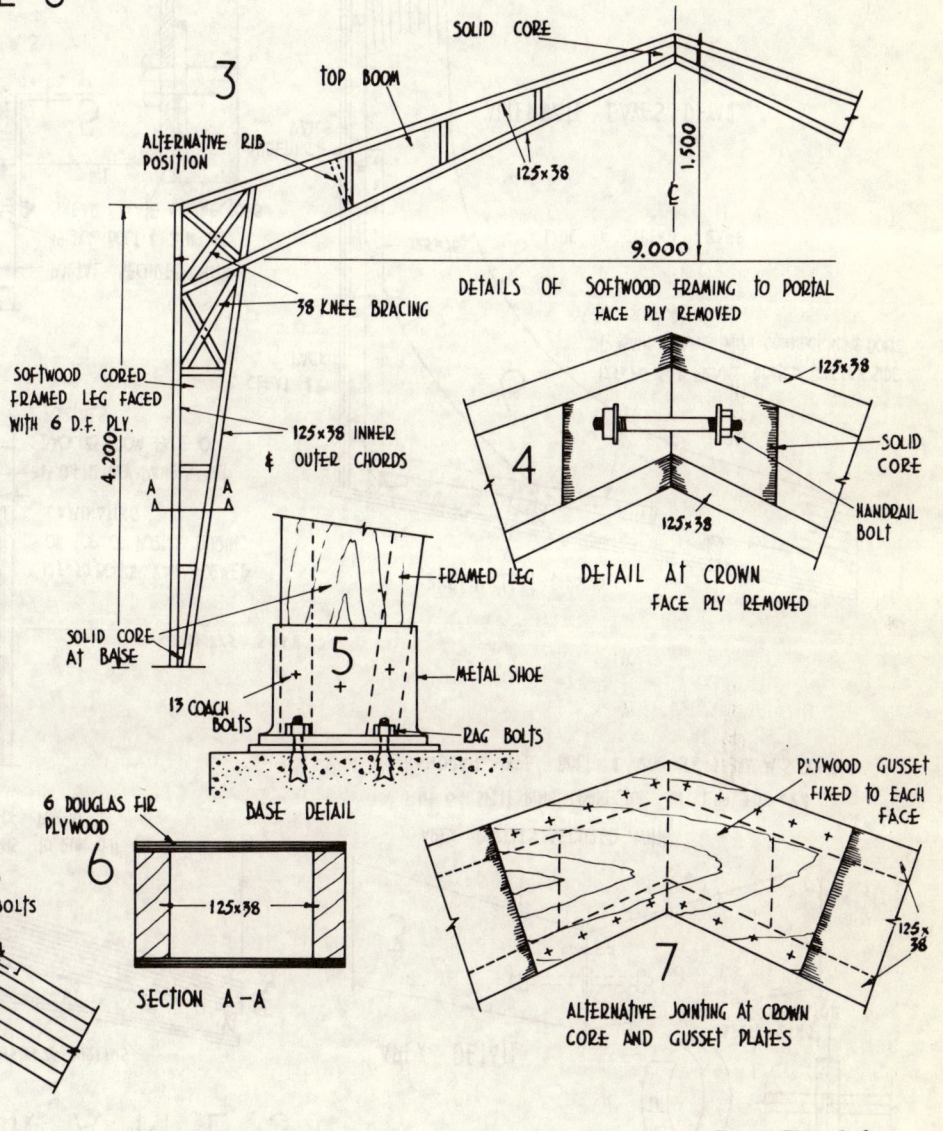

1

CROWN
℄
1,500
BOOM OR RAFTER
ALTERNATIVE FINISH AT KNEE JOINTING
KNEE
9.000
LEG
4.200
PLYWOOD FACED PORTALS
BASE

2
125×50 RAFTERS
250×100 PURLINS
3,900
EX 25 LAMINAE DOUGLAS FIR FRAME FABRICATED IN TWO SECTIONS
13.500 SPAN
℄
5,550
2,550 RAD.
LEG 150 WIDE
PART ELEVATION OF GLUED LAMINATED PORTAL
METAL SHOE BOLTED TO FOUNDATION CONCRETE
150 CONCRETE
150 HARDCORE

3
SOLID CORE
TOP BOOM
ALTERNATIVE RIB POSITION
125×38
1,500
9.000
DETAILS OF SOFTWOOD FRAMING TO PORTAL FACE PLY REMOVED
38 KNEE BRACING
SOFTWOOD CORED FRAMED LEG FACED WITH 6 D.F. PLY.
125×38 INNER & OUTER CHORDS
4.200
SOLID CORE AT BASE
A A
A A
FRAMED LEG

4
125×38
SOLID CORE
125×38
HANDRAIL BOLT
DETAIL AT CROWN FACE PLY REMOVED

5
METAL SHOE
13 COACH BOLTS
RAG BOLTS
BASE DETAIL

6
6 DOUGLAS FIR PLYWOOD
125×38
SECTION A-A

7
PLYWOOD GUSSET FIXED TO EACH FACE
125×38
ALTERNATIVE JOINTING AT CROWN CORE AND GUSSET PLATES

8
M.S. STRAP
BOLTS
JOINT AT CROWN

SHEET 14

PORTAL FRAMES

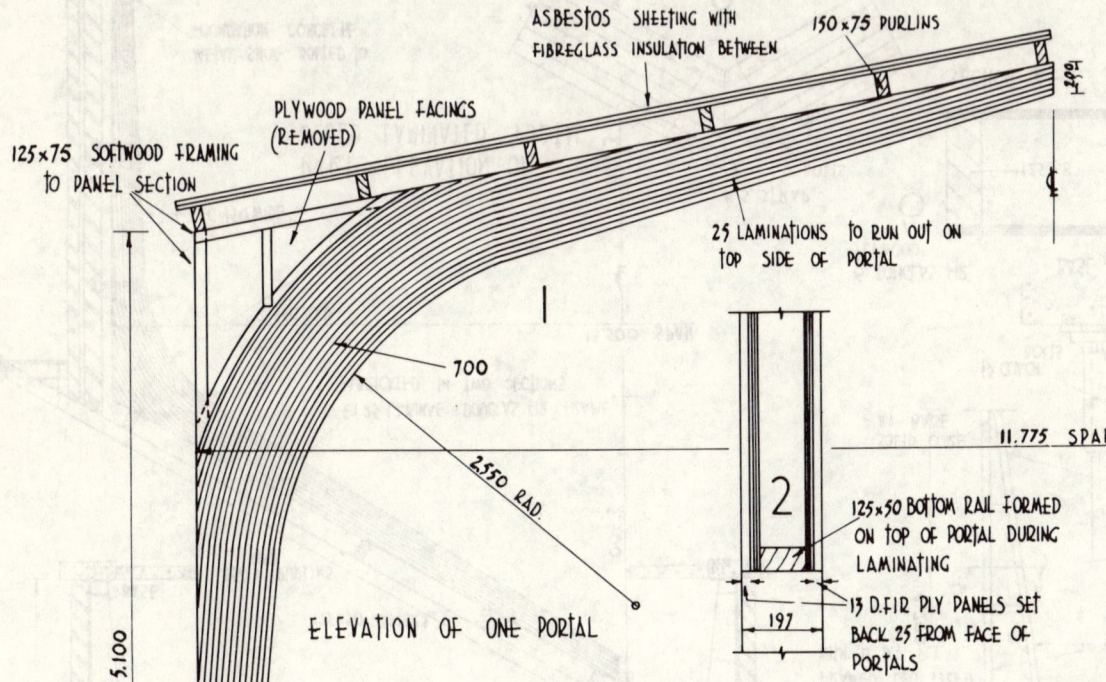

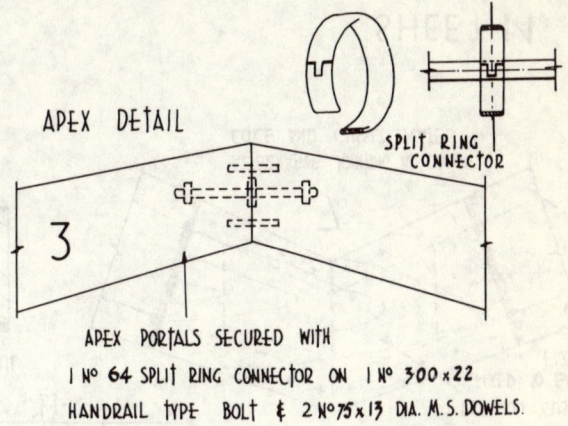

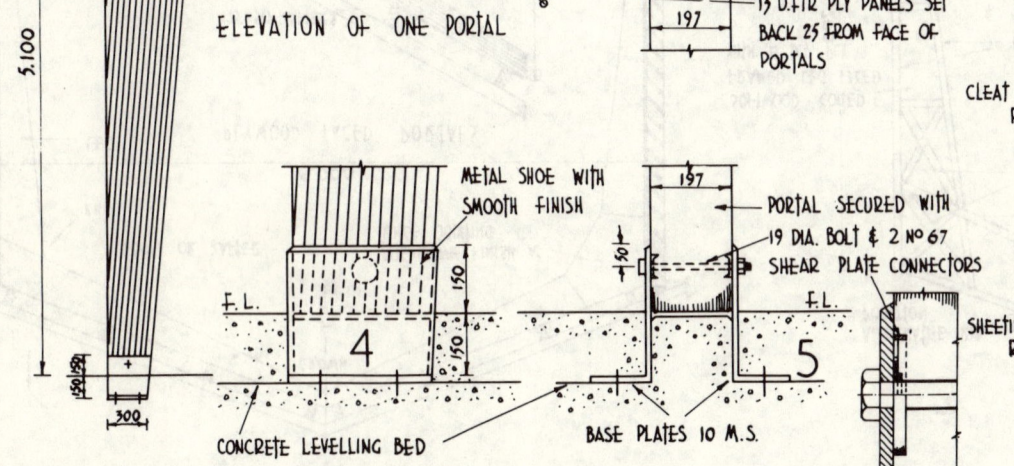

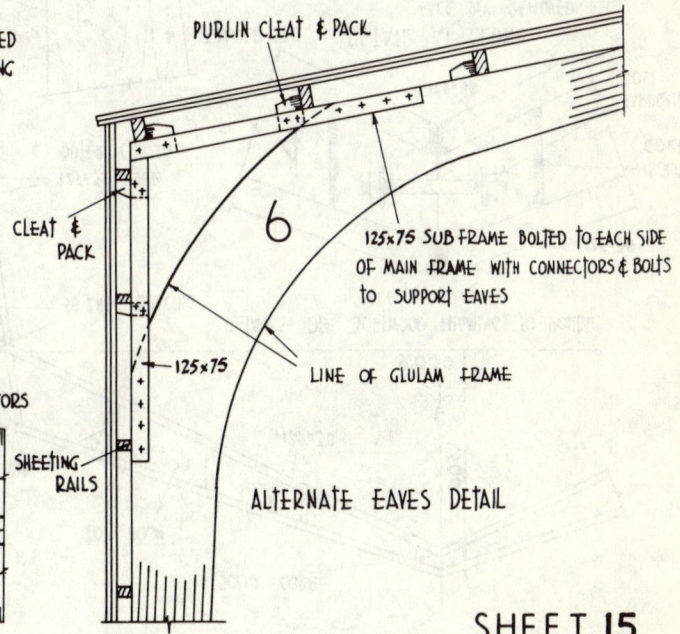

ASBESTOS SHEETING WITH
FIBREGLASS INSULATION BETWEEN

150 x 75 PURLINS

PLYWOOD PANEL FACINGS
(REMOVED)

125 x 75 SOFTWOOD FRAMING
TO PANEL SECTION

25 LAMINATIONS TO RUN OUT ON
TOP SIDE OF PORTAL

700

2550 RAD.

11.775 SPAN

1

5.100

ELEVATION OF ONE PORTAL

2

197

125 x 50 BOTTOM RAIL FORMED
ON TOP OF PORTAL DURING
LAMINATING

13 D.FIR PLY PANELS SET
BACK 25 FROM FACE OF
PORTALS

METAL SHOE WITH
SMOOTH FINISH

F.L.

150

4

150

300

CONCRETE LEVELLING BED

BASE DETAILS

197

50

PORTAL SECURED WITH
19 DIA. BOLT & 2 No 67
SHEAR PLATE CONNECTORS

F.L.

5

BASE PLATES 10 M.S.

APEX DETAIL

3

SPLIT RING
CONNECTOR

APEX PORTALS SECURED WITH
1 No 64 SPLIT RING CONNECTOR ON 1 No 300 x 22
HANDRAIL TYPE BOLT & 2 No 75 x 13 DIA. M.S. DOWELS.

PURLIN CLEAT & PACK

CLEAT &
PACK

6

125 x 75

SHEETING
RAILS

125 x 75 SUB FRAME BOLTED TO EACH SIDE
OF MAIN FRAME WITH CONNECTORS & BOLTS
TO SUPPORT EAVES

LINE OF GLULAM FRAME

ALTERNATE EAVES DETAIL

SHEET 15

LAMINATED WORK PROJECT
DETAILS

LAMINATED PORTAL FRAMES / GLAZED WALL PROJECT
DESIGN 1

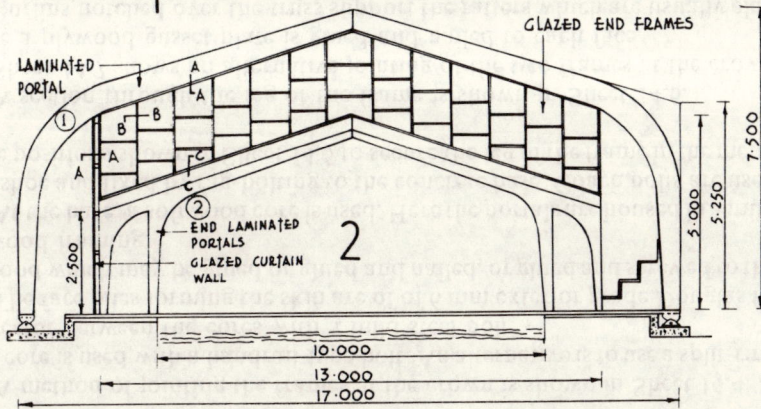

OUTSIDE LEG OF PORTAL

2

LAMINATED PORTAL

GLAZED END FRAMES

END LAMINATED PORTALS
GLAZED CURTAIN WALL

7.500
5.250
5.000
2.500

10.000
13.000
17.000

DETAILS — GLAZED END FRAMES

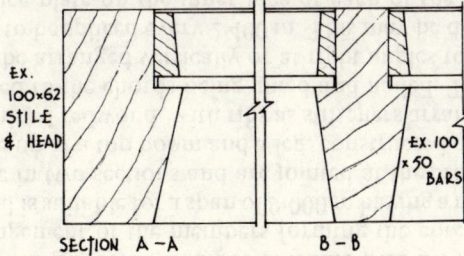

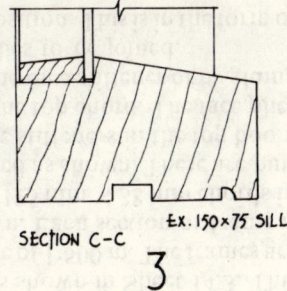

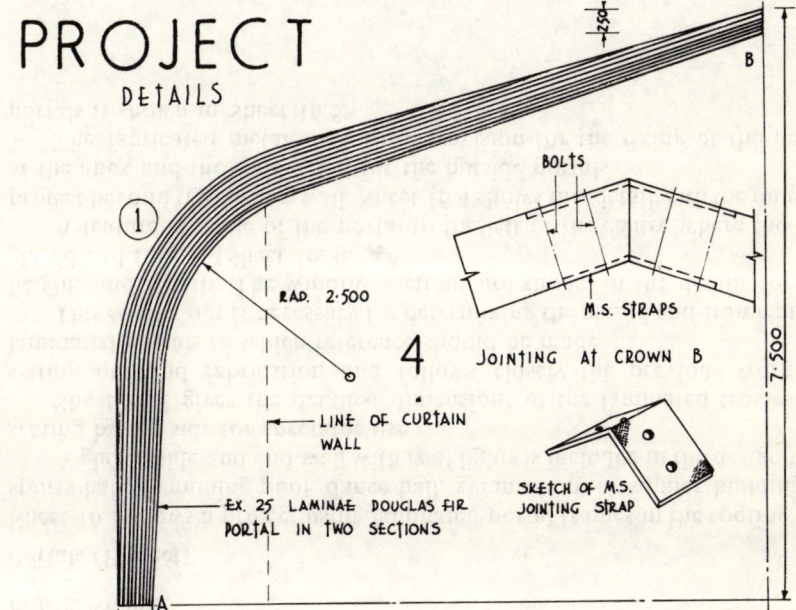

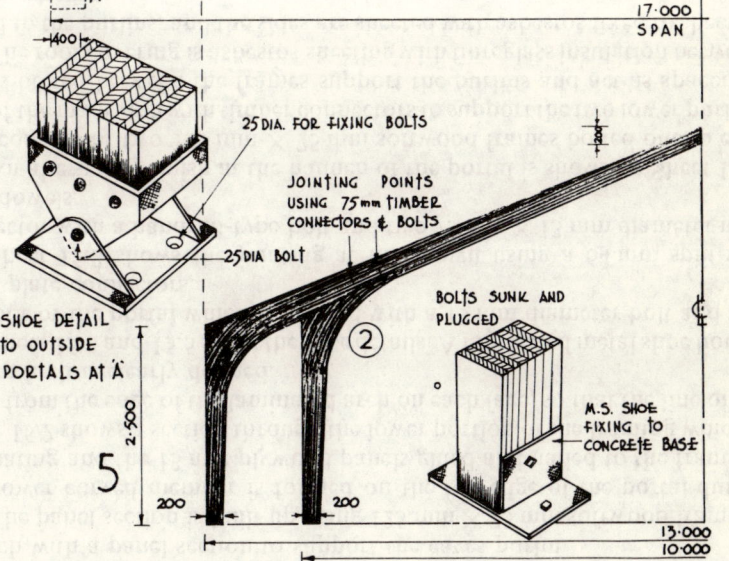

EX. 100×62 STILE & HEAD

EX. 100 ×50 BARS

SECTION A—A

B—B

3

SECTION C-C

EX. 150×75 SILL

4

RAD. 2.500

LINE OF CURTAIN WALL

EX. 25 LAMINAE DOUGLAS FIR PORTAL IN TWO SECTIONS

BOLTS

JOINTING AT CROWN B

M.S. STRAPS

SKETCH OF M.S. JOINTING STRAP

7.500

A

17.000
SPAN

B

400

25 DIA. FOR FIXING BOLTS

JOINTING POINTS USING 75mm TIMBER CONNECTORS & BOLTS

25 DIA. BOLT

SHOE DETAIL TO OUTSIDE PORTALS AT A

2

BOLTS SUNK AND PLUGGED

M.S. SHOE FIXING TO CONCRETE BASE

5

2.500

200   200

13.000
10.000

SHEET 16

the face plies to the boom or rafter are allowed to run through to the outside edge of the leg at the knee, with the plies to the leg jointing to them. To the right, an alternative finish at the knee is shown.

A larger detail of a portal frame with the face ply removed to show the arrangement of the members forming the core is shown in Sheet 14.3. This portal is suitable for a span of 9·000 m having a rise of 1·500 m. The frames are made in two sections and are jointed at the crown. Each section or frame is made up of a top boom and a leg, constructed of 125 mm × 38 mm chords in machined redwood, with ribs as stiffeners arranged as shown. These are butt jointed to the chords being glued and nailed. The stiffeners in the top boom may be arranged vertically or at right angles to the top chord. The face plies have to be spliced every 2·400 m. This may be done on a stiffener or by gluing a splice plate on the inner face of each of the plies to be joined.

Additional bracing is necessary at the knee position. This is in the form of diagonal knee bracing as shown.

A method of jointing the frames at the crown is shown in Sheet 14.4. A solid core is used with a handrail-type bolt. An alternative is to use a split-ring connector between the cores with a mild steel bolt.

The face plies forming the skin are of of 6 mm exterior grade Douglas fir plywood which may be glued or glued and nailed, or glued and screwed to the softwood framing.

At the base, a softwood core is used. Here the portals are housed in a mild steel shoe and fixed by rag-bolting to the concrete base. Coach bolts are used at the positions shown in Sheet 14.5 to secure the leg of the frame in the metal shoe.

A section through the leg of the frame is shown in Sheet 14.6.

Sheet 14.7 shows an alternative jointing of the two frames at the crown where a plywood gusset plate is glued and nailed to each face.

Purlins notched over the truss support the rafters which are usually clad on the underside with matchboarding or insulation board, the purlins being left exposed.

Sheet 14.2 shows the part elevation of a glued laminated timber portal frame of 13·500 m span suitable for a church or hall.

When choosing glued laminated construction for buildings there are certain practical considerations to bear in mind. One is the question of cost. A structural unit made by lamination is more costly than a solid member, but lamination serves many functions which cannot as readily be met by other methods. The cost may therefore be justified either by greater efficiency or improved appearance. The attractive profiles of laminated components have often encouraged designers to leave them exposed to the interiors of buildings. This can result in a saving of the total cost of construction since it has made the use of suspended ceilings unnecessary. Laminated arch forms of different types are designed to suit a variety of requirements. These include two- and three-pin segmental and semicircular arches, parabolic, and Gothic arches for spans up to 75·000 m.

The portal frames illustrated on Sheets 14 and 15 are in the latter group and are three-pinned arches, being hinged at the crown and at the springings or abutments. Arches in roof structures are normally spaced from 3·000 m to 6·000 m centres.

In Sheet 14.2 the portal is laminated from 25 mm Douglas fir laminates and jointed at the crown using purpose-made mild steel straps which are let into the top and bottom edges of the arches and secured with mild steel bolts as shown in Sheet 14.8.

The purlins are 250 mm × 100 mm with 125 mm × 50 mm rafters spaced at 600 mm centres, being clad on the underside leaving the purlins exposed.

Sheet 15.1 shows the elevation of a portal frame which is built up at the haunch with a panel section to support the eaves purlin.

The panel section is built up using 125 mm × 75 mm softwood framing. The lower curved member is formed on the top edge of the portal during laminating and the 13 mm plywood panels glued and nailed to the framing. Sheet 15.2 shows a section through the lower portion of the framing which is set in from the edge of the laminated arch on each face, so that the line of the glulam arch is clearly defined.

Sheet 15.4 and 15.5 show the base details. A fabricated metal shoe houses the foot of the portal which is secured with a 19 mm diameter bolt and two shear plate connectors.

Sheet 15.3 shows the jointing at the crown using a 64 mm split-ring connector with a handrail-type bolt and two 75 mm × 13 mm diameter mild steel dowels.

An alternative finish at the haunch of the portal is shown in Sheet 15.6. This consists of two 125 mm × 75 mm softwood frames bolted one to each side of the main arch with timber connectors to support the two lower purlins. Cleats bolted between the frames support the purlins and act as spacers.

The roof covering is asbestos sheeting with fibreglass insulation between, nailed to the purlins, and the sides are sheeted with asbestos fixed to sheeting rails as shown.

Portals (Project)

Sheet 16.1 shows a project using laminated portal frames in the roofing of a sports hall, swimming pool, dance hall, gymnasium, or similar building.

A glazed side and end wall with roof lights is included in the design with seating by the side for spectator use.

Sheet 16.2 gives the detailed dimensions of the laminated trusses for setting out and fabrication and follows closely the previous work on laminated portals to which reference should be made.

This setting out is necessary for determining the glazed end-frame pitch, height, and length. The window sections are shown in the details for the glazed end frames (Sheet 16.3).

A feature is made of the portal to the left of the centre where the legs project beyond the curtain wall. Sheet 16.4 shows this detail with the jointing at the apex and the shoe detail for the outside portals.

The fabricated metal shoe with provision for the fixing of the inside portals is shown in Sheet 16.5.

Chapter 14

Dormers

Sheet 17.1 shows the key front and side elevations of a segmental headed dormer window in a pitched roof, inclined at 45 degrees.

In this problem the development of the dormer roof and the shape of the opening in the main roof are required to be found.

First the front and side elevations are drawn in Sheet 17.2. The segmental roof line is divided into twelve equal parts and numbered as shown. These points are projected onto the main roof line and numbered 6′, 7′, 8′, 9′, 10′, 11′, and 12′. The plan is next projected below the side elevation with the same ordinates used in the elevation transferred to the plan. These are numbered points 0 to 6. Horizontal projections from these points are drawn to intersect with vertical lines dropped from 6′, 7′, 8′, 9′, 10′, 11′, 12′ in the side elevation. Through these intersections in points 0′, 1′, 2′, 3′, 4′, 5′, 6′, etc. a fair curve is drawn to complete the plan.

The development of the dormer roof is carried out as follows:

Points 0 to 12 from the segmental roof line in elevation are stepped out on the vertical line shown in the development. These points are projected horizontally and the distance 0–0′ taken from the plan, stepped off on the outside edge. The distances marked a, b, c, d, e, f, and g in the plan are transferred to the development and lines drawn vertically from these points to intersect with the horizontals, give the points 0′, 1′, 2′, 3′ etc. through which a curve is drawn to complete the development of the roof surface.

In obtaining the true shape of the opening in the roof, only the shaped portion is shown projected at right angles to the main roof line. The points projected intersect with the ordinates taken from the plan to give the shape on the main roof.

Turrets

Sheet 17.3 shows the plan and part section of a small turret roof suitable for an out building. The building is octagonal in plan with 50 mm hips secured at the top of an octagonal finial. Here, 100 mm by 50 mm jack spars are used between the hips. Joists secured to the wall plates serve as ties and carry the lining to the ceiling. The roof is boarded as shown with the hip rafters backed.

To the right of the centre line, the necessary bevels for the backing of the hips and the mitring of the boards are shown. It will be seen that one-half of the dihedral angle gives the required edge bevel for cutting the boards. The side bevel requires the development of one of the roof surfaces to be done as follows:

A–B is the length of one of the sides in plan, and C′ the height of the hip. With C′ as centre and radius C′A, describe an arc to A′ in C′ produced in the section. Draw horizontal lines from the side A–B in plan to intersect with a vertical line drawn from A′ in elevation giving points A′, B′. Join these points to C to give the development of the roof surface ABC which gives the required side bevel for cutting the boards.

Timber spires

Spires may be described as pyramidal roofs having a base outline square, octagonal or circular in plan. Being subject to considerable wind pressure, special care must be taken with the construction of the spire in bracing and anchoring to the main structure.

Sheet 18.1 shows the elevation of a timber spire 6·000 m high, octagonal at the base and covered in copper. The spire is mounted on a stone tower 2·700 m diameter.

A section through the tower and spire is shown in Sheet 18.2. A 275 mm × 75 mm wall plate or curb rests upon the top of the tower and is rag-bolted down flush with its outer face. The angles are secured by halving and bolting. Two 230 mm × 75 mm diagonial ties cross at right angles and are halved together at their intersection and notched and bolted to the wall plates, thus acting as a tie to the walls. From the centre of these a 125 mm × 125 mm octagonal mast or finial rises to a conical end piece at the top of the spire. The finial is housed into the conical end piece at the top, and the two bored to receive the cross. The foot of the finial is stub-tenoned and bolted to the ties. The faces of the octagonal pyramid are inclined at an angle of 80 degrees. The 125 mm × 50 mm hips are bird's-mouthed into the angles of the octagonal base and secured with iron angle brackets coach bolted to the rafters and wall plates. The upper ends of the hips are nailed to the finial and end piece. Two series of collars are framed between the hips as shown, and 100 mm × 50 mm sprockets, nailed to the hips and the wall plates, complete the framing. Diagonal boarding 16 mm thick with a 125 mm × 25 mm fascia and soffit complete the spire.

Sheet 18.3 shows the plan of the roof timbers to the spire, with the details at the apex and at the eaves shown in Sheet 18.4 and 18.5.

Domes and pendentives

Sheet 19.1 shows the plan and elevation of a hemisphere, part of which is covering a square plan. The square is exactly contained within the

DORMER

1

1.350
700
1.200

ELEVATION

45°

SIDE ELEVATION

2

0 1 2 3 4 5 6 7 8 9 10 11 12

ELEVATION

7 6
8
9
10
11
12

12

45°

SIDE ELEVATION

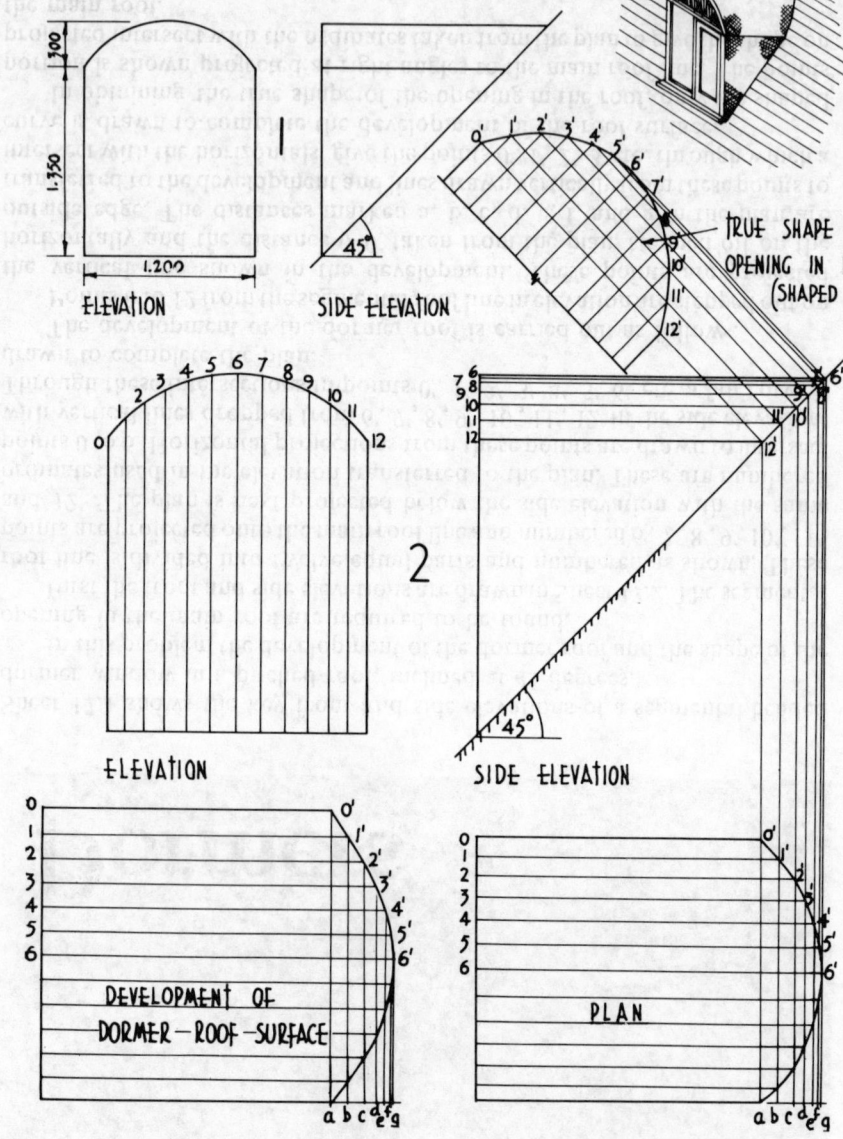

0 0'
1 1'
2 2'
3 3'
4 4'
5 5'
6 6'

DEVELOPMENT OF DORMER-ROOF-SURFACE

a b c d e f g

0 0'
1 1'
2 2'
3 3'
4 4'
5 5'
6 6'

PLAN

a b c d e f g

TURRET ROOF

3

0 1' 2' 3' 4' 5' 6' 7' 8' 9' 10' 11' 12'

TRUE SHAPE OF OPENING IN ROOF (SHAPED PORTION)

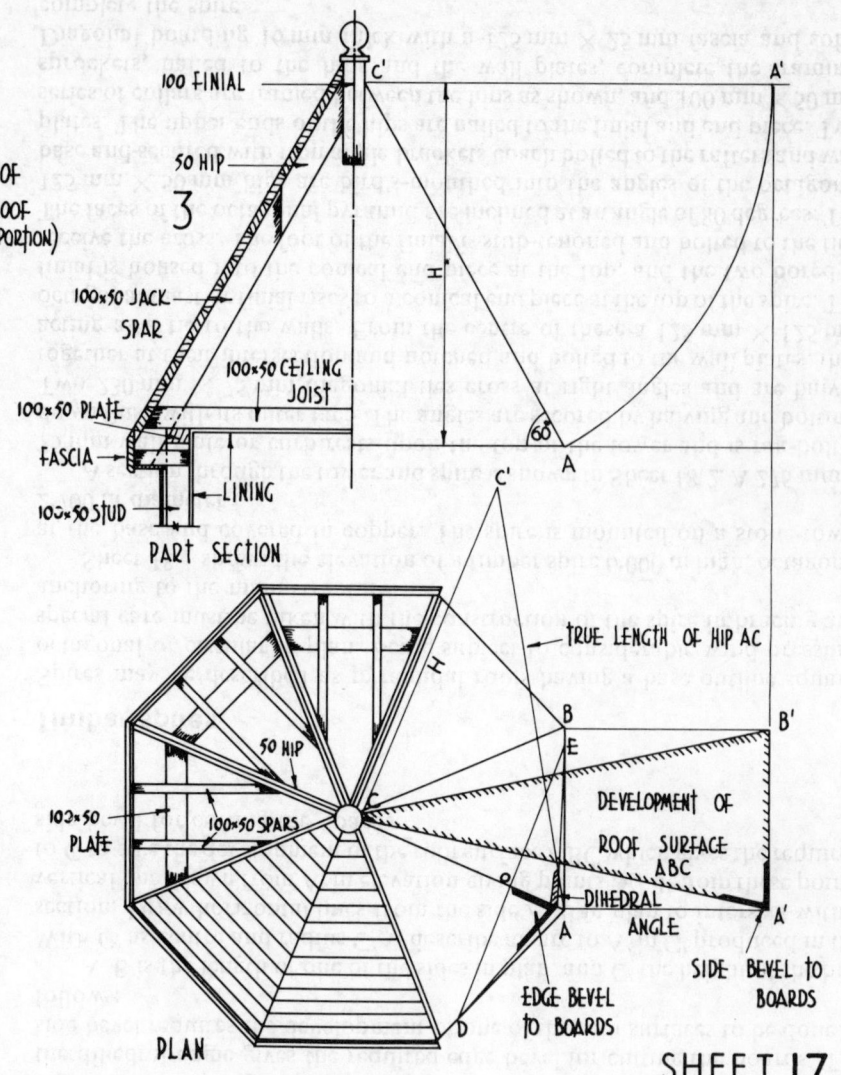

100 FINIAL

50 HIP

100×50 JACK SPAR

100×50 CEILING JOIST

100×50 PLATE

FASCIA

LINING

100×50 STUD

PART SECTION

C'

A'

H'

60°

A

TRUE LENGTH OF HIP AC

C'

H

B

E

B'

50 HIP

100×50 SPARS

102×50 PLATE

DEVELOPMENT OF ROOF SURFACE ABC

A

DIHEDRAL ANGLE

A'

SIDE BEVEL TO BOARDS

D

EDGE BEVEL TO BOARDS

PLAN

SHEET 17

TIMBER   SPIRE

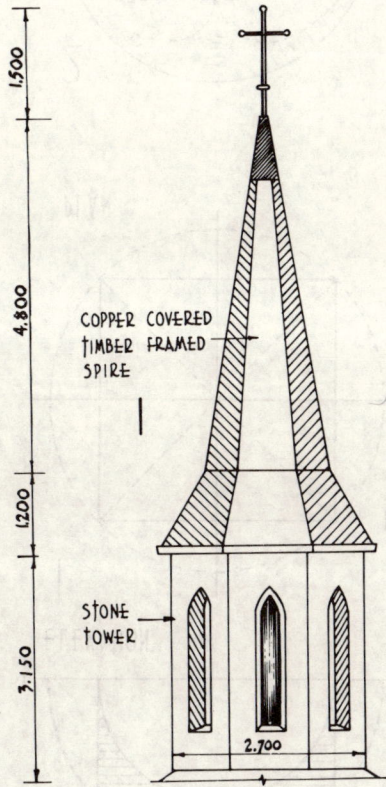

1500
4800
1200
3150

COPPER COVERED
TIMBER FRAMED
SPIRE

STONE
TOWER

2.700

I

ELEVATION

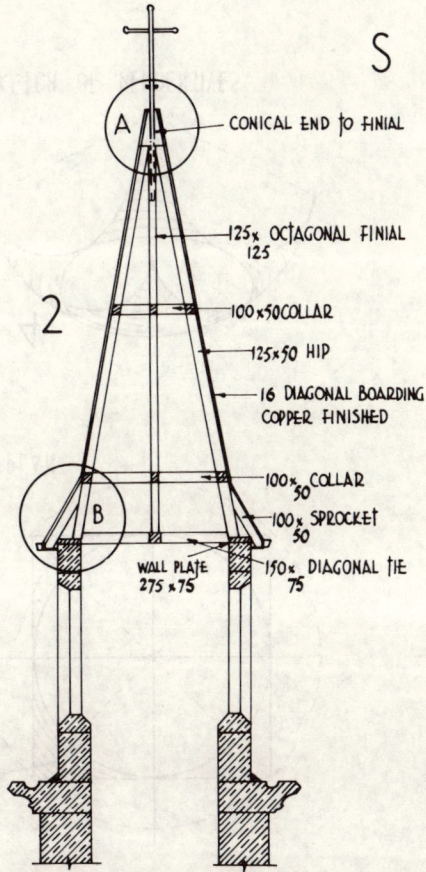

CONICAL END TO FINIAL

125 x OCTAGONAL FINIAL
/125

100 x 50 COLLAR

125 x 50 HIP

16 DIAGONAL BOARDING
COPPER FINISHED

100 x COLLAR
/50
100 x SPROCKET
/50
150 x DIAGONAL TIE
/75

WALL PLATE
275 x 75

2

SECTION THROUGH TOWER & SPIRE

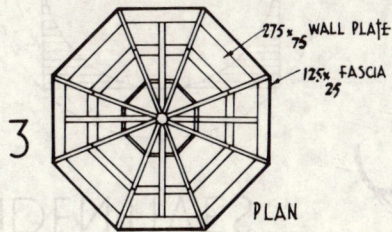

275 x WALL PLATE
/75
125 x FASCIA
/25

3

PLAN

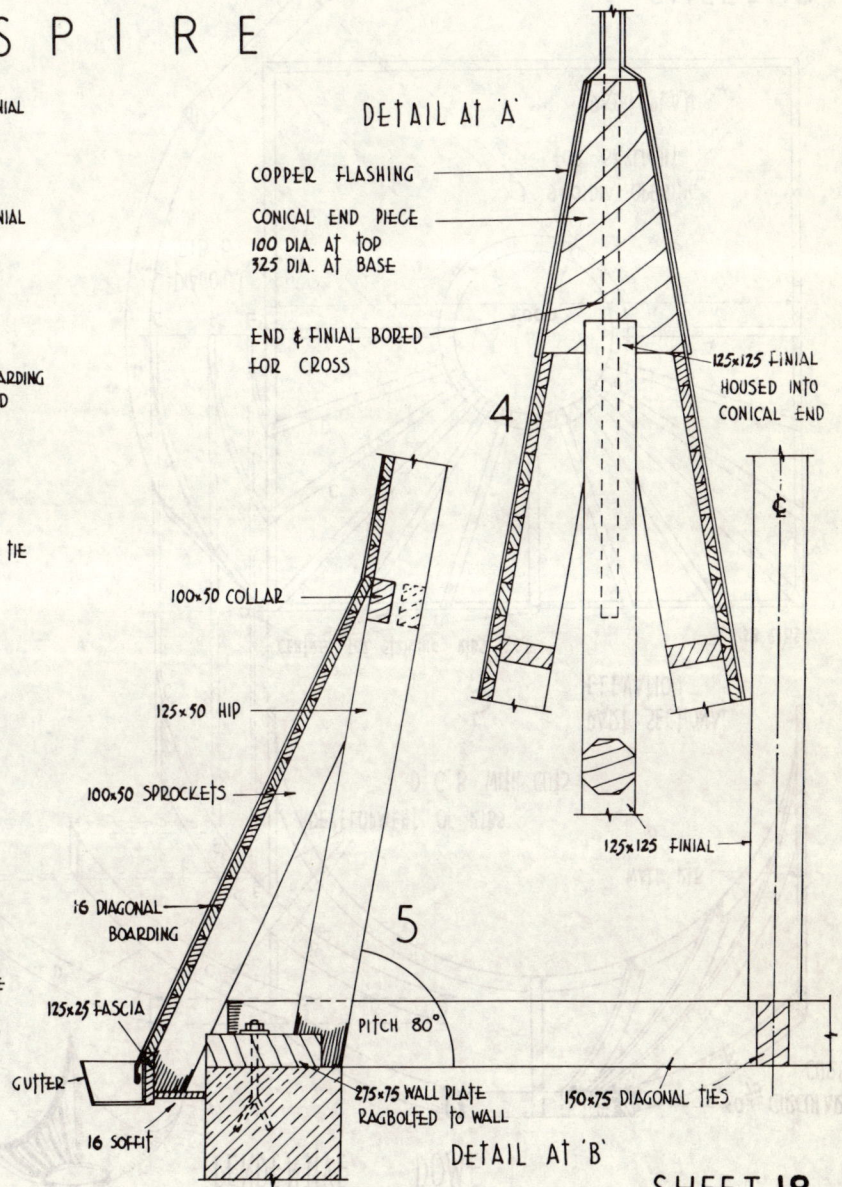

DETAIL AT 'A'

COPPER FLASHING

CONICAL END PIECE
100 DIA. AT TOP
325 DIA. AT BASE

END & FINIAL BORED
FOR CROSS

4

125 x 125 FINIAL
HOUSED INTO
CONICAL END

125 x 125 FINIAL

100 x 50 COLLAR

125 x 50 HIP

100 x 50 SPROCKETS

16 DIAGONAL
BOARDING

125 x 25 FASCIA

GUTTER

16 SOFFIT

5

PITCH 80°

275 x 75 WALL PLATE
RAGBOLTED TO WALL

150 x 75 DIAGONAL TIES

DETAIL AT 'B'

SHEET 18

137

DOMES & PENDENTIVES

PENDENTIVE DOME

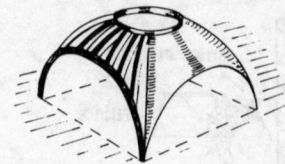

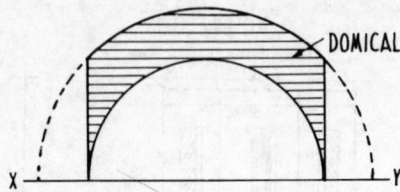

DOMICAL VAULT

ELEVATION

1

PLAN

PENDENTIVES

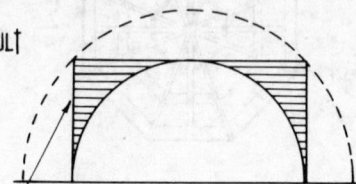

ELEVATION

3

PLAN

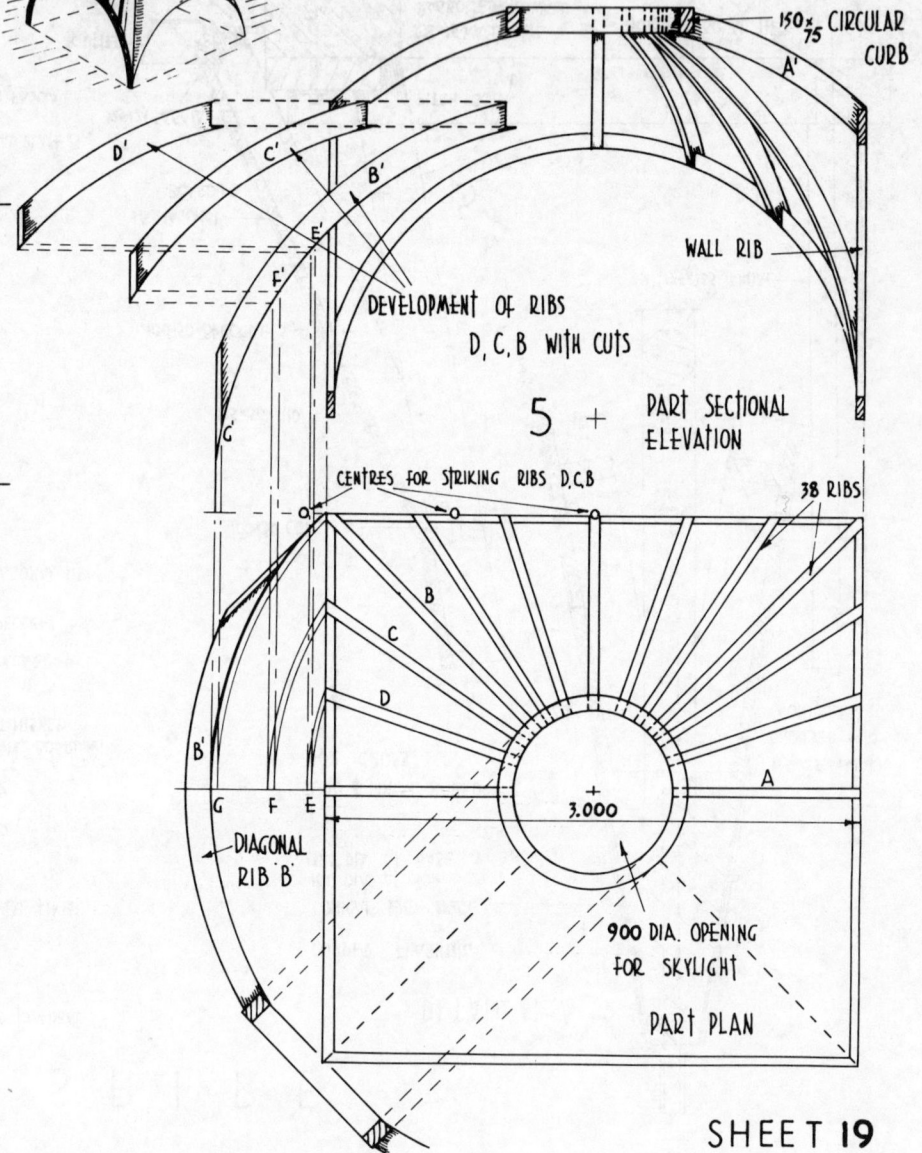

150×75 CIRCULAR CURB

A'

D'   C'   B'

WALL RIB

B'

E'

F'

DEVELOPMENT OF RIBS
D, C, B WITH CUTS

G'

5 +   PART SECTIONAL
        ELEVATION

CENTRES FOR STRIKING RIBS D, C, B

38 RIBS

B
C
D

B'

G F E

DIAGONAL RIB 'B'

A

+
3.000

900 DIA. OPENING
FOR SKYLIGHT

PART PLAN

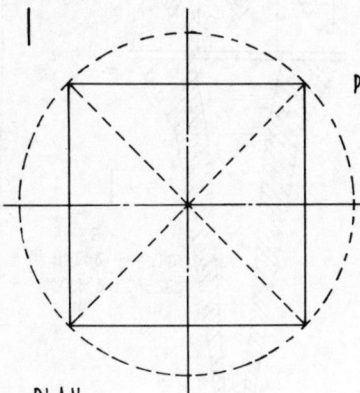

2

SKETCH OF DOMICAL VAULT

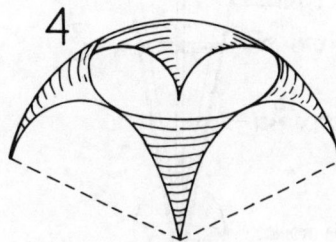

4

SKETCH OF PENDENTIVES

SHEET 19

hemisphere. The four vertical walls, which become vertical planes, remove four half-zones leaving the hemisphere as shown in the sketch, (Sheet 19.2). This is termed a domical vault.

Sheet 19.3 shows the top zone of the sphere removed from the previous figure, leaving a circular ring on plan. The spherical spandrels above each of the angles in plan are termed 'pendentives' and are shown in the sketch (Sheet 19.4).

Sheet 19.5 shows the part plan and part sectional elevation of a pendentive dome, having a circular opening for a skylight in the centre. The diameter of the dome is equal to the diagonal of the square in the plan.

One half of the plan shows the ribs in position, with the shape of the diagonal rib B developed to the left. This is struck from the centre with a radius equal to that of the dome and is laminated similar to those shown in the domed roof on Sheet 20. The spacing of the ribs will depend on the nature of the finish. To obtain the sectional elevation, the springing line of the dome is drawn and from the mid-point as centre, and radius equal to one-half of the diagonal, describe the inner curve of the rib A'. Four such ribs are required. From the same centre the semicircle, which is the intersection of the hemisphere and the wall, is drawn. The position of the curb for the skylight is projected vertically from the plan. The other ribs are drawn in the elevation by first positioning the ends at the side rib and at the curb. In this view, these ribs will be elliptical curves.

The ribs have different lengths, but the same curvature. The lengths of ribs B, C, and D are found by turning them to the centre line in plan at points E, F, and G. The diagonal rib B has been redrawn clear of the elevation at B' as shown. Vertical projections from points E and F on to the inner curve of the developed rib B', are drawn at E' and F'. Projections drawn from these points and from the head of the rib B' give the head and foot cuts of the ribs C' and D'. Four such ribs as C' and D' are required.

Domed roof

A dome is a vaulted roof having a circular, elliptical, or polygonal plan.

Sheet 20.1 shows the plan and elevation of an octagonal dome, the covering being omitted. To the left of the centre, the projections of two hips and one rib are shown, and to the right the method of obtaining the true shape of one of the hips.

The method of determining the elevation of the hips and rib is shown in the next example and fully described.

To obtain the shape of one of the hips, the outer curve of the elevation is divided into any convenient number of parts. Draw vertical projectors from these points on to the centre line of the hip member 0–6 in plan. Projections at right angles from these points and clear to the right, with corresponding heights from the elevation, enable the true outline of the hip to be determined.

The development of the hip rafter shows the member laminated in two thicknesses, with the joints of each layer falling in the middle of the adjacent layer as in small centring. The number of sections making up the hip depends upon the span and the available width of material. Likewise, the construction of the dome depends upon its size and its location.

Sheet 20.2 is an enlarged plan of the top ends of the hips, showing their connection to the finial. The tenons are each nailed from the side in turn. Purlins, at a convenient position, are housed into the hip rafters and provide a fixing for the head of the ribs.

Sheet 20.3 shows the true shape of the rib AB projected from plan. This is struck with a radius equal to the radius of the dome from the centre O. The construction of the ribs is similar to that for the hips.

When the hip rafter is cut to the true shape and square to the face, it still requires backing so that the roof boarding will seat properly. This is done by placing a templet cut to the developed shape on the hip and sliding it sideways a distance equal to C–D marked in plan. The hip is again marked to the templet and the process repeated on the other side. When cut to the lines the correct bevel will be found.

It should be noted from the enlarged plan of the hip that the amount of slide is dependent on the thickness of the hip.

Hemispherical dome

The sketch (Sheet 20.4) shows a hemispherical dome illustrating two approximate methods of covering this figure. To the left of the centre, gores are used. In this case the section planes cutting the sphere are all vertical and pass through the centre.

To the right of the centre, zones are used. This method of development is based on the assumption that the sphere is made up of a large number of conical surfaces, forming horizontal rings in elevation.

As it is not possible to develop accurately the surface of the sphere, approximations have to be resorted to in order to produce a practically accepted result in covering domes in timber, as a basis for further covering in lead or copper.

Sheet 20.5 shows the plan and elevation of a hemispherical dome. To the left of the centre the elevation of the ribs is determined. Divide the elevation as shown and project the points vertically downwards to the diameter in plan. With point 6 as centre, arcs are drawn from the points on the centre line, to cut the ribs in points 6–12' and 6–12". These points are projected back into the elevation and where they intersect with horizontal lines drawn from their opposite numbers will give the points to draw the elevations of the ribs.

To the right of the centre, the development of the sphere is shown. The points 0–6 in elevation are set out as horizontal sections. Draw vertical projectors from these points to the diameter in plan from which the concentric plan circles of six zones can be drawn. A line drawn through points 2 and 3 of zone A in elevation is produced to cut the centre line to give the centre C. With C as centre and C2 and C3 as radii, draw the arcs for 2 and 3. Similarly, from E as centre and radii E3 and E4, draw the smaller zone. The stretch-out for the zones is seen in plan, the distances stepped off and transferred to the development give the stretch-out for one-quarter of the complete zone. Each zone is treated in a similar way.

The approximate development of a gore is shown in the plan. With centre 6 and radii 6–1, 6–2, etc. draw arcs across the section to be developed, to give 1–1', 2–2', etc. The centre line is drawn clear of the plan and distances equal to

DOMED ROOF

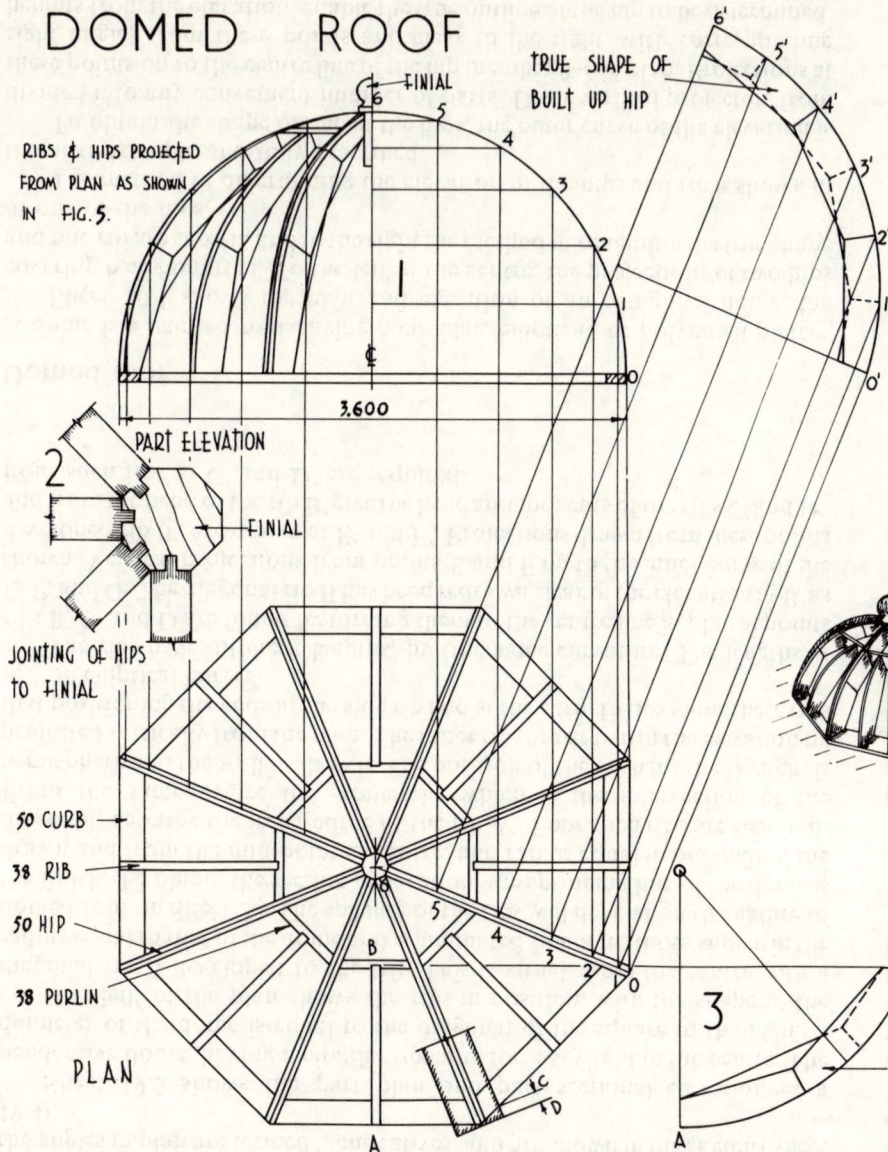

HEMISPHERICAL DOME

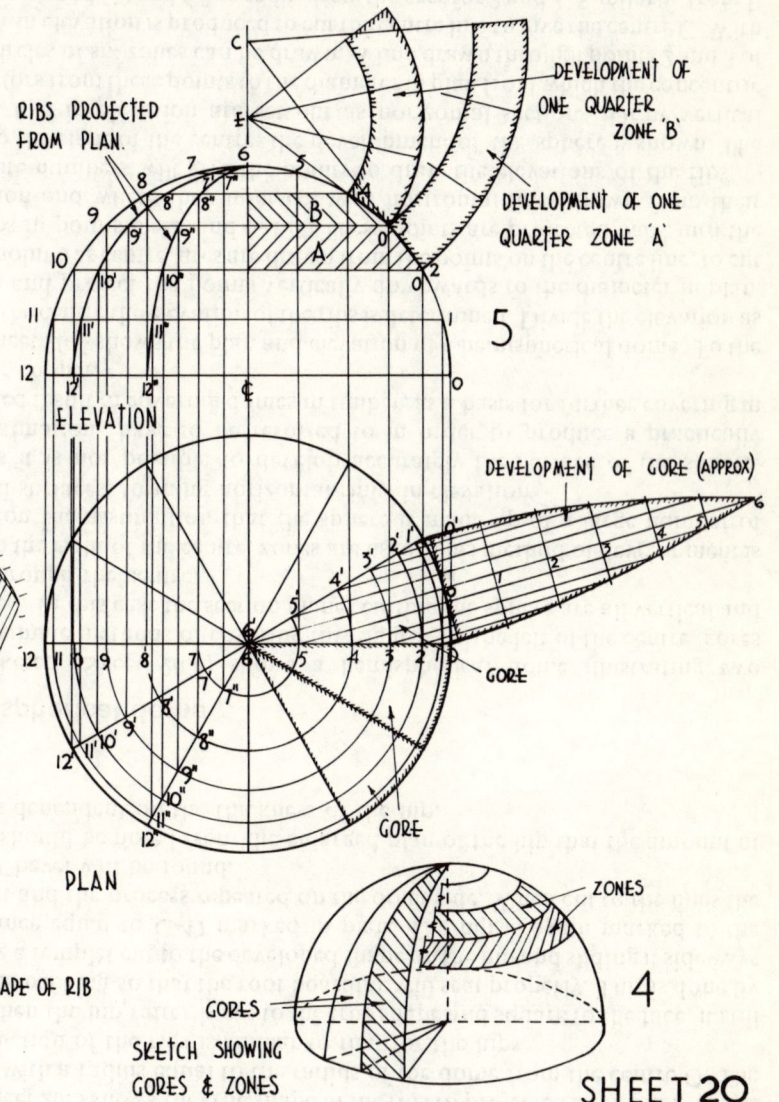

TRUE SHAPE OF BUILT UP HIP

FINIAL

RIBS & HIPS PROJECTED FROM PLAN AS SHOWN IN FIG. 5.

1

PART ELEVATION

3.600

2

FINIAL

JOINTING OF HIPS TO FINIAL

50 CURB

38 RIB

50 HIP

38 PURLIN

PLAN

3

TRUE SHAPE OF RIB AB

RIBS PROJECTED FROM PLAN.

DEVELOPMENT OF ONE QUARTER ZONE 'B'

DEVELOPMENT OF ONE QUARTER ZONE 'A'

5

ELEVATION

DEVELOPMENT OF GORE (APPROX)

GORE

PLAN

GORE

ZONES

GORES

SKETCH SHOWING GORES & ZONES

4

SHEET 20

those round the elevation marked off. Projections from the plan points to intersect with the vertical lines through 0, 1, 2, 3, 4, 5, 6 will give points through which to draw the approximate development.

Intersecting vaults

In vaulting there are three main types: barrel, groined, and ribbed.

When a barrel vault is intersected by one or more vaults, as in the corridors of public buildings, the line of intersection is called a groin, and the vault a groined vault.

Sheet 21.1 shows the intersection at right angles of two vaults of different span, but of the same height. The cross-section of the smaller vault is semicircular and the groins are straight lines on the plan. The shape of the section of the larger vault is found by dividing the semicircle into a number of parts, 1, 2, 3, etc. and projecting these points vertically downwards on to the plan line of the intersection. On projections from these points drawn horizontally, the heights, equal to those in the original section, are marked giving points 1', 2', 3', etc. The curve drawn through these points will give the required cross-section. It is semi-elliptical, its major axis being the span, and its semi-minor axis the rise of the vault.

The points on the plan line of the intersection are again used to obtain the true shape of the groin ribs, the method being similar to that already described. The ribs are built up or laminated, the spacing depending on the nature of the covering. A rebated section shows the method of obtaining the lengths and side cuts of the rib AB. The double lines at A' and B' show the bevel at the ends where they meet the groin rib.

Another method of framing for a vault of this type is for the centring for one of the vaults to continue across the intersection using ribs placed and fixed at intervals. The laggings run through to complete the centring for this vault. The ribs for the cross-vault are then fixed with one rib close up to the centre already in position and lagged. The laggings are run over those of the first set and scribed to fit the curve, the overhang being supported by part ribs and fillets. In this method groin ribs are not required.

Niches

A niche is a recess formed in a wall with the head usually semispherical. Sheet 21.2 shows the part elevation and plan of such a niche, framed in timber, it being assumed that the finish is to be plaster. The plan is semicircular and the face of the wall is flat.

The ribs of niches may be placed either in vertical planes as illustrated, or in horizontal planes to form the spherical surface.

In Sheet 21.3 the ribs forming the niche head stand in vertical planes which would pass through the centre if produced. The ribs are of similar curvature because the head of the niche is formed by a quadrant of a circle rotating about the centre. The lengths of the ribs will vary because of the thickness of material. The front or face rib is semicircular, the remaining ribs, all arcs of circles, are struck with the same radii as the front one. Additional

length for the trenching of the ribs into the curb must be allowed.

The shapes of ribs A, B, and C are shown projected at right angles to their positions in plan. The joints at the top of the ribs are also positioned from the plan as shown.

Ventilators

Sheet 22.1 shows the elevation and section through a triangular louvre ventilator. The frame is jointed together using dovetails at the two base angles, these being formed on the rail, and a mitred bridle joint at the top. The louvres are inclined at an angle of 45 degrees and are arranged so that the top edge of each is covered by the bottom edge of the one immediately above it. This cover is necessary so that one should not be able to see through to the other side and to prevent driving rain from passing through to the inside. The louvre boards are trenched into the frame as shown, so that it is necessary to develop the shapes of the boards and also one of the sides of the frame to enable the trenchings to be set out, the sides being set out as a pair.

To find the bevel for setting out the trenchings, project from the section the points a, b, c, and d onto the inside face of the frame. Rebate the side of the frame AB so as to show its width and project the points a, b, c, and d over at right angles to the developed side in a', b', c', and d'. Join these points to give the bevel and the setting out of the trenchings.

The shapes of the louvres can be developed in either of two ways. Both methods are shown. The first requires a templet on which the louvres can be marked (Sheet 22.2).

Points A and O in elevation are projected over to the right of the section in A' and O'. A line is drawn from O' at the same inclination as the louvre boards, in this case 45 degrees, to give the centre line. A line drawn at right angles to this through O' in B' and C', equal to BC in elevation, gives the base of the templet.

A louvre board marked E is projected from the section on to the centre line of the templet in points 1', 2', 3', 4'. From the centre line the points are projected at right angles to the edges of the templet to give the true shape of the top and bottom faces of the louvre E and the side bevel.

The second method of finding the bevels to cut the louvre boards is shown in Sheet 22.3.

Draw a vertical line through 2 to represent the vertical plane. With centre 2 and radius 2–1 draw an arc cutting the line in 1'. Draw a horizontal line from 1' to give 1" on a vertical line from point 1 in elevation. Join 2'–1" to give the side bevel to the louvre board. The edge bevel is obtained in exactly the same way.

A circular louvre ventilator is shown in Sheet 22.4. The frame in this case is built up in four sections and jointed with either handrail bolts and dowels, or double hammer-headed keys. The method of finding the shapes of the louvre boards is similar to the previous example, and will not be repeated here. The templet in this case is elliptical. The setting out of the trenchings in the frame requires special treatment and a good deal of skill. A rough frame is made to receive the circular one as shown in Sheet 22.5. The positions of the louvres as given by the section are marked on the outer sides of the rough

VAULTS & NICHES

INTERSECTING VAULTS

NICHE WITH DOMED HEAD

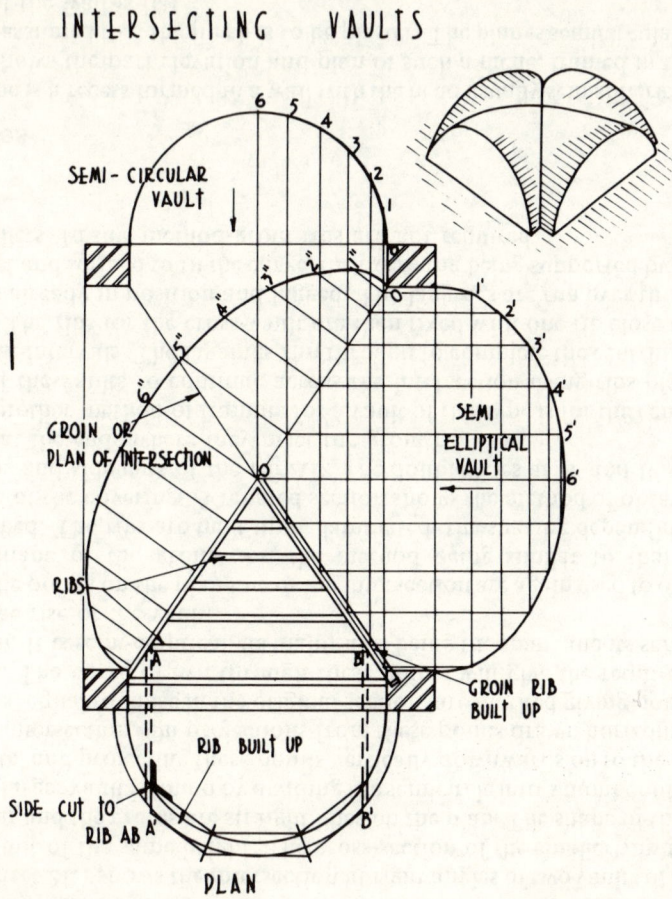

SEMI - CIRCULAR VAULT

SEMI ELLIPTICAL VAULT

GROIN OR PLAN OF INTERSECTION

RIBS

GROIN RIB BUILT UP

RIB BUILT UP

SIDE CUT TO RIB AB

PLAN

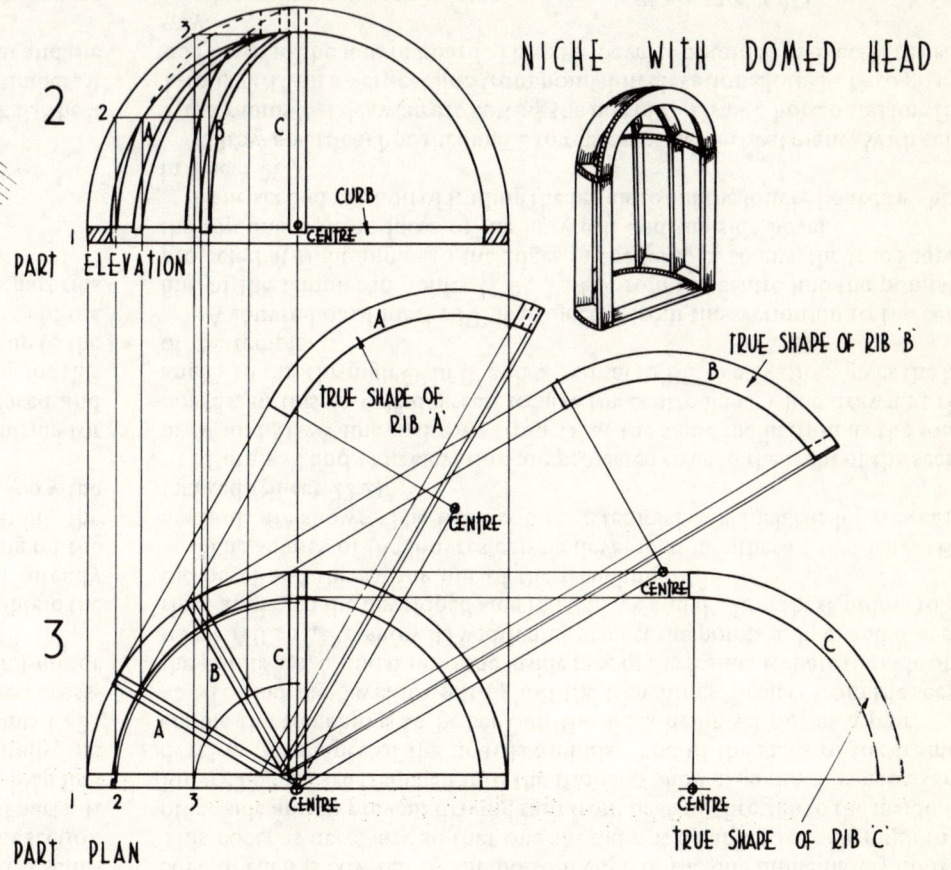

2

A B C

CURB

CENTRE

PART ELEVATION

TRUE SHAPE OF RIB 'A'

TRUE SHAPE OF RIB 'B'

CENTRE

A

B

C

CENTRE

3

A B C

CENTRE

1 2 3

CENTRE

CENTRE

TRUE SHAPE OF RIB 'C'

PART PLAN

SHEET 21

LOUVRED   VENTILATORS

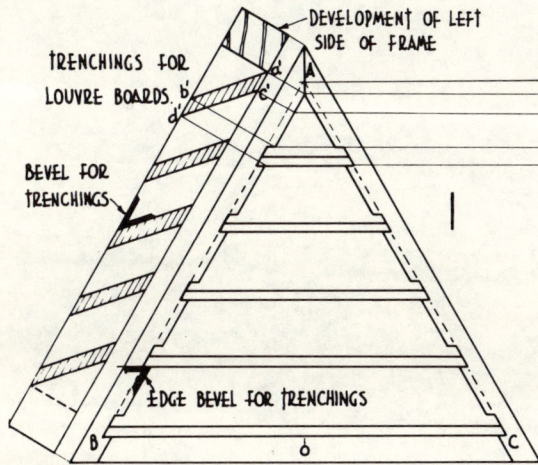

DEVELOPMENT OF LEFT SIDE OF FRAME

TRENCHINGS FOR LOUVRE BOARDS

BEVEL FOR TRENCHINGS

1

EDGE BEVEL FOR TRENCHINGS

ELEVATION

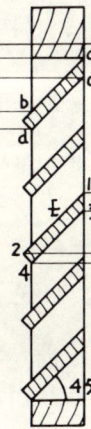

SECTION

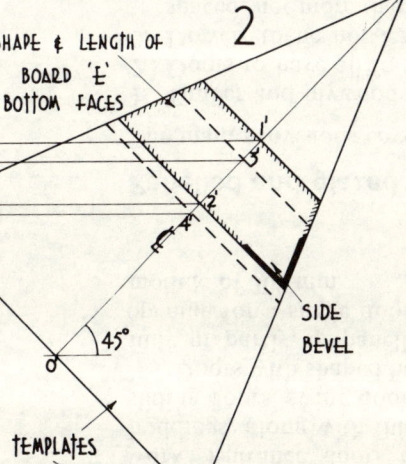

2

TRUE SHAPE & LENGTH OF LOUVRE BOARD 'E' TOP & BOTTOM FACES

45°

SIDE BEVEL

TEMPLATES

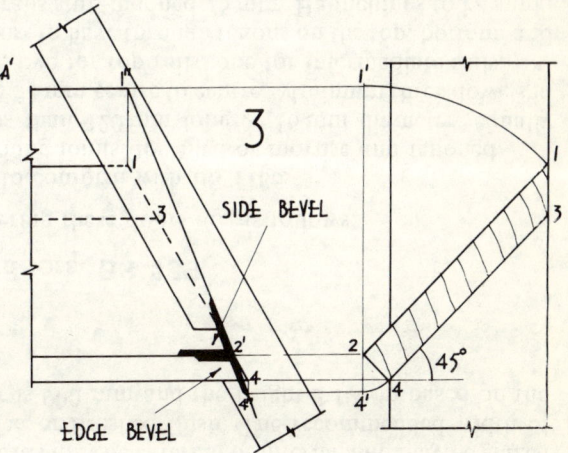

3

SIDE BEVEL

EDGE BEVEL

45°

LOUVRE BOARD BEVELS

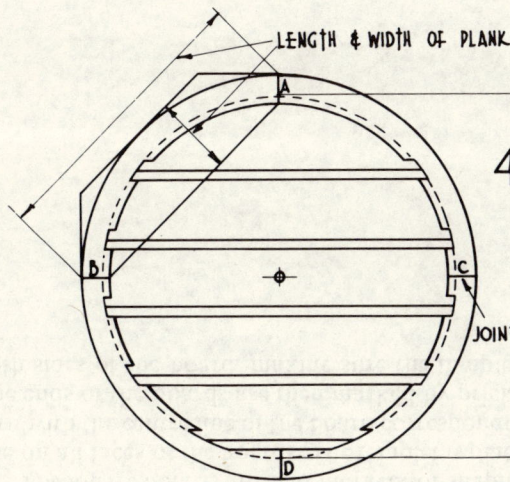

LENGTH & WIDTH OF PLANK

4

JOINT

ELEVATION

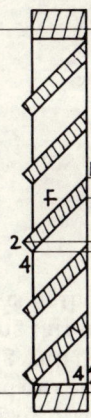

SECTION

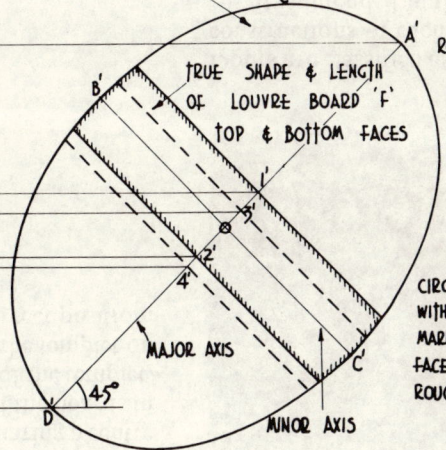

TRUE SHAPE & LENGTH OF LOUVRE BOARD 'F' TOP & BOTTOM FACES

MAJOR AXIS

45°

MINOR AXIS

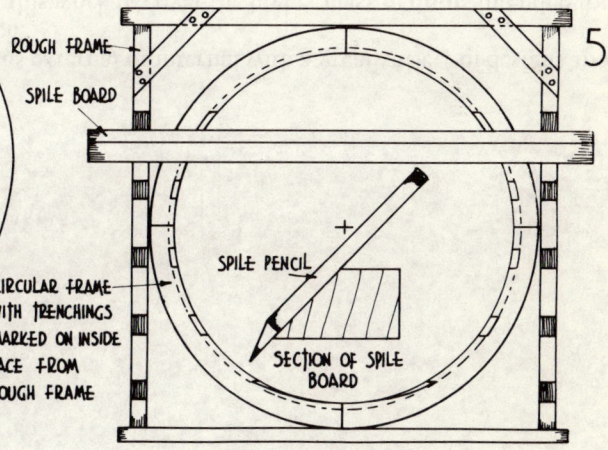

ROUGH FRAME

SPILE BOARD

5

CIRCULAR FRAME WITH TRENCHINGS MARKED ON INSIDE FACE FROM ROUGH FRAME

SPILE PENCIL

SECTION OF SPILE BOARD

SETTING OUT TRENCHINGS

SHEET 22

frame shown shaded. A spile board is then used to mark the positions of the trenchings on the front face of the frame. A section through the spile board is shown, having the front edge splayed at the same pitch as the louvre boards. This allows a pencil, sharpened so that the lead is in the same plane as the spile board edge, to be used to mark the housings on the inside face of the circular frame.

The louvre boards are next marked for shaping by first squaring a centre line on all faces of the boards. The templet is placed on each of the boards in turn with the centre line of the board corresponding with that of the templet. The ends of the boards are then marked by pricking through the templet on both sides of the board, making sure the templet is in the correct position.

Chapter 15

Doors

Doors are classified as external or internal with a great variety of designs and constructions of each.

It is intended in this work to cover the higher class of doors intended for particular purposes in specialised buildings requiring high-class joiner's work. Entrance doors to civic buildings, banks, churches, and public buildings, along with flush and fire-check doors; internal revolving doors; sliding doors, swing doors and vestibule doors fall into the two categories.

Doors with shaped heads may be external or internal and may be glazed fully or partly, or panelled or finished flush. The recommended width of opening for a single door is 900 mm and the height 2·100 m based on the module of 300 mm.

Panelled and glazed doors. BS 459

The main provisions covering these doors are as follows:

1. Timber and plywood to conform with BS 1186.
2. Doors to have all framing joints dowelled or mortise and tenoned.
3. Dowels to be not less than 120 mm long × 16 mm diameter, equally spaced, not more than 54 mm centre to centre. Minimum three dowels in bottom and lock rails, two for top rails, one for intermediate rails.
4. Mortise and tenon doors to have through tenons on the top, bottom, and one other rail. Other rails stub-tenoned 25 mm. Haunchings to be sunk, minimum 10 mm.
5. Through tenons to be wedged.
6. Plywood panels to fill the grooves to the full depth, less 2 mm.
7. Mouldings shall be worked on the solid and scribed. Rebates in glazed openings sunk in the solid.

Framed and panelled doors consist of a stable skeleton frame with glued joints, the openings of which are filled with thin panels of either solid timber

PANEL & LOUVRED DOORS

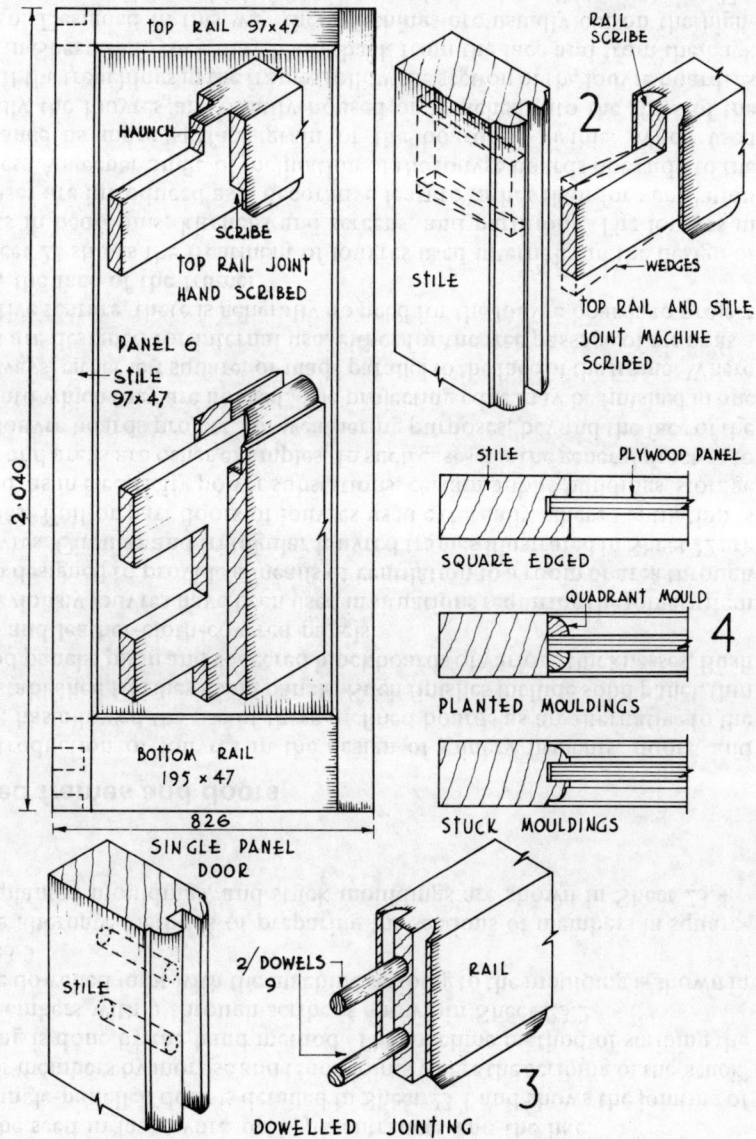

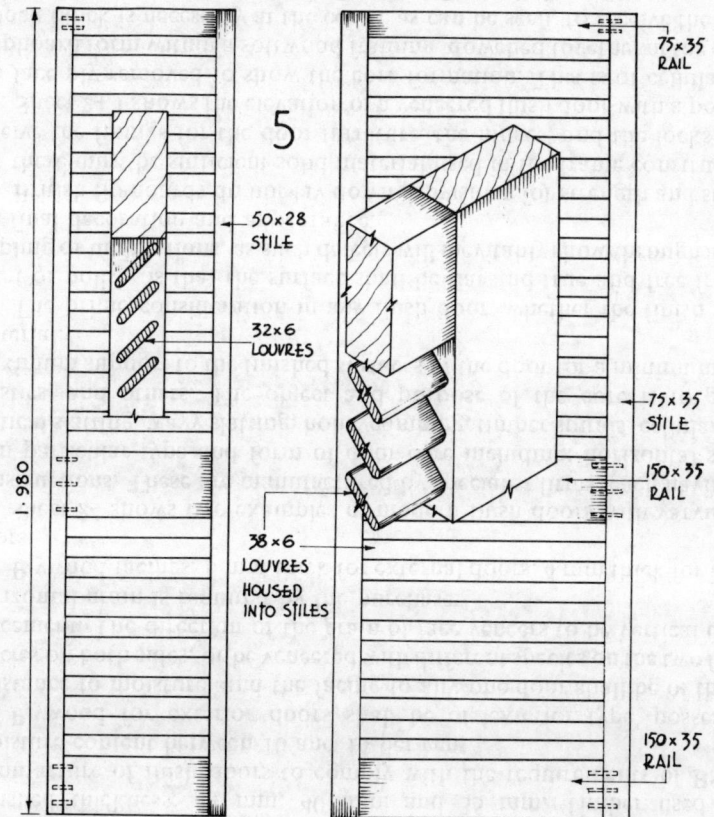

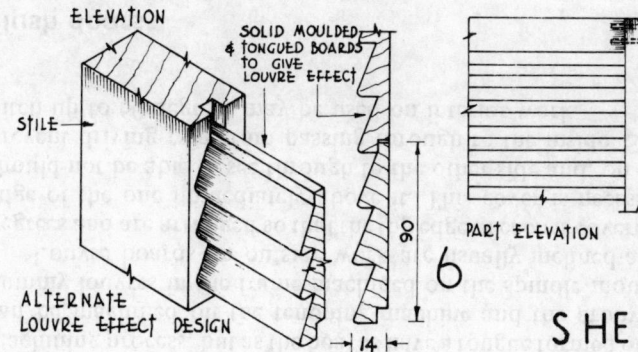

1

TOP RAIL 97 × 47

HAUNCH

SCRIBE

TOP RAIL JOINT
HAND SCRIBED

PANEL 6

STILE
97 × 47

2.040

826

BOTTOM RAIL
195 × 47

SINGLE PANEL
DOOR

2

RAIL
SCRIBE

STILE

WEDGES

TOP RAIL AND STILE
JOINT MACHINE
SCRIBED

4

STILE   PLYWOOD PANEL

SQUARE EDGED

QUADRANT MOULD

PLANTED MOULDINGS

STUCK MOULDINGS

3

STILE

2/DOWELS
9

RAIL

DOWELLED JOINTING

5

75 × 35
RAIL

50 × 28
STILE

32 × 6
LOUVRES

75 × 35
STILE

150 × 35
RAIL

38 × 6
LOUVRES
HOUSED
INTO STILES

1·980

150 × 35
RAIL

ELEVATION

6

STILE

SOLID MOULDED
& TONGUED BOARDS
TO GIVE
LOUVRE EFFECT

PART ELEVATION

ALTERNATE
LOUVRE EFFECT DESIGN

90

14

SHEET 23

or plywood. This type of door can be divided into any number of panels, separated by horizontal members or vertical members. In door construction the horizontal members are either lock rails or frieze rails and the vertical members, muntins. Usually panelled doors of this type are associated with more expensive work, and their construction allows considerable elaboration as can be seen in later work on bank entrances and the like.

A single-panelled door is detailed in Sheet 23.1 and shows the jointing of the door members by mortise and tenon joints where the scribing of the 'stuck' moulding is done by the hand method. The machine method of scribing the same members with a through scribe is shown in Sheet 23.2.

The dowelled joint with the machine scribing to the moulding is shown in Sheet 23.3.

The alternate methods of preparing the sections of members in square-edged, planted mouldings, and stuck mouldings are shown in Sheet 23.4.

Louvred frames and doors

The introduction of louvres in the design of joinery fitments, doors, and screens, has allowed the use of these inclined boards as an alternative to the more established finishes using panels. Such finishes include solid panel, thin plywood panels, plain and veneered blockboards of various thicknesses, flush panels, and leather-cloth-covered panels.

Previously louvres have been used in situations requiring the joinery item to be so designed to provide a means of ventilation to a room or area through the louvres. Circular and triangular louvred frames illustrated in Sheet 22 are examples. Full or part doors of louvres used externally where ventilation is required, as in electricity power substations, certain public buildings, storage rooms, and areas are other examples. In such cases it is the general practice to let the louvre boards project, for weathering purposes, beyond the face of the frame into which they are housed. The projecting edge may be finished in one of two ways, either left square, or made parallel to the face of the frame. Where louvres are designed for internal use, either for the free passage of air or as a decorative feature, there is generally no need for the louvre boards to project beyond the face of the frame.

Sheet 23 shows the treatment of louvres used internally in the design of fitments in bedrooms, kitchens and screens, and partitions. The louvres in such cases are introduced as a decorative feature rather than for ventilation purposes. A steeper angle of inclination of the louvre boards does add to the appearance as more surface grain of the boards is visible. When used internally the louvres are usually housed or trenched into the stiles of the doors. If the trenchings in the frames follow the section of the louvre board, as shown in Sheet 23.5, they may be set back from the face and from the back side also. Designed in this way the trenchings are usually cut on the high-speed routing machine which forms a rounded end on the trenching. The louvre boards require to be moulded to the same radius to add to the finished decor.

The louvre boards cannot be inserted into their trenchings if the frame is glued up. They must be a part of the general assembling of the frame, having first been cleaned up ready for the finishing specified.

An alternative design of infilling using suitably grained timbers to produce a louvred effect is shown in Sheet 23.6. It will be seen that the decorative effect is produced by boards moulded in section to represent inclined boards in the finished work. In this method there is much waste in the machining process, but as the boards have a tongue formed on each end, these can be machined on the tenoning machine and the groove to receive the dummy louvres in the frame machined on the spindle moulder.

Louvre boards on outside work are usually inclined at an angle of 45 degrees and are arranged so that the top edge of each is covered by the bottom edge of the one immediately above it. This cover is necessary so that one should not be able to see through to the other side and, on outside work, to prevent driving rain from passing through to the inside. Steeper angles of pitch up to 60 degrees may be used on interior work.

Flush doors

Finished thickness: 44 mm, 40 mm, and 35 mm. Timber used in the manufacture of flush doors to comply with the requirements of BS 1186. Moisture content between 10 and 14 per cent.

Plywood for exterior doors shall be of 'exterior type' possessing a resistance to moisture, and the facing to any one door shall be of the same species on both sides, or be veneered with different species on the two faces by agreement. The direction of the grain of face veneers to be vertical unless a horizontal grain is required by the purchaser.

Plywood facings: 6 mm thick for external doors, 4 mm thick for internal doors.

Sheet 24 shows two examples of modern flush doors with varying core constructions. These are manufactured by specialist firms, each having their own particular type and form of door core including: horizontal slatting, vertical slatting, wavy slatting, honeycombing, timber spirals, cellular paper, plastics, and others. The object and purpose of the core is to give the maximum support to the finished surfaces of the door for a minimum cost in material.

The prime consideration in any flush door, whether the finish is to be paint or polish, is that the surface shall be flat and true and free from any rippling or undulations, as such defects will inevitably show through and ruin the final decoration and appearance.

British Standards do not lay down provisions for strength and stability, but there must be sufficient solid material used in the frame construction to receive the fixings for the door furniture, the hinges, and the locks.

Sheet 24.1 shows the elevation of a veneered flush door with a portion of the face ply removed to show the core formation. This is of cellular paper chipboard form within a softwood framing, dowelled together at the corners. A lock block is necessary at the centre as can be seen, to receive the mortise lock. A 16 mm hardwood lipping to the two vertical edges of the door protects and receives the 4 mm face plies and gives a margin for fitting the door in width.

Sheet 24.2 shows a similar sized and finished flush door having a core of timber spirals with the lipping tongued into the framing at the door edges. The lippings should match the veneered faced ply in hardwood doors and the core

FIRE-CHECK & FLUSH DOORS

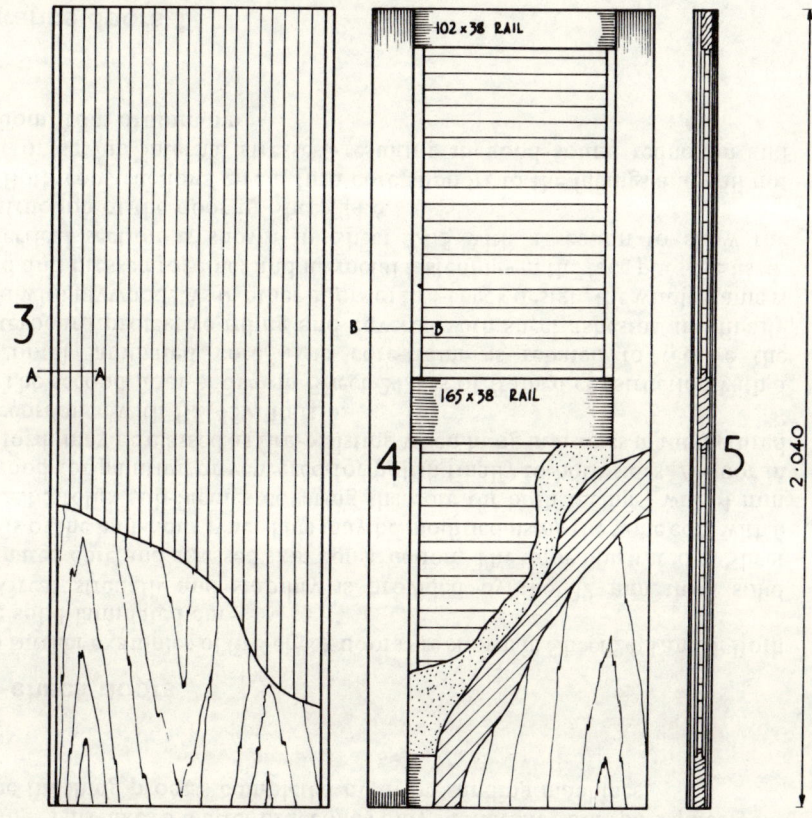

102 × 38 RAIL

165 × 38 RAIL

726

2·040

ELEVATION

SECTION

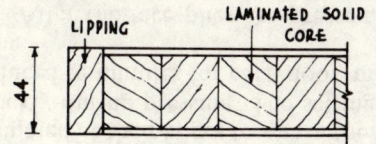

LIPPING

LAMINATED SOLID CORE

44

SECTION A-A

HALF HOUR TYPE

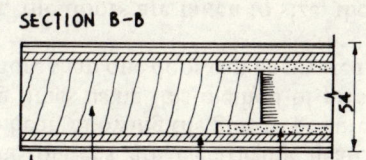

SECTION B-B

54

102 × 38 STILE
4 FACE PLY

10 PLASTERBOARD

4 ASBESTOS

ONE HOUR TYPE

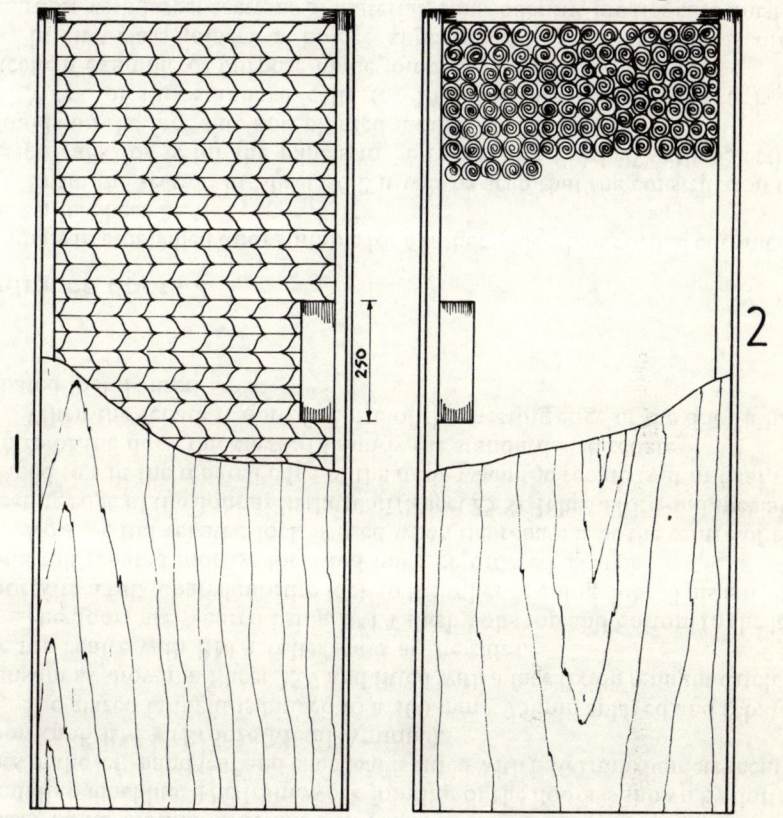

250

826

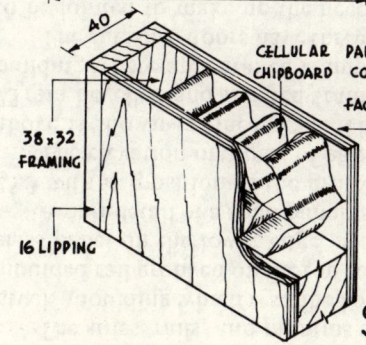

40

CELLULAR PAPER
CHIPBOARD CORE

FACE PLY 4

38 × 32
FRAMING

16 LIPPING

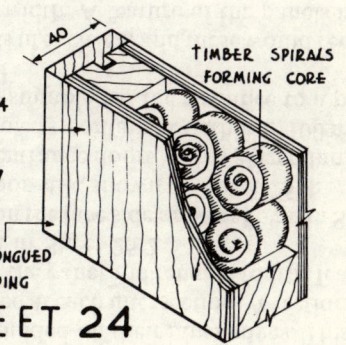

40

TIMBER SPIRALS
FORMING CORE

16 TONGUED
LIPPING

SHEET 24

cavity should be ventilated.

Openings may be required in flush doors for glass observation panels or louvres for ventilation and may be square, rectangular, or circular in outline.

Production of flush doors in quantity by specialist firms requires that the assembly of the frames and cores with the gluing of the face plies be carried out with modern machinery. This will include hydraulic presses operated by a motorised pump where a single large-diameter ram raises each of three pressing tables independently or all three locked together. A second type is an electric platen hydraulic press. The platens are electrically heated and insulated at their edges to retain the heat, allowing the glue to be cured very quickly during pressing. The setting times using this method of pressing is reduced to minutes for each door, more than one door can be pressed at one time.

After the face plies are pressed, the doors are taken to size, the edges prepared for the lippings which are then glued, positioned on the doors, and placed in a power-heating press for curing the glue line by radio-frequency heating. This takes a matter of seconds only so that the doors pass quickly on to the finishing process completed by drum sanding machines.

Fire-check doors

Two further examples of fire-check doors are shown in Sheet 24.3 and 4. Both have solid laminated cores.

Great strength and stability is provided by the 32 mm thick solid laminated core and stressed skin construction. The door illustrated in Sheet 24.3 is of the half-hour type. This may be modified using the same core with 6 mm asbestolux sub-facings covering the core on both facings, with 4 mm plywood for painting or veneered for polish finally to both sides. A door in this form may be classed as fire-resisting when hung in frames manufactured in accordance with BS 459, Part 3.

The second door shown in Sheet 24.4 is of framed construction with a horizontal laminated core. The core-frame is rebated to receive the plasterboard protective filling and covered with sheet asbestos and finally faced with plywood. As in other doors of fire-check design, the whole frame is glued and pressed together and no metal fastenings in the face plies are used. A vertical section is shown in Sheet 24.5 with a section to show the construction of the door in Sheet 24.6.

Both constructions give a firm core support to the facings and will not twist or set up internal stresses, resulting in good sound reduction and freedom from drumming.

Entrance doors 1

Sheet 25.1 shows the elevation of a more expensive class of entrance door. Panelled doors of this type, traditional in design, have again found favour in construction by many designers. They have a particularly pleasing appearance when framed in hardwood with a polish finish.

The design shows the door formed from framing enclosing nine raised panels with a centre 'bull's-eye' glass panel. Using a number of panels in this way reduces the width of individual panels which are more readily available than greater widths. The jointing of the various members in the framing is shown to the right of the centre line by dotted lines.

The stiles, rails, and muntins are moulded on their inner edges. This is a 'stuck' moulding which is scribed or mitred at each intersection. A section of a moulded rail grooved on the top edge for the panel and rebated for the feature glass panel on the lower edge, is shown in Sheet 25.2.

A part detail of a solid panel raised on the face side only is shown in Sheet 25.3 with its position in the grooved door stile shown in Sheet 25.4.

The elevation of a pair of panelled entrance doors with a glazed fanlight above is shown in Sheet 25.5. The whole is contained within a 100 mm × 75 mm bevelled and rebated frame and suitable for the entrance to a public building where appearance is important.

The entrance doors have three panels in each leaf and these would require to be jointed to make up the necessary width. A feature of the panels is the raised panel moulding detailed in Sheet 25.6 and formed on the spindle moulding machine. The framework jointing of the doors is shown by dotted lines in the left-hand leaf and each leaf is fitted with a 100 mm diameter centre door knob BMA (Bronze Metal Antique).

The glazed fanlight is hinged to a 100 mm × 75 mm splayed and rebated transom as shown in Sheet 25.7 and fitted with a face fixing fanlight catch at the top centre with BMA roller stays at the sides.

The doors are secured using BMA flush bolts top and bottom to the left door with a full rebated mortise lock to the other opening door. This is a left-hand full rebated mortise lock and must be ordered as such.

Note: A full rebated lock is used when the rebate is at the centre of the meeting edge of the door as in the plan (Sheet 25.8). It has a fore-end specially shaped to suit the rebated edge of the door when the rebate is at or near the centre of the door thickness. 13 mm is the standard size rebate.

When the rebate is out of centre of the meeting edge of the door a half rebated lock is used.

Entrance doors 2

A single five-panelled door suitable for a rather wider than normal entrance is shown in Sheet 26.1.

From the section in Sheet 26.2 it will be seen that the construction the design calls for is for the panels to be raised and finished with bolection moulds on the face side and planted moulds on the inside.

Work of this standard calls for first-class workmanship and is an excellent example of purpose-made joinery.

In first-class joinery a bolder appearance to the finishings is often desirable with larger sections of material; consequently, greater care must be taken to prevent any damage by shrinkage. A method of overcoming any movement by shrinkage in the entrance door illustrated, is by forming a subframe within the main frame. This allows for movement of the material without disturbing the joinery unit as a whole.

Sheet 26.3 shows the preparation of the main and subframe members in

ENTRANCE DOORS

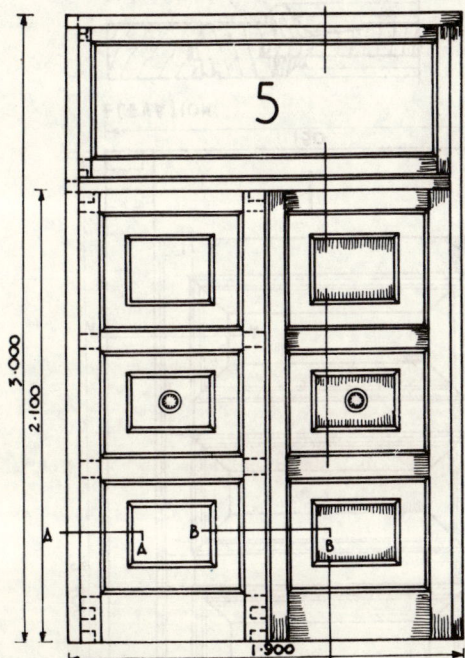

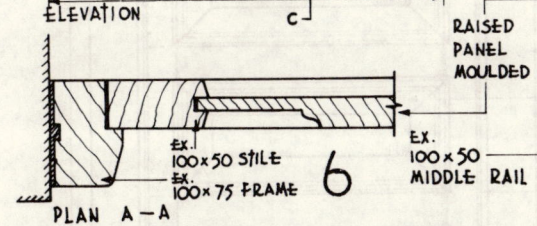

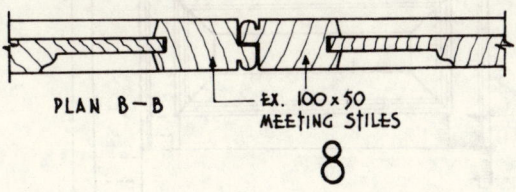

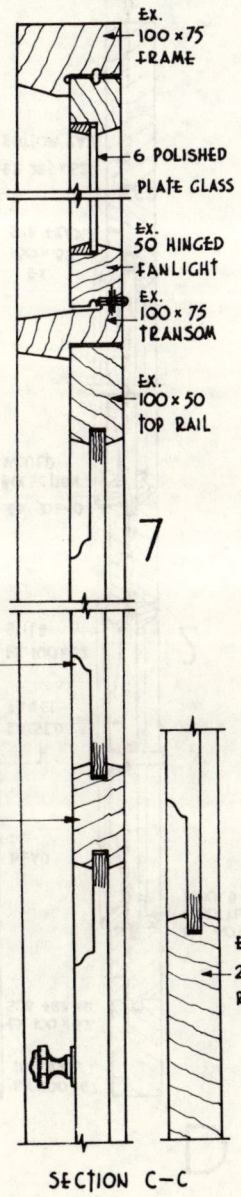

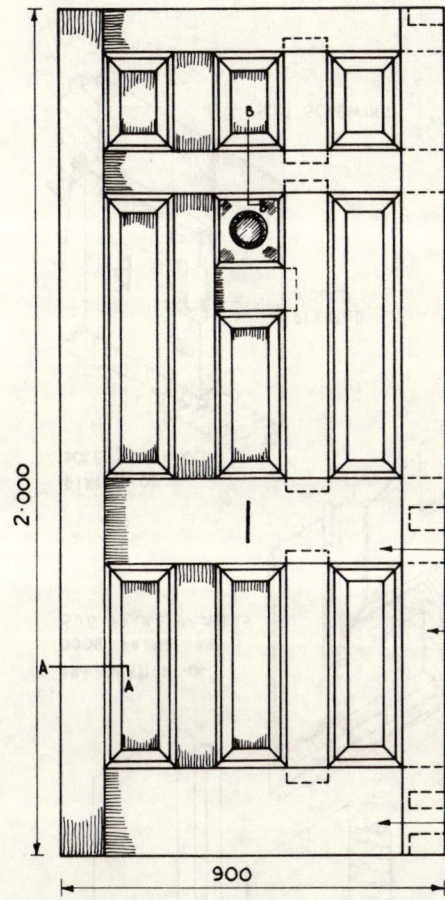

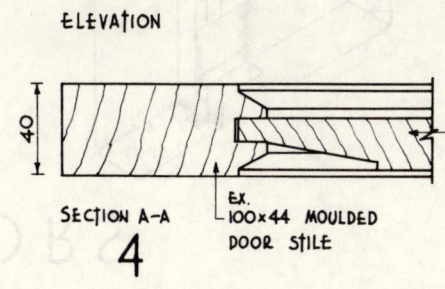

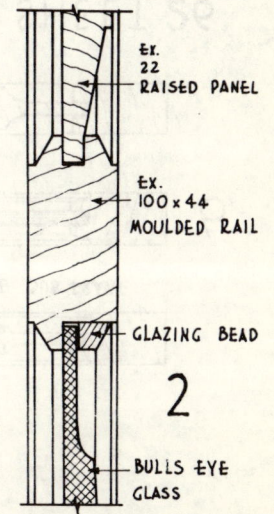

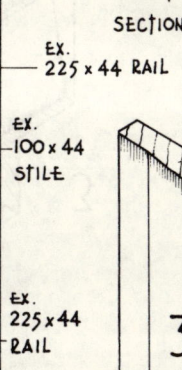

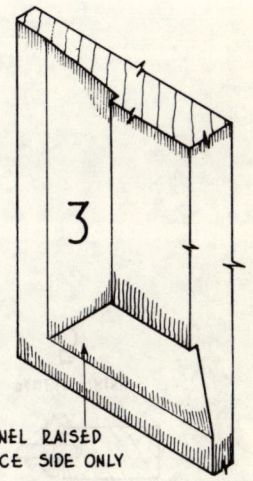

5

ELEVATION

3.000
2.100
1.900

A
B
A
B

EX.
100 x 75
FRAME

6 POLISHED
PLATE GLASS

EX.
50 HINGED
FANLIGHT

EX.
100 x 75
TRANSOM

EX.
100 x 50
TOP RAIL

7

RAISED
PANEL
MOULDED

EX.
100 x 50
MIDDLE RAIL

EX.
225 x 50
RAIL

SECTION C-C

PLAN A-A
EX.
100 x 50 STILE
EX.
100 x 75 FRAME
6

PLAN B-B
EX. 100 x 50
MEETING STILES
8

ELEVATION

2.000
900

A
B
A

1

EX.
225 x 44 RAIL

EX.
100 x 44
STILE

EX.
225 x 44
RAIL

EX.
22
RAISED PANEL

EX.
100 x 44
MOULDED RAIL

GLAZING BEAD

SECTION B-B

2

BULLS EYE
GLASS

3

PANEL RAISED
FACE SIDE ONLY

SECTION A-A
40
EX.
100 x 44 MOULDED
DOOR STILE
4

SHEET 25

PANELLED DOORS

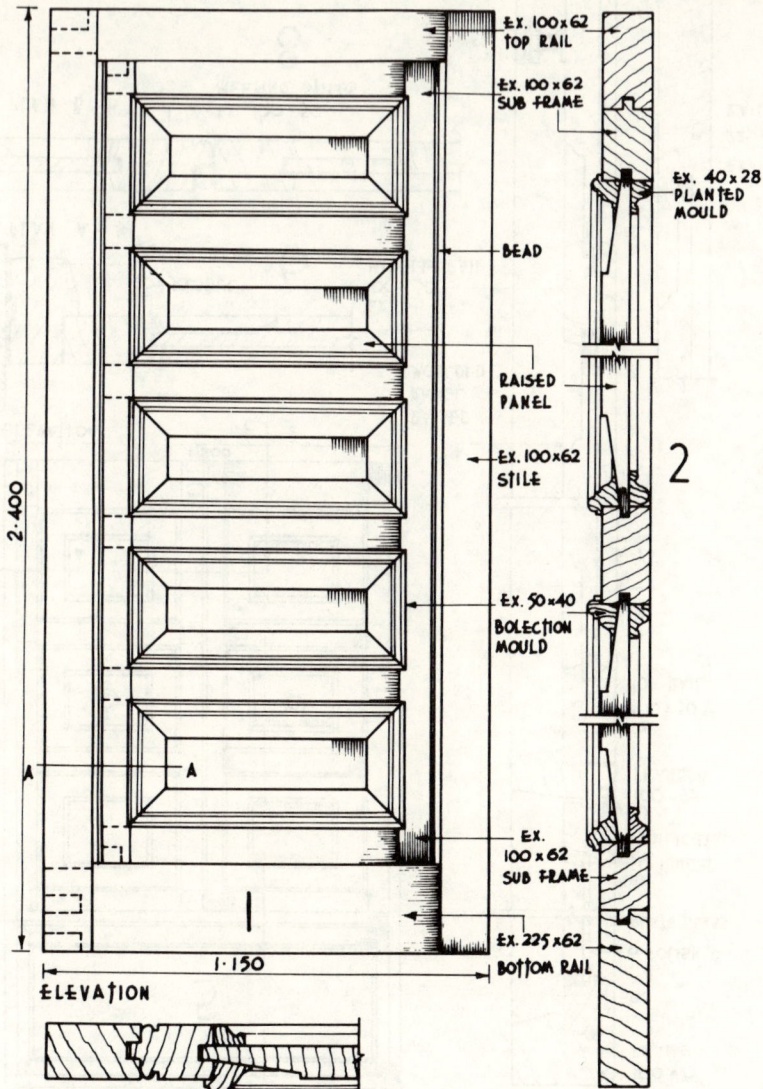

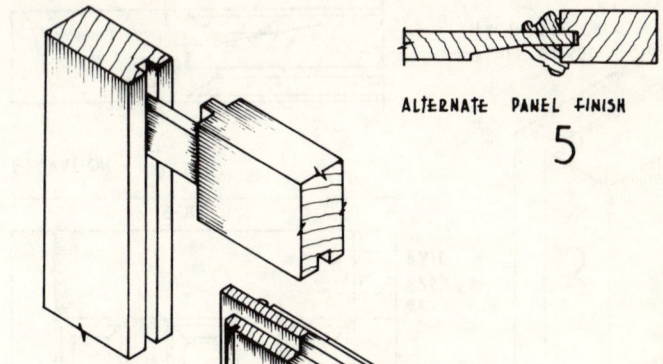

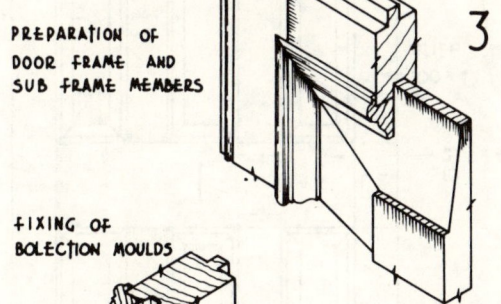

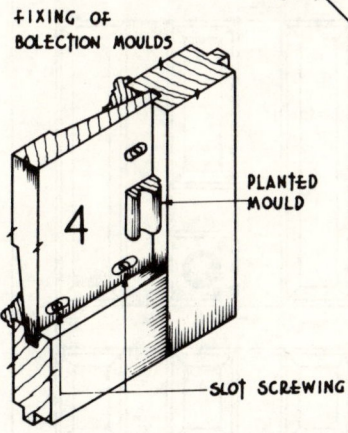

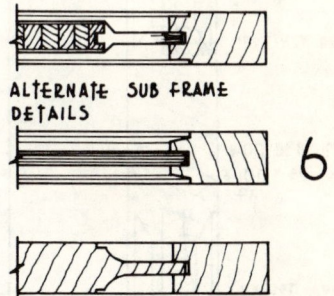

ELEVATION

2·400

1·150

1

A — A

PLAN A-A

EX. 100 x 62 TOP RAIL

EX. 100 x 62 SUB FRAME

EX. 40 x 28 PLANTED MOULD

BEAD

RAISED PANEL

EX. 100 x 62 STILE

EX. 50 x 40 BOLECTION MOULD

EX. 100 x 62 SUB FRAME

EX. 225 x 62 BOTTOM RAIL

2

PREPARATION OF DOOR FRAME AND SUB FRAME MEMBERS

3

ALTERNATE PANEL FINISH

5

FIXING OF BOLECTION MOULDS

4

PLANTED MOULD

SLOT SCREWING

ALTERNATE SUB FRAME DETAILS

6

SHEET 26

BANK ENTRANCE DOORS

56 FOLDING
DOORS

ELEVATION

1

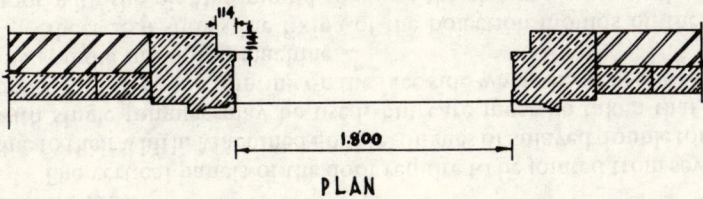

1.800

PLAN

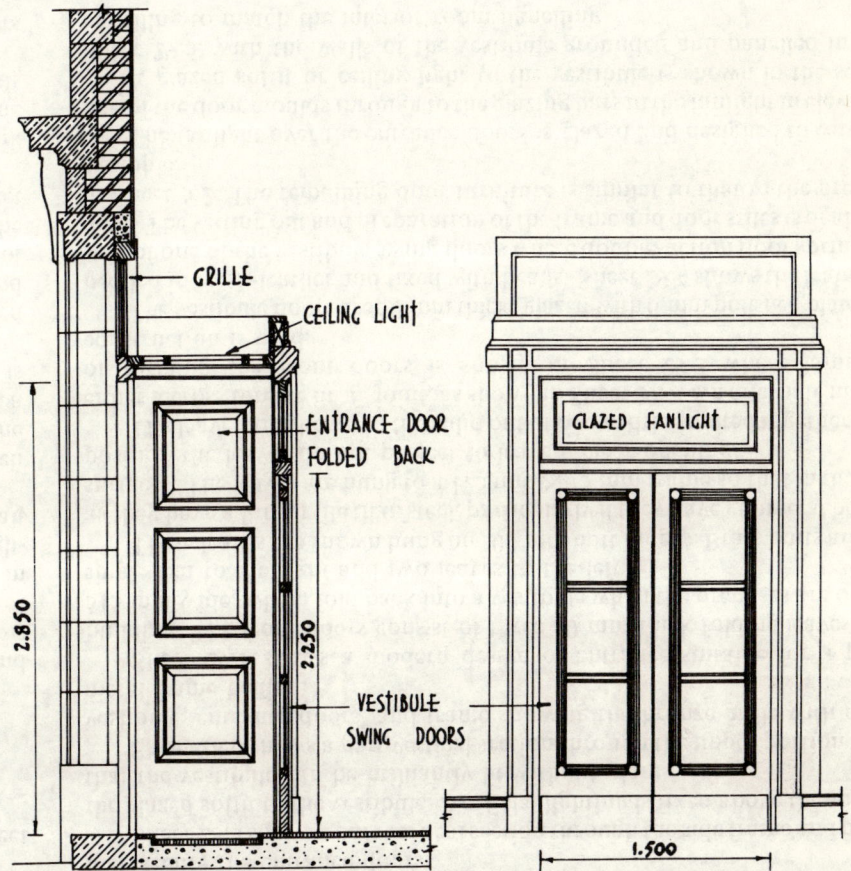

GRILLE

CEILING LIGHT

ENTRANCE DOOR
FOLDED BACK

2.850

2.250

VESTIBULE
SWING DOORS

SECTION

2

GLAZED FANLIGHT.

1.500

INTERNAL ELEVATION

3

SHEET 27

sketch form. It will be seen that a bead is stuck on the outer edges of the subframe stiles to help break the joint, and add to the appearance. This bead butt could be returned across the subframe at the inner edges of the top and bottom rails.

The vertical panels of the door require to be jointed from several boards due to their width. Machined double tongues or splayed double tongues along with single tongues may be used, but care must be taken that there is no breaking out of the jointing on the face side when the panels are raised using the spindle moulding machine.

Sheet 26.4 shows the fixing of the bolection moulds on the face of the door with the planted mould to cover the slot-screwing on the inside.

Bolection mouldings should not be fastened through the face of the panel but by slot-screwing from the reverse side as shown to allow movement of the panel.

Good mitre lines both on the panels and following through the bolection mouldings are important for a first-class finish.

Alternate details showing a panel raised and fielded are shown in Sheet 26.5 with alternate subframe finishes in Sheet 26.6.

Bank entrance doors and vestibule

Sheet 27.1 shows the plan and external elevation of the entrance doors and surrounding stone dressings to a bank. The joiner's work in this example is often termed 'traditional work', but even in new bank buildings this type of entrance is still to be found, apart from restoration work and work in extensions or additions to existing buildings. Certainly it represents high-class joiner's work in polished hardwood, and for this reason is being dealt with here.

The front doors consist of a pair of 56 mm thick folding leaves each 900 mm wide and 2.850 m high, which fold back into a vestibule when the premises are open. Above the doors can be seen a bronze grille which has a 32 mm metal frame behind glazed with 6 mm plate glass. The whole is contained within a 133 mm×100 mm frame with a 175 mm×100 mm transom. Sheet 27.2 shows a section through the vestibule with an entrance leaf folded back and a ceiling light above. Vestibule doors are usually swing doors and generally swing or open both ways on pivots or spring hinges. Examples of these can be seen in the chapter dealing with special ironmongery. The elevation of the vestibule doors with a glazed fanlight over is shown in Sheet 27.3.

Sheets 28.4, 28.5 and 28.6 show the details of the joiner's work for the bank entrance doors and vestibule. It will be seen that this is all moulded on the solid, with the exception of the entrance doors, which have bolection moulds on the face side.

The horizontal section in Sheet 28.4 shows one of the entrance doors folded back so that when open it is flush with the frame, the joint being masked by the roll moulding on the stile. In this case the door rotates on centres at the bottom and top. The side screens to the vestibule are framed and filled with raised panels. These are fixed in rebated end and corner frames with fillets. Moulded skirtings and architraves finish the vestibule as shown.

The inner swing doors, of which only a part is shown, are fitted into a solid frame. The hanging stiles of the doors are worked to a segment of a circle which fits into a corresponding hollow on the frame. The centre of rotation should be kept half the thickness of the door from the face of the frame, to allow the doors to open at right angles to the frame. The meeting stiles also are rounded; this curve is struck from the centre of rotation. The doors swing on centres, the lower one a floor spring and the upper one made to rise and fall to enable the door to be mounted, or on helical spring hinges. The swing doors have wide built-up stiles consisting of a reeded sub frame tongued within the main frame. This allows for any swelling or shrinking of the material without disturbing the joinery as a whole. Plate glass is bedded in wash leather and fixed within the subframe with beads. This method is used to prevent breakage of large sheets of glass through any movement in the door which may cause binding or the doors to foul each other when in use. Flush bolts with a mortise lock secure these doors while flush bolts and a rebated mortise lock secure the entrance doors.

Sheet 28.5 shows a part vertical section through the side frame and part of the glazed soffit to the vestibule. Artificial lighting is fixed above the soffit so that the vestibule can be brilliantly lit without glare.

Sheet 28.6 shows a part vertical section through the upper portion of the vestibule, entrance doors, and frame showing the bronze grille with glazed metal frame behind.

Sheet 29.1 shows a modern design of entrance suitable for a public building. The front doors consist of three 50 mm thick folding leaves, each 515 mm wide, which fold back into a vestibule when the premises are open, a single leaf to the right and two leaves to the left.

These leaves are shown hung on 100 mm butt hinges. Brass butts are used as they have a longer life than steel, particularly if they have gunmetal bearing surfaces. The leaves are hung to a 125 mm × 75 mm frame so that in the open position the leaves do not project to leave a clear opening.

The leaves are shown in the solid out of two widths jointed together using either a cross tongue or 'F' joint, as shown in Sheet 29.2. An alternate method of forming the front doors is shown in Sheet 29.3 where laminated construction is used.

The vestibule doors are 50 mm thick, glazed with 6 mm polished plate glass bedded in wash-leather and fixed with beads. Sheet 29.4 shows the frame and stile of one of the vestibule swing doors where double-action floor springs are used. The setting out and preparation of the frame and door stiles is dealt with in Sheet 5.2. The remaining door furniture is similar to that in the previous example.

The fanlight over the entrance doors is glazed and designed to carry the line of the door moulds through to the glazing bars in the fanlight in elevation.

A glazed soffit or ceiling light to the vestibule is shown in the section Sheet 29.5, with the walls of the vestibule grounded and panelled in flush panelling to match the interior room panelling.

DETAILS OF VESTIBULE

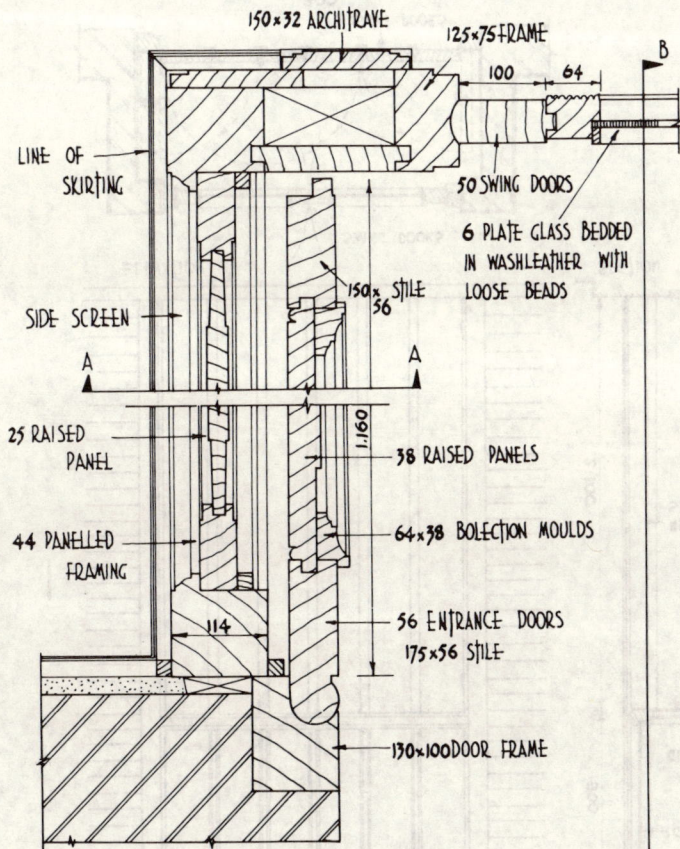

150×32 ARCHITRAVE

125×75 FRAME

100   64

LINE OF SKIRTING

50 SWING DOORS

6 PLATE GLASS BEDDED IN WASHLEATHER WITH LOOSE BEADS

SIDE SCREEN

150× STILE 56

A   A

25 RAISED PANEL

38 RAISED PANELS

64×38 BOLECTION MOULDS

44 PANELLED FRAMING

56 ENTRANCE DOORS 175×56 STILE

114

1,160

130×100 DOOR FRAME

4

HORIZONTAL SECTION THROUGH SIDE SCREEN & DOOR

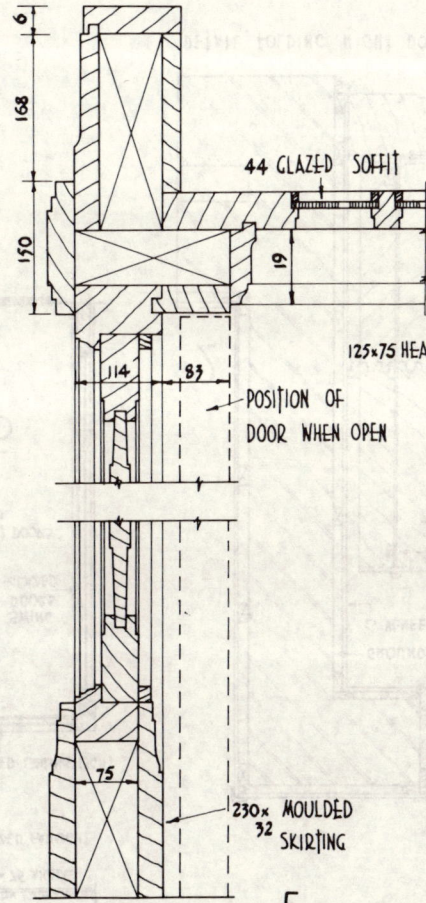

B

6

168

150

44 GLAZED SOFFIT

19

114   83

125×75 HEAD

POSITION OF DOOR WHEN OPEN

75

230× MOULDED 32 SKIRTING

5

VERTICAL SECTION A-A

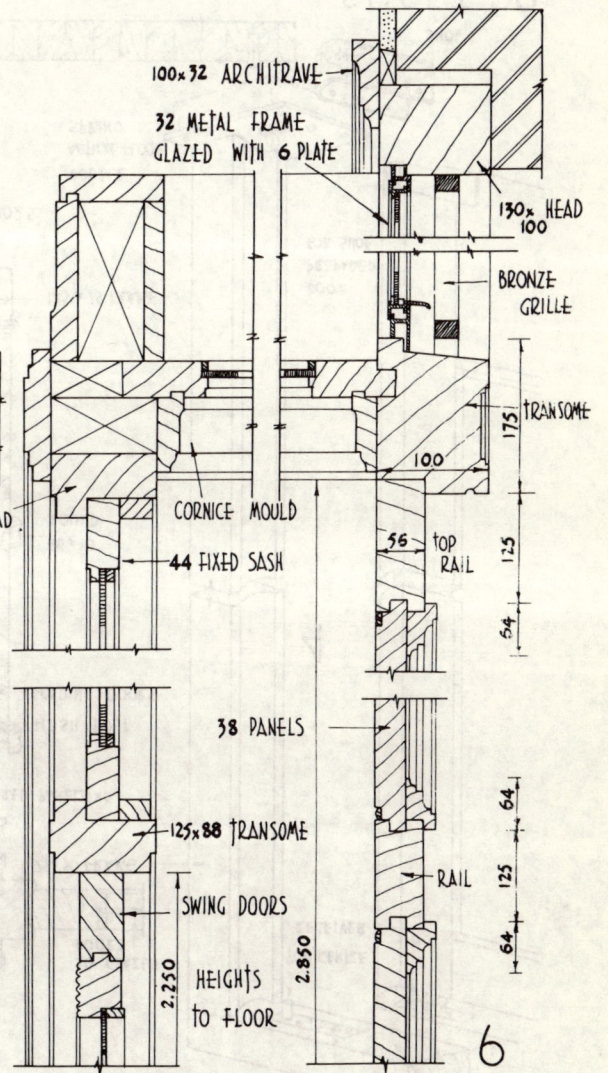

100×32 ARCHITRAVE

32 METAL FRAME GLAZED WITH 6 PLATE

130× HEAD 100

BRONZE GRILLE

TRANSOME

175

100

CORNICE MOULD

44 FIXED SASH

56 TOP RAIL

125

64

38 PANELS

125×88 TRANSOME

SWING DOORS

2.250

2.850

HEIGHTS TO FLOOR

RAIL

64

125

64

6

VERTICAL SECTION B-B

SHEET **28**

153

ENTRANCES & VESTIBULES

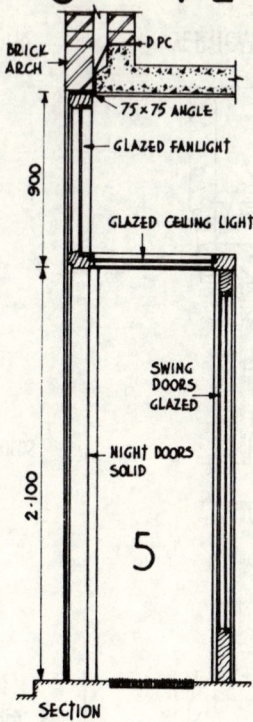

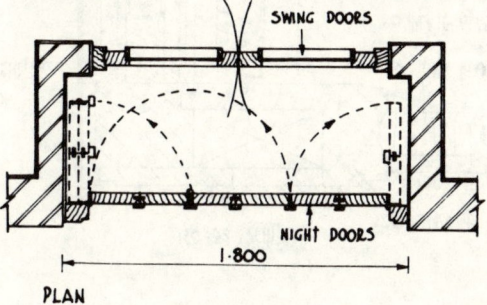

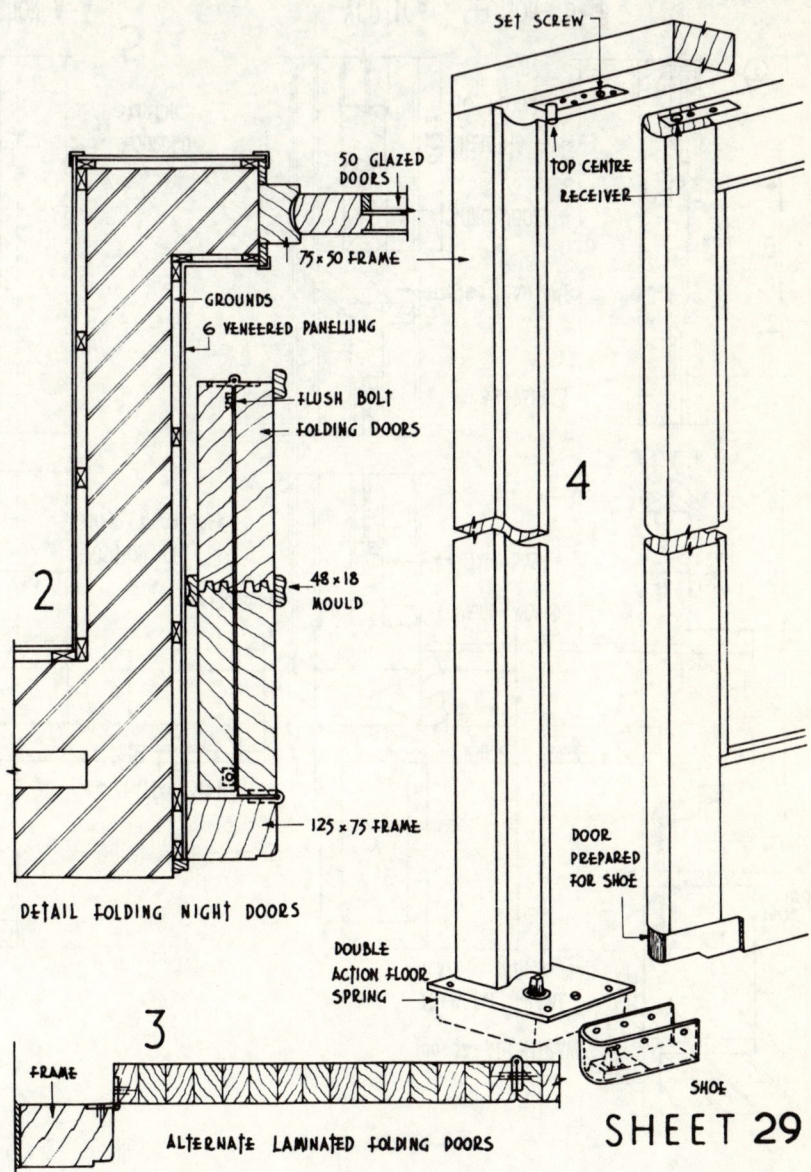

1

ELEVATION

SWING DOORS

NIGHT DOORS

1·800

PLAN

BRICK ARCH

DPC

75 × 75 ANGLE

GLAZED FANLIGHT

900

GLAZED CEILING LIGHT

SWING DOORS GLAZED

2·100

NIGHT DOORS SOLID

5

SECTION

50 GLAZED DOORS

75 × 50 FRAME

GROUNDS

6 VENEERED PANELLING

FLUSH BOLT
FOLDING DOORS

48 × 18 MOULD

125 × 75 FRAME

2

DETAIL FOLDING NIGHT DOORS

SET SCREW

TOP CENTRE RECEIVER

4

DOOR PREPARED FOR SHOE

DOUBLE ACTION FLOOR SPRING

SHOE

FRAME

3

ALTERNATE LAMINATED FOLDING DOORS

SHEET 29

Swing doors

Single and double swing doors are used extensively in public buildings, offices, hospitals, shops, and in libraries. In the main, such situations require doors to close automatically, and for this purpose some kind of spring hinge is used or alternatively a type of door closer.

It is intended to deal here with items of purpose-made joinery employing spring hinges or floor springs as the closing mechanisms, but first the student should consult the work on these forms of ironmongery on pages 179 and 180.

Spring hinges can be either single or double action for doors opening in one direction or both, and also be of varying types.

Sheet 30.1 shows a joinery example where in the elevation the appearance is that of a pair of glazed swing doors. In fact the right door only, which is fitted with a finger plate, is hinged with the matched glazed frame to the left fixed. As the door rotates through 180 degrees a double-action spring hinge is used at the top hinging position as shown, with a blank hinge at the lower position.

The details of the frame and door are shown in the plan in Sheet 30.2. It will be seen that the frame to the left is rebated with rebated mock top and bottom rails to line up with those of the door. No rebates are used on the frame at the door position.

The fitting of the hinges to the door and frame requires special care in positioning as shown in Sheet 30.3.

A sketch of the top helical hinge is shown in Sheet 30.4. The two barrels contain the springs. These are regulated by inserting a steel pin in the collar, when the hinge is fixed, one flap to the door and the other flap to the frame.

Sheet 30.5 shows a pair of swing doors rotating through 180 degrees termed 'double-action' as the doors open in both directions. Floor springs are used in this case.

The springs for double-action or swing doors are provided with shoes and top centres. Sheet 30.6 shows the shoe fixed to the heel of the door. A square hole in the bottom of the shoe fits over the pivot which is actuated by a spring in the box. The adjustable shoe is fitted with an adjustable bush which can be regulated to align doors which may not be straight. The details of the top centre which is fitted into the underside of the head or transom of the frame, with the top centre receiver fitted into the top edge of the door is shown in Sheet 30.7.

Deep bottom rails should be formed from two pieces jointed together to make up the depth. A bead may be used at the joint line on each face to break the joint, alternatively a sinking at this point may be used as shown in Sheet 30.8.

Sheet 30.10 shows a method of sealing the hanging and closing edges of swing doors to prevent draughts, using a patented seal. This consists of two aluminium strips each having neoprene inserts. One is housed into the door-frame and the other into the door-edge as shown; the fixing is by screwing. Where a pair of swing doors are used a seal is housed in the closing edge of each door.

Sheet 31.1 and 31.2 shows two further examples of varying designs in hardwood of entrances suitable for offices, restaurants, and shops. Sheets 31.3 shows frames using EX. 100 mm × 40 mm stiles and head, mortised and tenoned at the corners with a through sill member EX. 75 mm × 56 mm.

Mullions EX. 88 mm × 32 mm tenoned into the head and sill provide for infilling, using 6 mm polished plate glass above the middle rail with 12 mm exterior grade plywood panels below, using planted beads. A twelve-panel polished entrance door having planted mitred beads and hung using 100 mm brass butt hinges completes the modern elevation.

Sheet 31.4 shows a glazed entrance using double-glazed sealed units with facings to the framing panels in black plastic, matching the facing to the middle rails of the doors.

The jointing of the various members is shown by dotted lines in the detail in Sheet 31.4 where only one door is shown.

Showroom doors

The elevation of a set of doors suitable for a showroom is shown in Sheet 32.1. Two are sliding and two are fixed. Although these are classified as doors they may be grouped with windows.

The track and gear arrangement is similar to the details in later examples.

Sheet 32.2 shows the detail of the frame and door stile of one of the sliding doors. The detail at the meeting stiles of two doors is also shown.

Sheet 32.3 shows the vertical section through the doors.

Revolving doors

This type of door is very suitable for the entrance halls to such buildings as hotels, banks, restaurants, etc. where it is necessary to exclude draught. The arrangement consists of four leaves connected in various ways according to the patent at their centre. They are constructed so that openings can be thrown open by folding the leaves to the centre or to one side, to leave an unobstructed entrance for admitting goods. To do this, various mechanical devices are used, the leaves being hinged to fold up.

Sheet 33.1 shows a four-compartment plan of revolving doors within the entrance to a building. The four leaves collapse and fold to one side as shown.

Sheet 33.2 shows a three-compartment plan, with side lights between the circular enclosing screen and the vestibule walls. The three leaves collapse and fold against one side giving a clear opening. They are suitable for entrances where small doors only can be accommodated.

The plan and elevation, Sheet 33.3, show 38 mm thick doors 2.100 m high, contained within a flush circular enclosing screen, 1·800 m diameter and 35 mm thick. The doors, normally made in oak, walnut, or mahogany and polished, are constructed in the normal way and fitted with plate glass above the middle rail, push bars, and kicking plates. The hard wear to which these doors are subject requires mechanism of a substantial character. They are suspended from a trolley mounted inside two steel joists and this part of the mechanism is mounted out of sight, in the head soffit. A 13 mm spindle running on a large ball-bearing thrust race projects through the soffit and is mortised between the two stiles of the doors marked 'A' (Sheet 33.4). The doors are held together by means of the hinge wings, the centre block being

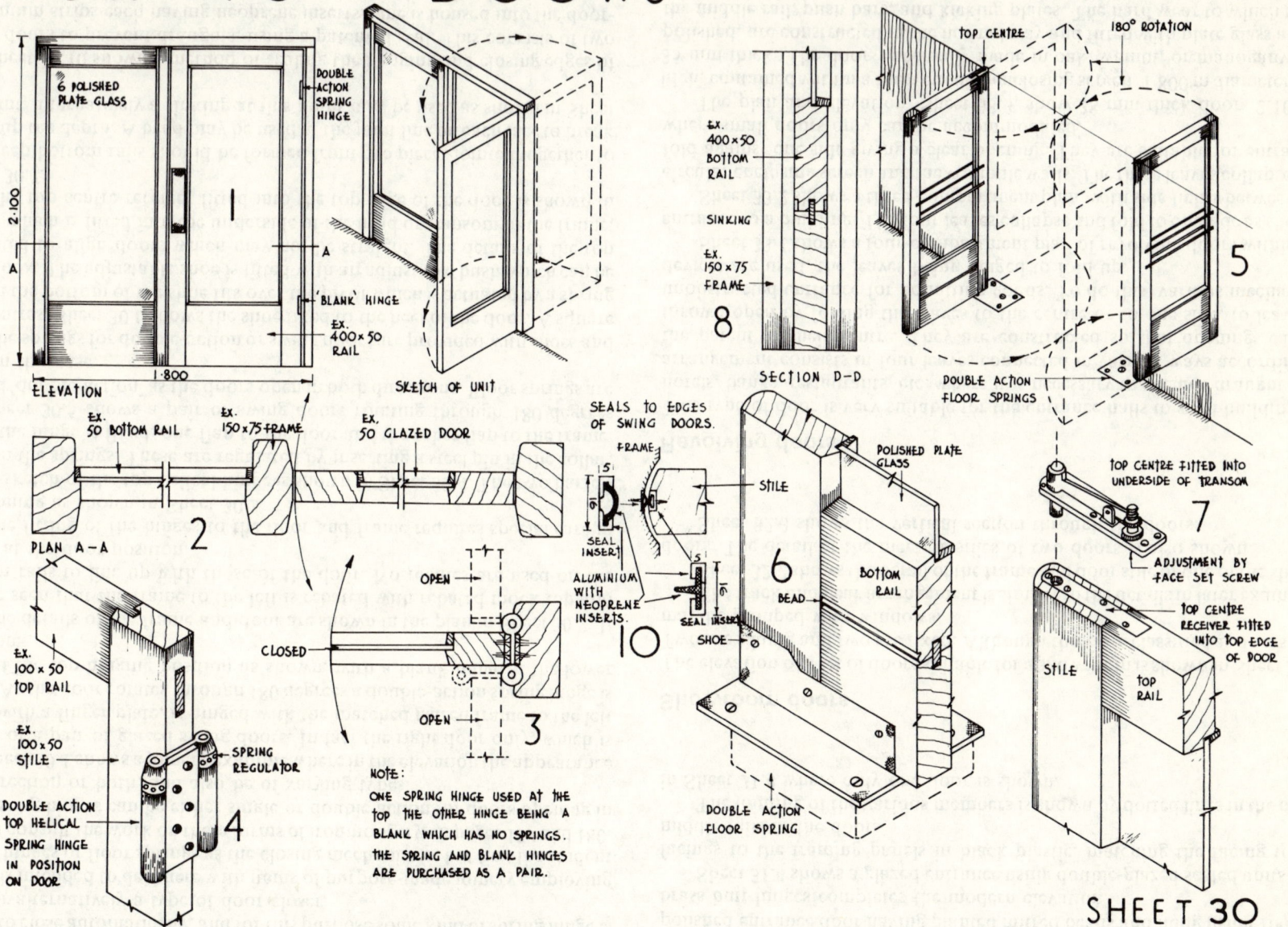

SWING DOORS

1
6 POLISHED PLATE GLASS
2·100
A
1·800
ELEVATION
DOUBLE ACTION SPRING HINGE
'BLANK' HINGE
EX. 400 × 50 RAIL

SKETCH OF UNIT

2
EX. 50 BOTTOM RAIL
EX. 150 × 75 FRAME
EX. 50 GLAZED DOOR
PLAN A-A

3
OPEN
CLOSED
OPEN
NOTE:
ONE SPRING HINGE USED AT THE TOP THE OTHER HINGE BEING A BLANK WHICH HAS NO SPRINGS. THE SPRING AND BLANK HINGES ARE PURCHASED AS A PAIR.

4
EX. 100 × 50 TOP RAIL
EX. 100 × 50 STILE
DOUBLE ACTION TOP HELICAL SPRING HINGE IN POSITION ON DOOR
SPRING REGULATOR

8
EX. 400 × 50 BOTTOM RAIL
SINKING
EX. 150 × 75 FRAME
SECTION D-D

5
TOP CENTRE
180° ROTATION
DOUBLE ACTION FLOOR SPRINGS

10
SEALS TO EDGES OF SWING DOORS.
FRAME
SEAL INSERT
ALUMINIUM WITH NEOPRENE INSERTS.
SEAL INSERT

6
STILE
POLISHED PLATE GLASS
BOTTOM RAIL
BEAD
SHOE
DOUBLE ACTION FLOOR SPRING

7
TOP CENTRE FITTED INTO UNDERSIDE OF TRANSOM
ADJUSTMENT BY FACE SET SCREW
TOP CENTRE RECEIVER FITTED INTO TOP EDGE OF DOOR
STILE
TOP RAIL

SHEET 30

ENTRANCES

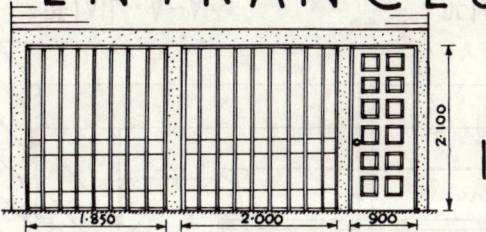

1

DESIGN

1·850   2·000   900

2·100

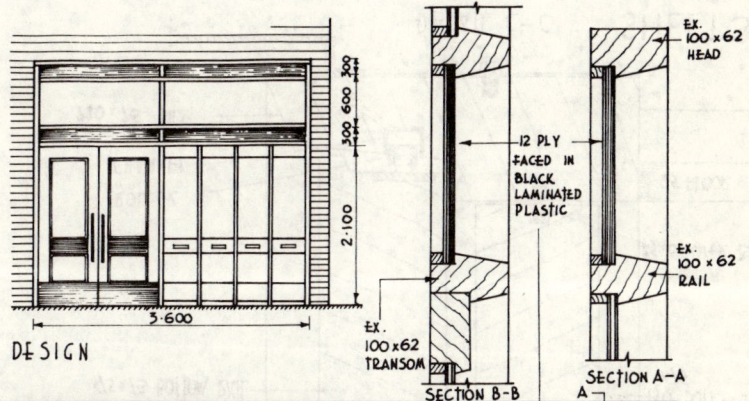

2

DESIGN

3·600

2·100

300 600 300

EX. 100 x 62 HEAD

12 PLY FACED IN BLACK LAMINATED PLASTIC

EX. 100 x 62 RAIL

EX. 100 x 62 TRANSOM

SECTION B-B

SECTION A-A

A

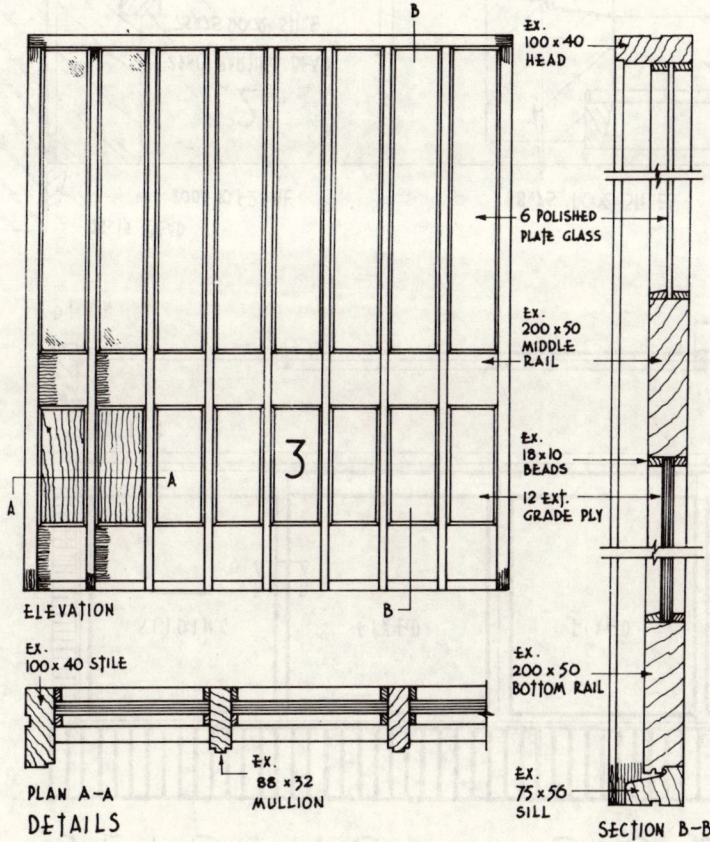

3

B

A

A

B

ELEVATION

EX. 100 x 40 STILE

PLAN A-A

DETAILS

EX. 88 x 32 MULLION

EX. 100 x 40 HEAD

6 POLISHED PLATE GLASS

EX. 200 x 50 MIDDLE RAIL

EX. 18 x 10 BEADS

12 EXT. GRADE PLY

EX. 200 x 50 BOTTOM RAIL

EX. 75 x 56 SILL

SECTION B-B

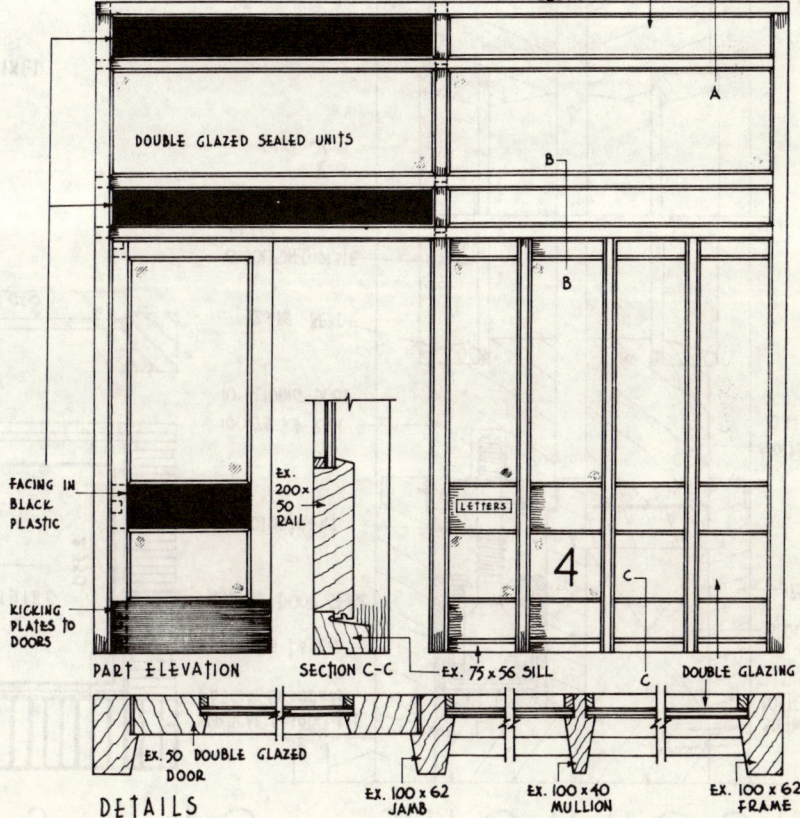

DOUBLE GLAZED SEALED UNITS

A

B

B

C

4

C

FACING IN BLACK PLASTIC

KICKING PLATES TO DOORS

PART ELEVATION

EX. 200 x 50 RAIL

LETTERS

SECTION C-C

EX. 75 x 56 SILL

DOUBLE GLAZING

DETAILS

EX. 50 DOUBLE GLAZED DOOR

EX. 100 x 62 JAMB

EX. 100 x 40 MULLION

EX. 100 x 62 FRAME

SHEET 31

SLIDING DOORS TO SHOWROOM

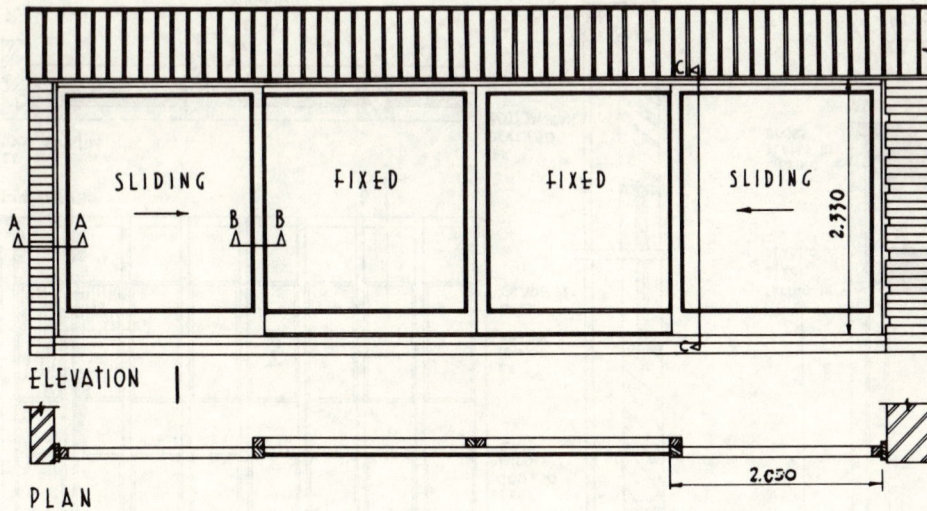

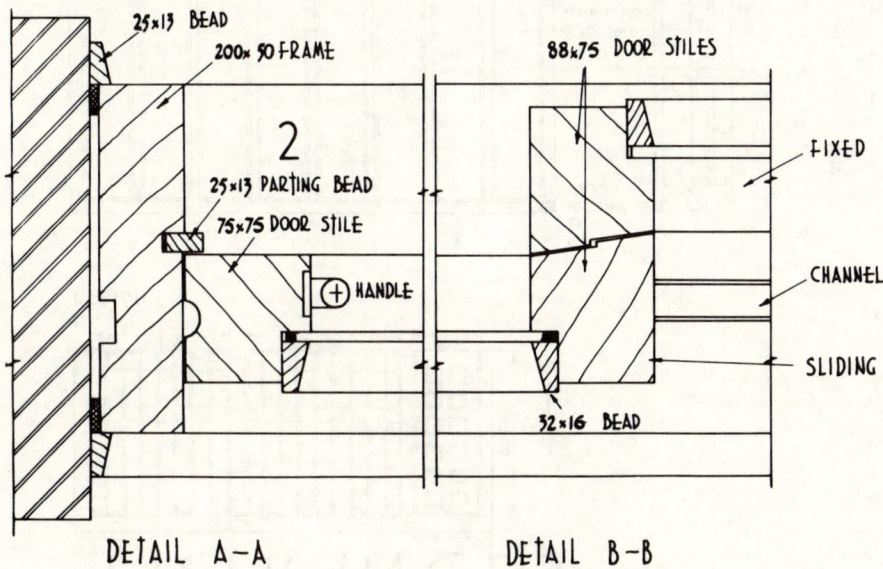

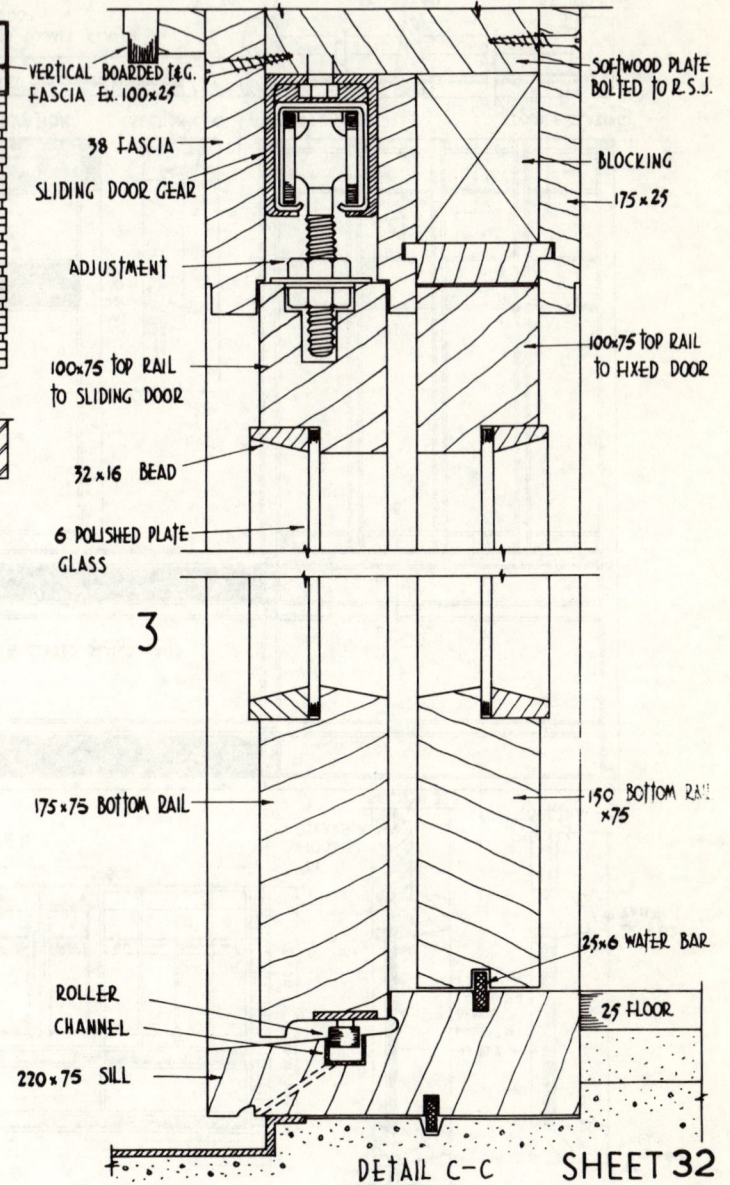

SLIDING FIXED FIXED SLIDING

2.330

ELEVATION 1

PLAN

2.650

DETAIL A-A

25x13 BEAD
200x50 FRAME
25x13 PARTING BEAD
75x75 DOOR STILE
HANDLE

2

DETAIL B-B

88x75 DOOR STILES
FIXED
CHANNEL
SLIDING
32x16 BEAD

3

VERTICAL BOARDED T&G. FASCIA Ex.100x25
38 FASCIA
SLIDING DOOR GEAR
ADJUSTMENT
100x75 TOP RAIL TO SLIDING DOOR
32x16 BEAD
6 POLISHED PLATE GLASS

SOFTWOOD PLATE BOLTED TO R.S.J.
BLOCKING
175x25
100x75 TOP RAIL TO FIXED DOOR

175x75 BOTTOM RAIL
150 BOTTOM RAIL x75
25x6 WATER BAR
ROLLER
CHANNEL
220x75 SILL
25 FLOOR

DETAIL C-C **SHEET 32**

REVOLVING DOORS

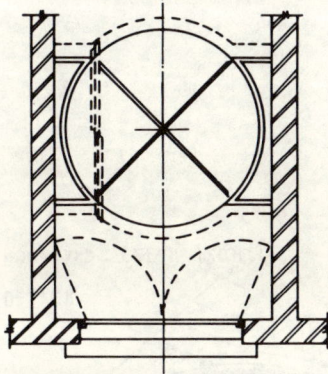

a) FOUR COMPARTMENT

1

PLANS SHOWING METHODS OF FITTING REVOLVING DOORS TO VARIOUS ENTRANCES

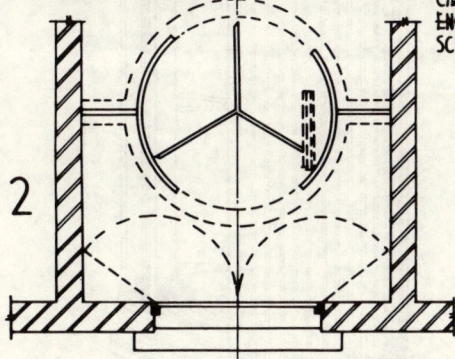

b) THREE COMPARTMENT

2

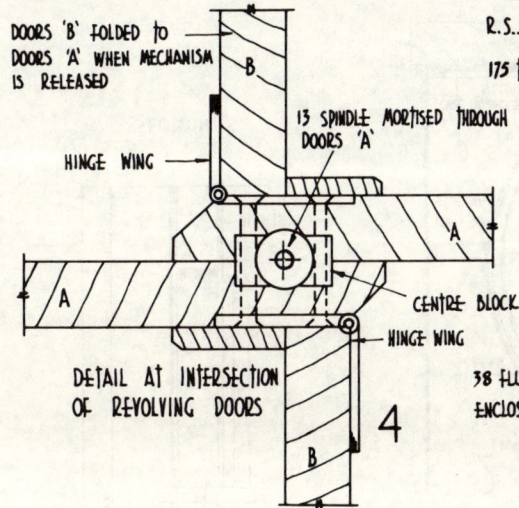

DOORS 'B' FOLDED TO DOORS 'A' WHEN MECHANISM IS RELEASED

B

HINGE WING

13 SPINDLE MORTISED THROUGH DOORS 'A'

A

A

CENTRE BLOCK

HINGE WING

B

DETAIL AT INTERSECTION OF REVOLVING DOORS

4

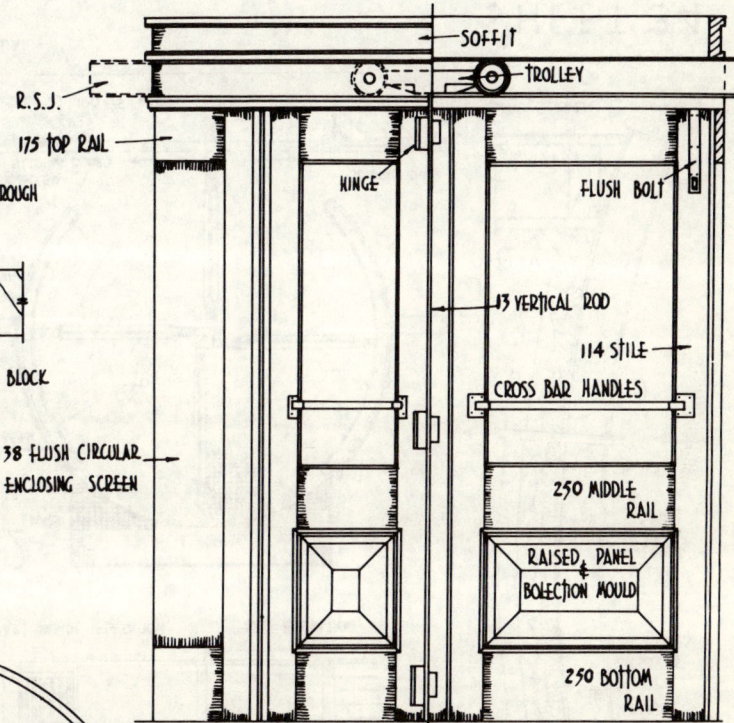

SOFFIT

TROLLEY

R.S.J.

175 TOP RAIL

HINGE

FLUSH BOLT

13 VERTICAL ROD

114 STILE

CROSS BAR HANDLES

38 FLUSH CIRCULAR ENCLOSING SCREEN

250 MIDDLE RAIL

RAISED PANEL & BOLECTION MOULD

250 BOTTOM RAIL

ELEVATION LEFT OF ℄   SECTION RIGHT OF ℄

3

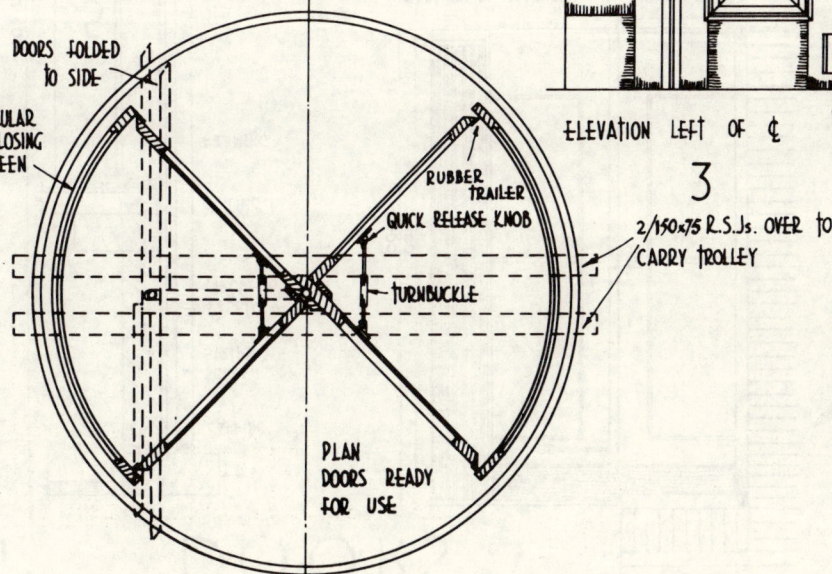

DOORS FOLDED TO SIDE

CIRCULAR ENCLOSING SCREEN

RUBBER TRAILER

QUICK RELEASE KNOB

TURNBUCKLE

2/150x75 R.S.Js. OVER TO CARRY TROLLEY

PLAN DOORS READY FOR USE

SPECIFICATION —
EXTERNAL DIAMETER 1.800
HEIGHT 2.100 UNDER SOFFIT.
FLUSH CIRCULAR ENCLOSING SCREEN.
DOORS 38 MOULDED WITH PLATE GLASS, PUSH BARS & KICKING PLATES FOUR COMPARTMENT.

SHEET 33

REVOLVING DOORS

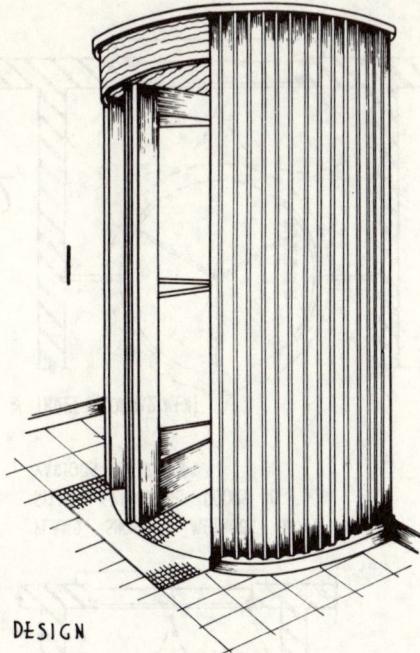

DESIGN

DOOR AND SCREEN PROJECT

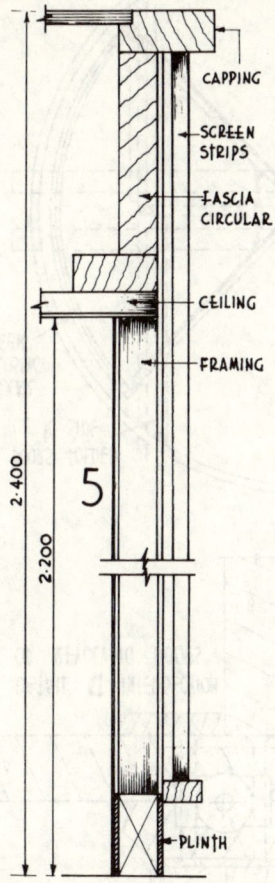

5

2·400

2·200

CAPPING

SCREEN STRIPS

FASCIA CIRCULAR

CEILING

FRAMING

PLINTH

SECTION

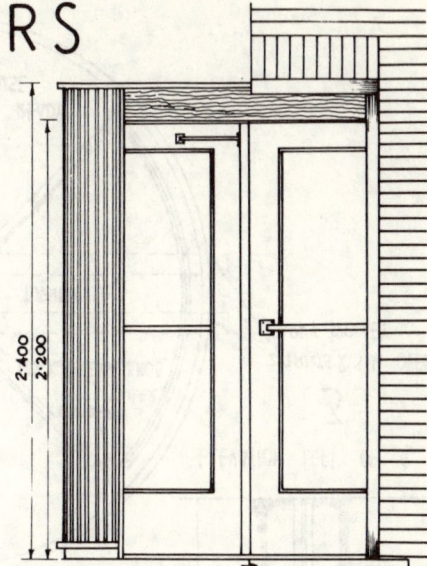

2·400

2·200

PART INSIDE ELEVATION   PART ELEVATION

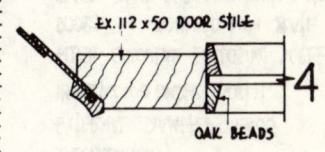

EX. 112 x 50 DOOR STILE

OAK BEADS

4

DETAILS

EX. 44 x 32 NOSING

EX. 50 x 25

32 FRAMING

EX. 50 x 25 OAK STRIPS

6 OAK VENEERED PLY

6 PLY

3

EX. 75 x 50

SIDE SCREEN

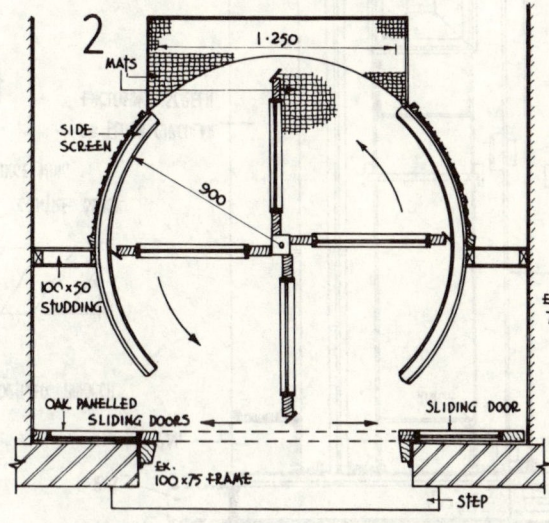

2

MATS

1·250

SIDE SCREEN

900

100 x 50 STUDDING

OAK PANELLED SLIDING DOORS

SLIDING DOOR

EX. 100 x 75 FRAME

STEP

PLAN

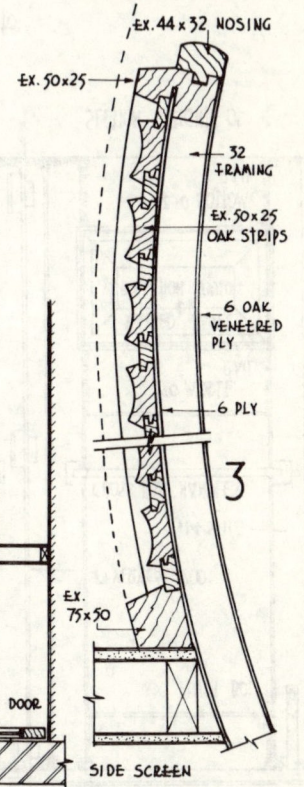

SHEET 34

tapped for this purpose. There are three centre blocks in the height of the doors, which are drilled vertically for the 13 mm rod to pass through for the purpose of lifting from the floor socket, the bottom pivot releasing the trolley from the head track.

The hinged doors are held to the fixed doors by cross connections, having a turnbuckle and quick-release knob shown in the plan. When folding the doors to the side, the connections are first released, doors 'B' are folded against doors 'A' when an automatic catch engages and locks the doors together. The centre rod is next lifted by the projecting handle and the whole assembly pushed to one side. Flush bolts lock the doors when not in use.

Between the doors and screen there should be a clearance of about 13 mm along the vertical edge for safety, but this space is closed by flexible rubber trailers fixed to the outer stiles and rubbing against the flush-faced enclosing screen.

Revolving doors/screen project

Sheet 34.1 shows a sketch of a revolving door having a side casing or screen within the entrance to a bank, hotel, or similar type of building.

As in the previous example this is high-class specialist joinery completed in hardwood and polished. Contrasting hardwoods may be used, the door leaves in mahogany with lighter coloured edging strips to match possibly with the glazing beads and moulded strip to the screen.

Sheet 34.2 shows the plan and elevation of the door. The inside elevation is seen to the left of the centre and the normal elevation to the right. The entrance doors are top-hung to slide clear of the opening on the inside face of the outer wall.

The detail of the circular screen faced on the inside with flush veneered ply and moulded strip finished with a nosing to the foyer side is shown in Sheet 34.3. The screen requires to be set out full size to determine the templets for marking out and machining the curved softwood rails.

The detail of the flexible rubber trailers fixed to the leading edges of the stiles of the four leaves is shown in Sheet 34.4 and the vertical section in Sheet 34.5.

Double-margin door

Sheet 35.1 shows the elevation of a double-margin door 2·100 m × 1·350 m. This type of door is made wide and in such a manner as to match with ordinary folding doors.

The door is first framed in two separate leaves which are secured together with a through top rail the full width of the door, brindle jointed at the centre stiles, and three pairs of hardwood fox wedges, shown by dotted lines in the drawing. A metal bar 38 mm × 10 mm is housed and screwed with brass screws into the bottom rails after the door has been fitted, for added stiffness. The wedges are positioned at distances shown, so that there may be sufficient timber between them and the ordinary mortises to resist the pressure when these wedges are tightened up.

The tenons are the usual thickness, one-third of the stiles, and the hardwood wedges are the same. The stiles and rails are prepared in the usual

way, with the two centre stiles mortised through the panel grooves for the extra keys. The doors are then glued, without the panels and top rail, care being taken that only the centre stile and rail ends that go into this are glued, cramped and wedged.

When both leaves have been dealt with in this way, the ends of the tenons are dressed off and the centre joint prepared. This is next glued and wedged, and any dressing off of wedges in the grooves done. Panels are then put in and the through top rail glued and dowelled in position, ready for the outside stiles to be glued and wedged. Panel moulds are then fitted, and the whole dressed off.

It should be noted that alternative panel finishings, suitable for this type of work, have been shown in the elevation.

Sheet 35.2 shows a sketch of the bridle jointing between the top rail and centre stiles. A bead stuck on the solid breaks the joint of the two leaves at the centre.

Alternative centre jointing is shown in Sheet 35.3 with centre stiles either rebated or tongued together with a loose tongue. When the latter is used, planted beads in a rebate break the joint.

Double-margin doors are very heavy because of their size and thickness, and should always be hung with a pair and a half of 100 mm butt hinges, half of the hinge sunk in the door edge and half into the door frame.

Double-faced door

In public buildings, where much high-class joiner's work is to be found, it is not unusual to have different wall finishings between adjoining rooms, and between rooms and corridors, where each may be finished in a different kind of wood. In such cases it is necessary that each side of a door shall match the other fixings, oak to oak, mahogany to mahogany, etc. If flush-veneered doors are to be used, then there is no difficulty, each side being finished to match the surrounding work. Where framed panelled doors with mouldings are required, then a door made of different kinds of wood on each face is used.

The elevation of a double-faced door, 2·100 m × 900 mm is shown in Sheet 35.4. The door, which is 56 mm thick, is made up of two parts, each 28 mm thick. These are secured together by keys in the positions shown in the drawing. The keys are tapered in length and have bevelled edges, being about 11 mm thick.

The keys are fitted into one member and then laid on the other, carefully marked and taken out, glued and screwed to the corresponding piece (Sheet 35.5). When this has been done the two members with the faces for the joint are glued and put together.

The panels, which are 19 mm thick, are kept apart by packing pieces, shown by the dotted lines in the elevation, and are inserted to keep the panels from warping. They are glued and pinned to the back of one panel, giving a solid bearing for the other panel.

When the door has been assembled, the edges should be veneered. The closing stile of the door should be finished with the same material as the fittings in the room, while the hanging stile corresponds to the fittings on the side to which the door leads off.

DOUBLE MARGIN DOOR.

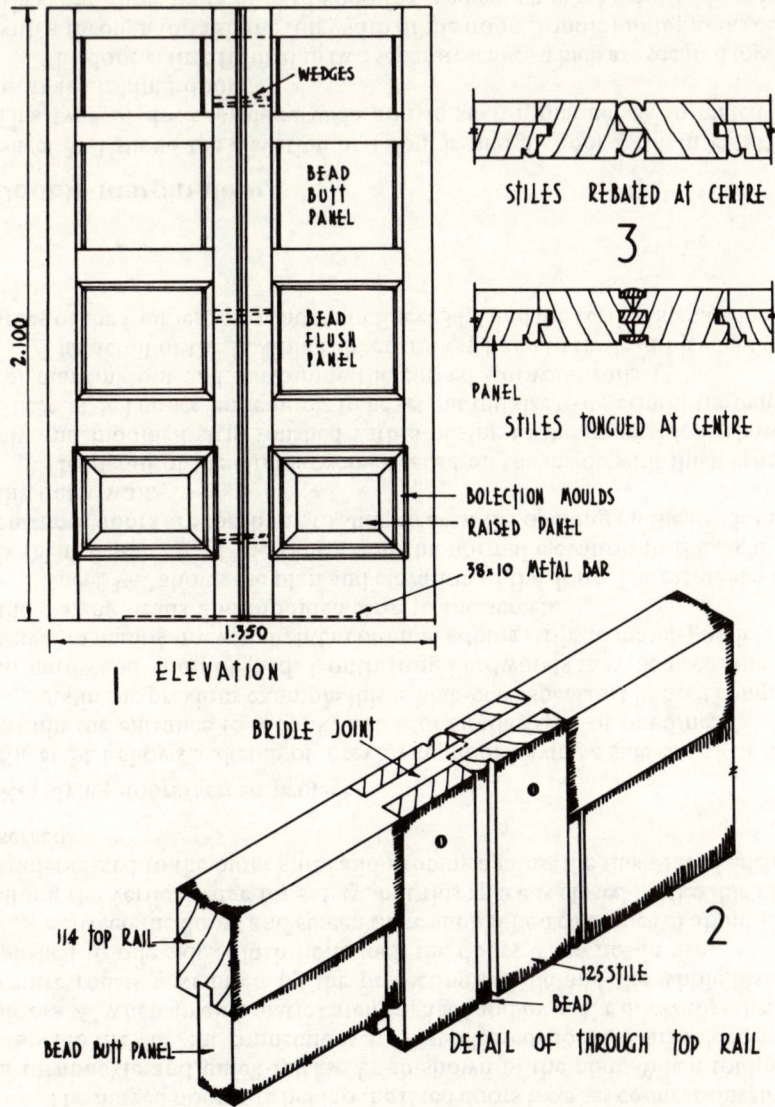

WEDGES

BEAD BUTT PANEL

BEAD FLUSH PANEL

2.100

BOLECTION MOULDS

RAISED PANEL

38×10 METAL BAR

1.350

1 ELEVATION

STILES REBATED AT CENTRE

3

PANEL

STILES TONGUED AT CENTRE

BRIDLE JOINT

114 TOP RAIL

125 STILE

BEAD

BEAD BUTT PANEL

2 DETAIL OF THROUGH TOP RAIL

DOUBLE FACED DOOR.

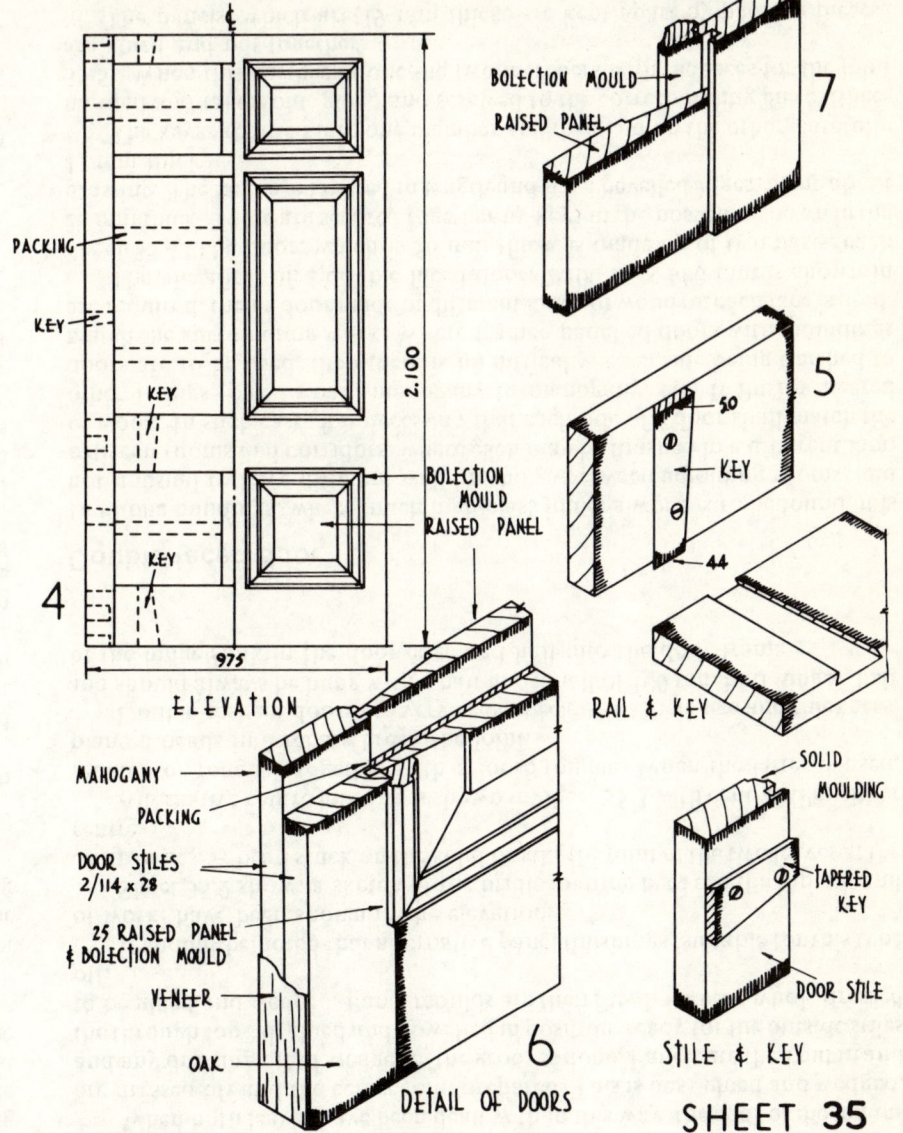

PACKING

KEY

KEY

KEY

2.100

BOLECTION MOULD RAISED PANEL

975

4 ELEVATION

BOLECTION MOULD

RAISED PANEL

7

50

KEY

44

5

RAIL & KEY

MAHOGANY

PACKING

DOOR STILES 2/114 × 28

25 RAISED PANEL & BOLECTION MOULD

VENEER

OAK

6 DETAIL OF DOORS

SOLID MOULDING

TAPERED KEY

DOOR STILE

STILE & KEY

SHEET **35**

162

Sheet 35.6 shows a sketch of part of the finished door with raised panels and bolection moulds on the outside, and raised panels with a solid moulding on the inside.

Sheet 35.7 shows a sketch of the stile, mould, and panel. It should be noted that the 10 mm groove is out of the centre of the stile, so that the 10 mm mortise and tenon joints between the muntin and rails and the stiles and rails will likewise be out of centre.

Fire-check doors

This type of door is being extensively used today in modern buildings. Hospitals, factories, boiler houses, warehouses, and workshops are examples of where they may be used. They may be constructed in a number of ways. Framed with solid panels in hardwood or as an alternative a solid door built up of three layers of tongue and groove boards, the surfaces then covered with facing sheets of aluminium or galvanized steel. Certain manufacturers have developed alternative constructions. One of these has a core-frame similar to normal flush door construction with an infilling of compressed straw slab. There are two types (a) half-hour resistance, (b) one-hour resistance, and they are highly efficient fire-barriers which fulfil fire-check requirements as laid down by BSS 459, Part 3. The facings are 6 mm plywood to half-hour type 44 mm finished thickness, and 6 mm metal-faced plywood to one-hour type 56 mm finished thickness.

Regulations concerning fire resistance in doors approve certain hardwoods – oak, teak, jarrah, karri, and others, provided the door is solid and the finished thickness is not less than 44 mm. All metal and metal-faced doors are approved.

Sheet 36.1 shows the elevation and vertical section of a solid panel door 48 mm finished thickness. The solid panels may have a 13 mm bead worked in the solid on the edges as bead flush panels, or be finished with loose beads and tongues. The panels may be built up in widths of 100–125 mm with a similar type of bead to each joint. This helps to reduce swelling and shrinking in the panels, but is more expensive. The top panel is prepared to receive a panel of fire-resisting plate glass as an observation panel, at a height of 1·500 m from the ground to the centre. Glazing beads fixed with cups and screws on both sides surround the panel.

Sheet 36.2 shows a section through the door-frame and stile of the door with a section through the meeting stiles shown in Sheet 36.3.

Sheet 36.4 shows the elevation of a flush fire-check door, one-hour type, with the facing cut away to show the construction. The core-frame is rebated to receive a plasterboard protective filling and covered with sheet asbestos and finally faced with plywood. The finished thickness is 56 mm as shown. The whole frame is glued and pressed together and no metal fastenings in the face plies are to be used. The timber should have a moisture content of 14 per cent, and all frames and exposed timber must be impregnated with a solution of 18 per cent mono-ammonium phosphate in water. A section through the door-frame and stile of the door is shown in Sheet 36.5. The maximum clearance allowed between the door and frame in this type of construction is 3 mm. Sheet 36.6 shows the vertical section through the door.

Sheet 36.7 shows a patent flush fire-check door, half-hour type, manufactured by Linden Doors Ltd. The construction is of stiles and cross-rails of timber with an infilling core of closely packed strips of compressed straw with lock blocks on each side. The middle rail allows for the fitting of a letter-box. Facings are of exterior grade plywood. The fire-resisting qualities of the compressed straw core make it possible to offer a 41 mm thick plywood-faced flush door that gives a 'half-hour' fire resistance, when hung in a suitable frame. This construction gives a firm core that supports the facings and will not twist or set up internal stresses and results in a properly stressed skin design. The core is of sufficient weight to give good sound reduction and freedom from drumming.

Sheet 36.8 shows the section through the door-frame designed to complete the unit. the door being left-hand hung.

Church doors

Sheet 37.1 shows the elevation of a pair of oak church doors and frame, suitable either for internal or external use. The stiles, muntins, and shaped rails are moulded on the solid. Typical sections are shown in Sheet 37.2 grooved for the panels. The bottom rail, (Sheet 37.3) is machined with a chamfer on the top edge over which the mouldings of the stiles and muntins are scribed. On external doors the chamfered bottom rail assists in weathering. The shaped top or head rails are cut out of the solid, and care should be taken when marking out the plank to see the run of the grain is such that as little short grain as possible occurs in the finished rail. The joints between the shaped rails and stiles in the doors and in the frame occur at the springing. Here, the jointing is made using a single hammer-headed key, worked on the stiles with cross tongues at the shoulders to prevent twisting, as shown in Sheet 37.4. The joint between the shaped heads of the door-frame is made using a dovetail key, being cut from hardwood and mortised into the shaped heads as shown in Sheet 37.5. It is glued and allowed to set before the remainder of the frame is assembled. Similarly, the joints between the shaped rails and stiles of the doors should be glued and allowed to set before the assembly of the doors, as some difficulty may be experienced in pulling up the joint. The most suitable joint at this point is a loose tenon inserted and well glued into the slotted rail and pinned as shown in Sheet 37.6. Normal wedging can then be carried out. A section through the rebated door stiles is shown in Sheet 37.7.

Sheet 37.8 shows a single door with shaped head suitable for the side entrance to a church or as an internal door. The construction is very similar to that in the previous example. At the head a hardwood dovetail key mortised into the shaped rails is used. A handrail bolt and dowels may be used at this point as an alternative method of jointing. The jointing at the springing between the shaped head and door stile is a hammer-headed key with hardwood wedges.

Sheet 37.9 shows a part section with solid panels beaded on the inside and tongued into the stiles and muntins.

A sketch of the joint between the chamfered bottom rail and door stile is shown in Sheet 37.10. It will be seen that here, as with the muntins, the mould

FIRE-CHECK DOORS

SOLID PANEL DOORS

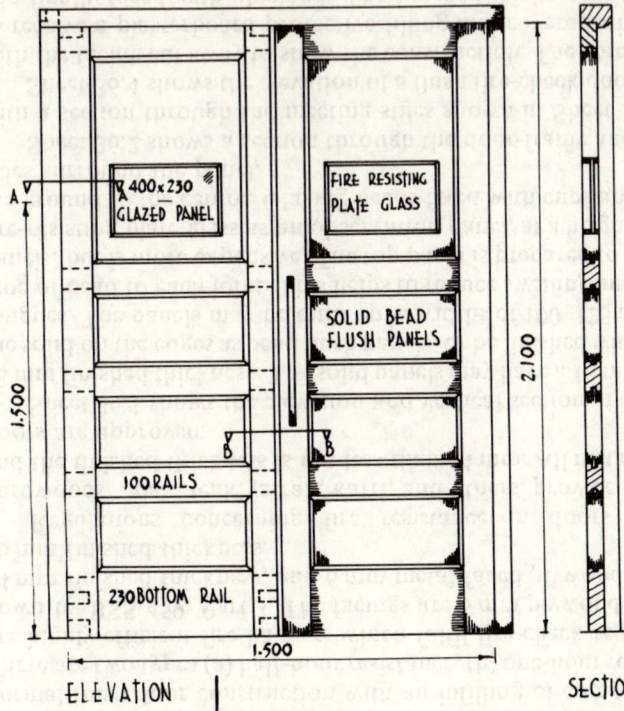

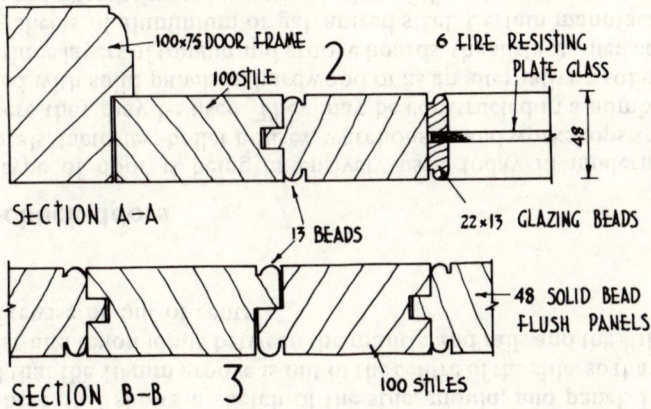

400 × 230 GLAZED PANEL

FIRE RESISTING PLATE GLASS

SOLID BEAD FLUSH PANELS

100 RAILS

230 BOTTOM RAIL

1.500

2.100

1.500

A

ELEVATION 1

SECTION

100×75 DOOR FRAME

100 STILE

2

SECTION A-A

13 BEADS

22×13 GLAZING BEADS

6 FIRE RESISTING PLATE GLASS

48

48 SOLID BEAD FLUSH PANELS

100 STILES

3

SECTION B-B

FLUSH FIRE CHECK DOORS

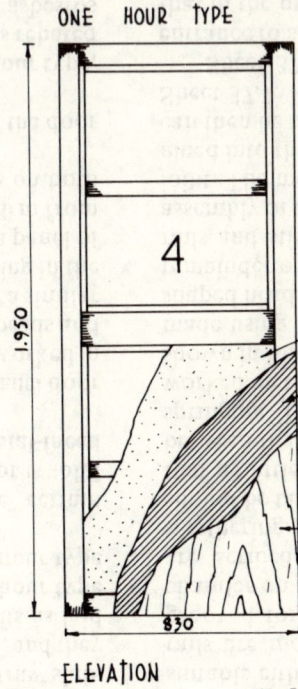

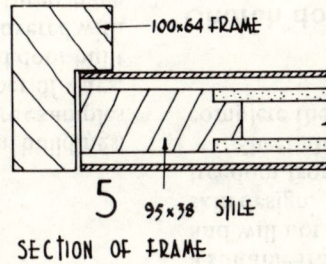

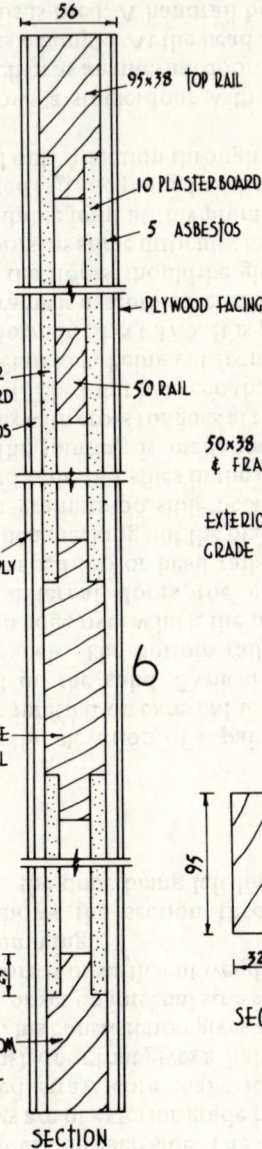

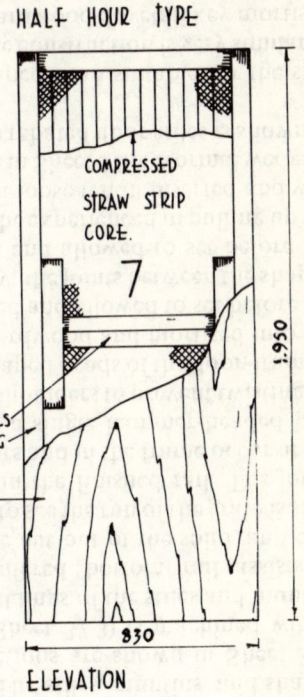

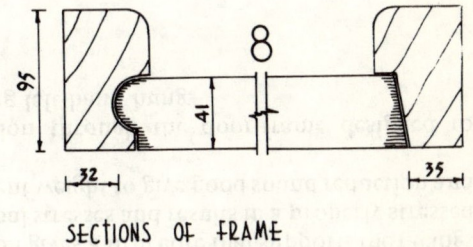

ONE HOUR TYPE

4

1.950

830

PLASTER BOARD

ASBESTOS

FACE PLY

ELEVATION

100×64 FRAME

5

95×38 STILE

SECTION OF FRAME

56

95×38 TOP RAIL

10 PLASTERBOARD

5 ASBESTOS

PLYWOOD FACING

50 RAIL

50×38 RAILS & FRAMING

EXTERIOR GRADE PLY

162×38 MIDDLE RAIL

6

25

95×38 BOTTOM RAIL

SECTION

HALF HOUR TYPE

COMPRESSED STRAW STRIP CORE.

7

1.950

830

ELEVATION

95

41

32

35

8

SECTIONS OF FRAME

SHEET 36

CHURCH DOORS

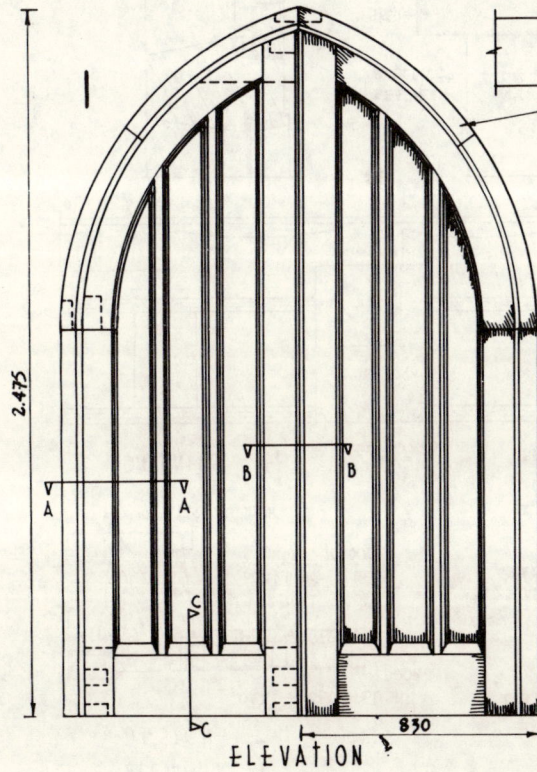

1

2.475

ELEVATION

830

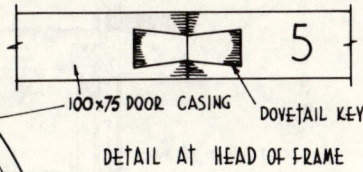

5

100x75 DOOR CASING DOVETAIL KEY

DETAIL AT HEAD OF FRAME

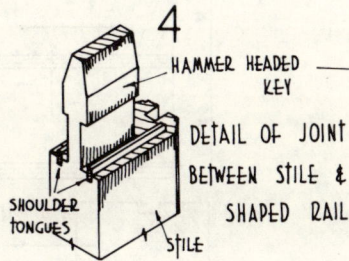

4

HAMMER HEADED KEY

DETAIL OF JOINT BETWEEN STILE & SHAPED RAIL

SHOULDER TONGUES

STILE

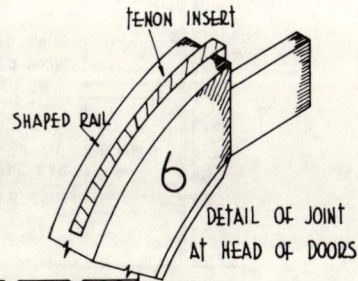

TENON INSERT

SHAPED RAIL

6

DETAIL OF JOINT AT HEAD OF DOORS

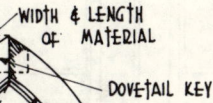

WIDTH & LENGTH OF MATERIAL

DOVETAIL KEY

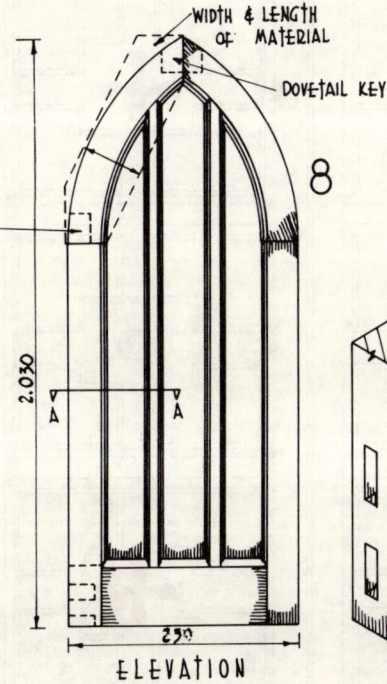

8

2.030

ELEVATION

230

A   A

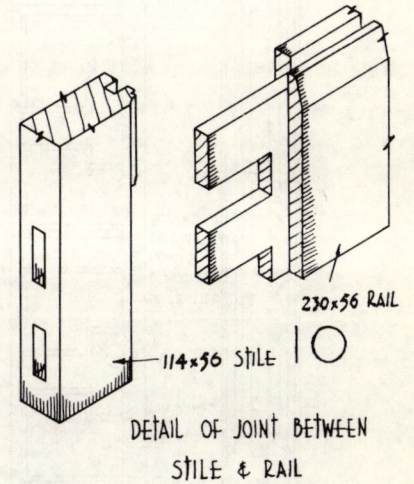

230x56 RAIL

114x56 STILE

10

DETAIL OF JOINT BETWEEN STILE & RAIL

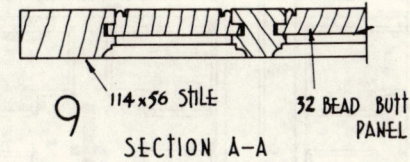

9

114x56 STILE   32 BEAD BUTT PANEL

SECTION A-A

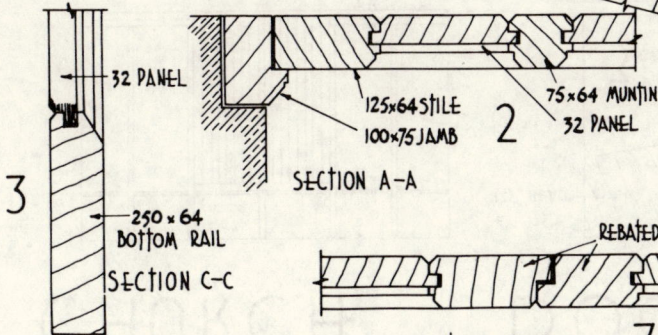

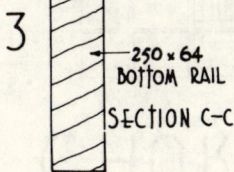

32 PANEL

3

250 x 64 BOTTOM RAIL

SECTION C-C

125x64 STILE

100x75 JAMB

2

75x64 MUNTIN

32 PANEL

SECTION A-A

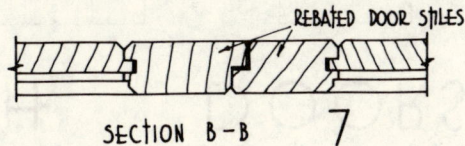

REBATED DOOR STILES

SECTION B-B   7

SHEET **37**

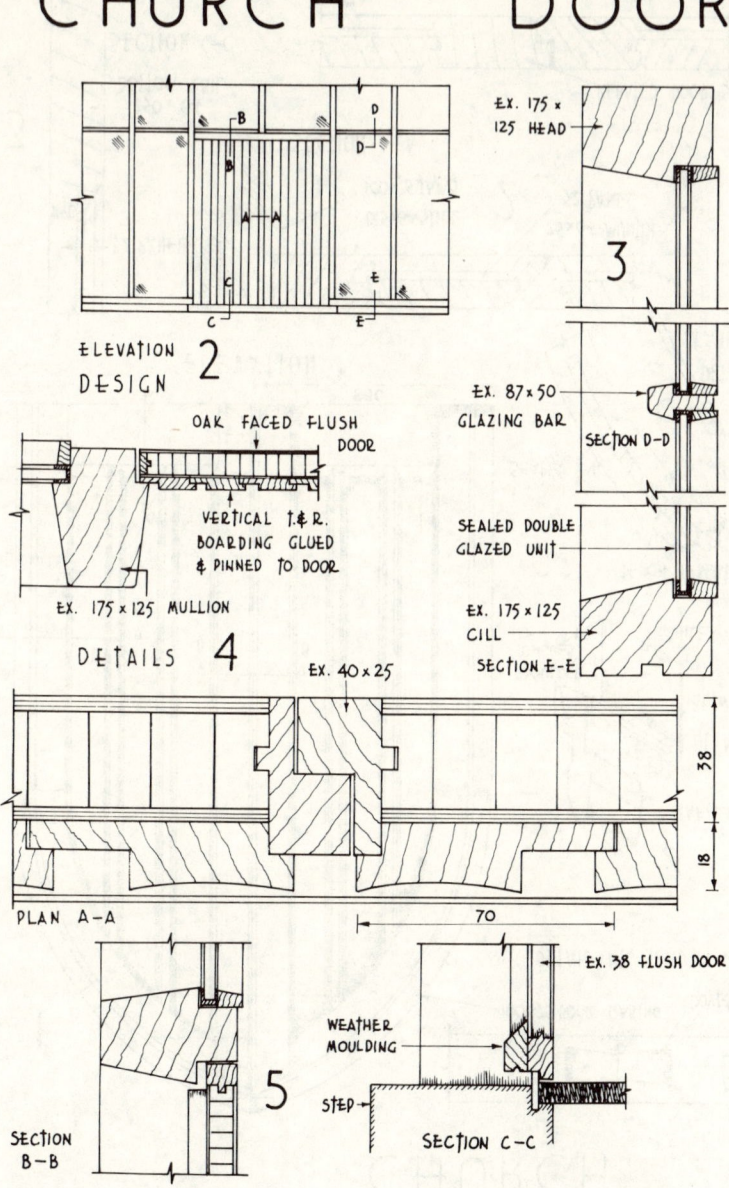

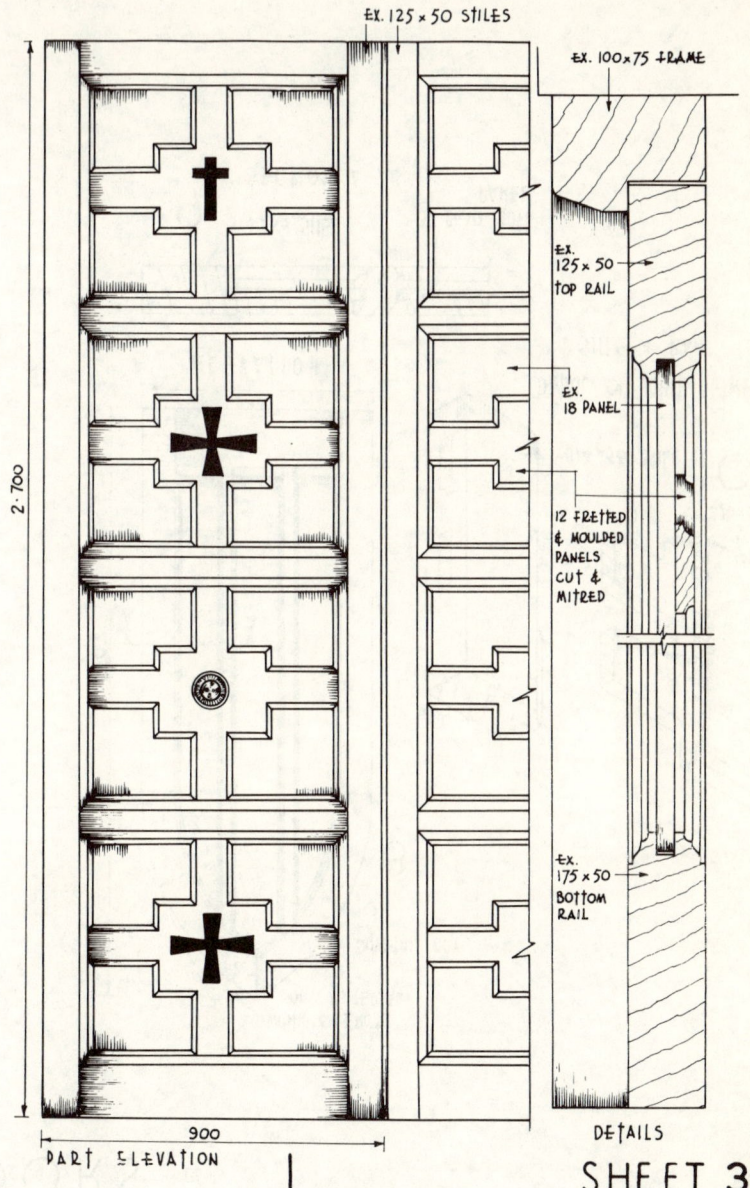

CHURCH DOORS

ELEVATION
DESIGN 2

DETAILS 4

OAK FACED FLUSH DOOR

VERTICAL T.&R. BOARDING GLUED & PINNED TO DOOR

EX. 175 × 125 MULLION

EX. 40 × 25

38

18

PLAN A-A

70

SECTION B-B

5

WEATHER MOULDING

EX. 38 FLUSH DOOR

STEP

SECTION C-C

EX. 175 × 125 HEAD

3

EX. 87 × 50 GLAZING BAR

SECTION D-D

SEALED DOUBLE GLAZED UNIT

EX. 175 × 125 CILL

SECTION E-E

EX. 125 × 50 STILES

EX. 100 × 75 FRAME

EX. 125 × 50 TOP RAIL

EX. 18 PANEL

12 FRETTED & MOULDED PANELS CUT & MITRED

EX. 175 × 50 BOTTOM RAIL

2.700

900

PART ELEVATION 1

DETAILS

SHEET 38

CHURCH DOORS

DOOR DESIGN

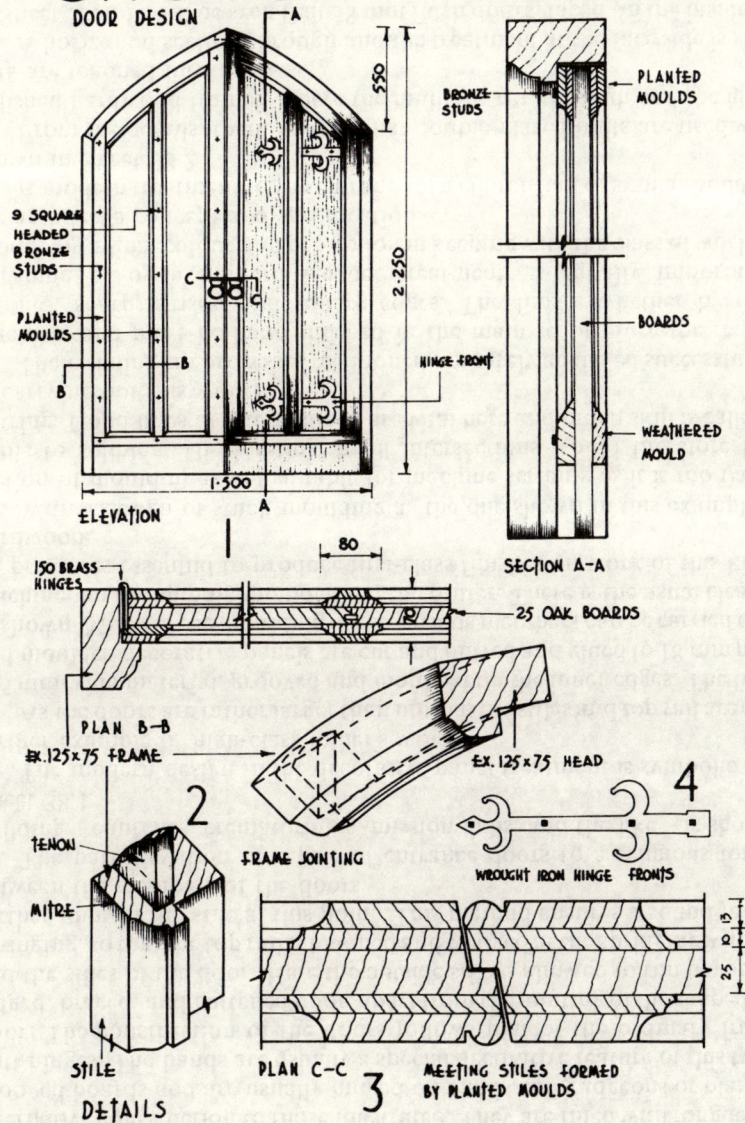

BRONZE STUDS

PLANTED MOULDS

9 SQUARE HEADED BRONZE STUDS

PLANTED MOULDS

BOARDS

HINGE FRONT

WEATHERED MOULD

550

2·250

1·500

ELEVATION

SECTION A-A

150 BRASS HINGES

80

25 OAK BOARDS

PLAN B-B

EX.125×75 FRAME

FRAME JOINTING

EX.125×75 HEAD

TENON

MITRE

STILE

DETAILS

2

WROUGHT IRON HINGE FRONTS

4

25 10 13

PLAN C-C

3

MEETING STILES FORMED BY PLANTED MOULDS

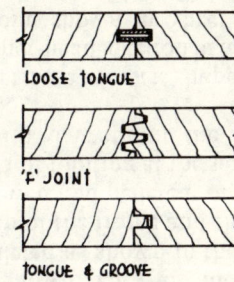

DESIGN 5

LOOSE TONGUE

'V' JOINT

TONGUE & GROOVE

9

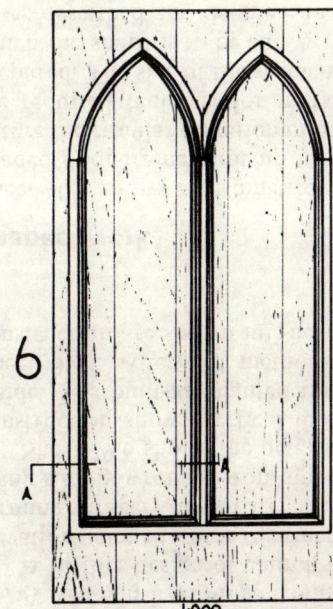

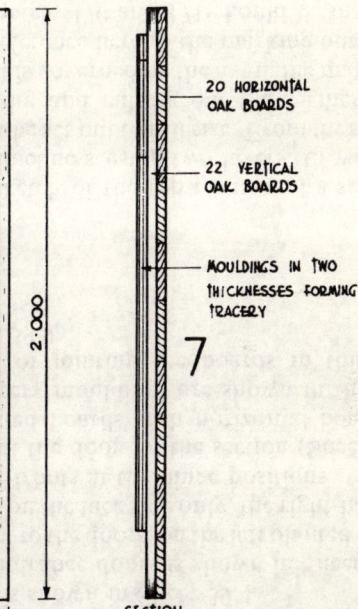

20 HORIZONTAL OAK BOARDS

22 VERTICAL OAK BOARDS

MOULDINGS IN TWO THICKNESSES FORMING TRACERY

6

7

2·000

1·000

ELEVATION

SECTION

200 × 20 BOARDS

200 × 22 BOARDS

PLAN A-A

STILE

MULLION

8

DETAILS

SHEET 39

is scribed over the chamfer as in the previous example. The muntins are stub-tenoned into the bottom rail and also into the shaped top rails where the moulds are mitred. The width and length of the material required to cut the shaped rail is indicated and also the jointing, shown by dotted lines.

Framed, ledged, and braced doors for external use may be used as an alternative construction to those illustrated. They are filled with tongued and grooved boards and are usually hinged on bands and gudgeons or plain cast butt hinges. The bands are usually a special decorative feature of this type of door. The construction of the doors follows that of the ordinary framed, ledged, braced, and battened door, apart from the jointing of the shaped head and the stiles of the door. Here the outside stile is allowed to run beyond the springing, so that the top rail is tenoned and wedged at the springing without a further joint in the stile at this point. This method ensures a sound jointing between the members of the doors.

The part elevation of a pair of entrance doors to a religious form of building – churches, crematoriums, mission halls, and the like – is shown in Sheet 38.1.

The modern design in the decorative panel treatment is symbolic and a further example of high-class joiner's work.

As the doors are rather larger than normal the stiles and top rail are out of 125 mm wide material, grooved and moulded on the inner edges. The fretted and moulded decorative panels are cut and mitred and glued to 18 mm panels as shown. Much of the moulding of the various members can be carried out by machine, on the spindle moulder and the router. There is the usual cleaning-up processes essential to produce first-class finishes on work of this kind in hardwood.

With a design of stuck moulding as the one shown in this example, the section of moulding is not suitable for machine scribing as it is too flat and liable to splinter. The jointing at all intersections would therefore be by mitring. Good lines of intersections are vital here and much skill is called for in carrying out this work.

The cleaning-up processes mentioned are rarely achieved successfully by machine, and must be hand finished in the main to produce the required profiles, sharp arrises, and square edges. The finish, whether by french polishing or by a modern lacquer treatment, is equally important in producing a final colour and protection in keeping with the class of work, and the architect's concept and specification.

A modern treatment to the entrance of a church hall or similar building is shown in Sheet 38.2.

From the details it will be seen that double-glazed units are used within polished hardwood frames, where the mullions run through and the glazing bars are tenoned into them.

A horizontal section through and the treatment at the face side is shown in Sheet 38.4. It will be seen that 38 mm flush doors, faced on the inside with oak-faced ply, are finished with tongued and rebated vertical hardwood boarding to the face-side.

Details of the finish at the base of the doors and the step are shown in Sheet 38.5, with vertical sections of the members in the framed units shown in Sheet 38.3 and 38.4.

Sheet 39.1 shows a pair of oak church entrance doors and frame for external use. The doors are formed of 25 mm thick jointed boards using cross-tongues. The stiles, muntins, and heavy moulded appearance is obtained by building up the mock sections using planted moulds.

Bronze studs, wrought-iron hinge fronts, and ring handles complete the traditional-type ironmongery in this particular design.

The frame is rebated and moulded to match the planted moulds of the door which is hung using 150 mm brass hinges.

Sheet 39.2 shows the frame jointing in sketch form, the two members being mitred together and tenoned.

The detail at the meeting edges of the doors is shown in Sheet 39.3 with the bevelled rebated edges finished with a bead on both faces.

A wrought-iron hinge front is shown in Sheet 39.4.

A further design of church entrance doors is shown in Sheet 39.5. The elevation gives alternate treatment to the doors; on the left planted mouldings forming a tracery design are used on the face side only, the right-hand design being simply wrought-iron hinge fronts at the hinge positions.

Sheet 39.6 shows the detail of the door. In the section (Sheet 39.7) the construction shows vertical jointed boards with horizontal boards to the inside. The built-up planted tracery mouldings are shown in the detail in Sheet 39.8. Alternate methods of jointing the boards in this form of construction are shown in Sheet 39.9.

Shaped work

Sheet 40.1 shows an isometric sketch of the top portion of a semicircular headed door-frame built up in thickness using two layers. In work of this nature the circular elevation must be set out to full size. From the setting out, the required templets for marking and cutting out the segments can be prepared. The size of the segment is governed by the available material from which the shape is to be cut. A reference here to the marking out of shaped work described and detailed on pages 170 and 171 should be made.

The segments forming the head are butt jointed with the joints arranged staggered between the layers as shown. They are then glued and screwed together with the screw-holes pelleted as shown in the section (Sheet 40.2).

The jointing between the jamb of the frame and the shaped head can be mortised and tenoned, glued, screwed and pelleted, or dowelled as shown in Sheet 40.3. An alternative method of jointing at the same point is by using a single hammer-headed tenon with or without the cross-tongues at the shoulders as shown in Sheet 40.4.

The jointing at the crown of a frame which is shaped from solid material may be made using a double hammer-headed tenon in hardwood with wedges which are used to tighten the joint shown in Sheet 40.5. An alternative method of jointing at the crown is shown in Sheet 40.6. Here a handrail bolt is used with hardwood dowels to prevent any twisting at the joint.

A handrail bolt is shown in Sheet 40.7. This, it will be seen, is a double-ended bolt with a square nut on one end and a circular slotted nut and washer on the other. The members to be joined require careful setting out and working to produce a satisfactory joint.

The rebating and moulding of the frame is usually done on the spindle

SHAPED WORK

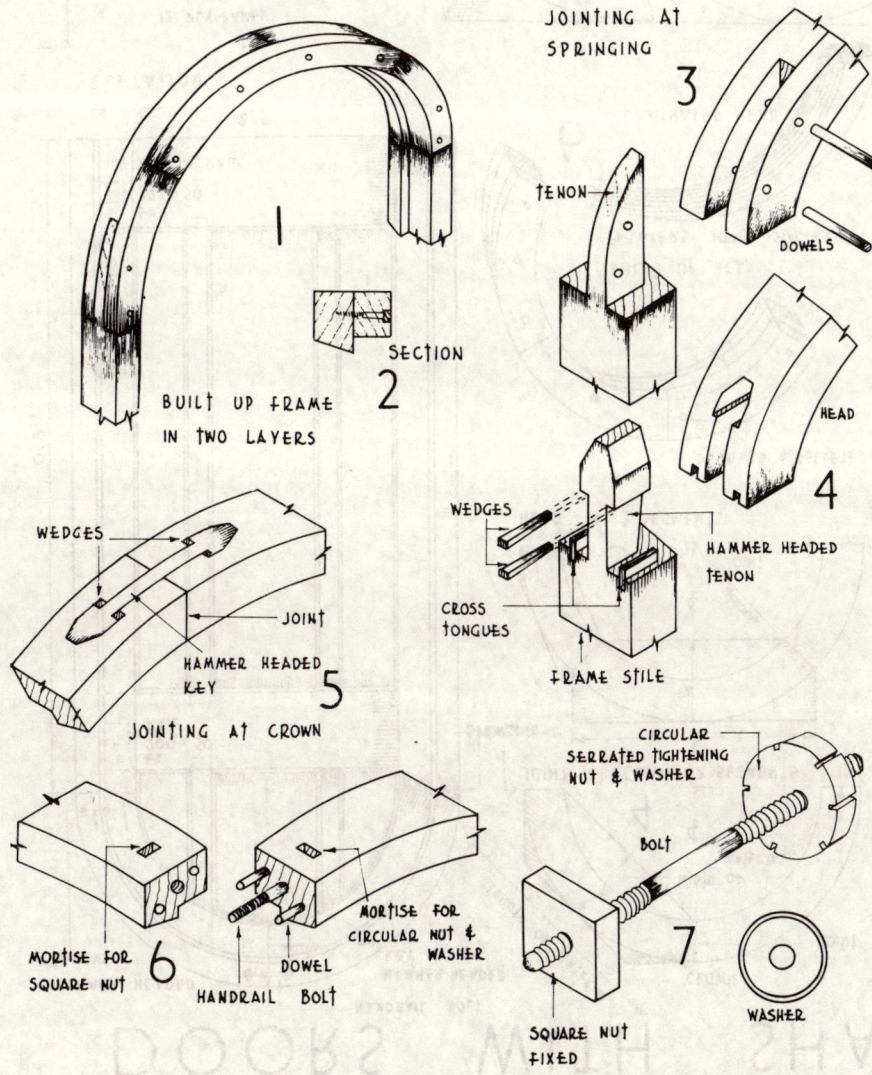

1 BUILT UP FRAME IN TWO LAYERS

2 SECTION

3 JOINTING AT SPRINGING

TENON

DOWELS

4 HEAD

WEDGES

HAMMER HEADED TENON

CROSS TONGUES

FRAME STILE

5 JOINTING AT CROWN

WEDGES

JOINT

HAMMER HEADED KEY

6

MORTISE FOR SQUARE NUT

HANDRAIL BOLT

DOWEL

MORTISE FOR CIRCULAR NUT & WASHER

7

CIRCULAR SERRATED TIGHTENING NUT & WASHER

BOLT

SQUARE NUT FIXED

WASHER

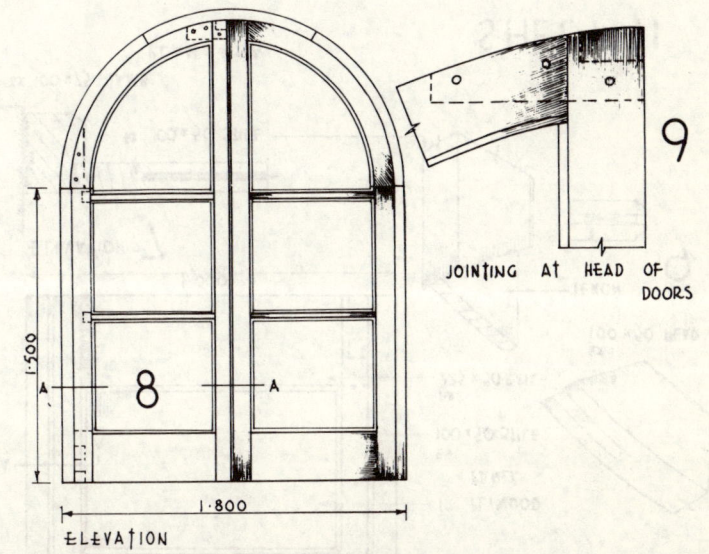

9 JOINTING AT HEAD OF DOORS

1·500

A — 8 — A

1·800

ELEVATION

SECTION A-A

EX. 100 x 50 MEETING STILES

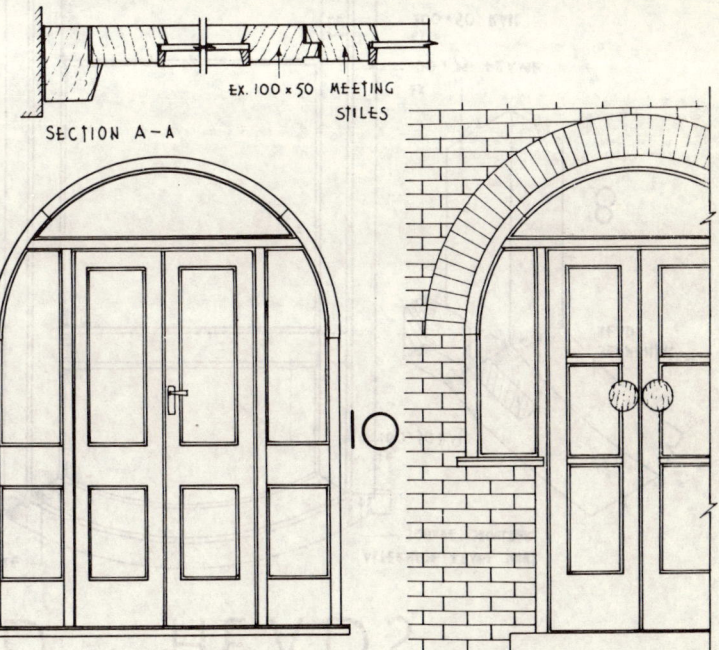

10

SHEET 40

DOORS WITH SHAPED HEADS

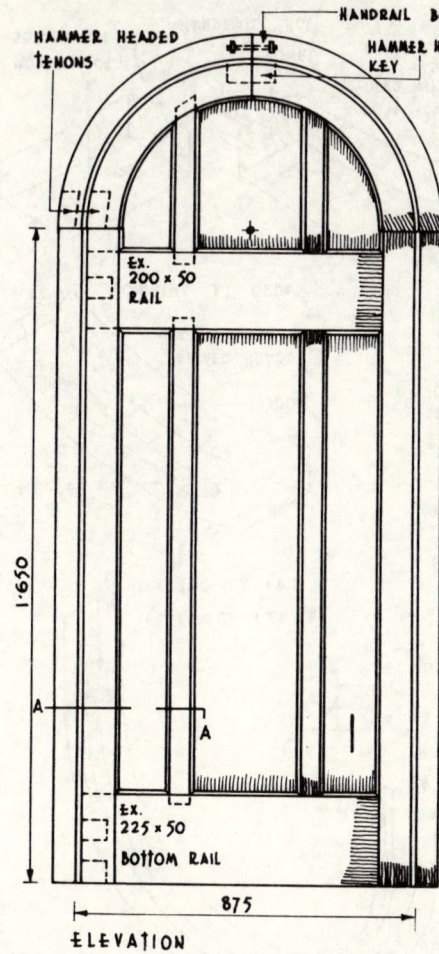

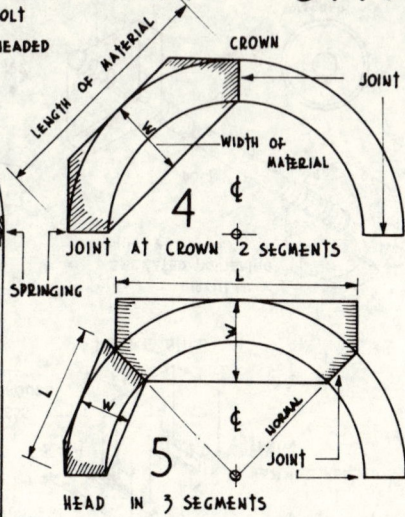

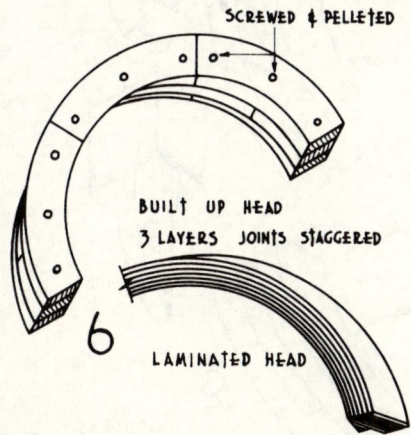

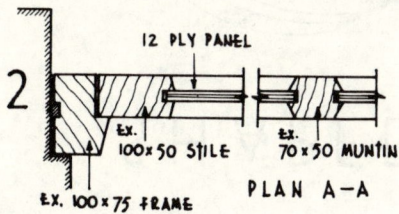

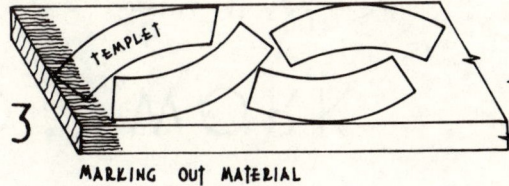

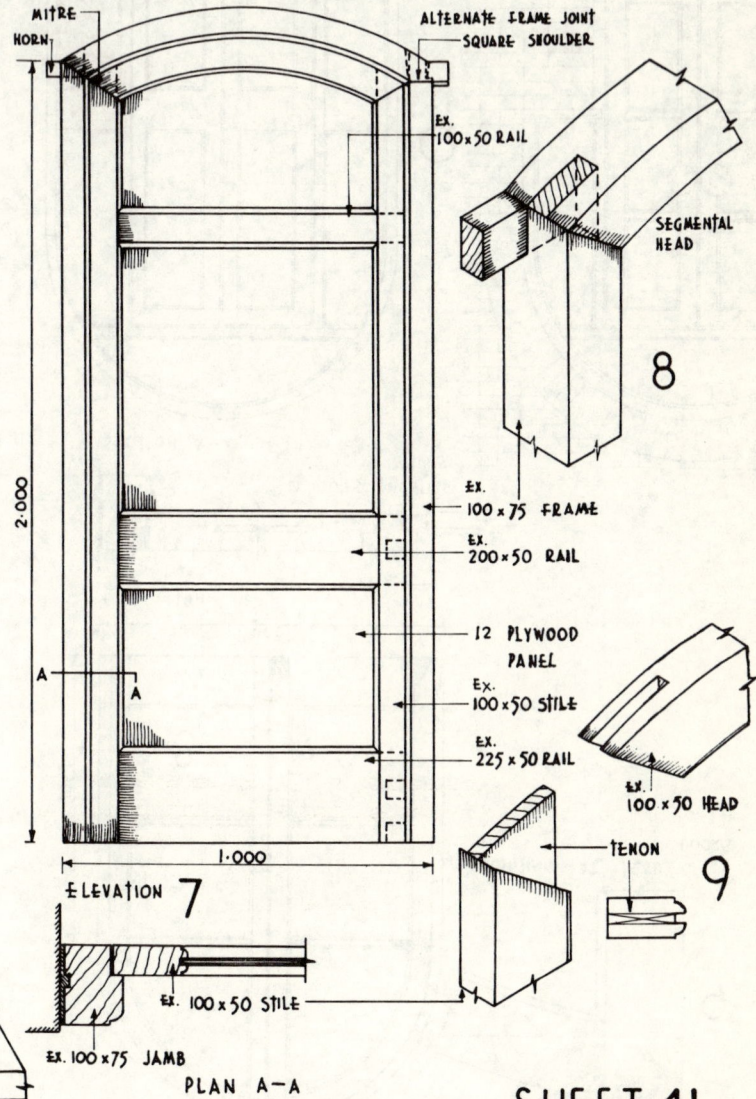

HAMMER HEADED TENONS

HANDRAIL BOLT

HAMMER HEADED KEY

EX. 200 x 50 RAIL

EX. 225 x 50 BOTTOM RAIL

1·650

875

ELEVATION

1

LENGTH OF MATERIAL

CROWN

JOINT

WIDTH OF MATERIAL

4 ℄

JOINT AT CROWN 2 SEGMENTS

SPRINGING

L

L

W

NORMAL

℄

JOINT

5

HEAD IN 3 SEGMENTS

SCREWED & PELLETED

BUILT UP HEAD
3 LAYERS JOINTS STAGGERED

6

LAMINATED HEAD

12 PLY PANEL

2

EX. 100 x 50 STILE

EX. 70 x 50 MUNTIN

EX. 100 x 75 FRAME

PLAN A-A

TEMPLET

3

MARKING OUT MATERIAL

MITRE

HORN

ALTERNATE FRAME JOINT SQUARE SHOULDER

EX. 100 x 50 RAIL

SEGMENTAL HEAD

8

EX. 100 x 75 FRAME

EX. 200 x 50 RAIL

12 PLYWOOD PANEL

EX. 100 x 50 STILE

EX. 225 x 50 RAIL

EX. 100 x 50 HEAD

TENON

9

2·000

1·000

ELEVATION **7**

EX. 100 x 50 STILE

EX. 100 x 75 JAMB

PLAN A-A

SHEET 41

moulding machine, using the ring fence and the shaping of the curved members on the same machine worked against the French head.

Semicircular headed door

Sheet 41.1 shows the elevation of a six-panel semicircular headed door and frame. The stiles, muntins, rails, and shaped heads are moulded on the solid. Typical sections are shown in Sheet 41.2 grooved for the panels.

The shaped top rail to the door and that of the frame may be cut out of the solid, and care should be taken when marking out the material to see the run of the grain is such that as little short grain as possible occurs in the finished member, shown in Sheet 41.3.

Sheet 41.4 shows the setting out of the door head. The jointing is arranged at the crown and at the springing so that the head is formed from two segments. The width and length of material required is clearly shown.

Sheet 41.5 shows the head set-out when the shaped work is formed from three segments with the jointing arranged at the springing and midway between the springing and the crown.

Alternate methods of forming the shaped members are shown by using laminations in Sheet 41.6.

The jointing at the head of the door and also the frame is shown in the elevation by dotted lines. Hammer-headed tenons and key with a handrail bolt are used. Details and descriptions with illustrations of these various methods of jointing are dealt with on pages 168 and 169 to which reference should be made.

Segmental headed door

Sheet 41.7 shows the elevation of a three-panel exterior door and frame having a segmental head.

As with any joinery item involving curved members this must be set out full size at the head, in this case on the setting-out rod.

The construction in this example follows very closely all other work previously described involving shaped or curved members.

Alternate jointing methods at the springing line of the frame are shown to the left and the right in Sheet 41.7. The left-hand side shows a mitred joint, and the right-hand side a square shoulder on the frame.

Sheet 41.8 shows a sketch of the joint in the frame and Sheet 41.9 the joint at the stile and head of the door.

Glazed entrance doors to shaped openings.

Various designs of glazed entrance doors for openings having semi-circular outlines are shown in Sheet 40.

Sheet 40.8 shows glazed doors and frame having a semicircular outline and a span of 1·800 m. The doors are rebated together as shown in the section and jointed at the crown position using a tenon insert, as shown in Sheet 40.9.

Narrower glazed entrance doors change the appearance and construction when used as stormproof doors with narrow sidelights. Two such units are shown in Sheet 40.10 incorporating transoms and muntins in a form of a combination frame.

Sliding and sliding–folding doors and partitions

There are three types of sliding doors:

1. Doors in one or more sections to slide to one or both sides.
2. Doors in hinged sections to slide and fold to one or both sides.
3. Doors in hinged sections on track which turns the doors to slide to one side and round the side wall inside the building.

The doors may be suspended from wheels running in a top track or fitted with swivel rollers running in a floor channel track. There is now a wide range of sliding and folding doors, but it is important to select the appropriate type to suit the plan, and the correct fittings for the size and weight.

Folding partitions work on the same principle as folding doors. Where there is wall space at each side of the opening, doors sliding along the face of the wall or in the cavity are suitable.

The illustrations *on the following pages* show different types of doors suitable for varying conditions, but the actual construction of the doors has not been dealt with, as this is done in the normal way and forms part of earlier work. The details of tracks and door gear are those supplied by P. C. Henderson Ltd.

Sheet 42.1 shows a two-leaf arrangement for garage doors with double top track sliding to one side, as group one. The doors lap the jambs and each other by 50–75 mm and are suspended by hangers, two to each leaf, each having four wheels and running in a steel track. An enlarged detail of the hanger or trolley is shown in the elevation. These are manufactured in various patterns and may be a single pair of wheels fixed centrally to the top of the hanger, or a double pair with the hanger between as shown. Brackets fixed to the lintel at not more than 900 mm apart carry the track with a closed end bracket at each end. The jointing of sections of track is made by positioning an open bracket to carry the two lengths at the joint. The top edge of the doors should cover the bottom edge of the lintel by about 25 mm, and a 50 mm clearance from the top edge of the doors to the underside of the track allows for vertical adjustment to be carried out if necessary. Corner protection plates are fixed to the outside bottom corners of the doors where they strike ground stops when opening and closing. The latter should be fixed securely with rag bolts to prevent hanger wheels hitting the end brackets.

The bottom of the doors is held in position by side guide rollers running in a steel channel let into the floor, bent fixing lugs bolted under the track at intervals firmly key the channel. Sheet 42.2 shows this detail with a 10 mm clearance between the bottom of the doors and the floor. Alternative designs of guides are shown in Sheet 42.3. To complete the work a filling piece is fixed on the inside of the pier to close the gap caused by the offset of the doors in using double track. The doors are normally secured by means of spring bolts into the floor with an outside fastening on the end leaf, using various types of locking bar used with a padlock or a cylinder lock.

If the doors are to be fixed on the outside face of a building, it is necessary to fix a canopy to protect the whole of the track and door gear from the weather, as shown in Sheet 42.4. One hanger is shown in section and part of the second in elevation in this detail. The steel machined wheels are supplied with ball-bearings or roller-bearings, grease packed, and lubricators which

SLIDING DOORS & DETAIL OF GEAR

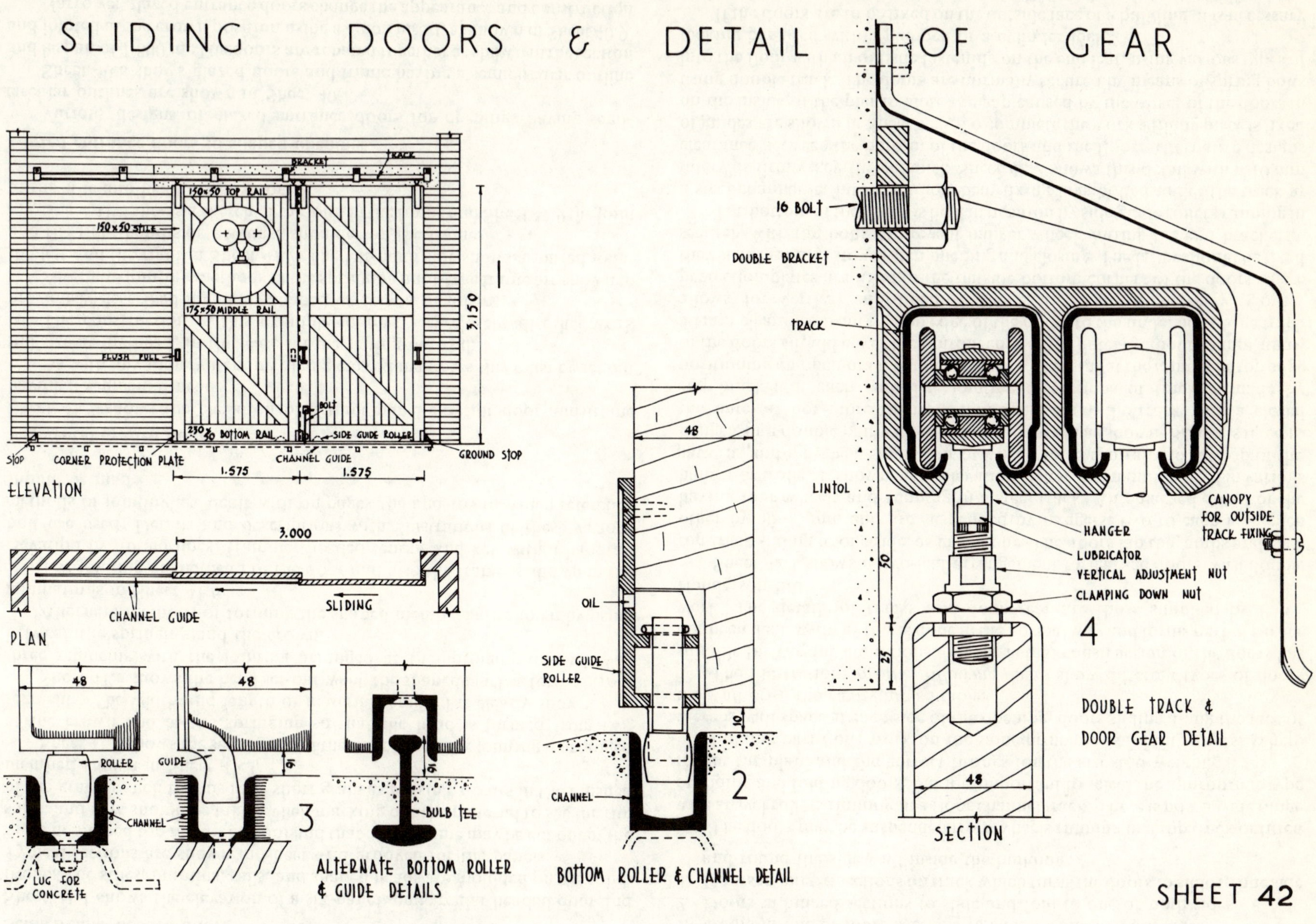

ELEVATION

150×50 STILE
150×50 TOP RAIL
BRACKET
TRACK
175×50 MIDDLE RAIL
FLUSH PULL
BOLT
230×50 BOTTOM RAIL
SIDE GUIDE ROLLER
STOP
CORNER PROTECTION PLATE
CHANNEL GUIDE
GROUND STOP
3.150
1.575   1.575

PLAN

3.000
CHANNEL GUIDE
SLIDING

3

48   48
ROLLER
GUIDE
CHANNEL
BULB TEE
LUG FOR CONCRETE

ALTERNATE BOTTOM ROLLER & GUIDE DETAILS

2

48
OIL
SIDE GUIDE ROLLER
CHANNEL
10

BOTTOM ROLLER & CHANNEL DETAIL

4

16 BOLT
DOUBLE BRACKET
TRACK
LINTOL
HANGER
LUBRICATOR
VERTICAL ADJUSTMENT NUT
CLAMPING DOWN NUT
CANOPY FOR OUTSIDE TRACK FIXING
50
25
48

SECTION

DOUBLE TRACK & DOOR GEAR DETAIL

SHEET 42

FOLDING DOORS

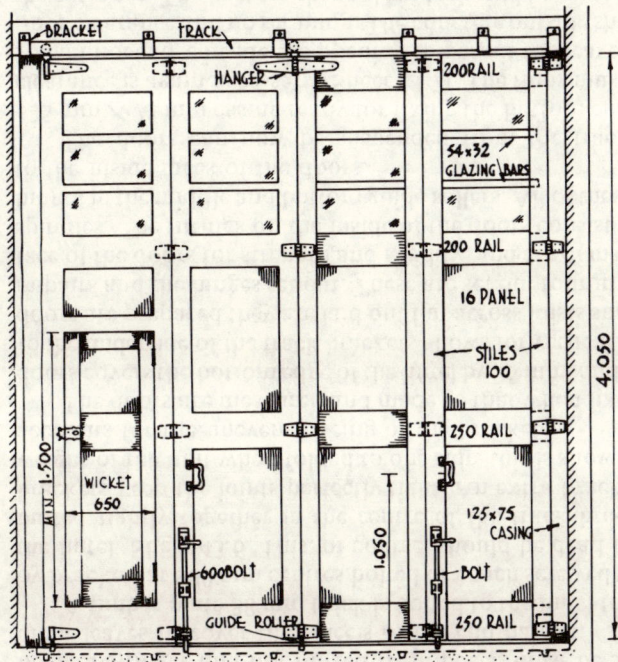

BRACKET   TRACK

HANGER

200 RAIL

54 × 32
GLAZING BARS

200 RAIL

16 PANEL

STILES
100

WICKET
650

MIN. 1.500

1.050

125 × 75
CASING

600 BOLT   BOLT

GUIDE ROLLER   250 RAIL

4.050

250 RAIL

ELEVATION 5

JOINTING BETWEEN LEAVES MAY
BE TONGUED OR REBATED

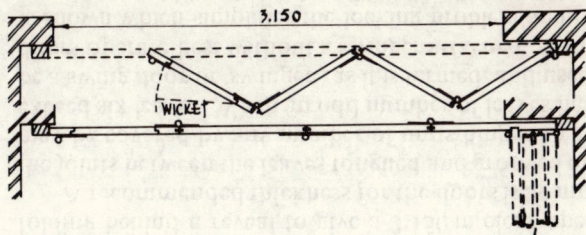

3.150

WICKET

PLAN

DOORS FOLDED TO SIDE

HENDERSON
TRACK & ROLLER
DETAIL
6

38   16 BOLT   FIXING
UNDER
R.S.J.

BRACKET

TRACK

PLATE   38

150

15   10

LUBRICATOR

19

HANGER   25

6

TOP RAIL   54

HINGE

ADJUSTMENT

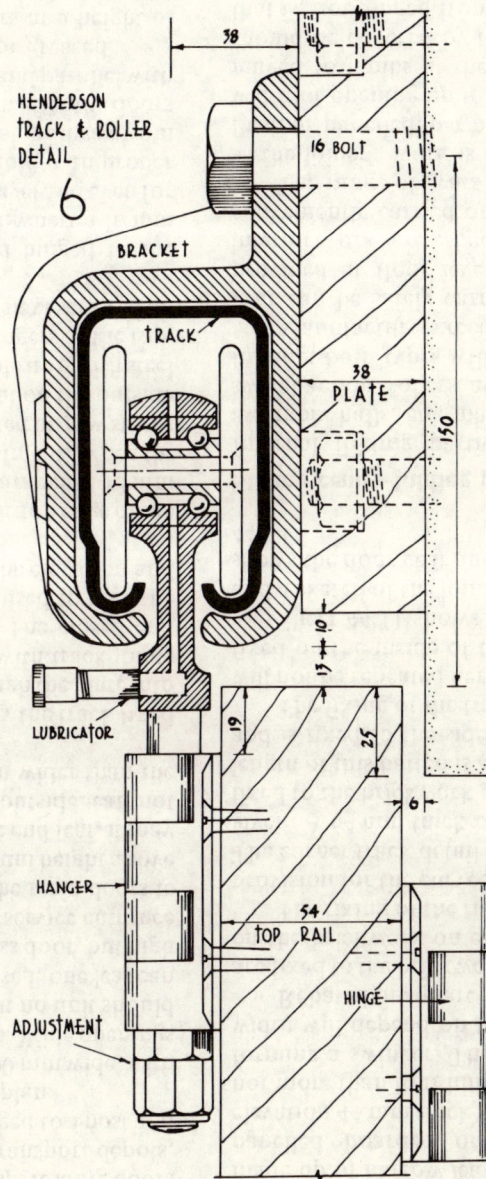

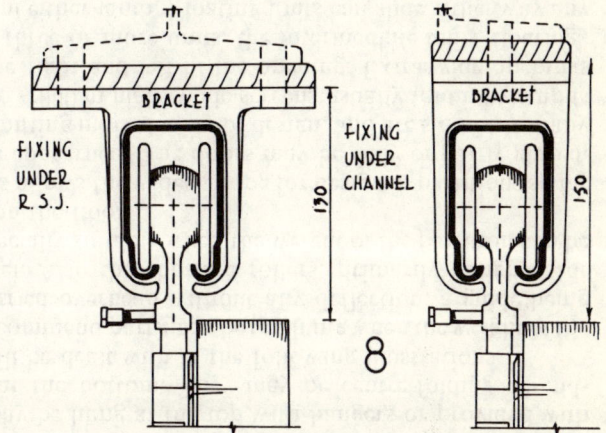

ALTERNATIVE HEAD FIXINGS 8

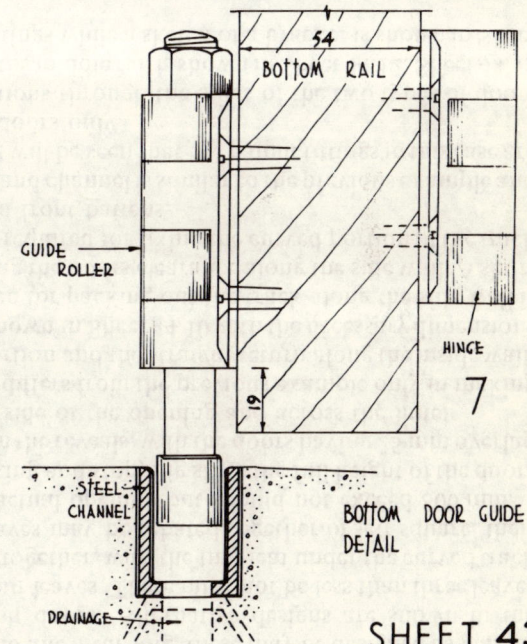

BRACKET   BRACKET

FIXING
UNDER
CHANNEL

130   150

54

BOTTOM RAIL

GUIDE
ROLLER

19

HINGE

7

STEEL
CHANNEL

BOTTOM DOOR GUIDE
DETAIL

DRAINAGE

SHEET 43

with proper maintenance and regular lubrication will ensure long and satisfactory life.

Sliding–folding doors

Sheet 43.5 shows the plan and elevation of a system of sliding–folding doors 4·050 m high, which fall in the second group, suitable for transport depots, warehouses, or garages. The doors consist of four leaves hinged to a post and folding behind a reveal to give a 3·150 m clear opening in plan.

A recommended thickness for the doors is 54 mm and 800 mm wide, with the joints between the leaves tongued and grooved or rebated. Wide openings may be covered by any number of units hinged to a post, but no unit should exceed six leaves. When an odd number of leaves has to be used, one leaf can be a swing door or 'swinger', as it is termed, and used as a pass door, but high doors operate best without a swinger attached. A wicket for service entrance is shown which simplifies the locking problem and permits the main doors to remain closed in bad weather. This should be 1·500 m minimum height above the bottom rail, and although it has been incorporated in the end leaf, it may be provided in any leaf. It should be noted that the end or outside leaf, not being a swinger and not hinged from posts, must be 38 mm wider than the other leaves to cover the wheels of the end hanger.

A timber plate 38 mm thick is bolted to the lintel to carry the track fixed by brackets at 900 mm centres bolted or coach screwed through the plate into the lintel, Sheet 43.6. This, of course, should be dead level with track joints butted tightly together in the centre of the track bracket. The closed end brackets keep the joints perfectly tight. An extra bracket is used to take the weight of the unit when folded to one side, this is shown in the elevation and accounts for the uneven spacing of the brackets.

The doors are measured and made so that when fixed, the top edge of the doors covers the bottom edge of the lintel by 25 mm, and a clearance of 13 mm to the underside of the track brackets allows for vertical adjustment. When the doors are prepared they are laid out flat across joists supported on saw stools in pairs and the hinges set out. These are set out four in height on the outside face of the doors for strength and rigidity, and are of malleable iron with steel spindles. The fittings on the inside of the doors consist of hangers at the top, hinges in the middle and bottom guide rollers. All of these are next screwed on to the inside faces of the doors.

The doors can now be suspended from the track and hinged to the 125 mm × 75 mm casing ready for fixing the bottom channel, where a 10 mm clearance is again necessary (Sheet 43.7). The floor must be level and even for 900 mm or more inside the opening to permit the leaves to fold with proper floor clearance and no scraping. The concrete outside should slope away from the entrance. The bottom channel guide should be cemented in after the doors have been tested when hung for plumb, and the guide is level and parallel with the top track. This is holed for drainage and should be kept greased.

Bow pull handles are screwed on the inside of the doors at a height of 1·050 m from the floor and 600 mm surface bolts let into the floor secure the doors. A cylinder lock is used on the wicket.

Sheet 43.8 shows alternative head fixings for the track, using special brackets where the doors are to be fixed under a rolled steel joist lintel or under a steel channel.

Sliding doors – around the corner

Sheet 44.9 shows the plan and elevation of a system of doors which fall into the third group – sliding to one side and round the side wall inside the building. They are ideal for domestic garages, and as they must of necessity be made up of narrow leaves to allow turning, these may be designed as glazed, panelled, boarded, or flush doors. Alternative designs are shown in the elevation 44 mm thick in four leaves. There must not be less than three leaves nor more than four hinged together, with the first leaf under the curved track forming a swinger. The leaves may be rebated together or left square, their width will depend on the actual opening but should not exceed 800 mm.

Rebated or square locking and stopping stiles the full height of the doors are fixed to the brickwork in the reveals, with the doors having 25 mm overlap on the brickwork on each side of the opening and across the lintel.

The fixing of the track differs from the previous example only in making provision for the curved portion and the straight return along the inside wall. The corner track detail is shown in Sheet 44.10 with the necessary dimensions given. A 64 mm thick batten for packing out the track along the side wall is fixed to the brickwork giving the doors clearance along the side wall. A short length of this batten is also required for fixing the curved portion of the track and is fixed to the side and front battens.

The fixing of the track and channel is similar to the previous example and will not be repeated here. It will be seen that the patent fittings in this case are fixed on the inside of the doors only.

Sheet 44.11 shows sections through the stiles of the two types of doors and a sketch of the joint at the middle rail is shown in Sheet 44.12. Sheet 44.13 shows the door gear and fittings while a sketch of a bracket is shown in Sheet 44.14.

Sliding centre-folding partitions

Internal folding partitions employed in schools, restaurants, churches, assembly halls, etc. may be hung at the top with hangers or provided with swivel-action rollers at the bottom. They may be centre-folding or end-folding; both types will be dealt with in the following illustrations.

Manufacturers recommend partitions be top hung when the weight of the gear can be safely carried overhead without any deflection, a guide being provided at floor level. Also that bottom rollers, primarily designed for interior work, are especially suitable when the weight of the partition can be conveniently carried on the floor.

The range of gears allows the widest scope for design in partitions so far as the joiner's work is concerned. The doors may be fully or partly glazed, flush or panelled, permitting individuality of design, and are suitable for any width of opening up to 4·800 m high. Wide screens usually fold in two units hinged to jambs at the right and left of the opening. Extra wide openings should be covered by three or more units, the intermediate units 'floating', that is, not hinged from either jamb. Floating units can slide either way any distance necessary to fold behind piers and clear of the opening.

For end-folding screens the number of leaves per unit should not exceed six. Floating units should always be four or six – never an odd number, because of the tendency to tilt and instability.

SLIDING DOORS

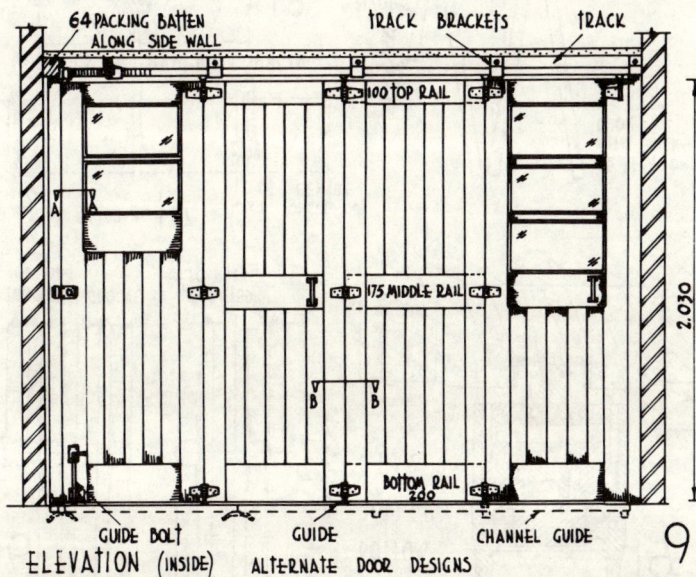

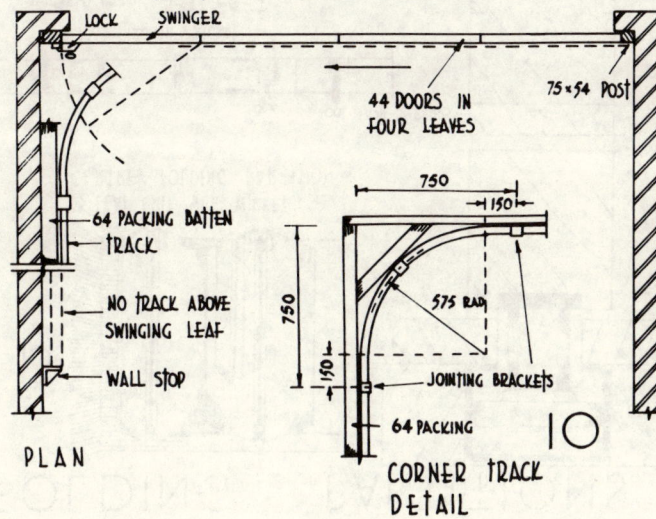

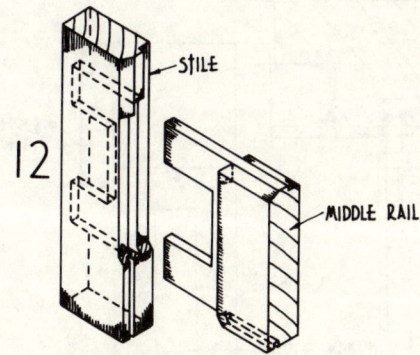

64 PACKING BATTEN ALONG SIDE WALL   TRACK BRACKETS   TRACK

100 TOP RAIL

175 MIDDLE RAIL

BOTTOM RAIL 200

2,030

GUIDE BOLT   GUIDE   CHANNEL GUIDE

9

ELEVATION (INSIDE)   ALTERNATE DOOR DESIGNS

LOCK   SWINGER

75 × 54 POST

44 DOORS IN FOUR LEAVES

64 PACKING BATTEN
TRACK

NO TRACK ABOVE SWINGING LEAF

WALL STOP

PLAN

750
150

750

575 RAD

150

JOINTING BRACKETS

64 PACKING

CORNER TRACK DETAIL

10

STILE

MIDDLE RAIL

12

SKETCH OF JOINT BETWEEN STILE & MIDDLE RAIL

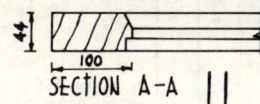

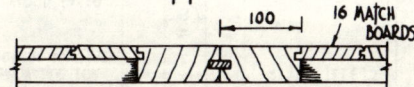

44
100

SECTION A-A   11

100   16 MATCH BOARDS

SECTION B-B

SKETCH OF BRACKET

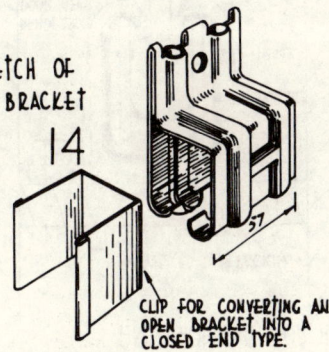

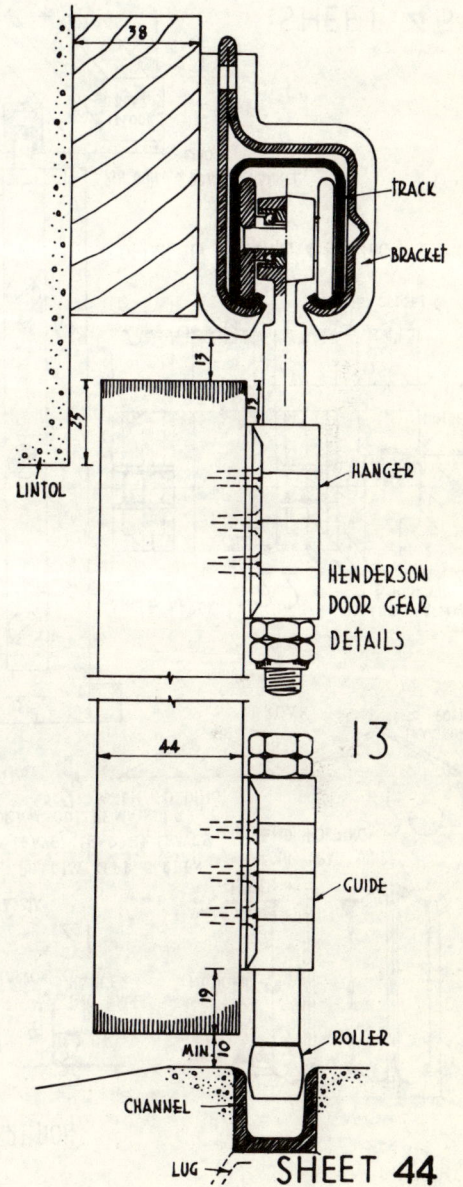

14

57

CLIP FOR CONVERTING AN OPEN BRACKET INTO A CLOSED END TYPE.

38

TRACK

BRACKET

13

LINTOL

25

HANGER

HENDERSON DOOR GEAR DETAILS

13

44

19

GUIDE

ROLLER

MIN. 10

CHANNEL

LUG

SHEET **44**

FOLDING PARTITIONS.

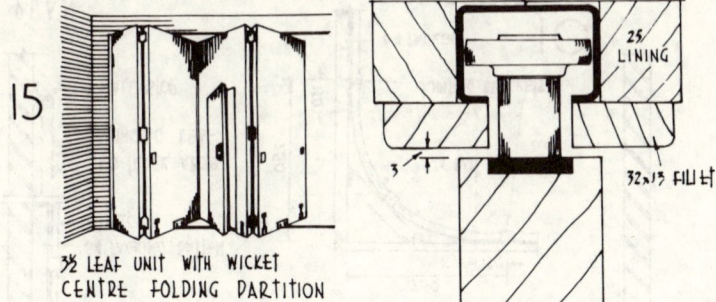

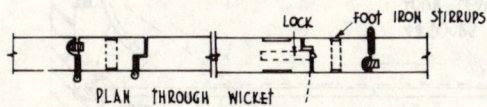

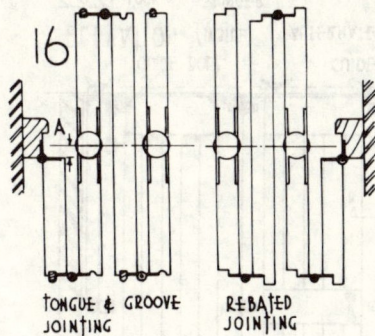

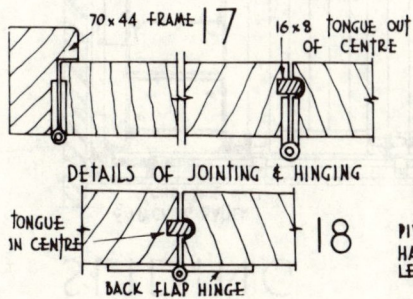

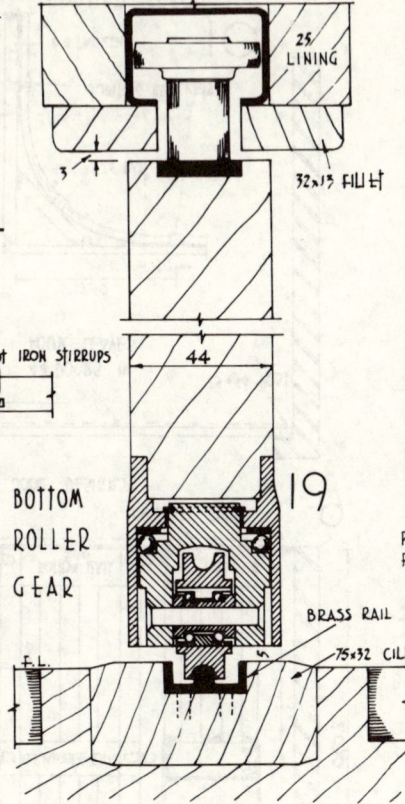

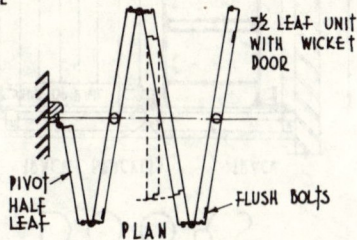

15 3½ LEAF UNIT WITH WICKET CENTRE FOLDING PARTITION

25 LINING
3
32×13 FILLET

LOCK · FOOT IRON STIRRUPS
PLAN THROUGH WICKET

16
A
TONGUE & GROOVE JOINTING · REBATED JOINTING

44

19 BOTTOM ROLLER GEAR
F.L.
BRASS RAIL
75×32 CILL

17 70×44 FRAME · 16×8 TONGUE OUT OF CENTRE
DETAILS OF JOINTING & HINGING

18 TONGUE IN CENTRE · BACK FLAP HINGE

PIVOT HALF LEAF · PLAN · FLUSH BOLTS
3½ LEAF UNIT WITH WICKET DOOR

END FOLDING TOP HUNG PARTITION

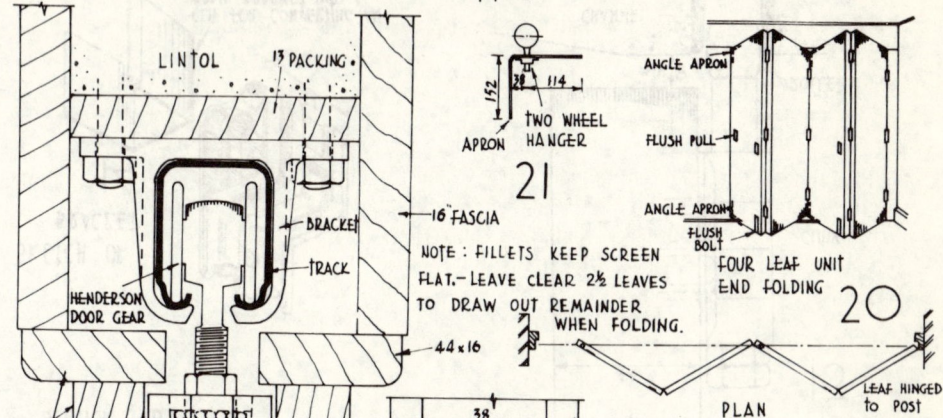

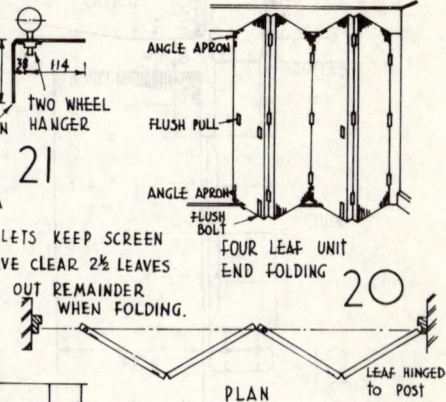

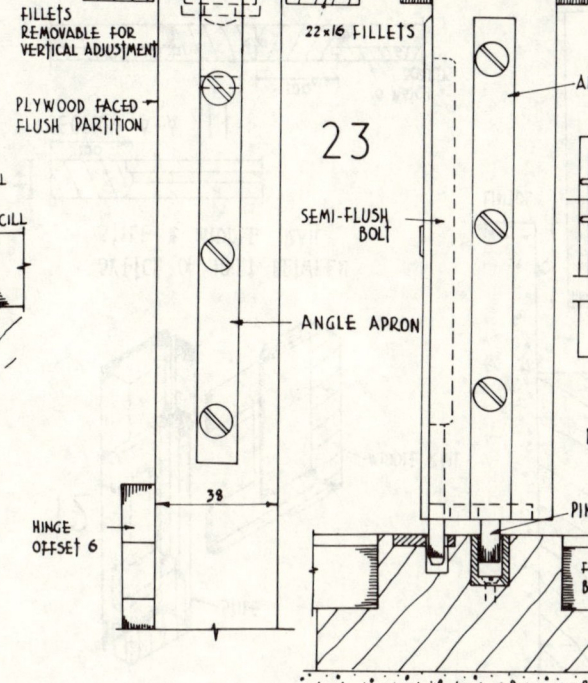

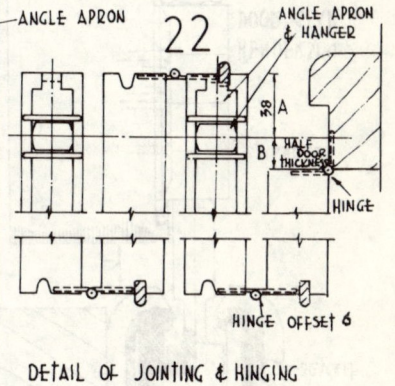

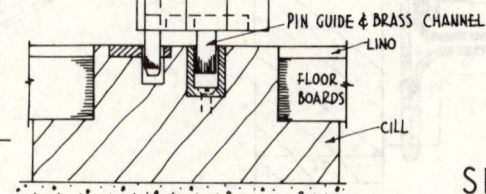

LINTOL · 13 PACKING
HENDERSON DOOR GEAR · BRACKET · TRACK
16 FASCIA
44×16
FILLETS REMOVABLE FOR VERTICAL ADJUSTMENT
PLYWOOD FACED FLUSH PARTITION
NOTE: FILLETS KEEP SCREEN FLAT.- LEAVE CLEAR 2½ LEAVES TO DRAW OUT REMAINDER WHEN FOLDING.
22×16 FILLETS

21 152 · 28 · 114 · APRON · TWO WHEEL HANGER

20 FOUR LEAF UNIT END FOLDING
ANGLE APRON · FLUSH PULL · ANGLE APRON · FLUSH BOLT
PLAN · LEAF HINGED TO POST

23 38
SEMI-FLUSH BOLT · ANGLE APRON · ANGLE APRON
HINGE OFFSET 6 · 38

22 ANGLE APRON · ANGLE APRON & HANGER · 38 A B HALF DOOR THICKNESS · HINGE
HINGE OFFSET 6
DETAIL OF JOINTING & HINGING

PIN GUIDE & BRASS CHANNEL · LINO · FLOOR BOARDS · CILL

SHEET 45

For centre-folding screens the maximum number of leaves hinged to a jamb should not exceed seven and a half leaves, while floating units should be in five or seven leaves.

Sheet 45.15 shows the small-scale key elevation of a partition with three full leaves and a half-leaf, which is always necessary in this type. All the full leaves are of equal width with the pivot half-leaf half the width of a full leaf, less the throw of the hinge shown at A in the plan (Sheet 45.16). The throw of the hinge is exactly half the finished thickness of the screen when the hinge is fitted with the centre pin in line with the face of the screen. A wicket door may be provided in any leaf not containing a bottom roller.

The formulae for the accurate width of the leaves for a screen 44 mm finished thickness, is as follows: add 22 mm to the width between the rebates of the end jambs for a screen with one pivot half leaf. Divide the result by twice the number of full leaves plus one for the half leaf. This gives half the width of a full leaf. Deduct 22 mm for the width of the pivot half leaf. Example:

2·733 m between rebates $\dfrac{2 \cdot 733\text{ m} + 22\text{ mm}}{7} = 394$ mm

394 mm less 22 mm = 372 mm width of pivot half leaf
394 mm × 2 = 788 mm width of three full leaves

The joiner's work in whichever type of design is used must be most accurately done so that the edges of all leaves are perfectly parallel with the ends square. The edges of the leaves may be rebated or tongued and grooved (Sheet 45.16). An enlarged detail of the latter is shown in Sheet 45.17 where it will be seen that to avoid cutting through the tongue and groove, the hinges as well as the tongues are offset. Broad hinges, fixed back-flap may be used as an alternative (Sheet 45.18), but this method does not give as neat a finish on the face side as hinges let in the edges of the leaves.

The fittings, apart from the hinges, are ball-bearing bottom rollers (Sheet 45.19) with guide rollers at the top, each of these is fixed centrally on flush plates let into the top and bottom edges of the leaves. A brass bottom rail is let in and screwed to a hardwood sill with a 5 mm clearance at the bottom. It is most important that the track is fixed level and perfectly straight. As can be seen, the top guide track is of a different section from any previously used and is fixed and cased in when the screens are in position.

Flush bolts with flush pulls over are fitted on alternate leaves and end leaves, with a rebated cylinder lock on the wicket to complete the partition.

Sliding end-folding partition

Sheet 45.20 shows the small-scale key elevation and plan of an end-folding top-hung partition with four leaves hinged to a jamb. An even number of leaves, not exceeding six and hinged together as a unit, should be used whenever possible. All the leaves are of equal width except the leaf hinged to the jamb. This is narrower by half the thickness of the screen plus 38 mm. Notice that the screen in this case is 38 mm thick.

The formulae for the accurate width of the leaves for such a screen is as follows:

Add 56 mm (A + B), (Sheet 45.22) to the width between the rebates, divide the total by the number of leaves to give the width of the equal leaves. The pivot or hinged leaf is 56 mm less in width.

The joinery work and preparation of the screen are similar to the previous example except that being top hung, special angle apron hangers (Sheet 45.21) and guide plates are fitted to the edges of the stile leaving clear flush faces to the screen.

Sheet 45.22 shows the plan of the doors folded back with the jointing and hinging of the leaves. A section showing the door gear with the fixing and finishings is shown in Sheet 45.23.

Spayed linings

Sheet 46.1 shows the part plan and elevation of a window with splayed soffit and jamb linings. The linings are tongued and grooved together. To obtain the bevel required for the shoulder of the jamb and the groove in the soffit, the lining should be developed. AB represents the face of the jamb in the plan splayed at 60 degrees to the face of the frame. The edges are projected into elevation and the head lining drawn in to intersect on the mitre line CD. With point A as centre and radius AB, describe the arc BB' bringing the edge B into the same plane as A. Draw a perpendicular from B' in plan into the elevation cutting the top edge of the soffit lining produced in C'. Join C'D and the contained angle is the bevel for the top of the jamb. As the soffit is splayed at the same angle as the jambs, the same bevel will answer for both. If the angle is different as in the next example, then the soffit also must be turned into the vertical plane to give the bevel for grooving the soffit.

The method of obtaining the angle which the edge cut makes with the face of the jamb lining is shown in Sheet 46.2. As a geometrical problem it is the determination of the angle between the two planes. CD is the elevation of the line of intersection of these two planes and CE its true length. From point 5 a line is drawn at right angles to CD, cutting the edges of the linings in points 2 and 3. With 5 as centre, an arc is drawn tangential to the true length CE cutting the intersection in point 4. From points 2 and 3 draw lines to point 4. Angle 2–4–3 is the dihedral angle. If the corners of the linings are mitred then angles 2–4–D and 3–4–D are the bevels required. If they are to be tongued and grooved then angle 3–4–F is the bevel to be applied to the vertical lining. In this example, only the side bevel for the vertical lining and the mitre bevel have been marked.

Sheet 46.3 shows the true shape of the jamb lining for a window or door opening having a segmental head. The jamb lining is splayed at 45 degrees. The method of obtaining the true shape is similar to the previous example and is clearly shown.

Sheet 46.4 shows the part elevation and plan of a window having a semicircular head with a framed splayed lining. The stiles to the head lining are worked in the solid in two pieces, joined at the crown and springings. The method of obtaining the face moulds for the head lining is shown to the right in plan. The edges of the stiles are drawn across to the centre line to give the centres E, 1, 2, and 3 for describing the moulds A, B, C, and D.

SPLAYED LININGS

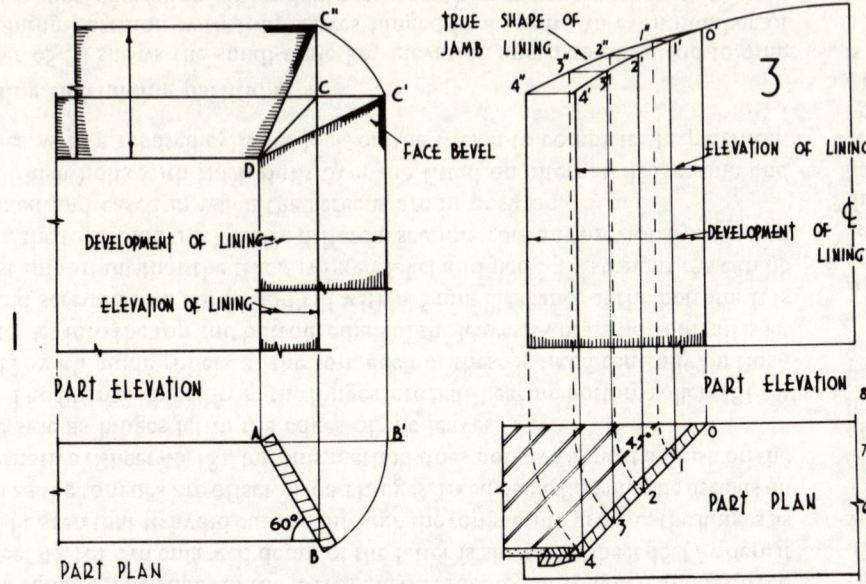

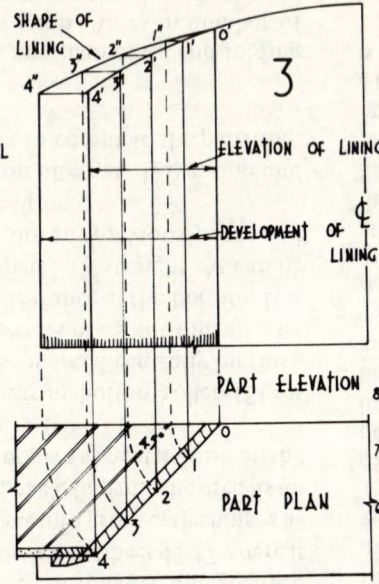

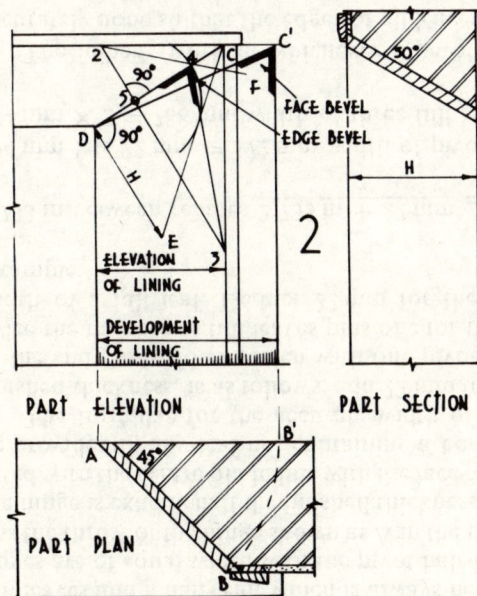

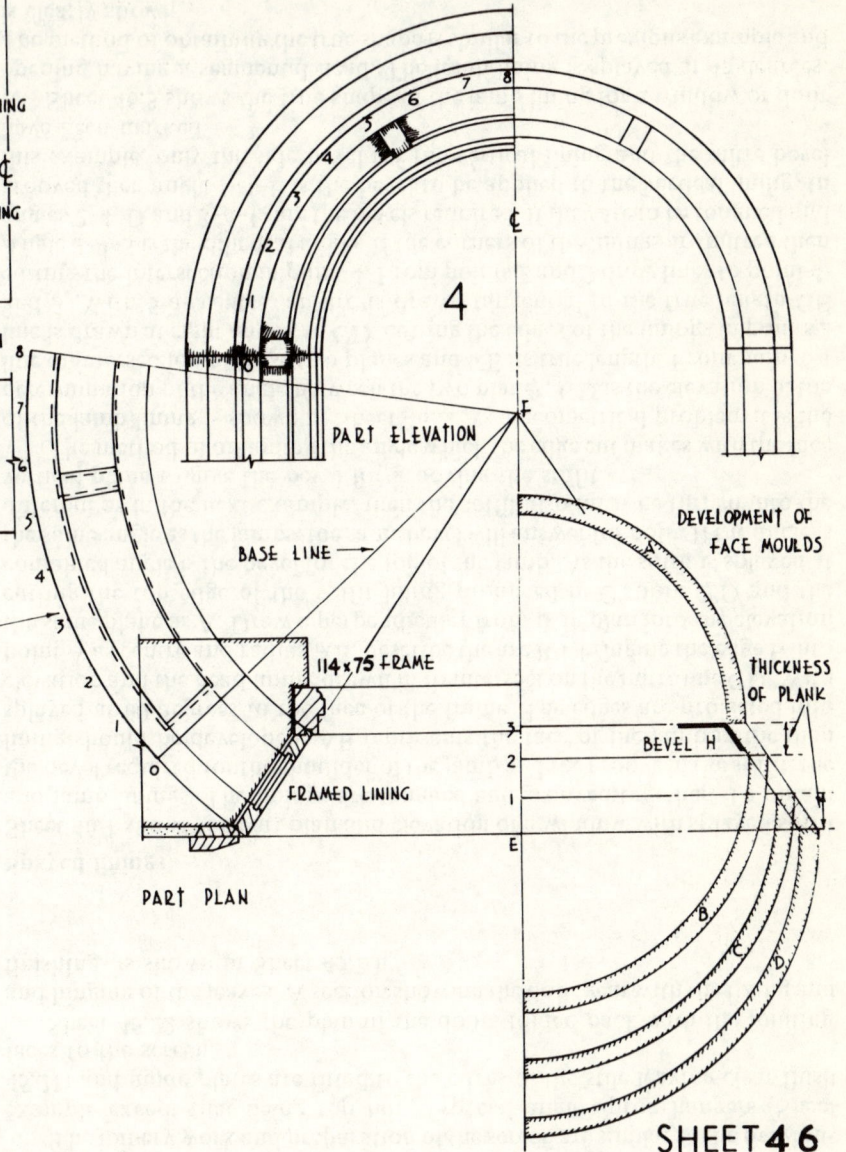

178

1
PART ELEVATION
PART PLAN

C″
C
C′
FACE BEVEL
D
DEVELOPMENT OF LINING
ELEVATION OF LINING
A   B′
60°
B

2
PART ELEVATION
PART PLAN
PART SECTION

2
90°
4
C
C′
5
F
D
90°
H
E
3
ELEVATION OF LINING
DEVELOPMENT OF LINING
A   45°   B′
B

50°
FACE BEVEL
EDGE BEVEL
H

3
TRUE SHAPE OF JAMB LINING
3″ 2″ 2″ 1″ 0
4″ 3 3 2′ 1′
4′
ELEVATION OF LINING
DEVELOPMENT OF LINING
PART ELEVATION
PART PLAN
8
7 0
6
5
1
4 2
3

4
PART ELEVATION
1 2 3 4 5 6 7 8
BASE LINE
PART DEVELOPMENT OF SPLAYED CURVED PANELS
114 x 75 FRAME
FRAMED LINING
PART PLAN
8
7
6
5
4
3
2
1
0
F
E
DEVELOPMENT OF FACE MOULDS
THICKNESS OF PLANK
BEVEL H
3 2 1
A
B
C
D

SHEET 46

The thickness of the plank is equal to the distances E–1, and 2–3 as shown on which the face moulds, along with the bevel, are applied. The width of the plank is determined by laying the two face moulds for each stile side by side, and measuring the width required. Mould A is applied to the face of the plank and the ends cut to the mould square across the face. The bevel H in plan is applied on the square ends working from the face side, giving points on the inside to which mould B is placed and marked. The piece is cut and worked to these mould lines, the inside edge being squared from the face ready for jointing at the ends. The other stile is dealt with similarly. The joints at the springing and crown are prepared for handrail bolts.

After setting out and mortising for the rails, the stiles are grooved for the panels. The development of a curved panel is shown in plan Sheet 46.4. A curved edge in elevation is divided into eight equal parts and numbered. Point F in plan is the intersection of the two linings produced and represents the apex of the semicone. With the apex F as centre, the edges of the stiles are struck, and from a base line the points 0–8 are marked off to give half the total length of the panels.

Thin veneered ply, which will easily bend to the curve, may be used for the panels.

Ironmongery

Windows designed so that they can be shut to exclude draughts must be fitted with the types of hardware which are adequate not only to close the window, but to keep it properly closed when so desired. A wide range of hardware is available including concealed espagnolettes, friction pivots, friction stays, casement stays, and espagnolette-handles of modern design.

Where windows are manufactured by specialist firms, normally the sashes are hung in their frames with all built-in hardware, such as pivots, espagnolettes, hinges, sash couplers, fitted in position. This particular range of hardware is rust-proof and therefore should not be painted as its efficiency may be impaired. Surface hardware such as handles, friction stays, and safety chains should be fixed after final completion of decoration.

Various types of hardware are illustrated on page 180. Sheet 47.1 shows a caulking lock with handle. These locks are fitted with a dead-locking lever and are particularly suitable for small windows, basement and stair windows, shutters, cupboard doors, etc. The lock is mortised into the stile or rail of the window, the receiver fixed to the frame with the handle screwed on the inside of the window. This handle is also used on horizontal pivot-hung windows. Sheet 47.2 shows a friction stay. The friction bearing is screwed to the sash with the pin mounting screwed to the frame. The stay makes use of friction to hold the sash in any desired position, the holding power of the stay is the same at all angles. The amount of friction can be adjusted by a screw so designed to prevent it being actuated by the movement of the arm.

The stay can be used on bottom-hung, top-hung, and pivot-hung sashes, as well as side hung.

Sheet 47.3 shows two types of espagnolette bolts. The type on the left is for windows and that on the right for doors. All bolts can be made to any desired length with the handle placed in any position. Normally it is placed in the centre of normal length units. The espagnolette is housed into the edge of the stile of doors or windows so that when closed only the handle is visible. The bar stiffens the stile and at a turn of a single handle its bolts lock simultaneously at several points, forcing the windows or doors against the frame, thus ensuring that they are safeguarded against warping and are completely draught-proof.

A key-operated lock can also be fitted to any espagnolette where the timber sections allow for a lock housing.

Sheet 47.4 shows an espagnolette with three side levers and a key-lock.

Sheet 47.5 shows a 'round closing' espagnolette with locking points at each corner for horizontal pivot-hung windows.

Sheet 47.6 shows a three-section brass coupling screw for fastening sashes in double glazed windows. An alternative sash coupler is shown in Sheet 47.7. The coupler ensures that all double sashes are secured in strict alignment, with the requisite amount of air space between them. The male half of the coupler is surface-mounted on one sash, whilst the female half is countersunk into the other sash. The female half incorporates a spring-loaded catch which greatly facilitates the separation or coupling of sashes.

Sheet 47.8 shows a friction sash pivot for side fixing on flush lights. A face-fixing pivot on flush lights is shown in Sheet 47.9. A semi-flush pattern for fixing on recessed lights is shown in Sheet 47.10. The pivot will hold the sash open, safely and securely, in any desired position. It is adjustable; the amount of friction can be easily increased or decreased by the adjustment nut and screw. Once the correct frictional strength is obtained, the pivot remains locked to this adjustment. It enables the window to be cleaned on both sides from inside the room.

A further type of pivoting window bearing for vertical pivoted sashes is shown in Sheet 47.11.

Sheet 47.12 shows another type of friction pivot for horizontal pivoted sashes rotating through 180 degrees.

Spring hinges

For doors which are required to close automatically, some kind of spring hinge is used. They are required to stand heavy wear, so their quality is most important. Various types of spring hinges are available for single-action doors and also for double-action or swing doors.

Sheet 48.13 shows a twin type of spring hinge for swing doors with a single type of the same hinge shown in Sheet 48.14. Two single-type hinges are used for doors weighing up to 22·70 kg in weight. For doors between 22·70 and 45·4 kg in weight, one single type and one twin type are used. Two twin-type hinges are used for doors weighing over 45·40 kg in weight.

A single-action helical spring hinge is shown in Sheet 48.15 suitable for doors up to 64 mm thick. A double-action hinge is shown in Sheet 48.16; this is also made in sizes to fit doors up to 64 mm thick and is suitable for swing doors. The barrels contain springs which are regulated by inserting a steel pin in the collar, when the hinge is fixed, one flap to the door and the other flap to the frame. In hanging a door, only one spring hinge is used at the top, the other hinge being a 'blank' (Sheet 48.17) which has no springs. The spring and blank hinges are sold together as a pair.

IRONMONGERY

CAULKING LOCK & HANDLE 1

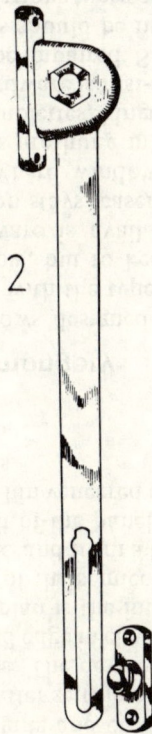

FRICTION STAY WITH PIN MOUNTING 2

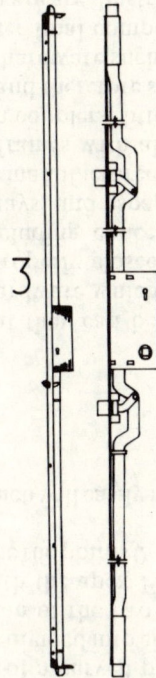

ESPAGNOLETTES FOR WINDOWS & DOORS 3

ESPAGNOLETTE FOR SINGLE SIDE HUNG SASH OR PIVOT HUNG SASH 4

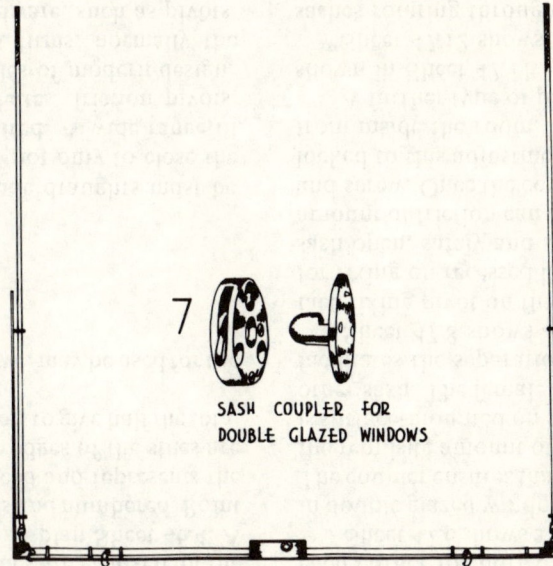

ESPAGNOLETTE FOR HORIZONTAL PIVOT-HUNG WINDOWS 5

COUPLING SCREW FOR DOUBLE GLAZED WINDOWS 6

SASH COUPLER FOR DOUBLE GLAZED WINDOWS 7

FRICTION SASH CENTRE SIDE FIXING ON FLUSH LIGHTS 8

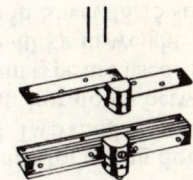

SASH CENTRE FOR FACE FIXING ON FLUSH LIGHTS 9

SASH CENTRE SEMI-FLUSH FOR RECESSED LIGHTS 10

VERTICAL PIVOTING WINDOW BEARING 11

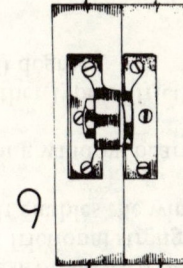

FRICTION SASH PIVOT 12

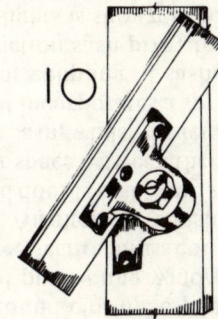

SHEET 47

IRONMONGERY

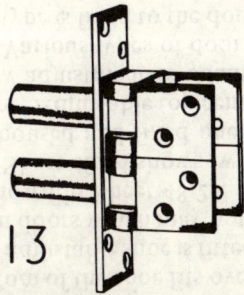

13 DOUBLE SPRING HINGE

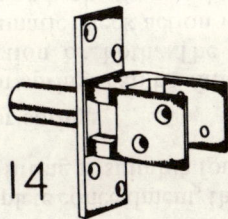

14 SINGLE SPRING HINGE

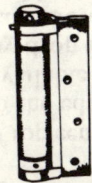

15 SINGLE ACTION SPRING HINGE

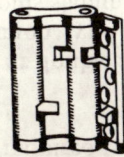

17 BLANK HINGE

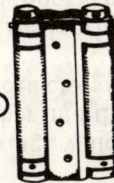

16 DOUBLE ACTION SPRING HINGE

19 DOOR CLOSER

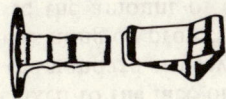

SPRING CLIP

DOOR HOLDERS

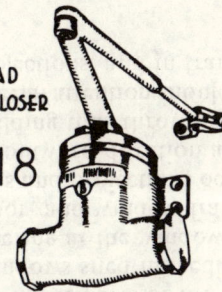

18 OVERHEAD DOOR CLOSER

20 CONCEALED DOOR CLOSER

TOP CENTRE

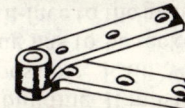

24 TOP CENTRE FOR SINGLE ACTION FLOOR SPRINGS

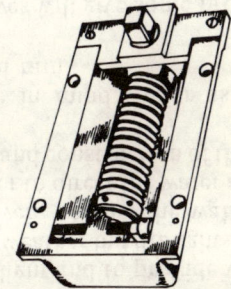

DOUBLE ACTION FLOOR SPRING

22 DOUBLE-ACTION SHOE

DOUBLE-ACTION STRAP

23 SHOE FOR SINGLE-ACTION FLOOR SPRINGS

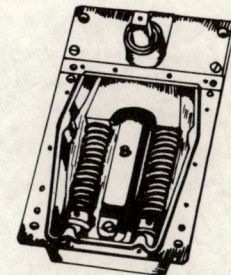

21 DOUBLE ACTION FLOOR SPRING WITH HYDRAULIC CHECK

SHEET **48**

181

Overhead door closers

Two examples of this type of fitting are shown in Sheet 48, figs. 18 and 19. They combine the features of a door check and spring for closing doors without noise, and are hydraulic. They may be obtained with either single or double action; single-action closers are made either to push or pull the door closed. Because it is sometimes necessary for the door to remain open for ventilation purposes, or while things are carried through the doorway, a closer which will hold the door open at 90 degrees may be fitted, being released with a slight pull.

The body of the closer is fixed to the top rail of the door and the arm bracket fixed to the face of the frame. With the door hung on normal butt hinges, the standard arms will allow the door to open through 120 degrees. If the door is hung on projecting hinges, the angle of opening will be reduced, subject to the amount of hinge projection. In some instances the door is required to open 180 degrees and longer arms become necessary.

A concealed door closer is shown in Sheet 48.20. These units are door closers specially designed for use where the first consideration is to maintain the aesthetic qualities of the door and surround. The model is primarily intended to control interior doors not exceeding 54·48 kg in weight. Apart from the slim arms of the fitting on the hinge side of the door, there is complete concealment, the body being recessed into the top edge of the door. The fitting is suitable for a door required to open to 90 or 180 degrees.

Floor springs

Floor springs can be either single or double action for doors opening in one direction or both. The single-action type have either a hydraulic or a pneumatic check action which avoids slamming. The springs are let into the floor and take the weight on the pivot (Sheet 48.21). The springs for double-action or swing doors are provided with shoes and top centres. The shoe (Sheet 48.22) is fitted and screwed to the heel of the door. A square hole in the bottom of the shoe fits over the pivot which is actuated by a spring in the box. The adjustable shoe is fitted with an adjustable bush which can be regulated to align doors which may not be straight. A shoe for a single-action floor spring is shown in Sheet 48.23.

Sheet 48.24 shows two types of top centre used with floor springs, which are housed and fixed one part into the edge of the door, the other into the frame. Adjustable top centres are available which are fitted with regulators to allow adjustment to counteract any drop in the door.

Various types of door holders are shown in Sheet 48.25. The steel spring clip type is fixed to the door-frame with the plunger screwed to the door. The other patterns are housed and fixed in the floor.

Chapter 16

Windows

There are a number of types of window. Some are complicated, and are usually employed in the more specialised buildings such as schools, hospitals, multi-storey flats, offices, and the like.

The purpose of any window is to admit light and to provide ventilation to the building. It must not only keep out the weather but draughts also so far as is possible. Rain which falls, and is driven, against the wall-face of a building has to be reckoned with, and in order to direct the water away from the wall-face to the ground area, good design and construction of the building is most important.

The following points should be borne in mind so far as design is concerned when considering the exclusion of rain:

1. A good overhang or projection of the eaves will give some protection to the upper parts of the outside walls.
2. Windows sited immediately under the projecting eaves will prevent rain entering at the window head.
3. Door- and window-frames are better set back from the wallface.
4. Sills and projections occurring on a building face should be weathered in an outward direction and where possible project beyond the face of the building thus throwing water clear of the face.
5. Special attention should be paid to throatings, or drips on the underside of projections, and in frames and opening sashes.

Window weatherings

The most vulnerable position on the exposed faces of any building is at window and door openings. It is pointed out here that the first consideration in the design and construction of any window or door is the shedding of rain as it streams down the vertical face. The weakest parts of a window where

WEATHERINGS

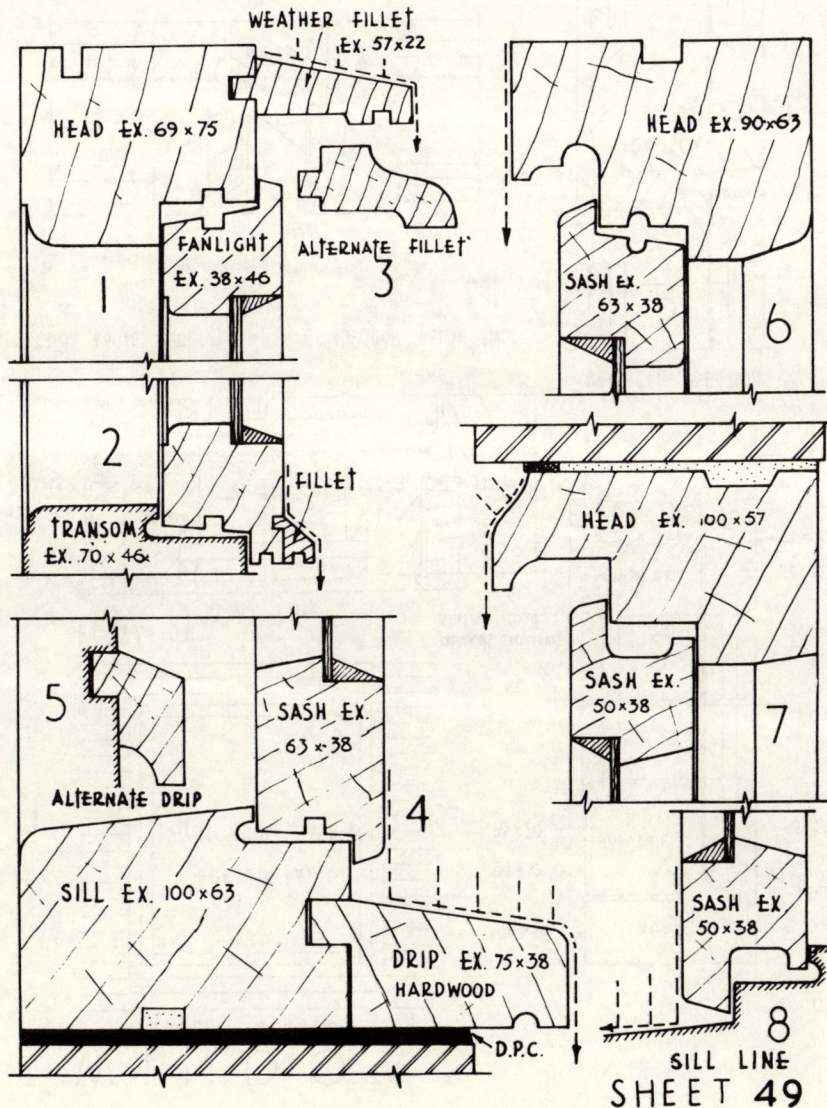

WEATHER FILLET
EX. 57 x 22

HEAD EX. 69 x 75

FANLIGHT
EX. 38 x 46

1

ALTERNATE FILLET

3

2

TRANSOM
EX. 70 x 46

FILLET

HEAD EX. 90 x 63

SASH EX.
63 x 38

6

5

ALTERNATE DRIP

SASH EX.
63 x 38

4

SILL EX. 100 x 63

DRIP EX. 75 x 38
HARDWOOD

D.P.C.

HEAD EX. 100 x 57

SASH EX.
50 x 38

7

SASH EX
50 x 38

8

SILL LINE

SHEET 49

penetration may occur, are the opening lights.

Sheet 49 shows various sections of timber windows with outward-opening casements where a number of alternative designs of frames, sashes, and sills with weatherings are illustrated. Sheet 49.1 shows the head of a frame prepared to receive a hardwood weather fillet. The purpose of this fillet is to prevent rain from sweeping in over the tops of the opening lights and down inside the window by shedding it well clear of the window as shown by the dotted line.

The method of shedding rain at the bottom rail of the sash and transom position is shown in Sheet 49.2. A drip moulding is used, housed, and bonded to the frame with synthetic resin, casein glue, or white lead paint and nailed into the bottom rail of the sash. Such a drip must be used where casements occur below transoms. An alternate fillet is shown in Sheet 49.3.

Water is thrown well clear of the face of the building at sill level by the hardwood drip shown in Sheet 49.4. This is tongued to the sill of the frame, weathered on the top side and throated on the underside. Fixing is done as with weather fillets at the head. An alternate drip is shown in Sheet 49.5. Sheet 49.6 shows the head of the window-frame profiled to form its own weather fillet on the solid. This prevents the rain from sweeping in over the lipping of the top rail of the sash. A similar form of moulding at the head of the window-frame to form a drip out of the solid is shown in Sheet 49.7.

Equally as important as the use of drips in weatherproofing wood windows is the construction and fitting of opening lights. As these have to open for ventilation purposes, there must be a clearance between the edge of the sash and the frame; this tolerance is also necessary to allow for a number of coats of paint. The lipping on the outer edges of the sash shields this clearance between sash and frame and does in fact force any weather attempting to penetrate to change its route.

Throating, or grooving, of sashes as shown on the edges of the sash further assists in preventing penetration. These grooves disrupt capillary action and direct any rain entering the window away to either sills or transoms without causing any dampness on the inside of the buildings. The throating in the frame is placed in the rebate opposite to those in the edges of the sash. A further throat placed in the angle of the rebate is used on a square arris casement. The edges of the grooves should not be less than 6 mm from the face of the member in which they occur, positioned so that the screws for the hinges enter a flat portion of the rebate or in the centre of a groove.

Sheet 49.7 shows the section through the head of a window-frame and the top rail of an opening sash. The throat, it will be seen, is much deeper and wider in section than any of the accompanying illustrations and is a channel. The effect of this is to increase the space between the edge of the sash and the rebate of the frame, preventing further capillary action.

The detail at the bottom rail and sill is shown in Sheet 49.8.

The types of windows in general use may be classified as follows:

1. Casement, with sashes either fixed or to open. Opening sashes may be hinged at the side, top or bottom, to open inwards or outwards. Inward-opening sashes require special attention, they are difficult to make watertight and with opening into a room tend to foul decorations like curtaining.

WINDOWS
DOUBLE GLAZED

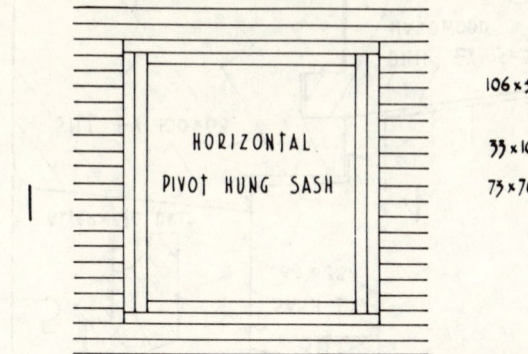

HORIZONTAL
PIVOT HUNG SASH

1

ELEVATION

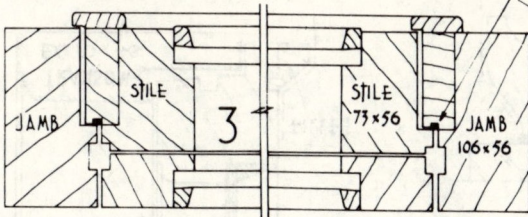

JAMB — STILE

3

STILE 73×56

JAMB 106×56

SECTION ABOVE PIVOT   SECTION BELOW PIVOT

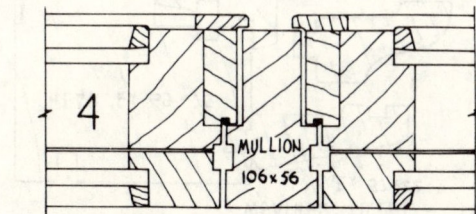

4

MULLION 106×56

DETAILS OF MULLION & TRANSOME SECTIONS
FOR WINDOWS WHERE THESE ARE EMPLOYED

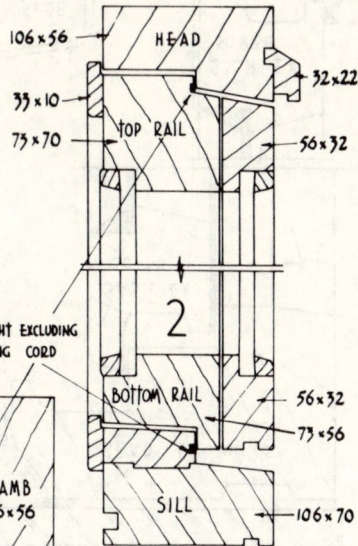

106×56   HEAD
33×10   32×22
73×70   TOP RAIL   56×32

DRAUGHT EXCLUDING
SEALING CORD

2

BOTTOM RAIL   56×32
                73×56

SILL   106×70

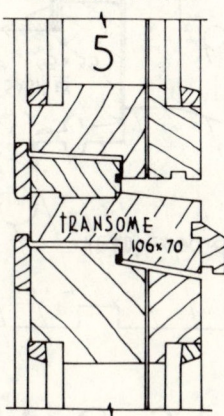

VERTICAL SECTION

5

TRANSOME 106×70

COMBINED PIVOT & SIDE HUNG CASEMENT WINDOW

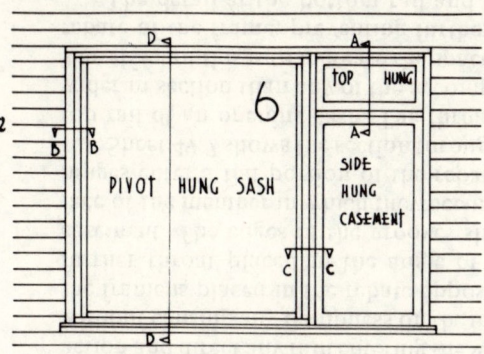

6

TOP HUNG

PIVOT HUNG SASH   SIDE HUNG CASEMENT

50×19

ELEVATION

SKETCH OF PIVOT HUNG SASH
OPENING APPROX 30°

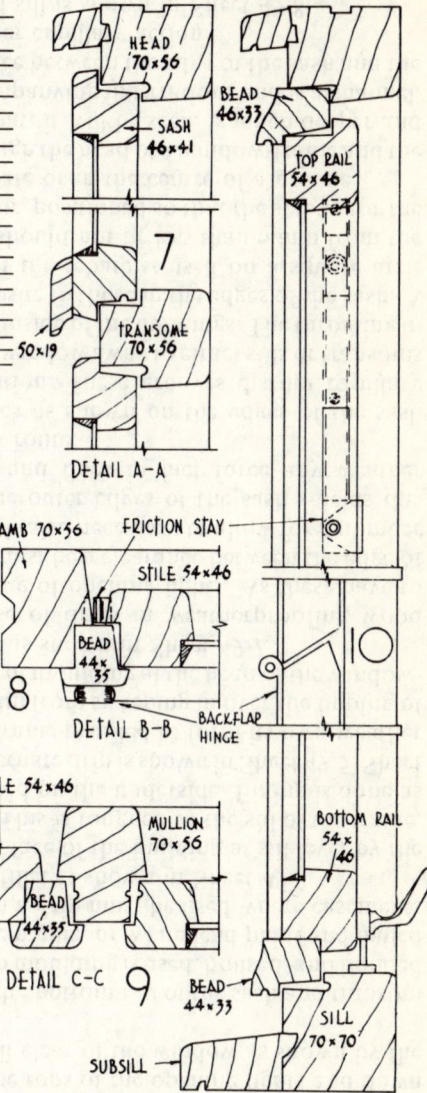

COMBED JOINTING

11

DETAIL D-D

HEAD 70×56
SASH 46×41
BEAD 46×33
TOP RAIL 54×46

7

TRANSOME 70×56

DETAIL A-A

JAMB 70×56   FRICTION STAY
STILE 54×46

8   BEAD 44×35

DETAIL B-B   BACKFLAP HINGE

STILE 54×46
MULLION 70×56
BEAD 44×35

DETAIL C-C   9

BOTTOM RAIL 54×46

10

BEAD 44×33
SILL 70×70

SUBSILL

2. Sliding, with sashes to slide vertically – when they are termed double-hung – or horizontally.
3. Pivot-hung, with sashes hinged on pivots, either vertical pivoted or horizontal pivoted.
4. Special windows to circular openings and the like.

The type and size of windows used in new construction work are designed and set out on a 300 mm module. Casements should have rebates of 10 mm deep to receive the glass, and the frame should have a rebate 12–15 mm deep to receive the casement. Throatings should be at least 6 mm in diameter.

The preparation or processing of window material sections will vary from works to works. In the small joinery unit, woodworking machines may be limited in both their scope and number. Rebates, moulds, and any other operations are carried out on the spindle moulding machine after surfacing and thicknessing. In the larger joinery manufacturing unit, this process of producing the finished sections is carried out at the one passing through the moulding machine, or four-cutter as it is known.

Any deterioration in timber window-frames usually occurs at or near to the jointing positions of the members. Where moulded sections are used, any scribing is perhaps better done on the tenoning machine rather than by hand. A through scribe at the shoulders of a moulded rail will allow any moisture entering at the joint to pass through where otherwise it could be trapped, resulting in possible decay and failure of the joint. The life of any window is prolonged where hardwood sills are used, as sills are, by the nature of their position, more subject to rot than other members in any window.

In the design and construction of windows, it should be pointed out that the timber members are subjected to changes in humidity due to their very position in the buildings. These changes have the effect of swelling and shrinking of members and such movement is more pronounced the greater the section of material used. Opening sashes in particular can be affected by this movement resulting in the difficulty of opening through swelling.

In factory-produced windows opening sashes are usually jointed at their corners by a comb joint shown in Sheet 50.11. This is an alternative joint to the mortise and tenon. It has been specially developed for machine manufacture being cut on the tenoning machine or spindle moulder. It relies for its permanence on the strength and durability of the weather-proof adhesive that is used to secure it. The joint gives proportionally greater gluing area than the mortise and tenon joint.

The range of modified standard casement windows following the basic principles set out in BS 644 is based on a module for basic spaces of 300 mm giving the following dimensions in length:
600 mm, 900 mm, 1·200 m, 1·800 m, 2·400 m
Heights of frames are:
600 mm, 900 mm, 1.050 m, 1.200 m, 1.500 m
Window types are identified by figures and letters, i.e. 5CV40 where:
 5 = five width modules = 5 × 300 mm = 1·500 m
 C = casement
 V = ventlight
 40 = four height modules = 4 × 300 mm = 1·200 m

Much research has gone into the design and production of this standard type of window, and has resulted in a casement window which is fairly draught-proof and weather-proof through the lipping of the casement over the frame. This does allow an opening light to be less subject to swelling and sticking through changes in humidity mentioned previously. Sheet 50.6 shows a combined pivot- and side-hung standard casement window.

Double-glazed windows

Restrictions imposed on the use of timber during the war years, and for some time afterwards, delayed the development in this country of double-glazed timber windows.

In addition to the basic requirements of admitting the optimum amount of light and conforming to certain aesthetic standards, double-glazed windows pay due regard to the ever-increasing importance of heat loss, sound transmission, the inclusion of pleated blinds of the most modern appeal, and methods of opening, the latter being influenced by ventilation requirements and the ease of indoor window cleaning.

Heat loss

There has been, until quite recently, a general disregard of thermal efficiency in some aspects of building design and construction. Deterioration in the supply of fuel, and its increased cost, have already compelled a far more serious attitude towards the problem of heat loss through roofs and walls. Logically, the same serious consideration should be given to heat loss through windows. Such losses can be controlled by paying due regard to the following factors: (1) fit between sash and frame; (2) closing mechanism; (3) double glazing.

Fit between sash and frame

While considerable heat loss takes place through a single pane of glass, as compared with double-glazed windows, an even greater loss occurs through gaps between the sashes and frames. Due to the natural movement of timber in changing moisture conditions, and to the need of space for paint, a fair clearance must be left between the edges of the sashes and frames. Air and moisture can best be prevented from entering by providing a form of packing at the joint; wind pressure and capillary action can be minimised by ensuring that grooves and rebates are of correct size and shape. If the packing is to serve as a draught excluder, then it is necessary for it to be sufficiently flexible to allow it to be compressed.

Closing mechanisms

The types of hardware fitted must be adequate not only to close the window but to keep it properly closed when so desired. Examples of the range of hardware available are detailed on Sheet 47, to which reference should be made.

Double glazing

Sheet 50.1 shows the elevation of a 'Tomo' double-glazed window. The

WINDOWS

DOUBLE HUNG SASHES

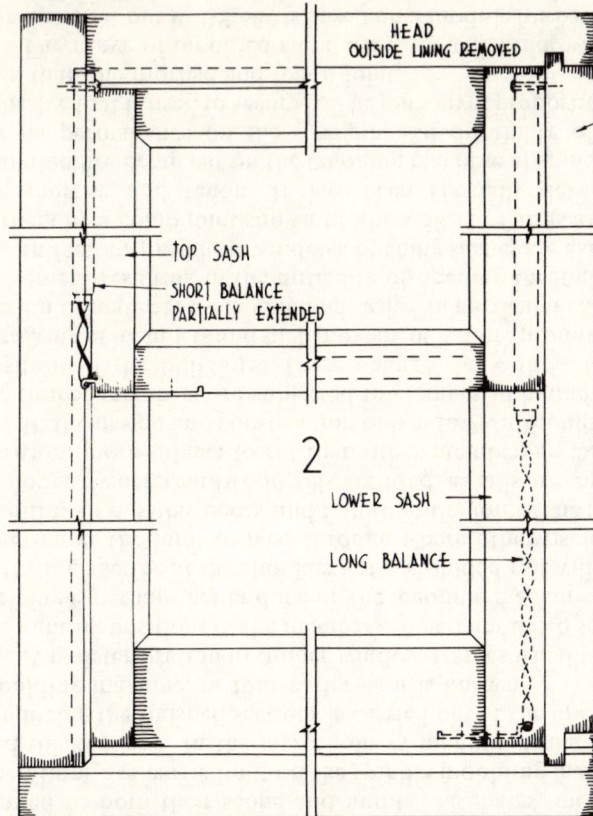

HEAD
OUTSIDE LINING REMOVED

106×30 STILE

48×38

TOP SASH

SHORT BALANCE
PARTIALLY EXTENDED

2

LOWER SASH

LONG BALANCE

PART ELEVATION

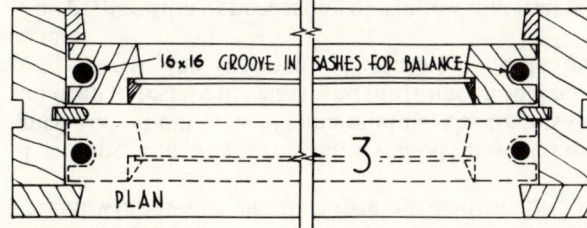

16×16 GROOVE IN SASHES FOR BALANCE

3

PLAN

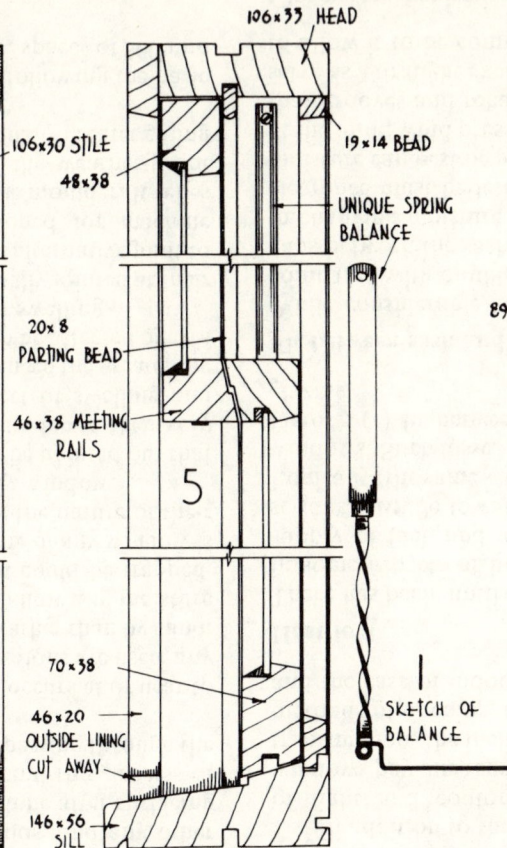

106×33 HEAD

19×14 BEAD

UNIQUE SPRING
BALANCE

20×8
PARTING BEAD

46×38 MEETING
RAILS

5

70×38

46×20
OUTSIDE LINING
CUT AWAY

146×56
SILL

SECTION

SKETCH OF
BALANCE

1

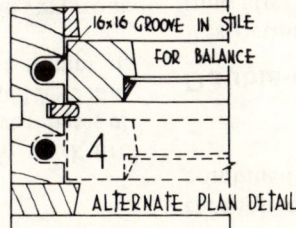

16×16 GROOVE IN STILE
FOR BALANCE

4

ALTERNATE PLAN DETAIL

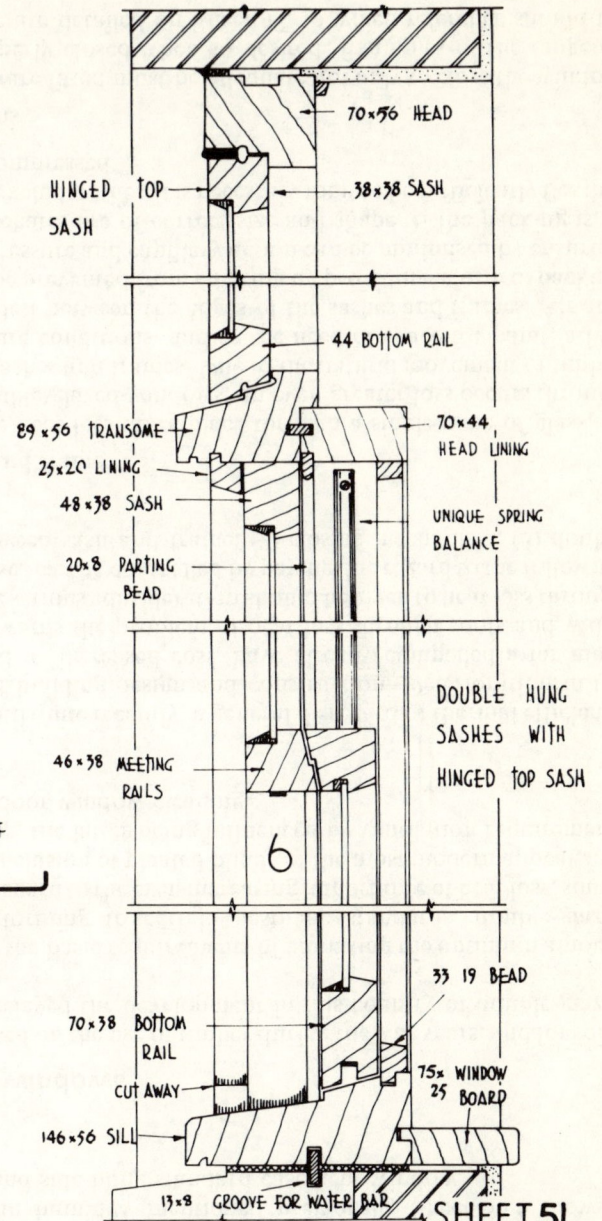

70×56 HEAD

HINGED TOP
SASH

38×38 SASH

44 BOTTOM RAIL

89×56 TRANSOME

70×44
HEAD LINING

25×20 LINING

48×38 SASH

UNIQUE SPRING
BALANCE

20×8 PARTING
BEAD

DOUBLE HUNG
SASHES WITH
HINGED TOP SASH

46×38 MEETING
RAILS

6

33 19 BEAD

70×38 BOTTOM
RAIL

CUT AWAY

75× WINDOW
25 BOARD

146×56 SILL

13×8 GROOVE FOR WATER BAR

SHEET 51

DORMER WINDOWS

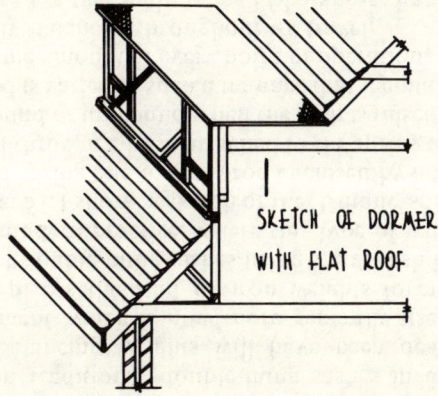

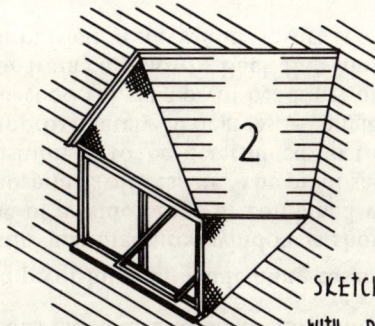

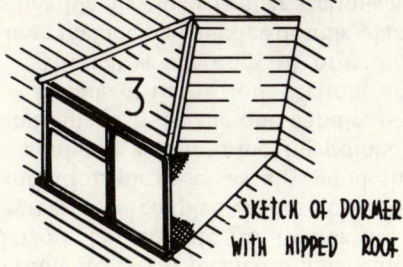

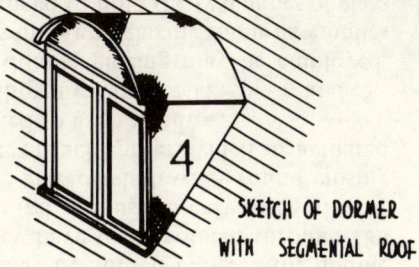

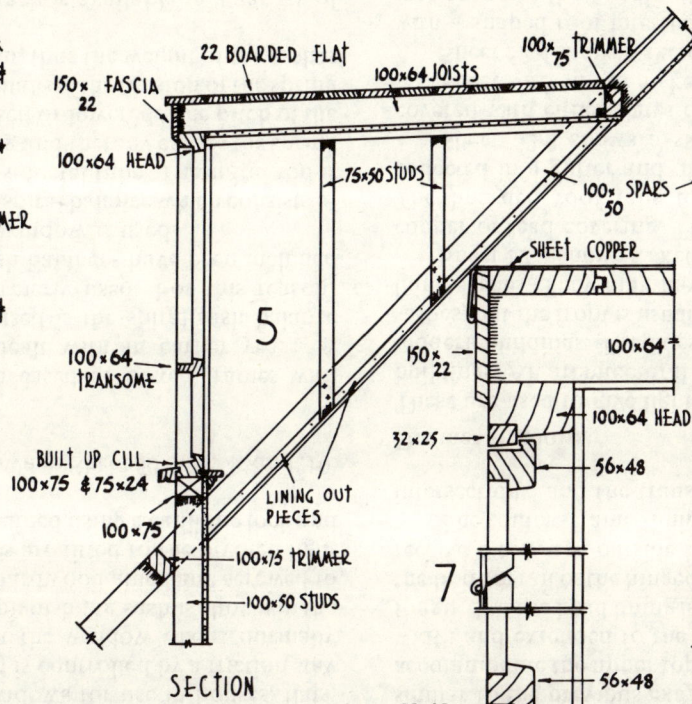

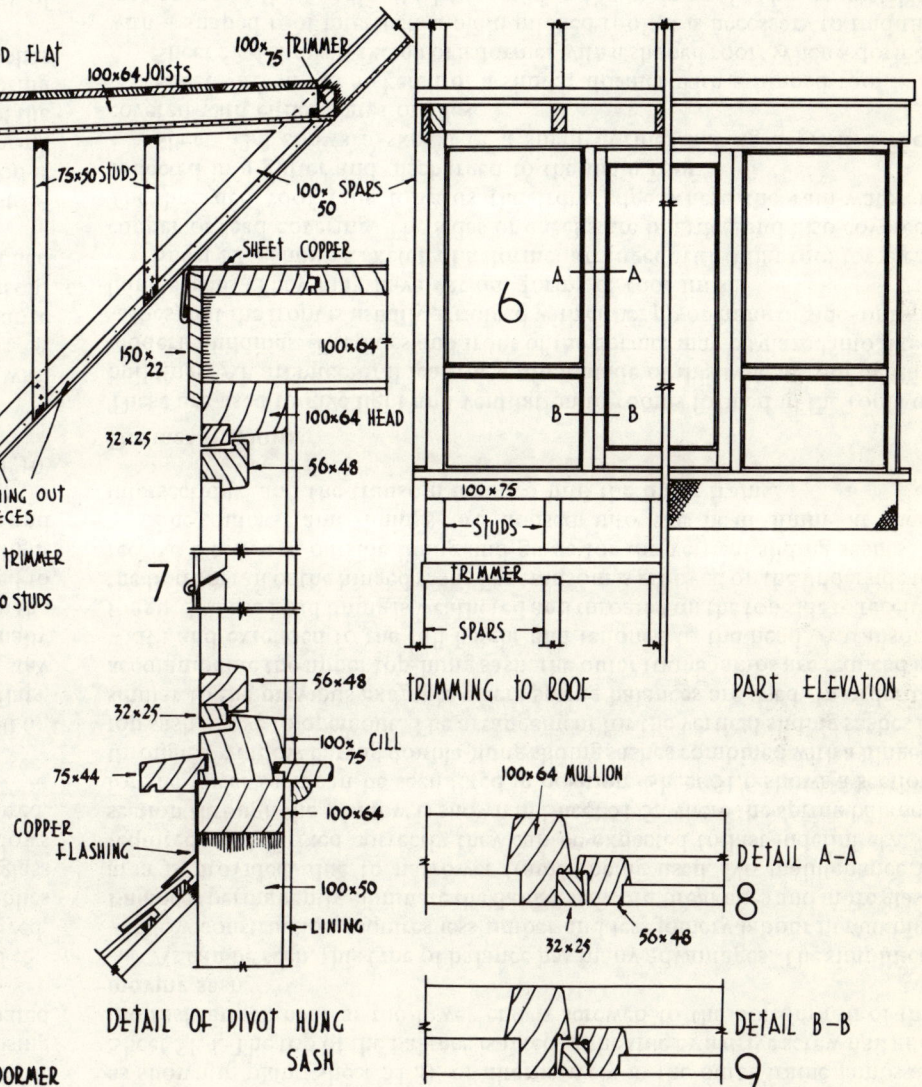

1 SKETCH OF DORMER WITH FLAT ROOF

2 SKETCH OF DORMER WITH PITCHED ROOF

3 SKETCH OF DORMER WITH HIPPED ROOF

4 SKETCH OF DORMER WITH SEGMENTAL ROOF

5 SECTION

22 BOARDED FLAT
150 × 22 FASCIA
100 × 64 JOISTS
100 × 75 TRIMMER
100 × 64 HEAD
75 × 50 STUDS
100 × SPARS 50
100 × 64 TRANSOME
BUILT UP CILL 100 × 75 & 75 × 24
100 × 75
LINING OUT PIECES
100 × 75 TRIMMER
100 × 50 STUDS

7 TRIMMING TO ROOF

SHEET COPPER
150 × 22
100 × 64
32 × 25
100 × 64 HEAD
56 × 48
56 × 48
32 × 25
100 × 75 CILL
75 × 44
100 × 64
COPPER FLASHING
100 × 50
LINING

DETAIL OF PIVOT HUNG SASH

6 PART ELEVATION

A — A
B — B
100 × 75
STUDS
TRIMMER
SPARS

100 × 64 MULLION

8 DETAIL A–A
32 × 25
56 × 48

9 DETAIL B–B

SHEET 52

187

second pane of glass is carried in a separate sash which is coupled to the other sash in such a way that adequate air filtration can take place to prevent condensation between the panes. This is obtained by using sash-couplers or coupling-screws which allow the windows to be cleaned between panes, and permits access to pleated blinds which may be housed between the double glazing. To separate the two sashes they have to be rotated through 180 degrees. The coupled sashes act as one and are just as easy to handle for ventilation purposes as any good single-glazed window. A round closing espagnolette with four locking points, one at each corner, which are actuated simultaneously by the one handle, is fitted to each window.

Details of the various sections are shown in Sheet 50.2, .3, .4, and .5.

Alternative methods of providing double glazing may be employed. These include sealed-edge double-glazing units and glazing frame assemblies in flexible plastic or rubber sections arranged to hold the two panes of glass with a 19 mm cavity between. They may be secured to wood or metal windows using various methods and the sections may be readily removed if required.

Combined pivot-hung and side-hung casement

This 'Austin' manufactured window is shown in the elevation (Sheet 50.6), and is one of a wide range of combined windows for use in houses, flats, schools, hospitals, offices, etc. The pivot sash is controlled by a friction stay whose position is to be determined so that the window opens normally through approximately 30 degrees. A single pivot-hung sash is shown in the sketch (Sheet 50.11). The pivot beads are of hardwood glued and screwed to the sash or frame as shown. Back flap hinges are fitted to the pivoted light with a pair of friction stays, and closure is obtained using a caulking lock and handle.

The details of the various sections are shown in Sheet 50.7, .8, .9, and .10.

Double-hung sashes

Work on traditional double-hung sashes in cased or boxed frames with counterbalancing weights will have been dealt with in earlier years. A development which is now more generally used is the spiral sash balance which replaces the lead or iron weights formerly used. For this reason, details of double-hung sashes using spiral sash balances have been included here as modern practice, where this type of window is used.

Sheet 51.1 shows a sketch of the 'Unique' spiral balance which consists of a torsion spring and a spiral rod enclosed by a metal tube. The spiral rod is threaded through a bush attached to the spring and thereby causes the spring to be wound or unwound when the sash is raised or lowered. The pitch of the spiral rod is varied in such a manner that it equalises the tension of the spring to the same amount at every point of operation, thus the weight of the sash is uniformly balanced throughout its travel.

There are two main types of balances. Each is available in a variety of lengths to suit different heights of sashes. Both can be adjusted within limits after installation to suit the actual weight of sash, but it is important that the correct size of balance, both for weight and length, is used. One type is suitable for domestic windows with sashes up to 13·60 kg glazed weight, the second type for all other installations where the weight of the glazed sash is up to 34·02 kg.

Sheet 51.2 shows a part external elevation of double-hung sashes with one short balance partially extended to the top sash. A similar balance will be required for the opposite side of the sash, but this is not shown. One of the long lower balances is shown on the right, and again the one for the opposite side has been omitted from the drawing.

Each balance is housed in a groove. This may be in the stile of the sashes as shown in plan (Sheet 51.3), or alternatively in the outer frame jambs as Sheet 51.4. The top of the balance is fixed to the stile by a drive screw nail and the sash attachment at the lower end is screwed to the bottom rail of the moving sash.

As can be seen, this type of balance has many advantages. The simplified window construction requires less timber and less joinery labour in making. Balances permanently eliminate the dangers of cord breakages and more glass area is provided, due to narrower frames being used. No maintenance is required, and if fixed correctly they can be expected to last indefinitely. A section through the window is shown in Sheet 51.5, where the spring balance for the lower sash can be seen fixed in position. Sheet 51.6 shows a section through a window having double-hung sliding sashes combined with a hinged top sash above to open out. The arrangement for the vertical sliding sashes is similar to the previous example where spring balances are used. In order to accommodate the upper top-hung sash, the outer frame jambs are reduced in width and extended to the full height and tenoned to the head. A transom tongued to the head lining is weathered and throated on the top side to receive the bottom rail of the hinged sash. The transom is grooved on the underside to receive a tongued outside lining and guide for the vertical sliding sashes.

The outer frame linings are housed into the head lining at their intersections, and the transom tenoned into the outer frame.

Dormer windows

These are used to give light and ventilation to rooms formed in the roofs of buildings. An architectural feature is often made of this type of roof light in modern buildings. The sides and front of the dormer may be glazed into fixed sashes, but the front is usually arranged with either pivot-hung or side- or top-hung sashes. They may have various forms of roof finish.

Sheet 52.1 shows a sketch of a dormer arranged with a flat roof for sheet copper or lead covering. The sides or cheeks are boarded and also covered. The boarded roof falls towards the front edge where the rain-water is collected in a gutter and discharged to the main roof.

Sheet 52.2 shows a sketch of a small dormer having a pitched roof covered with either slates or tiles.

Sheet 52.3 shows a sketch of a similar dormer with a hipped roof.

Sheet 52.4 shows a sketch of a dormer with a shaped roof. When a dormer with a shaped roof intersects a main pitched roof, it is necessary to find the true shape of the opening to be formed in the main roof. This geometrical construction is shown on Sheet 17.

Sheet 52.5 shows a section through a dormer with a flat roof having a casement with a pivot-hung opening light with the cheeks prepared for boarding and finished in sheet copper.

Sheet 52.6 shows a part elevation of the roof light to the right and the roof trimming to the left.

The opening in the main roof is formed between two 100 mm × 75 mm trimmers, with the feet of the roof spars bird's-mouthed over the top one and housed into the lower one and nailed. Two 100 mm × 75 mm trimming rafters are substituted for 100 mm × 50 mm common rafters at the sides, and the trimmer set vertically between them at the head of the opening. The trimmer at the foot of the opening may be placed as shown or a deeper trimmer used and placed vertically. This is allowed to stand 75 mm above the rafters to form a seating for the sill of the casement frame.

When the trimmer is fixed square to the rafters, 100 mm × 50 mm studs or uprights supported from the floor below carry a 100 mm × 75 mm head which supports the sill of the frame to the dormer front. The casement frame provides for a pivot-hung opening light in the centre, the others are fixed.

Sheet 52.7 shows an enlarged section through the front of the dormer showing details of the pivot-hung sash. Typical sections taken above and below the pivot of the sash are shown in Sheet 52.8 and .9.

The roof joists are fixed from the top trimmer to the head of the casement to carry a lined ceiling of insulation board, and a boarded flat which falls towards the front edge. A fascia is mitred round the face and cheeks of the dormer to finish the flat with a gutter to collect the rain-water and discharge it to the main roof.

Skylights

These are sashes fixed on inclined roofs to provide additional lighting to the space below. They may be fixed or made to open for ventilation purposes.

Sheet 53.1 shows a section through a framed skylight hinged at the top. The opening in the rafters is formed by trimming in the usual way to the required size. A curb or lining is framed, tongued and grooved at the angles, and fixed within the opening. The curb is 50 mm thick and 275 mm deep, standing high enough above the main roof to provide an adequate gutter at the top of the skylight. The curb is covered with a flashing on the outside to render it watertight.

The skylight rests on the curb and is hinged at the top. It is at this point where maintaining a watertight joint presents a problem. Alternative finishes will be dealt with later. A 50 mm projection on all four sides over the curb is necessary, with a throating on the underside, on all edges. The light consists of two stiles, top and bottom rails. The bottom rail of the skylight is kept below the level of the glass so that the glass runs over it. Bars are sometimes used to reduce glass sizes. A wide bottom rail is used, giving a wider projection over the curb on the inside to accommodate the opening gear. A sketch of this is shown in Sheet 53.2 which is of the quadrant type operated by cord gear and worm.

Sheet 53.3 shows a horizontal section through the skylight where glazing bars are shown. An alternative finish on the left is also shown. Here, the curb or lining is tongued and the sash correspondingly grooved. A fillet is also tongued on the underside of the projecting sash to assist in keeping out the driving rain. These are mitred at the corners. Haunched mortise and tenons are used between the top rail and stiles. A barfaced tenon is used for the bottom rail with glazing bars stub-tenoned at the top and bottom. It should be

noted that the top rail is grooved to receive the glass with all other members rebated.

As the insulation properties of glass are low, condensation which takes place on the underside of the glass in cold weather has to be allowed to escape. To allow this to happen the face of the bottom rail is kept below the rebate line of the stiles by 10 mm as shown, leaving a gap between the glass and bottom rail. Fillets fixed to the bottom rail will help to check any draughts and assist in channelling the condensation clear, as shown in Sheet 53.4. When skylights are hinged, non-corrosive butts should be used, fixed first to the curb and then to the sash.

Sheet 53.5 shows a section through a skylight in an asbestos-covered roof with steel trusses, using patent glazing bars. Although this work is not the work of the carpenter and joiner, Sheet 53.6 shows an example where he has to make provision for the fixing by others of this type of glazing in a timber roof.

The trimming is done as for a timber skylight. At the bottom, a filler piece having a bevelled top edge is fixed to the trimmer to the required height and a fascia or lining, for fixing the tops of the bars, fixed on the inside.

A section of a glazing bar is shown in Sheet 53.7. This consists of a steel tee bar galvanised and clothed with a lead sheath sealed at both ends. The two lead wings make a watertight joint with the glass seating on an imperishable, oiled asbestos cord cushion.

Lantern lights

This form of light is used on flat roofs to give top lighting to staircases and rooms. They also provide extra ventilation from vertical side lights. The outline of the plan may be square, rectangular, circular, or polygonal, and the roof covering may be formed with inclined lights or a lead flat.

Sheet 54.1 shows the half plan and half horizontal section of a lantern light having glazed hipped roof lights. A half sectional elevation and half elevation are also shown. The roof lights are constructed of four frames which are mitred and tongued and grooved at their intersections. The lantern is raised above the main roof on a curb, resting on the trimming joists of the lead flat. The side lights are 600 mm deep and may be top hung to open out or pivoted or fixed. The details of the vertical lights are similar to those for a square bay window.

The corner posts and mullions are rebated in the solid. The head is bevelled to the slope of the roof on the top edge, which may have a tongue worked on it with a corresponding groove in the bottom rail of the roof light. The sill is mitred at the corners and fixed by handrail bolts. The stiles are tenoned into the sill and head, the tenons being set in line with the edge of the rebates.

When the lantern is fixed, a lining, which may be framed and panelled or of hardboard, is fixed to mask the curb and trimming. To prevent condensation falling into the room below, condensation gutters are planted and flashed, with outlets bored through the curb as shown. Hip and ridge rolls are mitred together at their intersections and fixed ready for flashing.

Sheet 54.2 shows an enlarged detail through the lantern showing a fixed

ROOF LIGHTS

OPENING GEAR 2

BOTTOM RAIL LINING
CORD

175×38 BOTTOM RAIL

125×50 TOP RAIL

GUTTER BACK
HINGE

100×50 SPARS
100×75 TRIMMER
50 LINING

1.130

SECTION 1

FILLET

ALTERNATE FINISH TO CURB & SKYLIGHT

ALTERNATE FINISH
FILLET

50×32 BARS
125×50 STILE
FLASHING

830
50 LINING
100×75 TRIMMER

3

SECTION A—A

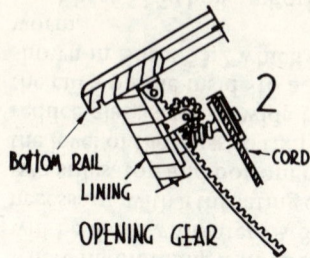

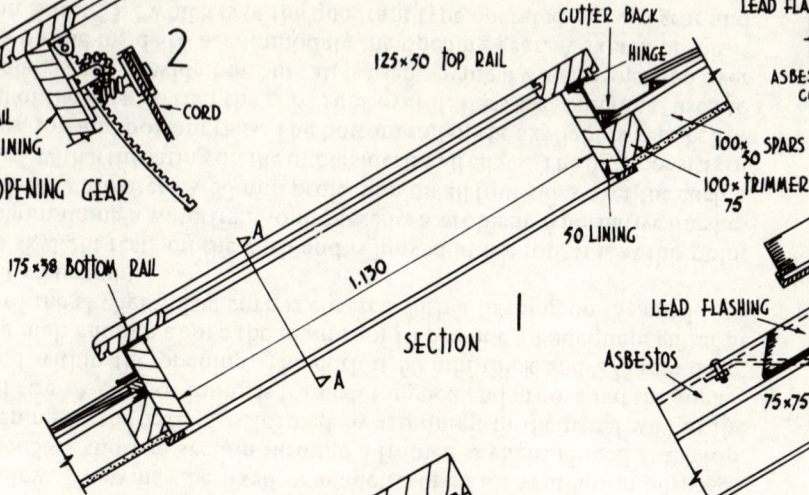

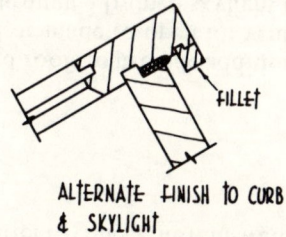

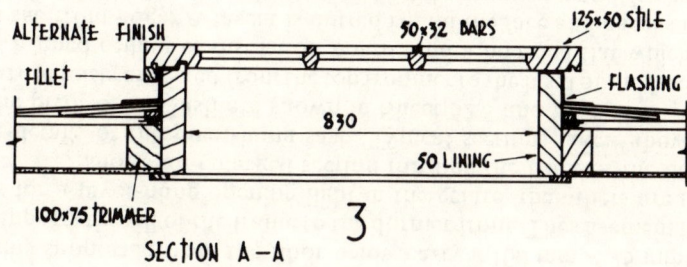

PATENT GLAZING BAR SECTION 7

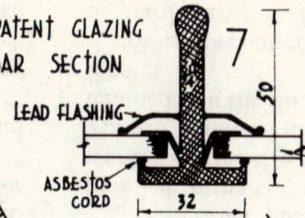

LEAD FLASHING
ASBESTOS CORD
50
6 CAST GLASS
32

PATENT GLAZING IN ASBESTOS ROOF WITH STEEL TRUSSES

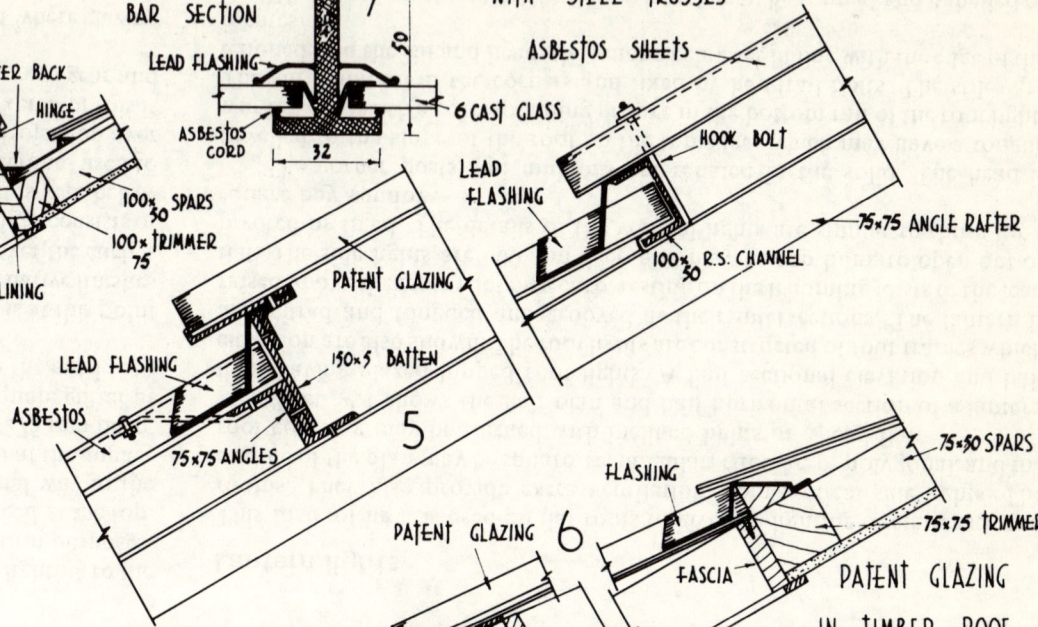

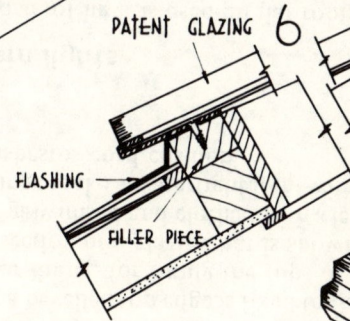

ASBESTOS SHEETS
HOOK BOLT
75×75 ANGLE RAFTER
100×50 R.S. CHANNEL
LEAD FLASHING

PATENT GLAZING
LEAD FLASHING
150×5 BATTEN
ASBESTOS
75×75 ANGLES

5

PATENT GLAZING
FLASHING
75×50 SPARS
75×75 TRIMMER
FASCIA

6

FLASHING
FILLER PIECE

PATENT GLAZING IN TIMBER ROOF

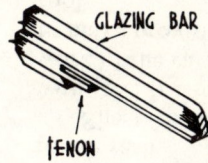

GLAZING BAR
TENON

CONDENSATION ESCAPE
STILE
BAREFACED TENON
BOTTOM RAIL

4

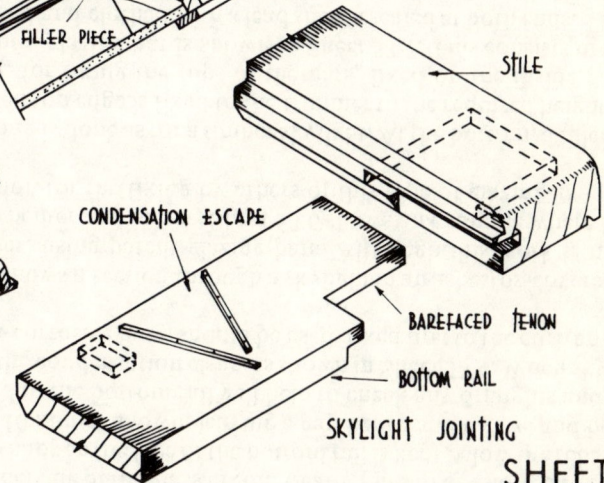

SKYLIGHT JOINTING

SHEET 53

LANTERN LIGHT

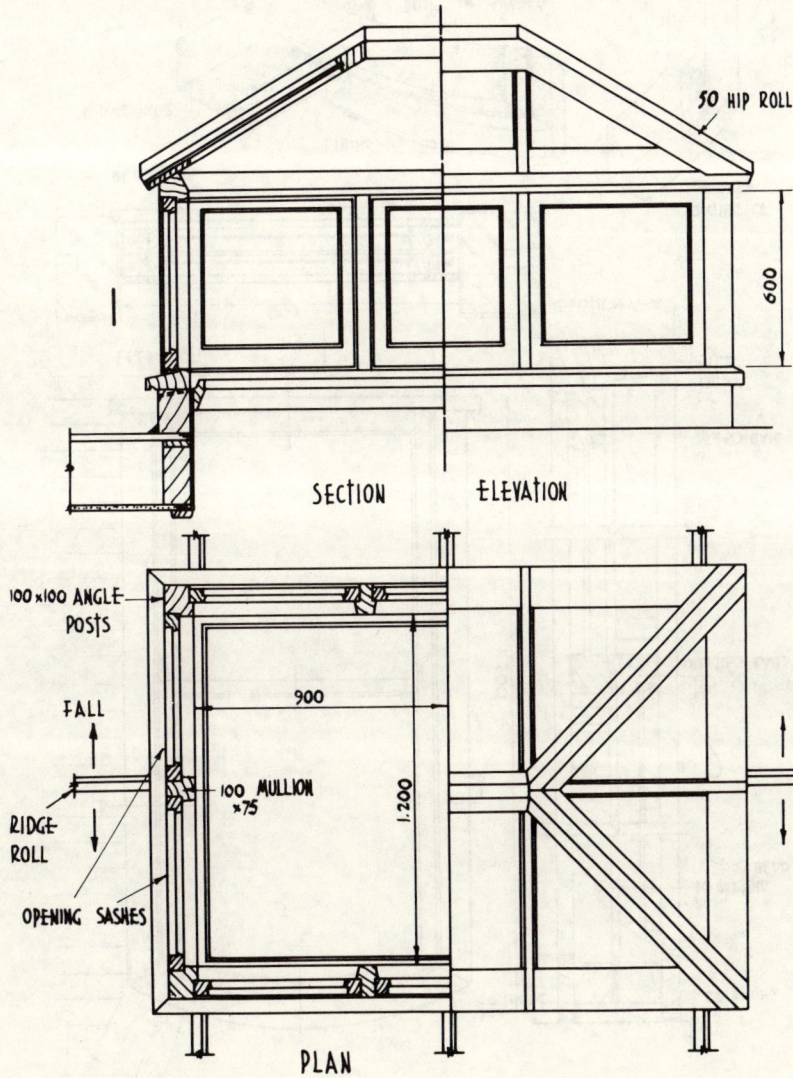

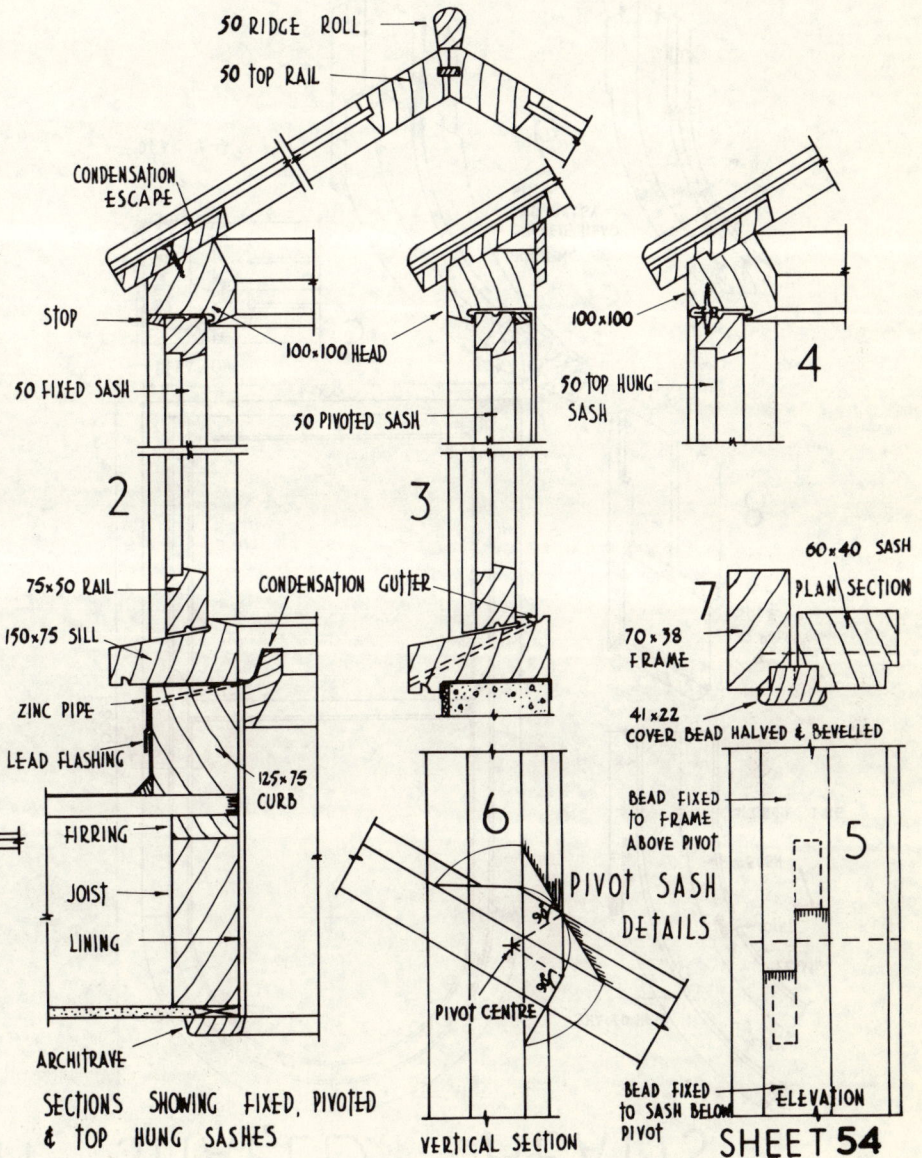

1

SECTION   ELEVATION

50 HIP ROLL

600

PLAN

100×100 ANGLE POSTS

FALL

RIDGE ROLL

OPENING SASHES

900

100 MULLION ×75

1.200

---

50 RIDGE ROLL

50 TOP RAIL

CONDENSATION ESCAPE

STOP

50 FIXED SASH

100×100 HEAD

50 PIVOTED SASH

2

75×50 RAIL

150×75 SILL

CONDENSATION GUTTER

ZINC PIPE

LEAD FLASHING

125×75 CURB

FIRRING

JOIST

LINING

ARCHITRAVE

SECTIONS SHOWING FIXED, PIVOTED & TOP HUNG SASHES

3

4

100×100

50 TOP HUNG SASH

6

PIVOT SASH DETAILS

PIVOT CENTRE

VERTICAL SECTION

7

60×40 SASH
PLAN SECTION

70×38 FRAME

41×22
COVER BEAD HALVED & BEVELLED

BEAD FIXED TO FRAME ABOVE PIVOT

5

BEAD FIXED TO SASH BELOW PIVOT

ELEVATION

SHEET **54**

WINDOWS WITH SHAPED HEADS

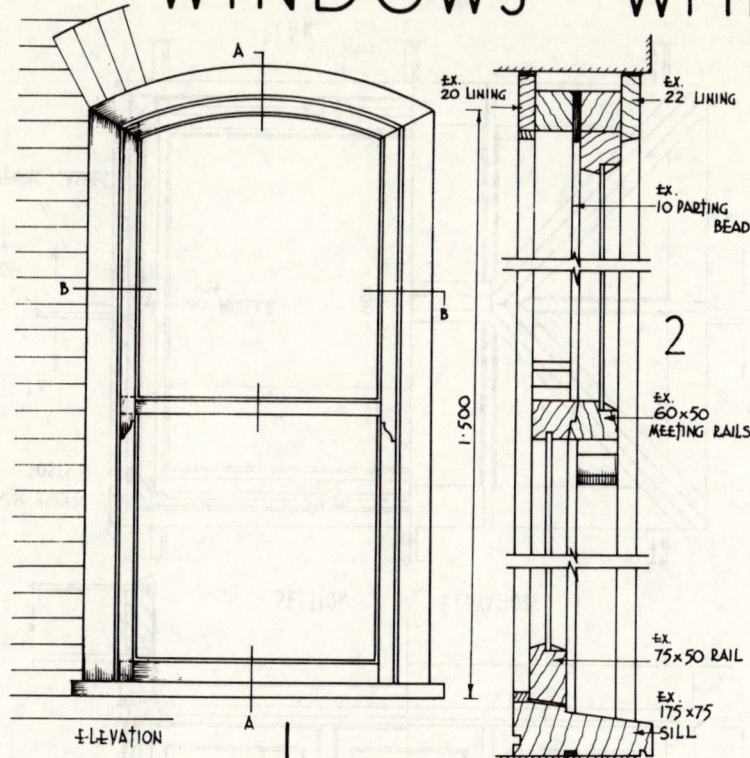

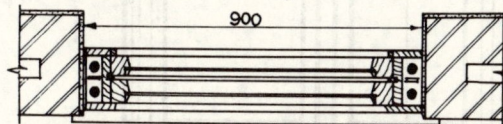

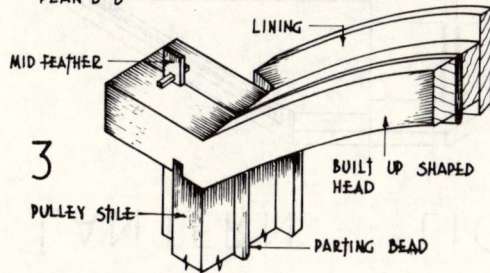

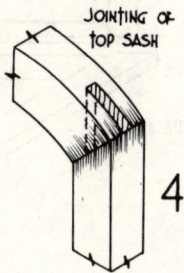

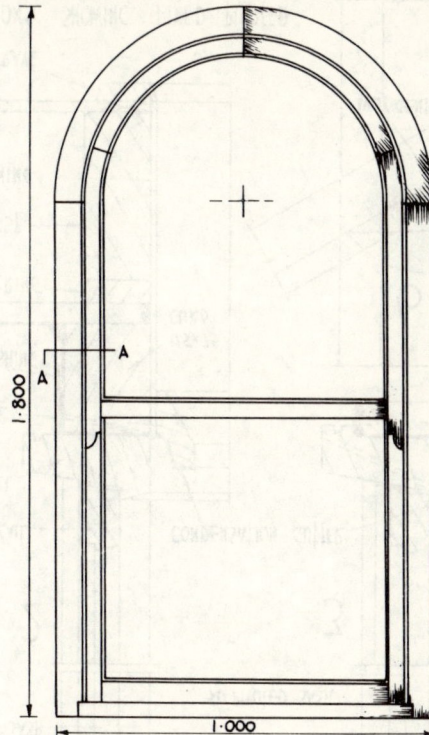

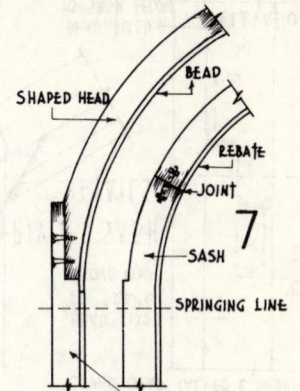

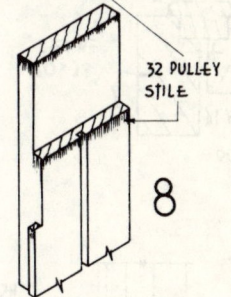

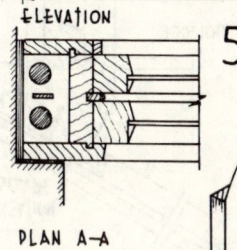

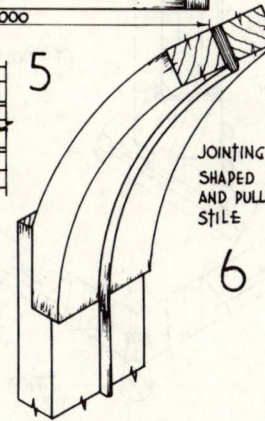

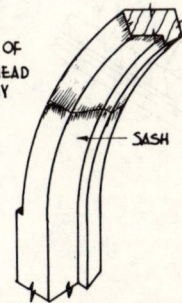

ELEVATION

A

B

B

A

1·500

900

PLAN B-B

1

EX. 20 LINING

EX. 22 LINING

EX. 10 PARTING BEAD

2

EX. 60×50 MEETING RAILS

EX. 75×50 RAIL

EX. 175×75 SILL

SECTION A-A

MID FEATHER

LINING

PULLEY STILE

PARTING BEAD

BUILT UP SHAPED HEAD

3

JOINTING OF TOP SASH

4

A

A

1·800

1·000

ELEVATION

PLAN A-A

5

SHAPED HEAD

BEAD

REBATE

JOINT

SASH

SPRINGING LINE

7

32 PULLEY STILE

8

JOINTING OF SHAPED HEAD AND PULLEY STILE

SASH

6

SHEET 55

sash in position.

Sheet 54.3 shows a section through a concrete flat roof with the curb formed also in concrete and the whole covered with asphalt. It will be noted that the sill shown has a condensation groove worked in the solid. The head is tongued with the bottom rail grooved to receive this, and a lining fixed on the inside.

Sheet 54.4 shows a detail of the head prepared for sidelights, top hung to open out.

When the lantern is roofed with four sashes or lights as shown in the illustration, the mitre bevel required is half the dihedral angle.

An alternative method of roofing a lantern is to use moulded bars. The hip, ridge, and other bars will be of different sections, owing to their varying inclinations. Reference should be made to the section dealing with the work on inclined mouldings to lantern on Sheet 96.

A further method of covering a lantern is by using a flat roof. In this case, joists, having a fall in two directions, are boarded and covered with lead and finished with a fascia and gutter all round.

Pivot sashes

Pivot-hung windows are now widely used in all types of modern buildings including offices, flats, hospitals, schools, and stores. In curtain-wall construction pivot windows designed for horizontal and vertical swinging often form a special feature of the architectural design. The kinds of pivots available are many and varied, the more recent of these friction-type pivots shown on Sheet 47 have been in use on the Continent some long time.

The construction of the window is governed by a number of factors:

1. The type of pivot.
2. The method of opening – horizontally or vertically.
3. The amount of swing.
4. The position of the sash in relation to the frame.

As there are various types of pivot, so there are varying timber sections used in the construction of the window. Manufacturers have their own sections and architects specify many others in which the joiner's work in making the window and in the hanging of the sashes requires special attention.

The section of such a window is shown in Sheet 54.7. In this detail the sash is flush with the frame on the outside. Both members are of stock sections and rebated on all edges to receive a rebated cover bead, screwed to the frame above the pivot and to the sash below the pivot. The bevels for cutting the beads are shown in Sheet 54.6, which as well as being bevelled are halved as seen in the elevation (Sheet 54.5). The sash in this example opens to an angle of 30 degrees to the horizontal.

Sash windows with shaped heads

Windows with shaped heads are usually of two types: (1) segmental and (2) semicircular.

Sheet 55.1 shows the plan and elevation of a segmental headed sash window having a height of 1·500 m to the springing of the brick arch over the opening which is 900 mm wide.

Sheet 55.2 shows a broken vertical section through the frame and the sashes on the centre line and may be used as a basis for setting out on the setting-out rod. The shaped head requires to be set out full size in order that templets and the lengths and widths of material may be determined.

There are a number of methods of building up shaped heads whatever the outline, and applies both to windows and doors. Reference to further shaped work on pages 168 and 171 should be made.

The method of building up the head in this example is shown in Sheet 55.3 and referred to as 'sandwich construction'. It will be seen that the parting bead is placed between the two shaped outside layers, the three members being glued together under pressure using a resin adhesive.

Sheet 55.4 shows a method of jointing the shaped top rail to the stile of the top sash to form a mitred joint in elevation.

Although the more traditional counterbalancing lead or iron weights have been shown in the plan detail of the window, it should be pointed out that spiral sash balances used in the previous work on double-hung sashes may also be used as an alternative here.

A sash and frame window with a segmental head may have only the outside lining cut to the curve of the arch, the inner side of the frame being left square. The top sash will then require a top rail with a straight upper edge and a curved lower edge.

Segmental-headed windows

Two further examples of windows with shaped heads of good-quality joiner's work are detailed on Sheet 58.

Sheet 58.1 shows the elevation of the window having a centre opening sash. This requires a transom with two mullions to be incorporated in the design to accommodate the sash which is hinged at the transom to open inwards. A vertical section is shown in Sheet 58.2.

Again the importance of accurate setting out is stressed. The head of the frame in this case is laminated and is set out on site dimensions of span 1·800 m and rise 300 mm.

Sheet 58.3 shows a similar window designed to have two opening sashes low down, at sill level. A transom and single mullion are necessary in this design. The head is laminated and the setting-out on site dimensions of 2·100 m span and 300 mm rise.

The frame jointing is shown by dotted lines to the right of the centre line in both examples.

Semicircular-headed sash window

Sheet 55.5 shows the elevation and part plan of a semicircular-headed sash window. A vertical section has been omitted as this is similar to the work in the previous example where identical sectioned material has been used. Again

WINDOWS & FANLIGHTS

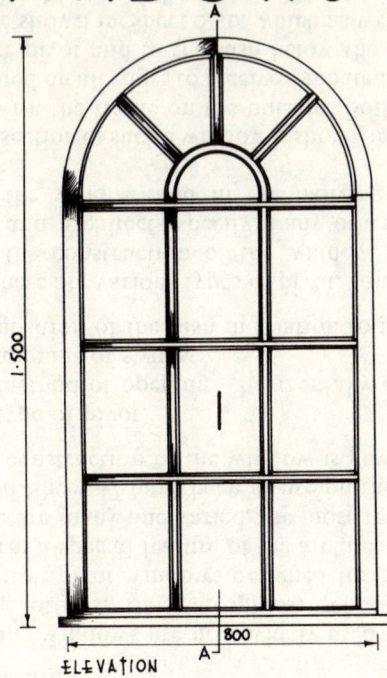

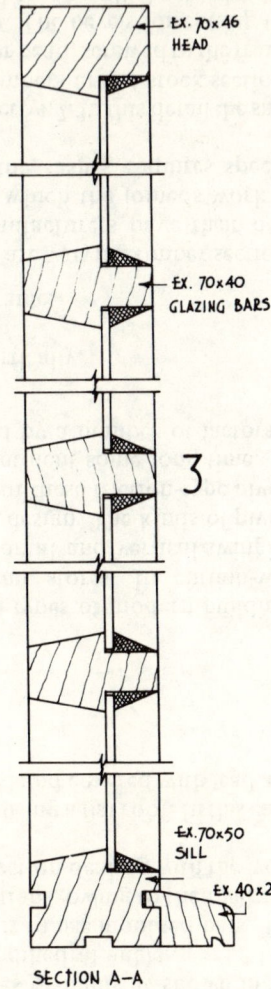

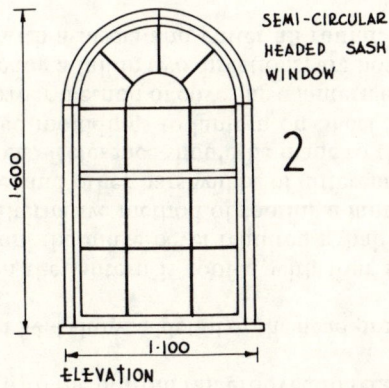

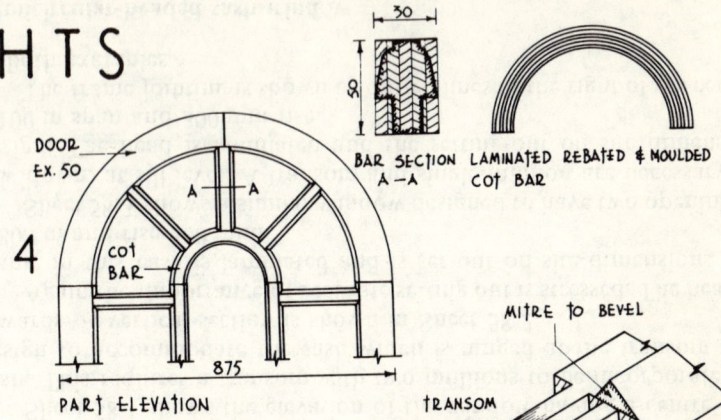

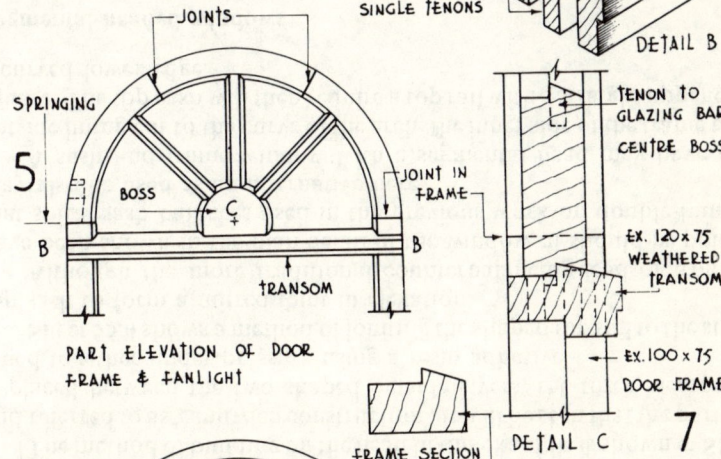

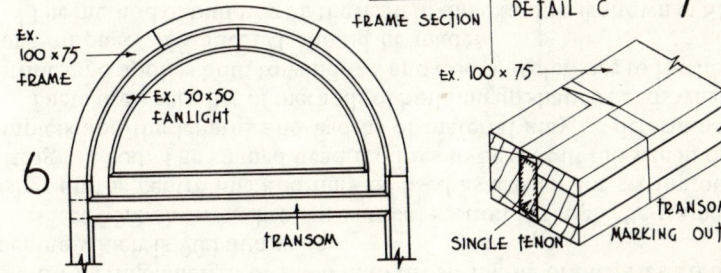

1

1·500

800

ELEVATION

A — A

2

SEMI-CIRCULAR
HEADED SASH
WINDOW

1·600

1·100

ELEVATION

3

EX. 70×46
HEAD

EX. 70×40
GLAZING BARS

EX. 70×50
SILL

EX. 40×28

SECTION A-A

4

DOOR
EX. 50

A — A

COT
BAR

875

PART ELEVATION

30

BAR SECTION
A-A

LAMINATED REBATED & MOULDED
COT BAR

MITRE TO BEVEL

TRANSOM
JOINTING

DOUBLE OR
SINGLE TENONS

DETAIL B

5

SPRINGING

JOINTS

BOSS

C

JOINT IN
FRAME

B — B

TRANSOM

PART ELEVATION OF DOOR
FRAME & FANLIGHT

TENON TO
GLAZING BAR
CENTRE BOSS

EX. 120×75
WEATHERED
TRANSOM

EX. 100×75
DOOR FRAME

FRAME SECTION

DETAIL C

6

EX.
100×75
FRAME

EX. 50×50
FANLIGHT

TRANSOM

7

EX. 100×75

SINGLE TENON

TRANSOM
MARKING OUT

SHEET 56

CIRCULAR WINDOWS

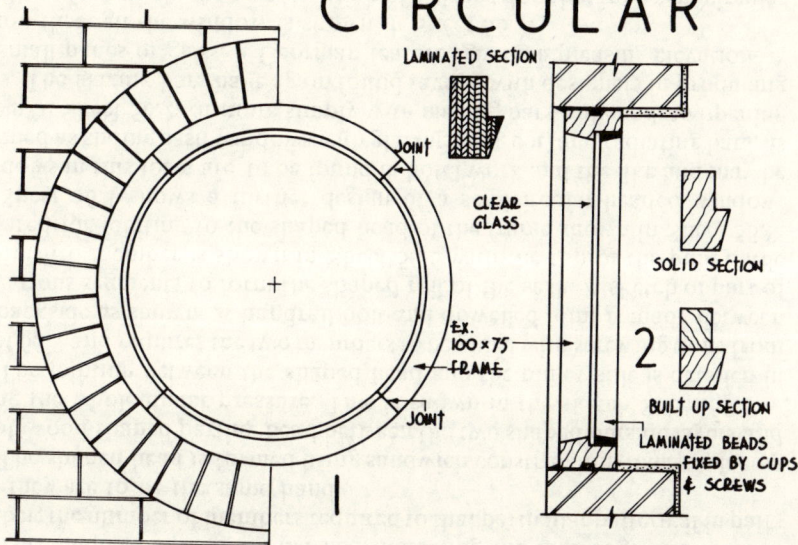

LAMINATED SECTION

JOINT

JOINT

ELEVATION

1

CLEAR GLASS

SOLID SECTION

EX. 100 × 75 FRAME

BUILT UP SECTION

LAMINATED BEADS FIXED BY CUPS & SCREWS

SECTION

2

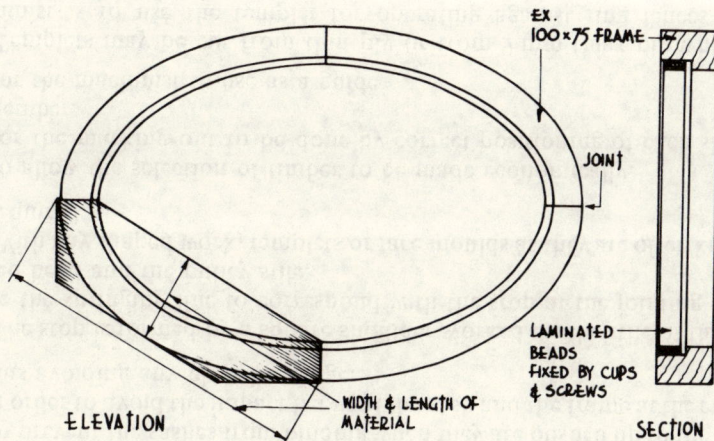

EX. 100 × 75 FRAME

JOINT

WIDTH & LENGTH OF MATERIAL

LAMINATED BEADS FIXED BY CUPS & SCREWS

ELEVATION

SECTION

ELLIPTICAL FRAME

3

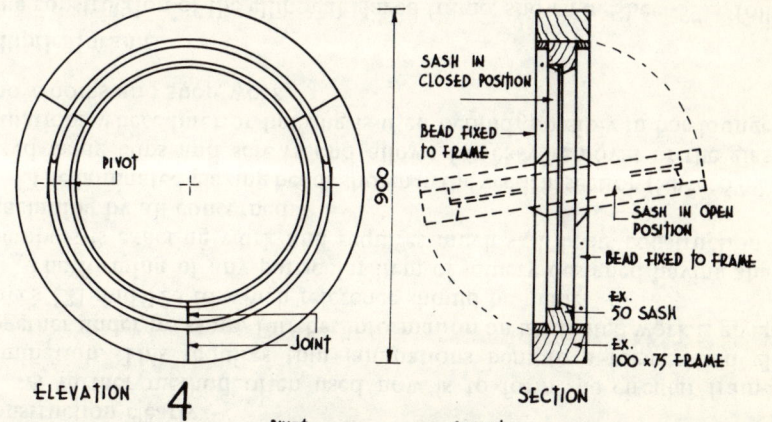

PIVOT

JOINT

ELEVATION

4

SASH IN CLOSED POSITION

BEAD FIXED TO FRAME

SASH IN OPEN POSITION

BEAD FIXED TO FRAME

EX. 50 SASH

EX. 100 × 75 FRAME

SECTION

900

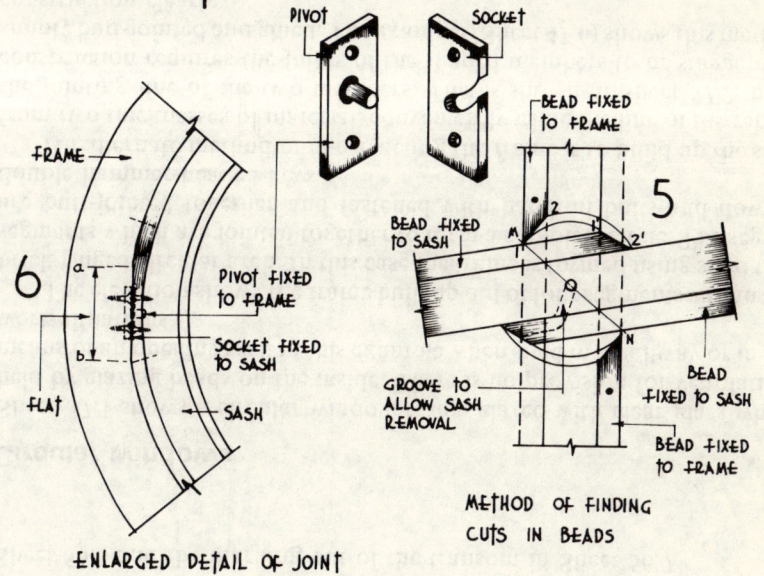

FRAME

FLAT

PIVOT FIXED TO FRAME

SOCKET FIXED TO SASH

SASH

ENLARGED DETAIL OF JOINT

PIVOT

SOCKET

BEAD FIXED TO FRAME

BEAD FIXED TO SASH

GROOVE TO ALLOW SASH REMOVAL

BEAD FIXED TO SASH

BEAD FIXED TO FRAME

METHOD OF FINDING CUTS IN BEADS

5

6

SHEET 57

this work must be set out full size on a board in order that plywood templets may be prepared for the shaped head of the sash, and for the shaped head of the frame and linings.

In setting out, as the overall site sizes in height and width are given, the centre for striking the shaped members can be determined and the springing line marked. This is indicated in Sheet 55.5. Here it should be noted that the joint of the head to the pulley stile is made between 50 and 75 mm above the springing to form a stop for the sash. The reasons for this are:

1. To prevent the sashes from binding when they are pushed up to the head.
2. In order to avoid the impact between the sash and the frame at the crown, thus avoiding any glass breakage.

The stop is formed by a square shoulder worked on the stile of the sash above the springing line to correspond with the stop at the jointing of the shaped head and the pulley stile.

With any shaped work, templets or face moulds as they are often known, are required:

1. To allow the selection of timber to be made economically.
2. For the marking out to be done by correct positioning of each shaped member.
3. For the machinist to use as a guide.

Templets may be cut from thin ply or from 9 mm thick ply when the machinist is to use the templet for operating against ring fences when moulding, rebating, and cleaning up the shaped work on the spindle moulding machine.

The templets should be marked indicating the face edge and the rod number; the number of members required to that particular pattern; if in pairs or if they are to be the same hand.

The shaped head is formed using sandwich construction, that is, placing the plywood shaped parting bead between the two shaped outside layers and gluing the whole under pressure. This is shown in the sketch Sheet 55.6.

The jointing between the shaped head and the pulley stile is detailed in Sheet 55.7 and requires the two members to be glued with screwing done from the backside as shown. A handrail bolt and dowelled joint is made between the various segments to form the shaped rail of the sash. A sketch of part of the finished member is shown in Sheet 55.6 with the top of the pulley stile prepared for jointing to the shaped head of the frame shown in Sheet 55.8.

Sheet 56.1 shows a further design of a semicircular-headed window. Windows in this form are to be found in hotel work and the like and may be designed as sliding sash windows with glazing bars, cot, and radiating bars, as shown in Sheet 56.2, or more simply with glazing bars only and no opening parts. The glazing bars have again found favour with designers in producing the small panes of glass, a Georgian feature, giving a pleasing elevation. A section through the window is shown in Sheet 56.3.

Sheet 56.4 shows the part elevation of a glazed door having a semicircular head. This is formed out of the solid, while the cot bar and radiating glazing bars in the design are laminated to give additional strength at the jointing positions.

An external door frame with a glazed fanlight is shown in Sheet 56.5. A centre boss cut, moulded and rebated on the solid is used at the centre with the radiating bars tenoned into it. The centre boss is itself tenoned into the weathered transom.

It will be seen that the jointing between the transom and the jambs of the frame is arranged out of the springing line. This is done to prevent any weakening at the jointing at this point. Single or double tenons may be used at the transom joint as shown in Sheet 56.7.

A semicircular-headed frame with opening glazed fanlight is shown in Sheet 56.6 and the marking out of the transom in Sheet 56.7.

Circular windows

Sheet 57.1 shows a circular window-frame glazed with clear glass which is held by glazing beads on the inside. There is no provision for ventilation by means of an opening sash in this example when sited in a hallway or in hotel-work situations.

The elevation shows the frame built up out of four segments within a one-brick ringed circular arch. In this case the frame is formed using solid timber segments which are jointed together to form a complete circle. The segments are butt-jointed together and fastened with handrail bolts and dowels or double hammer-headed keys.

An alternate method of constructing the frame is to build up the section from two thicknesses of material, conveniently using the line of the rebate as the jointing line of the two members. This is shown in Sheet 57.2, but the construction requires the joints of the shaped members to be staggered and simply butt-jointed and glued. The example (Sheet 41.6) shows this method of construction clearly.

A further method often used now is to form the circular frame by lamination. This requires thin laminations bent to a former and glued together under pressure. Further information on laminated work is given on pages 121 and 123 to which reference should be made.

The forming of any particular item of joinery designed having shaped members is exacting work and requires much skill in its construction and machining by all concerned.

The laminated glazing beads shown in the vertical section (Sheet 57.2) are fixed using cups and screws and allows for easy removal of the glass in situations where interior lighting is used behind, possibly in bar lounges or showrooms and shop work.

Elliptical frame

The construction of the elliptical glazed frame, shown in Sheet 57.3, follows that previously dealt with in the circular frame. The same alternative methods of construction apply, but, of course, the elliptical outline requires to be set out full size. There are a number of methods of setting out the ellipse using practical and geometrical constructions necessary to produce the required templets for marking out the material and shaping and jointing the segments. These methods are shown in Sheets 81–4.

WINDOWS WITH SHAPED HEADS

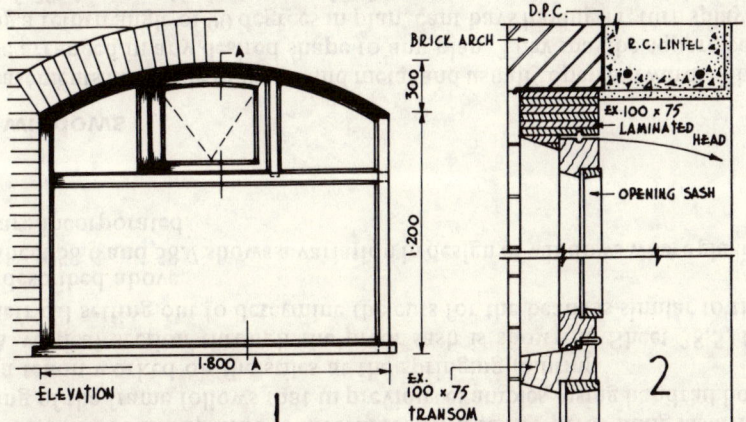

1

ELEVATION

SEGMENTAL HEADED

1·800

300

1·200

LAMINATION THICKNESS:

SOFTWOOD : $\dfrac{\text{RADIUS}}{100}$

HARDWOOD : $\dfrac{\text{RADIUS}}{150}$

2

D.P.C.

BRICK ARCH

R.C. LINTEL

EX. 100 x 75 LAMINATED HEAD

OPENING SASH

EX. 100 x 75 TRANSOM

32 WINDOW BOARD

EX. 150 x 75 SILL

D.P.C.

SECTION A-A

300

3

ELEVATION

2·100

1·800

SECTION A-A

EX. 100 x 62 TRANSOM

EX. 50 SASH

EX. 150 x 75 SILL

4

ELEVATION

2·100

JOINT

CENTRE PIVOTS

600

1·200

SECTION A-A

EX. 100 x 62 HEAD

EX. 50 SASH

6 GLASS

EX. 100 x 62 TRANSOM

5

6

MULLION

B

B

C

C

EX. 100 x 62

SECTION B-B

7

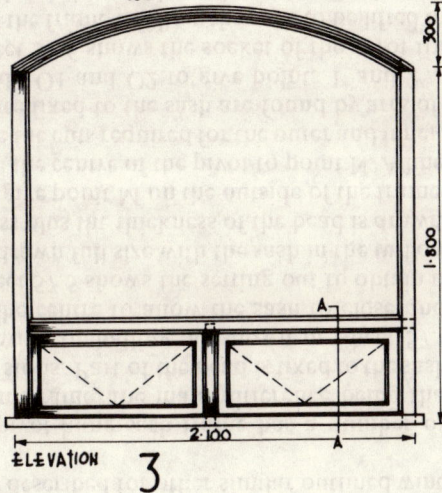

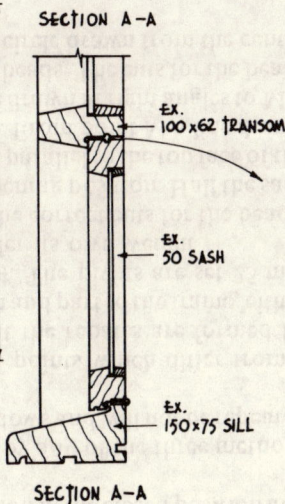

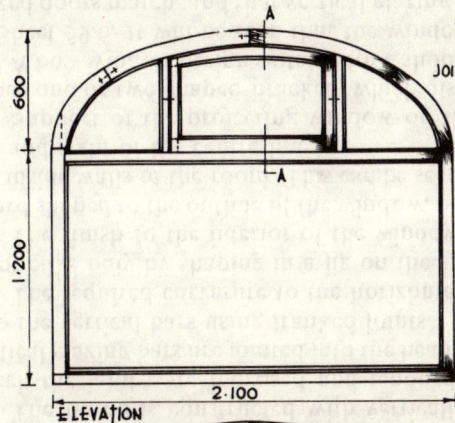

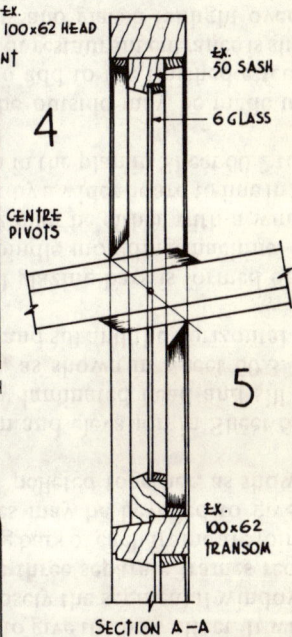

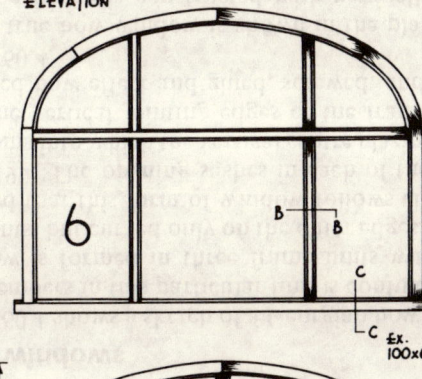

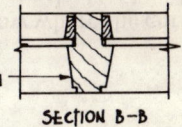

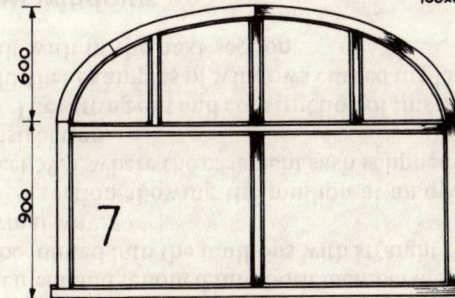

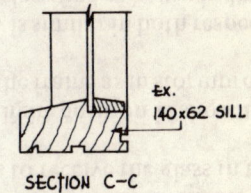

SEMI-ELLIPTICAL

600

900

EX. 140 x 62 SILL

SECTION C-C

DETAILS

SHEET 58

Bull's-eye window

Sheet 57.4 shows the elevation and vertical section of a bull's-eye window. The window consists of a circular frame with a pivot-hung casement or sash fitted into it. The sash is shown in the height section in the closed position and in the open position indicated by dotted lines.

Constructing the frame may be carried out by any of the three methods already described for other similar outlined windows and will not be repeated here.

A pivot-hung sash frame has a number of points which differ from a casement frame, the main difference being that the rebates are formed by beaded stops. Part of the bead is fixed to the sash and part to the frame, either by screwing or nailing, as shown in Sheet 57.4. The pivots are set 25 mm above the centre to allow the sash to close under its own weight.

Sheet 57.5 shows the setting out to obtain the correct cuts for the beads. This is drawn full size with the sash in the full opening position. Half the sash thickness plus the thickness of the bead is drawn parallel to the top face of the sash to give point M on the outside of the frame. From point M a line drawn through the centre of the pivot to point N. A line drawn at right angles to MN will give the cuts required for the outer and inner beads. The cuts for the beads which are fixed to the sash are found by arcs of circle drawn from the centre with radii O1 and O2 to give points 1' and 2'.

Sheet 57.6 shows the socket of the pivot fixed to the sash and the pivot fixed to the frame to allow the sash to be lifted out. When the sash is trenched and the bead slotted, the sash can be withdrawn by lifting up so that the pivot passes in the groove allowing the sash to be removed.

Semi-elliptical headed windows

Sheet 58 shows three semi-elliptical headed windows of varying designs. Sheet 58.4 has a built-up head to the frame with a centre-pivot-hung sash. The jointing of the frame follows that in previous examples, using handrail bolts with a tenon worked on the stiles at the springing jointing.

A vertical section through the pivot sash is shown in Sheet 38.5, the geometrical setting out to determine the cuts for the beads is similar to that fully described above.

Sheet 58.6 and 58.7 shows a variation in design of windows where glazing bars are incorporated.

Bay windows

Bay casements are made of wood and metal and usually open outwards. They can be arranged in any desired shape to any plan. They may be square bays having a return angle of 90 degrees in plan, cant bays having a return splay or cant of 30 degrees, 45 degrees or 60 degrees, or segmental or semicircular bays.

Sheet 59.1 shows a pictorial view of a cant bay window having a splay of 45 degrees.

The construction follows closely that of a casement window except at the angles. The angle posts may be solid or built up, as shown in Sheet 59.2, where two jambs are cross-tongued together to form a solid angle post. An alternate finish at the angle is shown in Sheet 59.3, where the front and cant casements come together at the intersection and an infill mould is fixed across the angle as shown. The sill is mitred with the angle posts framed into both pieces.

A pictorial view of a segmental five-light bay window is shown in Sheet 59.4. The joint lines of the head and sill radiate to a common centre. The head and sill are curved in plan with the mullions running the full height of the casement and tenoned into both members. This allows the transom members to be jointed into the mullions with straight rebates to receive the glass in the normal way.

A section showing the mullion at an opening light position is shown in Sheet 59.5, where the casement sash is lipped over the frame as in stormproof construction.

The setting out and construction of this window is similar in both respects to further examples of windows curved in plan and classified as bow windows dealt with in the next section.

Bow windows

Sheet 60.1 shows a sketch of a Georgian bow window where the curvature of the members in this particular unit is confined only to the head and sill. The window is formed in three frame units with the head and sill jointed in segments, but curved only on the outer edges to give the bow effect. It will be realised that this form of window follows closely the segmental window on page 199. The opening sashes in each of the three separate frames require transoms into which the vertical centre glazing bars of each frame are jointed.

The vertical jointing edges of the frames may be bevelled to give the required bow effect and glued, screwed, and pelleted together, as shown in Sheet 60.4.

A true bow window is shown in the plan and elevation in Sheet 60.2.

The frame is constructed with vertically laminated head and sill into which the jambs are mortised and tenoned, as shown in Sheet 60.3. The vertical glazing bars are jointed into the head and sill and the horizontal bars into the vertical bars using franked joints.

The required curvature to the horizontal glazing bars is formed on the front edge only by shaping in a jig on the spindle moulding machine.

The finish to the interior of the window may be either with a window board shaped to the outline of the window, or by a wider board to line in with the inside walls of the room. This can be seen in the plan in Sheet 60.2 to the left and right of the centre line.

Support of the projecting window on the outside may be made using either one or two shaped brackets which also add to the finished effect.

A bow window design suitable for a shop or restaurant entrance is shown in Sheet 59.6. It will be seen that the window and glazed fanlight over the glazed doors match, and that vertical glazing bars only are used in the design.

The setting out is based on dimensions taken from the site between cross-walls of span 3·000 m and projection 500 mm. The overall height of the window is 2·100 m and the finish from the pavement level to the underside of the sill is by vertical weather-boarding. This is fixed to studding which has building paper, acting as a moisture barrier, applied to it before the exterior cladding is fixed. Insulation is used between the inside and outside finishes.

BOW & BAY WINDOWS

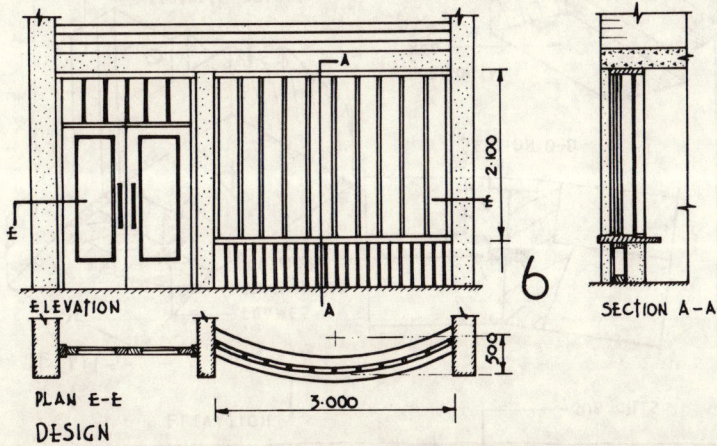

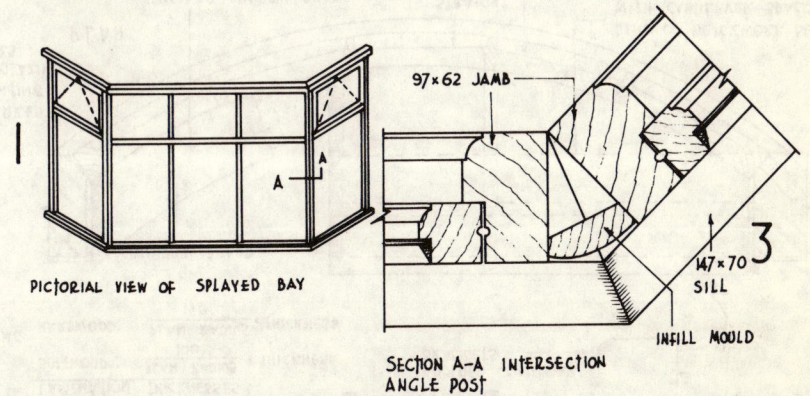

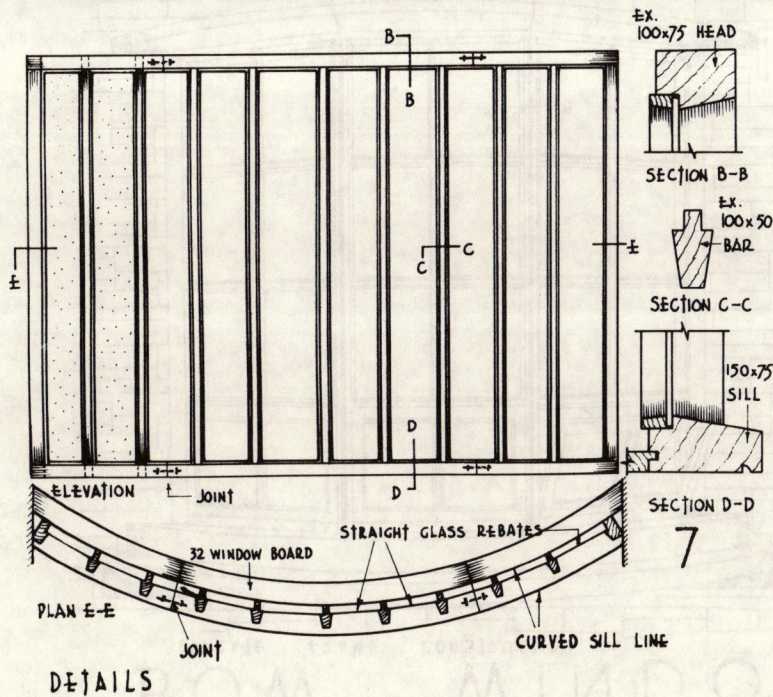

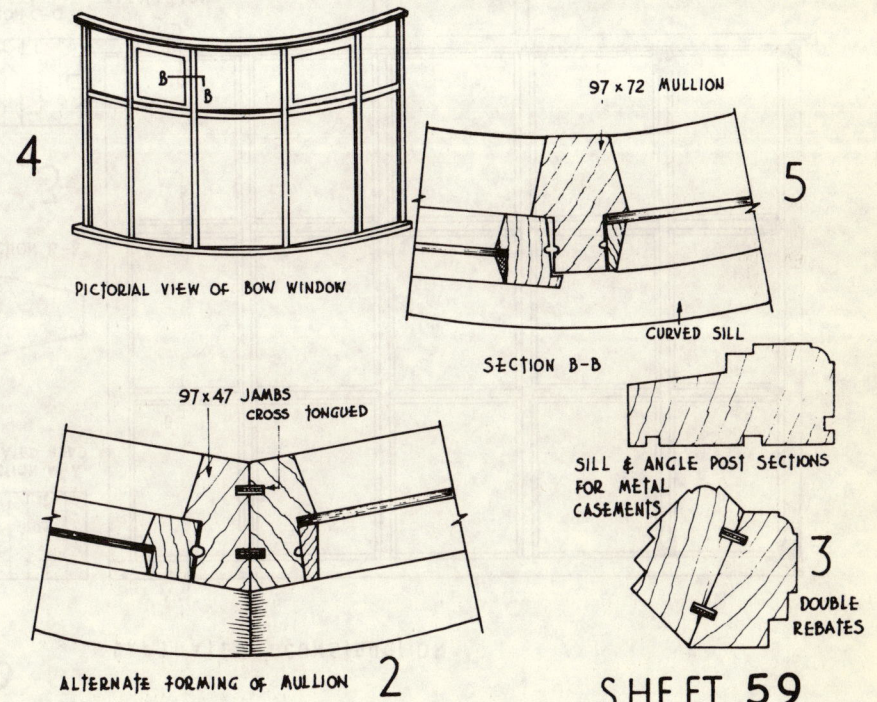

ELEVATION

2·100

E

PLAN E-E

3·000

500

DESIGN 6

SECTION A-A

PICTORIAL VIEW OF SPLAYED BAY 1

A

97 × 62 JAMB

SECTION A-A INTERSECTION
ANGLE POST

147 × 70
SILL

INFILL MOULD

3

EX.
100×75 HEAD

SECTION B-B

EX.
100 × 50
BAR

SECTION C-C

150×75
SILL

SECTION D-D

7

ELEVATION

JOINT

PLAN E-E

JOINT

32 WINDOW BOARD

STRAIGHT GLASS REBATES

CURVED SILL LINE

DETAILS

PICTORIAL VIEW OF BOW WINDOW 4

B

97 × 72 MULLION

5

CURVED SILL

SECTION B-B

97 × 47 JAMBS
CROSS TONGUED

ALTERNATE FORMING OF MULLION 2

SILL & ANGLE POST SECTIONS
FOR METAL CASEMENTS

DOUBLE
REBATES

3

SHEET 59

BOW WINDOWS

SEPARATE FRAME CONSTRUCTION

LAMINATED CONSTRUCTION

ELEVATION 1

SUB CILLS

SECTION A-A
LAMINATED HEAD

SECTION B-B

3

SECTION C-C
LAMINATED CILL

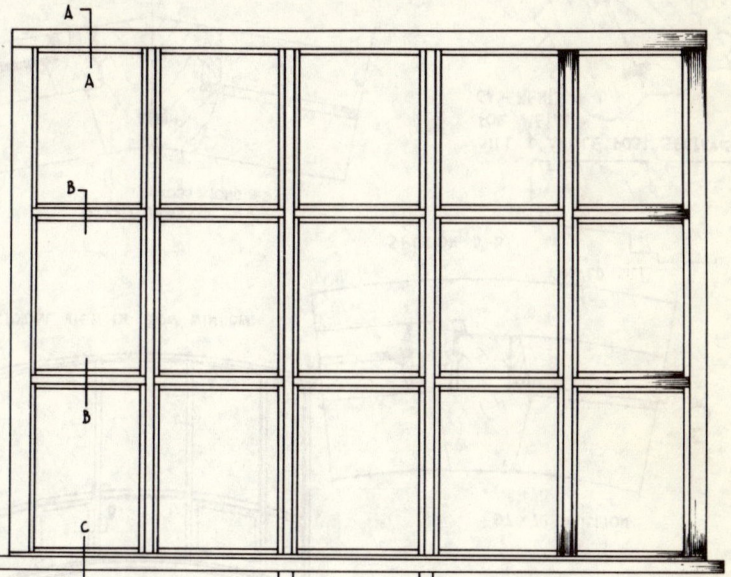

ELEVATION

MOULDED CANTILEVER
BRACKETS

LAMINATION THICKNESSES:

SOFTWOOD: $\frac{PLAN\ RADIUS}{100}$ = THICKNESS

HARDWOOD: $\frac{PLAN\ RADIUS}{150}$ = THICKNESS

DETAILS

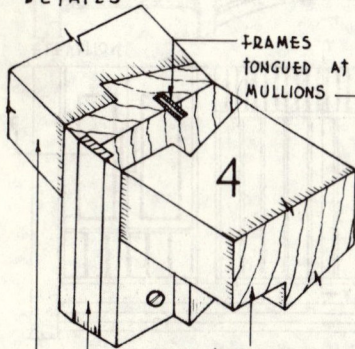

FRAMES
TONGUED AT
MULLIONS

4

SEPARATE FRAMES
SCREWED TOGETHER THROUGH
REBATES AND FINISHED WITH
COVER MOULD

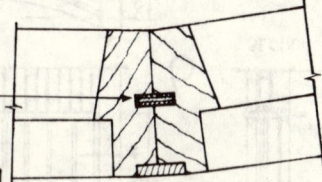

SECTION D-D

HORIZONTAL
BAR

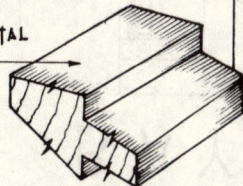

VERTICAL
GLAZING
BAR

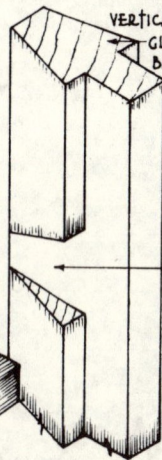

FRANKED
JOINTING
OF GLAZING
BARS

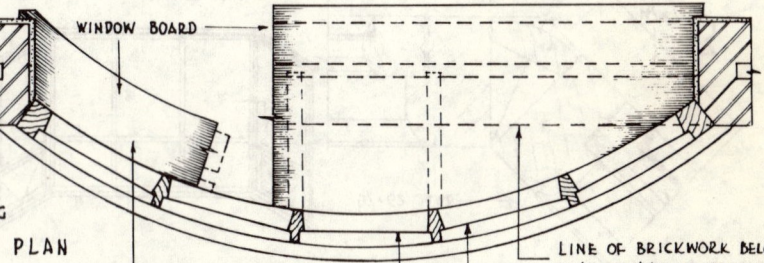

WINDOW BOARD

PLAN

CIRCULAR WINDOW BOARD
OVER CIRCULAR BRICKWORK

STRAIGHT
REBATES

LINE OF BRICKWORK BELOW
WITH CANTILEVER BRACKETS
TO PROJECTING BOW WINDOW

2

SHEET 60

BALCONIES

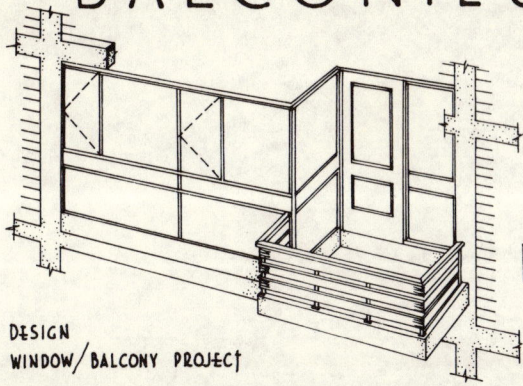

DESIGN
WINDOW/BALCONY PROJECT

1

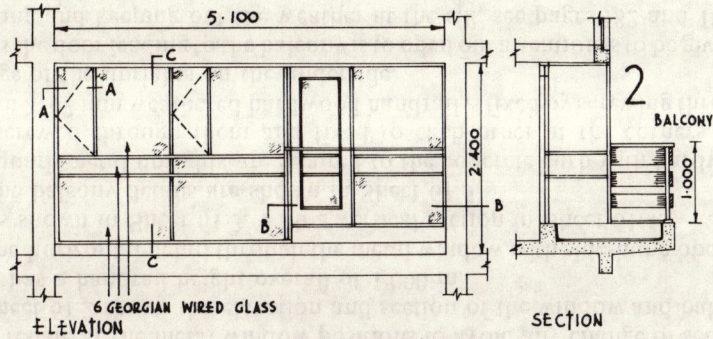

5·100

2·400

1·000

6 GEORGIAN WIRED GLASS

ELEVATION

BALCONY

SECTION

2

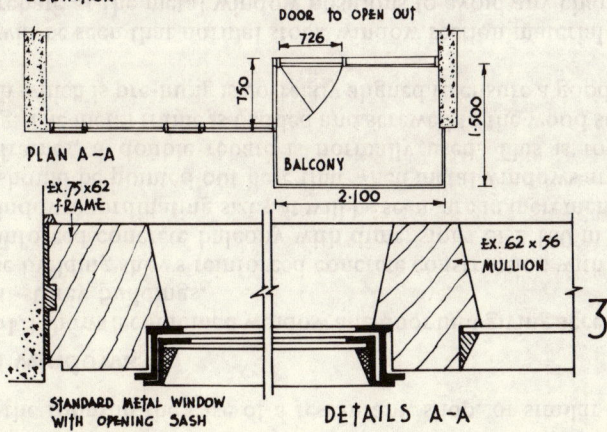

DOOR TO OPEN OUT
726

150

1·500

PLAN A-A

BALCONY

2·100

EX.75×62 FRAME

EX.62×56 MULLION

STANDARD METAL WINDOW WITH OPENING SASH

DETAILS A-A

3

USUAL TWICE REBATED TIMBER SECTION FOR METAL CASEMENTS

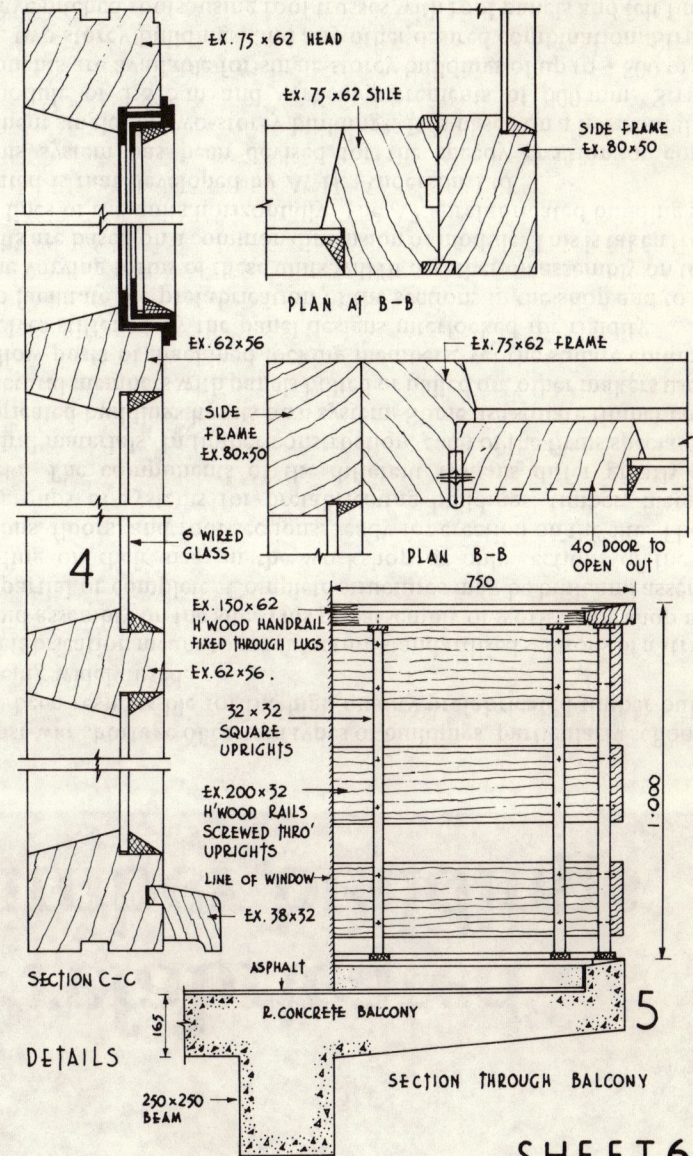

EX. 75×62 HEAD

EX. 75×62 STILE

SIDE FRAME EX.80×50

PLAN AT B-B

EX.62×56

SIDE FRAME EX.80×50

6 WIRED GLASS

EX.75×62 FRAME

PLAN B-B

40 DOOR TO OPEN OUT

750

EX. 150×62 H'WOOD HANDRAIL FIXED THROUGH LUGS

EX.62×56.

32 × 32 SQUARE UPRIGHTS

EX.200×32 H'WOOD RAILS SCREWED THRO' UPRIGHTS

LINE OF WINDOW

EX.38×32

1·000

SECTION C-C

4

DETAILS

ASPHALT

R.CONCRETE BALCONY

165

SECTION THROUGH BALCONY

250×250 BEAM

5

SHEET 61

202

The jointing of the shaped head and sill is similar to previous examples and the glazing is fixed using glazing beads from the inside as shown in the details in Sheet 59.7.

A curved window board is shown which allows additional seating area within the room in the case of a restaurant, shop, or similar use.

Metal windows

Sheet 61.1 shows a combined window and door unit giving access to a balcony in multi-storey buildings.

The building shows reinforced concrete construction with brick infilling and reinforced concrete balcony with dimensions of $2 \cdot 100$ m $\times$ $1 \cdot 500$ m.

Window coordinating sizes, it will be seen, are in increments of 300 mm, and it should be pointed out here that when metal windows are fitted within wood frames, a double rebate is normally used. This is to avoid excess packing. The metal frame is bedded and screwed to the wood surround while the sash which is pre-hung is correctly aligned to ensure a good seating when closed.

It will be seen that normal stock window section material is used with a single rebate at the metal window positions to avoid any change of section.

Sheet 61.2 shows the elevation and section of the window and balcony which has a handrail height overall of $1 \cdot 000$ m.

The horizontal detail through the metal window with side-hung opening light is shown in Sheet 61.3, with a vertical section in Sheet 61.4.

The balcony details are shown in Sheet 61.5.

Square metal uprights are secured to the concrete curb with hardwood rails screwed through them and fixed to each other at the corners. The 150 mm $\times$ 62 mm weathered hardwood handrail is fixed by screwing through the lugs of the uprights on the underside.

As the door leading to the balcony is to open out attention is to be given to throating and keeping out the weather at the sill, see pages 182 and 183.

Chapter 17

Prefabricated timber buildings

The post-war shortage of certain types of buildings, particularly schools, has largely been responsible for the high-quality prefabricated timber buildings now being widely used.

Prefabrication means to manufacture standardised sections of a structure for rapid assembly on the site. This pre-assembly of work in the shop may be either partial or complete. Complete structures may be built and assembled, depending on their size, in the workshop or only sections of the walls, partitions, floors, and roof sections, ready for erection on the site. There are three groups of systems for prefabricating buildings: timber, metal, and concrete. The components of the different systems differ greatly in the structural materials. In timber construction, each of the firms specialising in prefabricated buildings has its own system. Some use square timber columns as structural members with panels bolted or nailed on, other makers use built-up hollow posts of machined locking members, yet the square columns are themselves stiffened by the panel designs interlocked for rigidity.

To facilitate the prefabrication of the sections in the shop and to ensure that the varying forms of these units will fit together on assembly on the site, the units are based on a common dimension or module. This is taken from the centre lines of the units horizontally. The A75 prefabricated building system illustrated is that developed by A. H. Anderson Ltd.

This system has been devised for the speedy erection of complete permanent single- or two-storey buildings. It is based on a horizontal centre line module of $1 \cdot 875$ m and vertical increments of 600 mm. Structural components are available for single-storey buildings of up to $4 \cdot 800$ m ceiling height, two-storey buildings, and any other desired combination. Structures may have pitched roofs, using roof trusses with roof panels and felt finish, or flat roofs, using plywood box beams with roof panels and similar finish.

With this system there is unusual freedom of action. The system can be employed, if desired, in conjunction with other methods of construction. Considerable variations in elevational treatment and in finish are possible. Sheet 62 shows a sketch of a layout of single- and two-storey structures which

TIMBER BUILDINGS

1 LAYOUT OF PERMANENT SINGLE & DOUBLE STOREY STRUCTURES

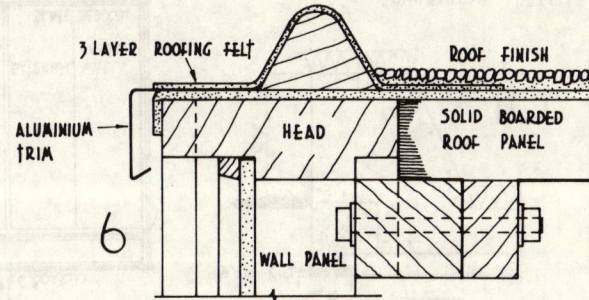

6

3 LAYER ROOFING FELT

ROOF FINISH

ALUMINIUM TRIM

HEAD

SOLID BOARDED ROOF PANEL

WALL PANEL

EAVES FLASHING DETAIL

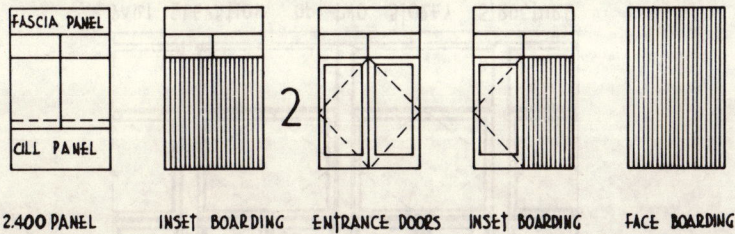

FASCIA PANEL

CILL PANEL

2

2.400 PANEL   INSET BOARDING   ENTRANCE DOORS   INSET BOARDING   FACE BOARDING

WALL PANEL RANGE 2.400 HIGH

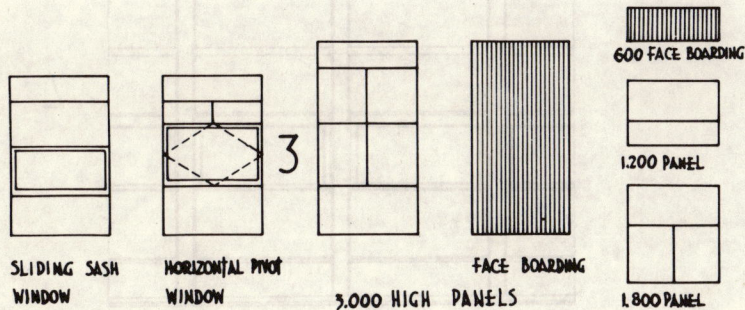

3

SLIDING SASH WINDOW   HORIZONTAL PIVOT WINDOW   3.000 HIGH PANELS   FACE BOARDING

600 FACE BOARDING

1.200 PANEL

1.800 PANEL

WALL PANEL FRAMING DETAILS

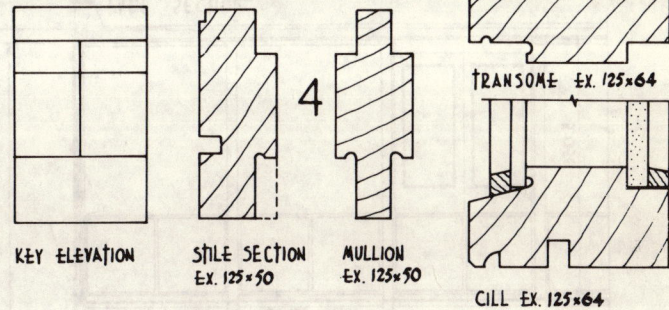

4

KEY ELEVATION   STILE SECTION EX. 125×50   MULLION EX. 125×50

TRANSOME EX. 125×64

CILL EX. 125×64

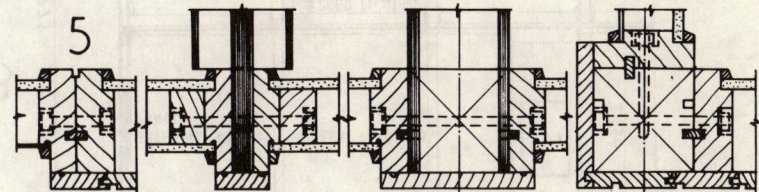

5

GLAZED PANEL TO BOARDED PANEL   GLAZED PANEL TO GLAZED PANEL   GLAZED PANEL TO GLAZED PANEL WITH BEAM ENTRY TO STRUCTURAL COLUMN   GLAZED PANEL TO BOARDED PANEL AT EXTERNAL CORNER

WALL PANEL JUNCTIONS

SHEET 62

TIMBER BUILDINGS

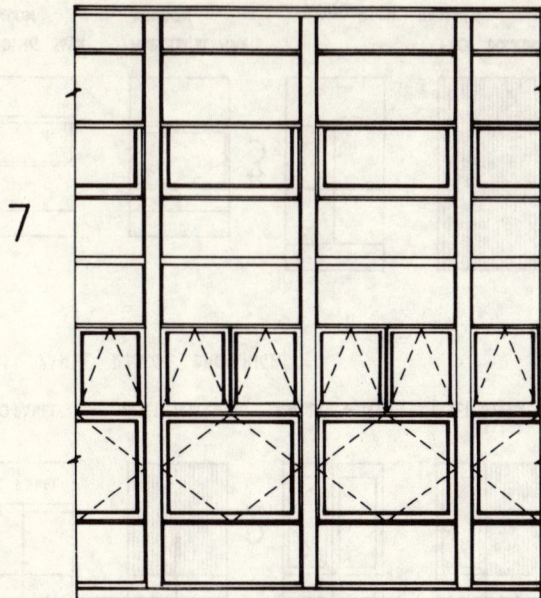

7

PART ELEVATION OF TWO STOREY STRUCTURE

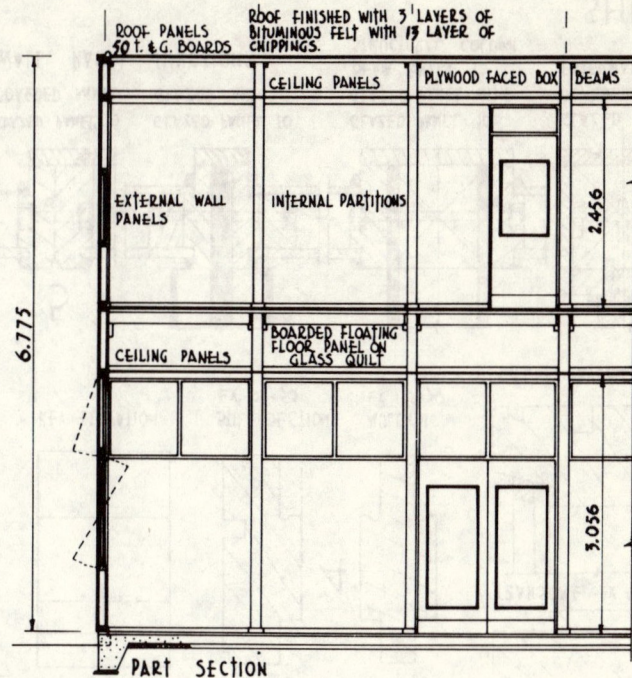

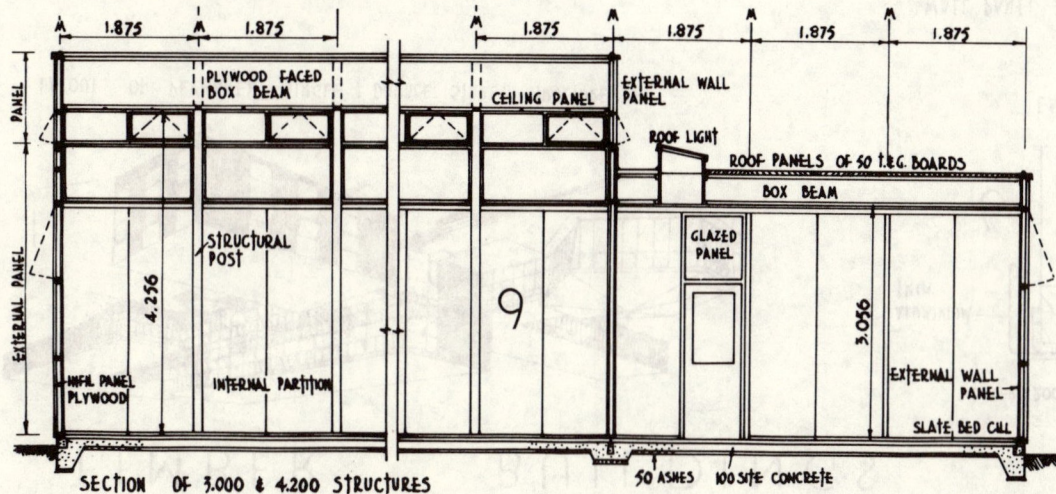

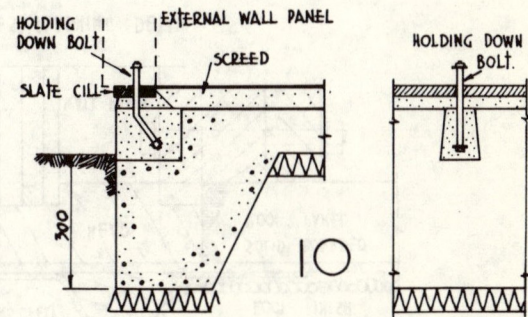

8

ROOF PANELS 50 T. & G. BOARDS

ROOF FINISHED WITH 3 LAYERS OF BITUMINOUS FELT WITH 13 LAYER OF CHIPPINGS.

CEILING PANELS	PLYWOOD FACED BOX	BEAMS

EXTERNAL WALL PANELS

INTERNAL PARTITIONS

2.456

6.775

CEILING PANELS

BOARDED FLOATING FLOOR PANEL ON GLASS QUILT

3.056

PART SECTION

9

1.875   1.875   1.875   1.875   1.875   1.875

PLYWOOD FACED BOX BEAM

CEILING PANEL

EXTERNAL WALL PANEL

ROOF LIGHT

ROOF PANELS OF 50 T.& G. BOARDS

BOX BEAM

PANEL

EXTERNAL PANEL

STRUCTURAL POST

4.256

GLAZED PANEL

3.056

INFIL PANEL PLYWOOD

INTERNAL PARTITION

EXTERNAL WALL PANEL

SLATE BED CILL

SECTION OF 3.000 & 4.200 STRUCTURES

50 ASHES   100 SITE CONCRETE

10

HOLDING DOWN BOLT

EXTERNAL WALL PANEL

SCREED

SLATE CILL

300

HOLDING DOWN BOLT.

FOUNDATION DETAILS

SHEET 63

TIMBER BUILDINGS

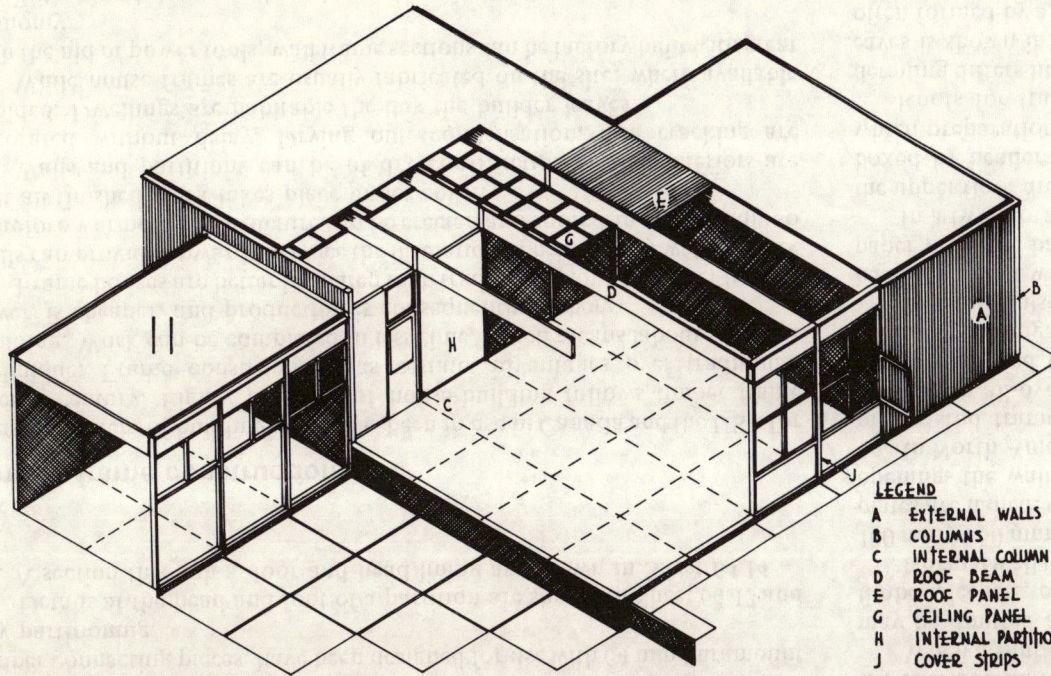

SKETCH OF TYPICAL SINGLE STOREY A 75 ASSEMBLY

11

LEGEND

A EXTERNAL WALLS
B COLUMNS
C INTERNAL COLUMN
D ROOF BEAM
E ROOF PANEL
G CEILING PANEL
H INTERNAL PARTITION
J COVER STRIPS

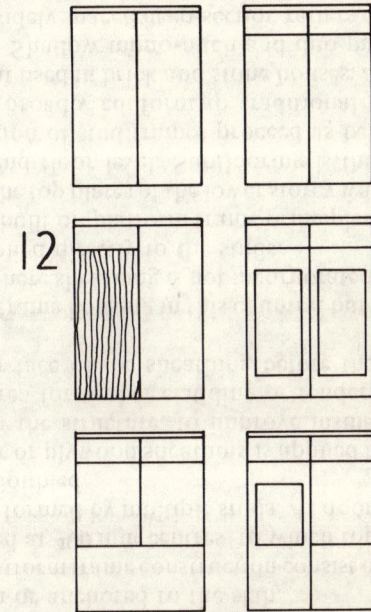

12

TYPICAL 2.400 PARTITION FRAMES

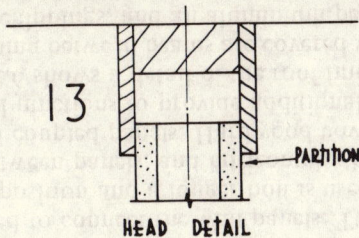

13

PARTITION

HEAD DETAIL

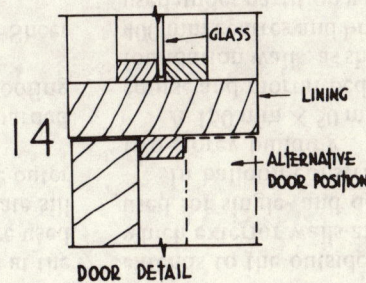

14

GLASS

LINING

ALTERNATIVE DOOR POSITION

DOOR DETAIL

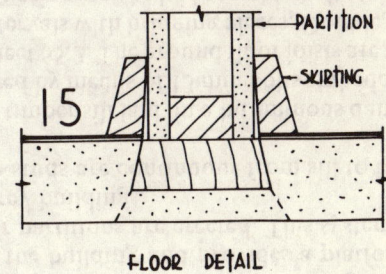

15

PARTITION

SKIRTING

FLOOR DETAIL

SHEET 64

205

can be altered or dismantled, and re-erected with little waste of materials.

External wall panels consist of a basic frame jointed in the usual way, with a variety of infilling subassemblies. The principal types in the 2·400 m and 3·000 m range are shown in Sheet 62.2 and 3. Details sections of the panel framing members are shown in Sheet 62.4.

Typical details of wall panel junctions are shown in Sheet 62.5, 13 mm bolts are used to connect the wall panels. The end stile is reinforced at the beam entry position and a longer bolt is used. Hardwood tongues are used vertically between panels, and horizontally between panel sill and slate sill and between coupled panels. Hardwood cover strips are fixed on the outer face of panel junctions to provide additional weathering protection.

Sheet 62.6 shows a detail of the roof finish, 50 mm thick solid boarded panels spanning between beams are covered with three layers of felt roofing and 13 mm chippings, and an aluminium perimeter trim.

A part elevation of a two-storey structure is shown in Sheet 63.7. Sheet 63.8 shows a part section of the same structure.

A combination of 3·000 m and 4·200 m structures is shown in section in Sheet 63.9. The details of the foundations are shown in Sheet 63.10.

Sheet 64.11 shows an isometric view of a typical single-storey assembly. 2·400 m partition frame range is shown in Sheet 64.12. These, together with timber connecting pieces, have been designed for use with 64 mm Paramount dry partitioning.

Details at the head and foot of a partition are shown in Sheet 64.13 and 15. A section through a door and head lining are shown in Sheet 64.14.

Timber frame construction

This is a system of building which has been in use in Canada and the USA for over a century. Eighty per cent of house-building follows timber frame technique. Frame construction has certain advantages over traditional building. Work can be completed in less time, which means labour costs are lower, is cheaper, and production is consequently higher.

Frame houses are better insulated than traditional houses, timber framed walls can provide upwards of twice the insulation of brick cavity walls and are therefore warmer. The structure can be erected and roofed-in very rapidly so that all finished work takes place under cover.

Walls and partitions can be of dry construction so that interiors are decorated without delay. Drying out, condensation, and cracking are avoided. Dwellings are habitable the day the builder leaves.

While house frames are usually fabricated on the site, where available with the aid of power tools, wall frame sections can be factory built with great economy.

With sound design and workmanship, timber frame houses will outlast many others. In frame construction all timber is isolated from brickwork and masonry and damp-proofed, so that the usual starting points of dry rot are eliminated.

In a timber frame house the basic framework of studding is a free-standing structure, acting as a braced and unified assembly which supports all loads. Since the exterior facing, brick or stone veneer, timber weather-boarding or cement rendering, performs no structural function other than that of providing additional bracing, it need not be started until the building has been closed in.

Two main forms of frame construction are in current use, platform and balloon.

Sheet 65.1 shows a section using platform construction. The ground floor extends to the outside edges of the building and provides a platform upon which exterior walls and interior partitions are erected. This system may be used for single- and double-storey buildings.

In balloon construction the studs are continuous from sill to eaves in a two-storey building.

A 150 mm × 50 mm treated timber sill laid on a bituminous damp-proof course and mortar bed is anchored by means of 13 mm bolts embedded in the foundation walls, as shown in Sheet 65.2. The ground floor joists are placed at 400 mm centres and braced at intervals with bridging pieces. Double joists are used under partition walls. Subflooring may be laid at right angles to the joists or diagonally using square-edged or tongued and grooved boards (Sheet 65.3). Plywood subflooring may be used in 2·400 m × 1·200 m sheets, with blocking pieces fixed to afford edge nailing. The finished floor is not laid over the subfloor until the building has been closed in.

When a timber frame house is to be built on a concrete slab, subflooring may be omitted and the finished flooring laid directly on pressure-treated timber sleepers, embedded in or anchored to the slab.

Normal wall frames in platform frame construction consist of storey-high 100 mm × 50 mm studs placed at 400 mm centres, to which top and bottom plates are nailed. Corners are formed by multiple studs. At door and window openings the wall studs are doubled.

In North America, timber or plywood sheathing is applied to the outside of the stud frames to stiffen the structure, to improve insulation, and to provide an all-over nailing area for timber cladding or rendering. Building paper is applied to the outer face of the sheathing before the addition of exterior cladding.

Sheathing also figures in frame housing in this country, but many houses have been built without it. Where sheathing is not incorporated the building paper moisture barrier is applied directly to the studs.

In a two-storey dwelling built on platform frame principles, the joists of the upper floor are spiked to the top plates of the lower storey wall frames and boxed by headers as at ground floor level. Subflooring is then laid, after which preparations and erection of stud frames proceed as before.

Roofs for frame houses broadly conform to traditional practice and detailing differs little from that used in brick and stone houses; a detail at the eaves is shown in Sheet 65.4. Shallow mono-pitch and duo-pitch roofs are often formed by a system of widely spaced deep-section rafters or beams and structural roof planking of 50 mm Western red cedar, a form of construction which ensures excellent thermal insulation. Roof planking and beams are left exposed for interior decorative effect and are overlaid above with additional rigid insulation and the outer weather skin. The standard pitched roof system designed by the T.R.A.D.A. can also be used in conjunction with frame house construction.

Most people who build timber frame houses like to have at least part of

TIMBER FRAME CONSTRUCTION

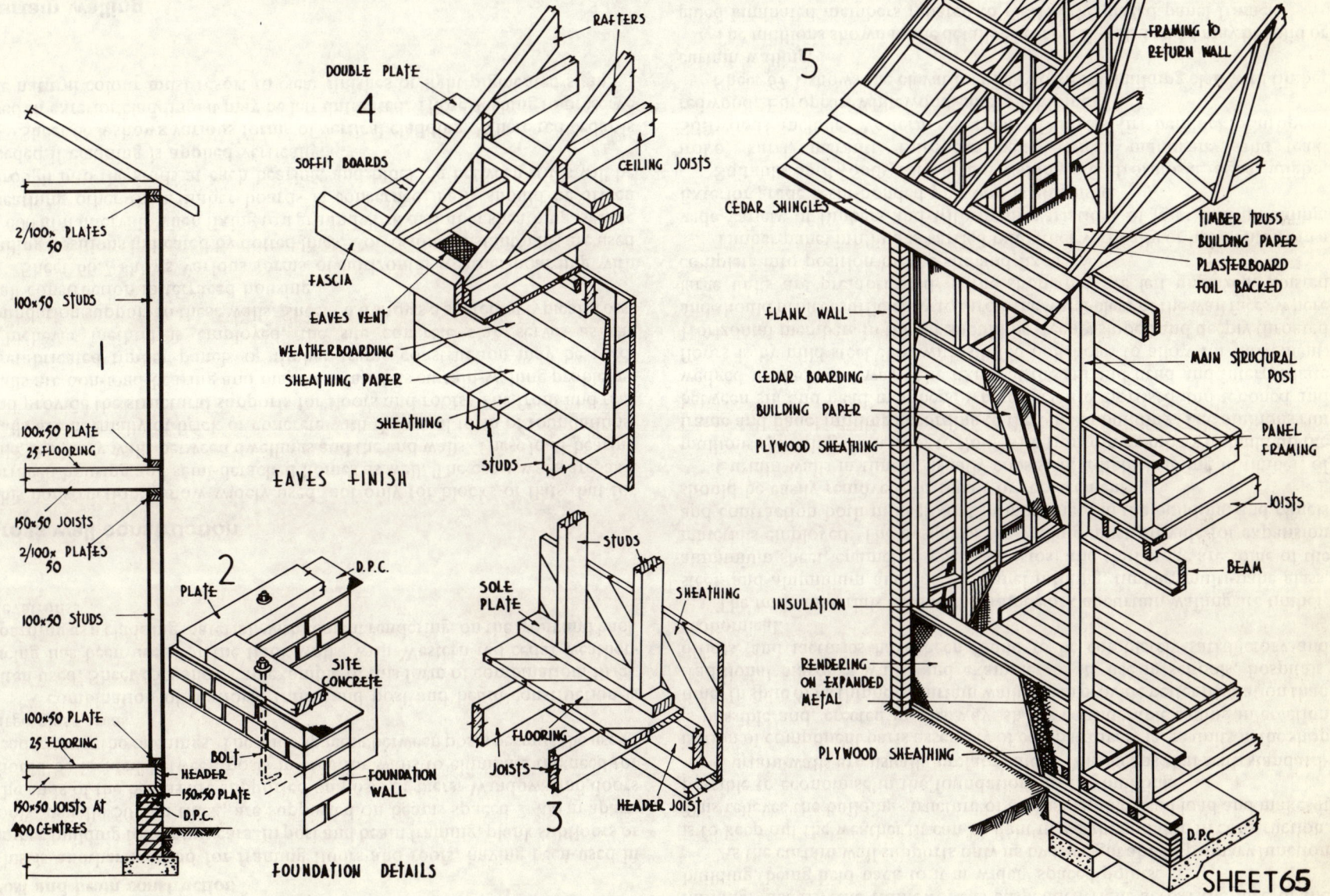

1

2/100x PLATES 50

100x50 STUDS

100x50 PLATE
25 FLOORING

150x50 JOISTS

2/100x PLATES 50

100x50 STUDS

100x50 PLATE
25 FLOORING

150x50 JOISTS AT 400 CENTRES

2

PLATE

D.P.C.

SITE CONCRETE

BOLT

HEADER

150x50 PLATE

D.P.C.

FOUNDATION WALL

FOUNDATION DETAILS

4

RAFTERS

DOUBLE PLATE

CEILING JOISTS

SOFFIT BOARDS

FASCIA

EAVES VENT

WEATHERBOARDING

SHEATHING PAPER

SHEATHING

STUDS

EAVES FINISH

3

STUDS

SOLE PLATE

SHEATHING

FLOORING

JOISTS

HEADER JOIST

5

FRAMING TO RETURN WALL

CEDAR SHINGLES

TIMBER TRUSS
BUILDING PAPER
PLASTERBOARD FOIL BACKED

FLANK WALL

CEDAR BOARDING
BUILDING PAPER
PLYWOOD SHEATHING

INSULATION

RENDERING ON EXPANDED METAL

PLYWOOD SHEATHING

MAIN STRUCTURAL POST

PANEL FRAMING

JOISTS

BEAM

D.P.C.

SHEET 65

207

the structure clothed externally with wood as well. Brick, stone, rendering, tile, and shingle can all be used alone or in combination with vertical or horizontal timber weather-boarding.

Post and beam construction

This is another method for framing floors and roofs, having been used in timber building for many years. In post and beam framing, plank subfloors or roofs, usually 50 mm thick, are supported on beams spaced 2·400 m apart. The ends of the beams are supported on posts or piers. Windows and doors should be located between posts in exterior walls to eliminate the need for headers over the openings. The wide spacing between posts permits the use of large glass areas.

A combination of platform frame and post and beam construction is often used. Sheet 65.5 shows an example of this form of combination. Brick facing has been used for the flank walls, with Western red cedar weather-boarding as a cladding material, with cement rendering, on the front and back elevations.

Cross wall construction

This construction is now widely used, not only for blocks of flats, but for terraced housing and semi-detached homes as well. The cross walls are, as a rule, the party walls between dwellings and the end walls. These load-bearing walls are normally of brick or concrete with the usual form of foundations, and provide the structural supports for floors and roofs. The front and rear walls are non-load-bearing and may be treated as curtain-walling problems. Prefabricated timber panels or timber frame construction may be used. Whichever method is employed, the site concrete slab serves as the foundation support to these walls. Sheet 66.6 shows a sketch of typical cross-wall construction in terraced housing.

Sheet 66.7 shows various forms of horizontal weather-boarding with nailing positions indicated by dotted lines. Corrosion-resistant nails are used at 600 mm intervals when fixing to a groundwork of timber sheathing. Where sheathing other than timber boards is concerned, nails should be driven through into the studs at each bearing, and blocking between studs will be needed if cladding is applied vertically.

Sheet 66.8 shows various forms of vertical cladding. Where red cedar is used in exterior cladding, it may be left untreated. Those wishing to preserve the natural colour must resort to clear finishes or light pigmented stain.

Curtain walling

A curtain wall is a wall of windows and spandrel panels in a framework which is not build up within the main structure of the building. These walls are not load-bearing, therefore any form of walling which is not load-bearing is a curtain wall. The illustrations in frame and cross-wall construction on Sheets 65 and 66 show this type of construction where the enclosing walls are curtain walls.

Recent years have seen considerable changes in the design of buildings with the replacement of brick and masonry walls by thin sheet materials in large panels, or by light timber or metal frameworks with various forms of infilling. Usually the framework is suspended right across the face of the building, being held back to it at widely spaced points.

As the curtain wall supports only its own weight and its primary function is to keep out the weather, it can be light in weight and thin in construction. This relieves the building structure of a considerable dead load and makes it possible to economise in the foundations to the building.

Curtain walls are usually prefabricated. This means that with standardisation of component parts assembly of comparatively large units in the shop is possible and, erected in this way, shows considerable saving in erection time. In spite of its thinness, curtain walling can provide better insulation than traditional masonry walls, and examples of its use in schools, hospitals, offices, and factories have been found to be completely satisfactory and economical.

The main materials used in the framework of curtain walling are timber, steel, and aluminium and for the panel infilling, timber, multi-pane glass, aluminium sheet, enamelled glass, asbestos, and thin stone are some of the materials employed. The design of the walling must provide for expansion and contraction both in itself and in movement in the building, and panels should be easily removed for repair or replacement.

Curtain walls in timber usually consist of a light framing in timber, of mullions spanning from floor to floor with transom members to build up the frame and panel infilling according to the type of building. The mullions run between sill and head members, where they are mortised and tenoned and wedged in the usual way. The fixing at the sill and head and intermediate floors is by mild steel clips with slotted bolt holes to allow for movement. Horizontal members in timber should be steeply sloped and deeply throated and should project sufficiently to throw the water clear of the wall face. Where large units are prefabricated in the shop they are left unglazed, hoisted complete into position on the site and fixed.

Timber panel infilling is carried out either vertically or horizontally in a wide variety of timbers and with many variations of joint and moulding. Exterior grade resin-bonded plywood is also used.

Suitable hardwoods for curtain walling are afrormosia, agba, idigbo, iroko, karri, meranti, European oak, African mahogany, and teak. Softwoods include Western red cedar, Douglas fir, hemlock, European redwood, European whitewood, and pitch pine.

Sheet 67.1 shows the elevation of a two-storey building clad with timber curtain walling.

The mullions shown in the details Sheet 67, figs. 2, 3 and 4 may be solid or glued laminated members rebated to receive the glazed panel frames.

Vertical cedar boarding is used between the two floors and also at the roof level, and finished with linseed oil or a clear lacquer or varnish.

A vertical section through the lower storey of the curtain walling is shown in Sheet 67.5.

It will be seen that stock single and double rebated window sections are used which require fillets to be planted in the rebates of the vertical members where the cedar weather-boarding is fixed. The entrance doors are not in line

CROSS WALL CONSTRUCTION & CLADDING

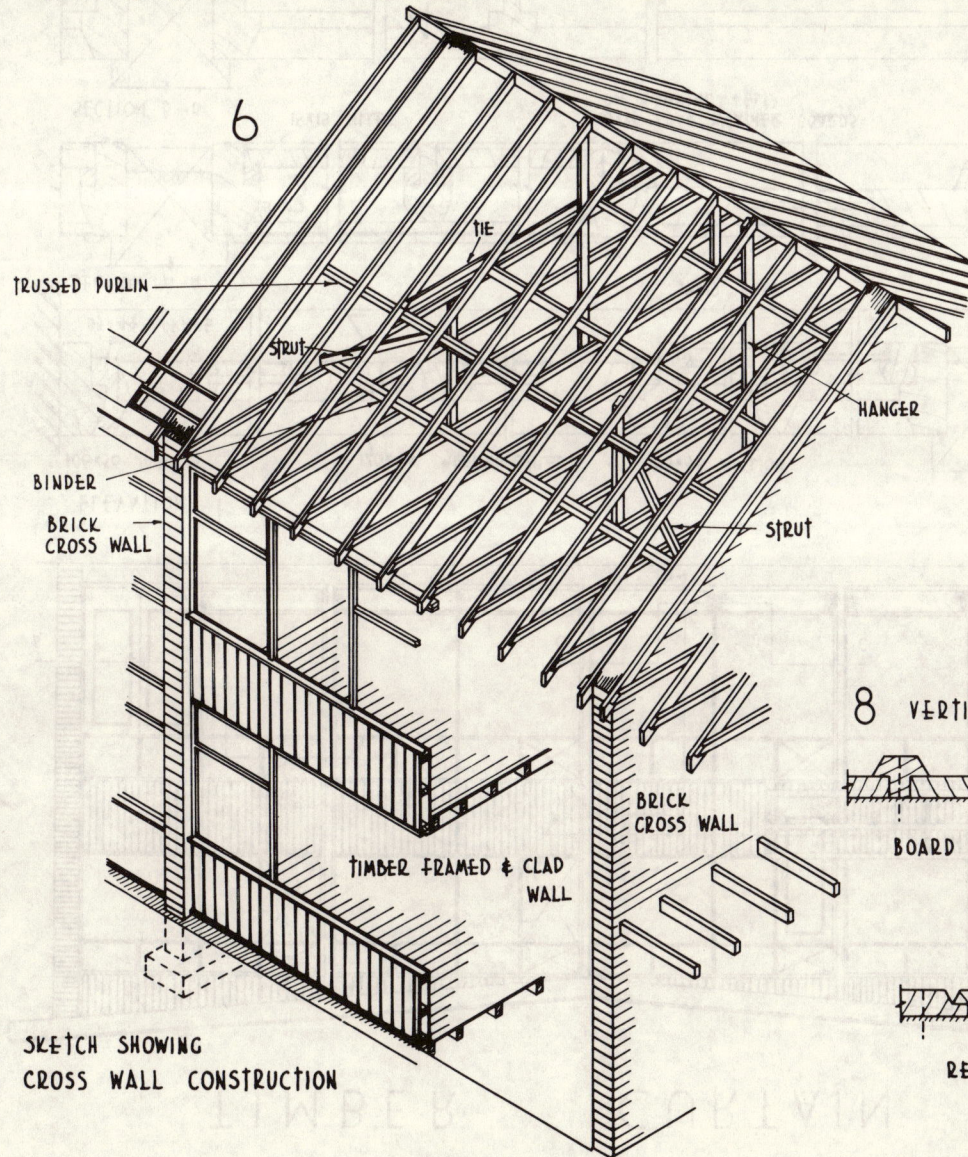

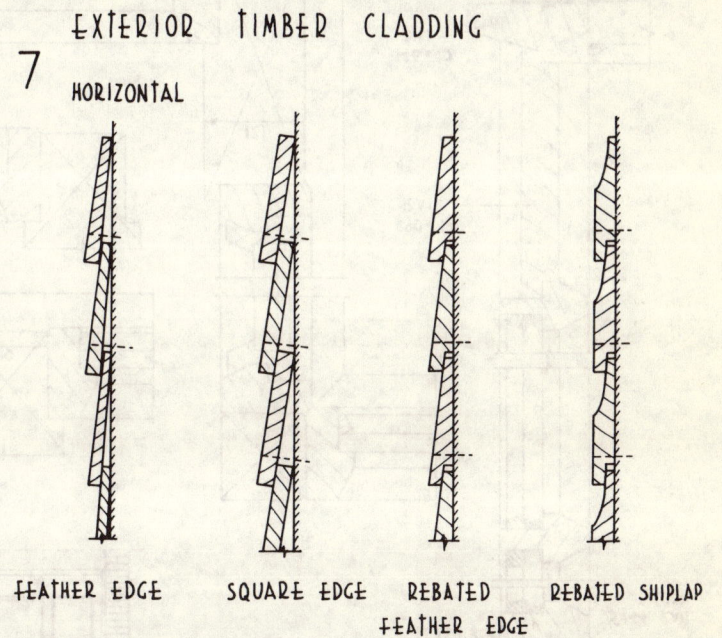

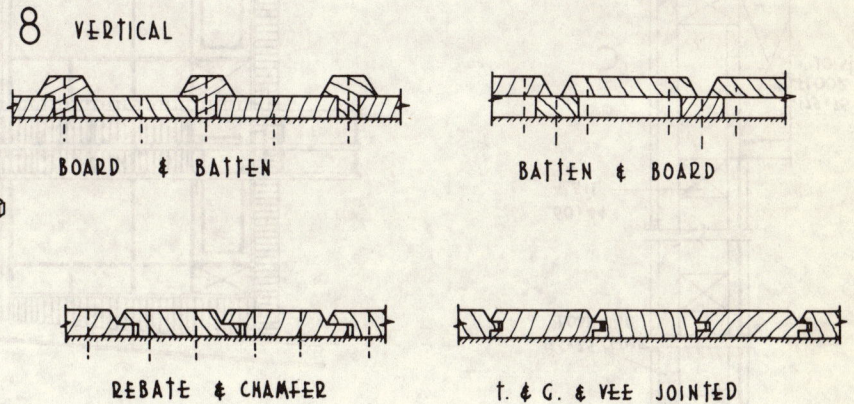

6

TIE

TRUSSED PURLIN

STRUT

HANGER

BINDER

STRUT

BRICK
CROSS WALL

TIMBER FRAMED & CLAD
WALL

BRICK
CROSS WALL

SKETCH SHOWING
CROSS WALL CONSTRUCTION

7 EXTERIOR TIMBER CLADDING

HORIZONTAL

FEATHER EDGE   SQUARE EDGE   REBATED   REBATED SHIPLAP

FEATHER EDGE

8 VERTICAL

BOARD & BATTEN   BATTEN & BOARD

REBATE & CHAMFER   T. & G. & VEE JOINTED

SHEET **66**

TIMBER CURTAIN WALLING

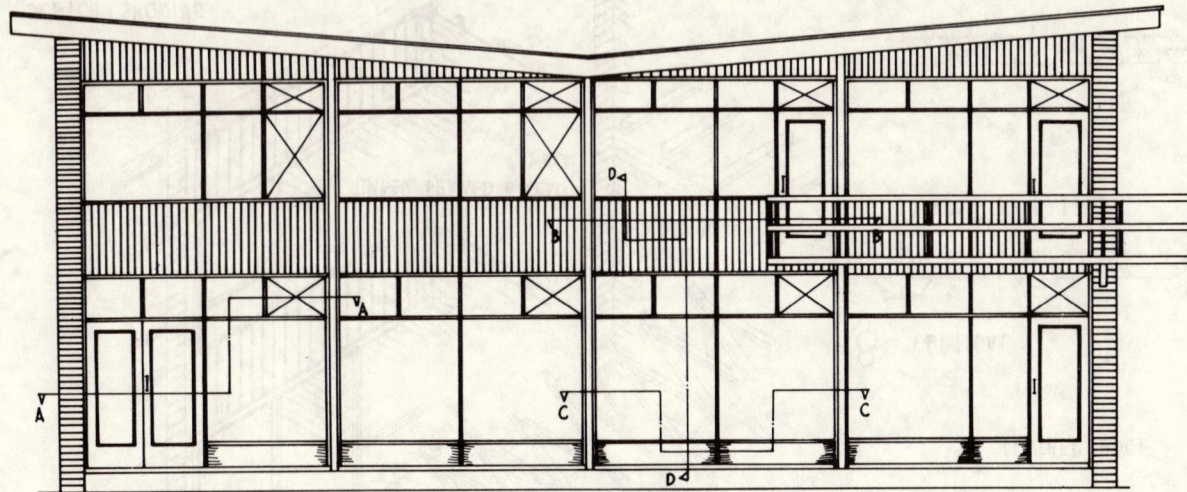

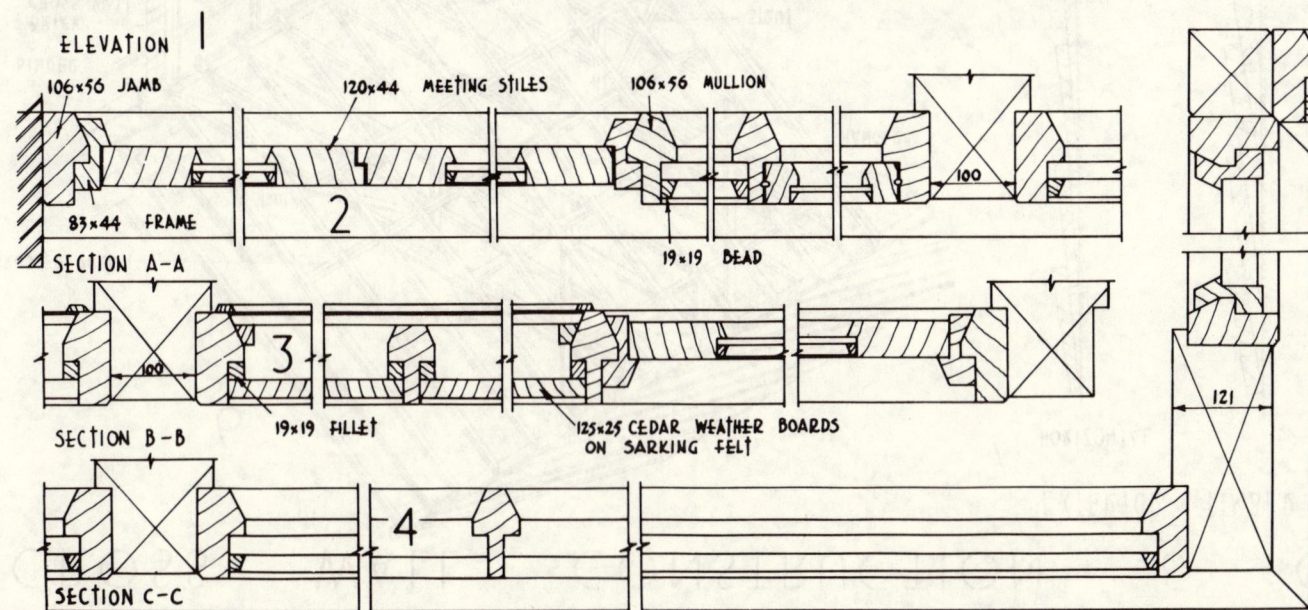

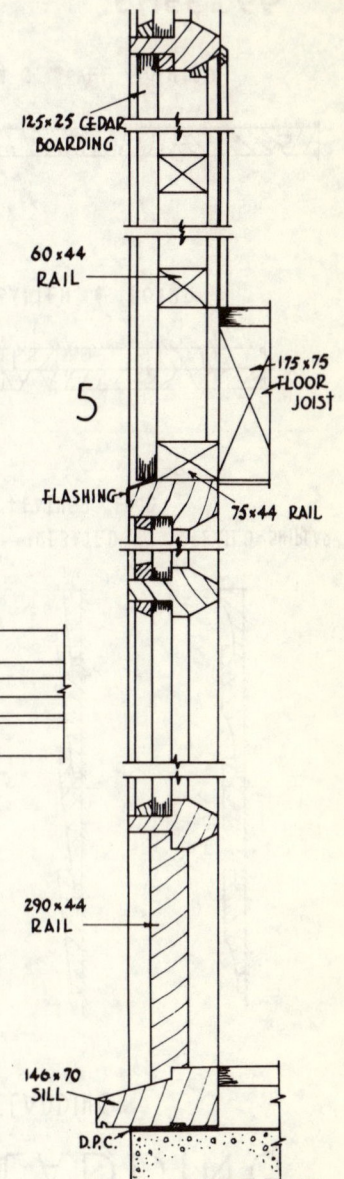

ELEVATION 1

125x25 CEDAR BOARDING

60x44 RAIL

175x75 FLOOR JOIST

5

FLASHING

75x44 RAIL

290x44 RAIL

146x70 SILL

D.P.C.

106x56 JAMB

120x44 MEETING STILES

106x56 MULLION

2

100

83x44 FRAME

19x19 BEAD

SECTION A-A

3

100

19x19 FILLET

125x25 CEDAR WEATHER BOARDS ON SARKING FELT

SECTION B-B

121

4

SECTION C-C

SHEET 67

T I M B E R   F O O T B R I D G E

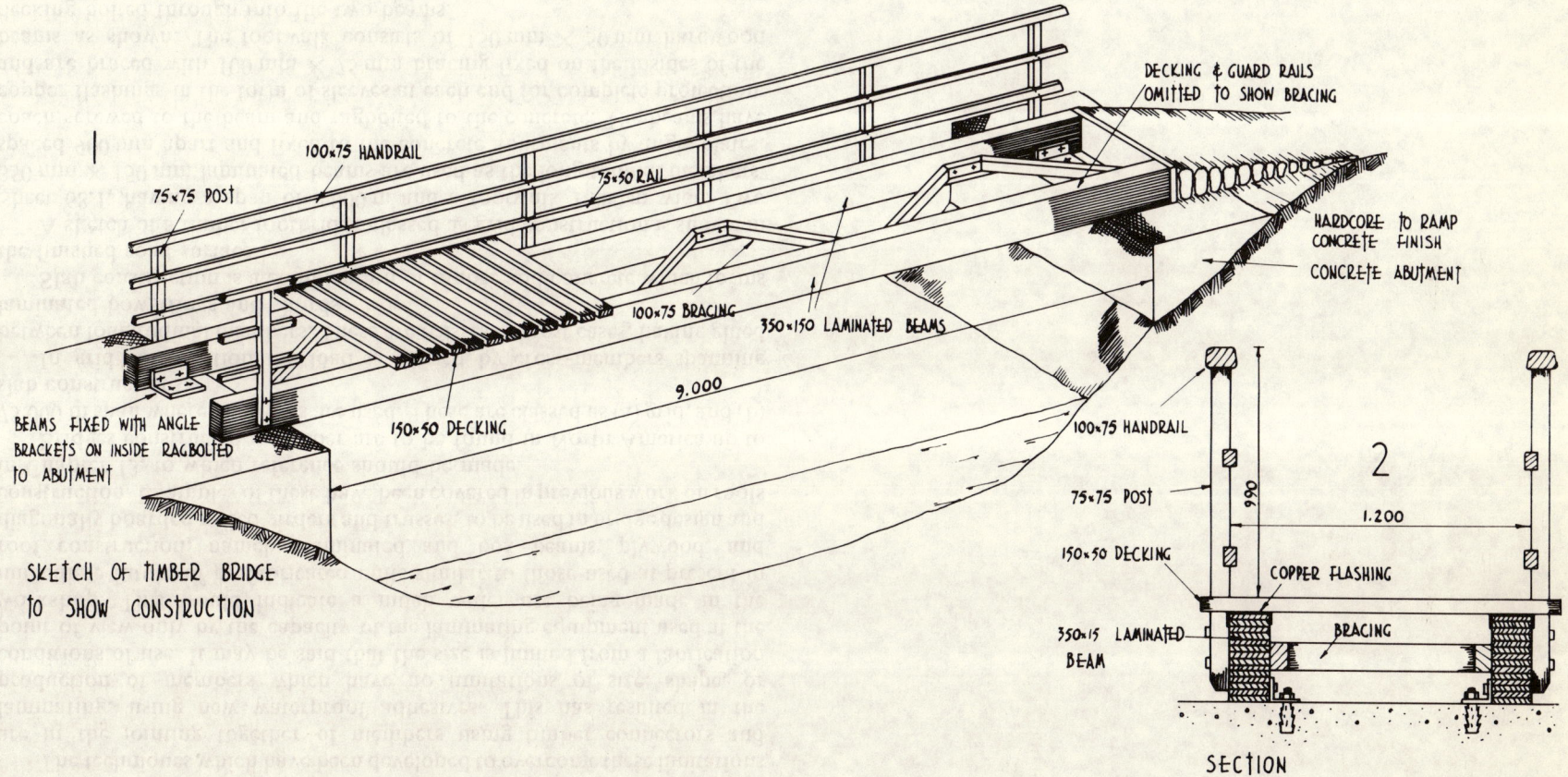

DECKING & GUARD RAILS
OMITTED TO SHOW BRACING

100×75 HANDRAIL

75×50 RAIL

75×75 POST

HARDCORE TO RAMP
CONCRETE FINISH

CONCRETE ABUTMENT

100×75 BRACING

350×150 LAMINATED BEAMS

9.000

150×50 DECKING

BEAMS FIXED WITH ANGLE
BRACKETS ON INSIDE RAGBOLTED
TO ABUTMENT

SKETCH OF TIMBER BRIDGE
TO SHOW CONSTRUCTION

100×75 HANDRAIL

2

75×75 POST

990

1.200

150×50 DECKING

COPPER FLASHING

350×15 LAMINATED
BEAM

BRACING

SECTION

SHEET 68

211

with the glass rebates, requiring the subframes to be used at these points.

Timber bridges

The use of timber as a structural engineering material has only become accepted in this country since the Second World War. Progress made both in North America and the Continent has been much more advanced in this field where timber is used in bridge construction as well as in methods of roofing.

Timber bridges were in use by the railways in this country until the 1930s, where the components were limited in size, shape, and quality available. Reinforced concrete and steel constructions replaced timber in this type of construction due to these limitations.

The techniques which have been developed to overcome these limitations are in the jointing together of members using timber connectors and laminating, using new waterproof adhesives. This has resulted in the production of members which have no limitations of size, shape, or conditions of use. It may be said that the size is limited from a fabrication point of view only by the capacity of the laminating equipment used at the workshop. This would indicate a much wider use being made in the immediate future of prefabricated units similar to those used at present in roof construction, namely, laminated and box beams, plywood, and diagonally boarded nailed girders and trusses, to be used in bridge design and construction. Examples of these have been covered in previous work on roofs in Chapter 13, to which reference should be made.

Bridges constructed in timber are to be found in North America up to 75·000 m span where two types are used. These are classed as (a) grid, and (b) slab constructions.

In grid construction the load is carried by cross-members spanning between longitudinal members. These are trusses in most cases, having glued laminated bowstrings and chords.

Slab construction is a combination of timber and concrete which forms the finished road surface.

A sketch of a timber footbridge classed as grid construction is shown in Sheet 68.1, having a span of 9·000 m and a footwalk 1·200 m wide. Two 350 mm × 150 mm laminated beams are used as the longitudinal members spaced 900 mm apart and fixed to the concrete abutments by angle plates, coach screwed to the beam and ragbolted to the concrete. The beams have copper flashings in the form of sleeves at each end for complete protection, and are braced with 100 mm × 75 mm bracing fixed on the insides of the beams as shown. The footwalk consists of 150 mm × 50 mm hardwood decking bolted through into the two beams.

A filler piece is used against the beam ends to receive the hardcore for the ramps which are finished in concrete.

The section through the bridge is shown in Sheet 68.2. The 100 mm × 75 mm handrail is carried on 75 mm × 75 mm posts bolted to the laminated beams with intermediate rails housed as shown. The decking is cut round the posts where these occur to allow firm fixing of the posts to the outside faces of the beams.

All the timber used in bridge construction should be treated with a preservative of some kind. The treatment of timber has been dealt with in Chapter 3. Likewise all ironwork, bolts, angles, etc. should be treated with a rust-preventing coat.

Chapter 18

Stairs

Building Regulations

The new Building regulations made by the Minister for the Environment have now replaced the Model By-Laws and apply to all building applications lodged with the relevant Authority, except Scotland which has its own regulations, and Inner London, where the London Building Acts still operate.

The existing by-laws did not include requirements for the construction of stairs, staircases, and handrails.

Interpretation

The Building Regulations differentiate between stairs for domestic use and stairs in all other classes of buildings. The regulations are concerned with four groups of buildings:

a.   A stairway within a dwelling or serving exclusively one dwelling.
b.   A stairway in common use in connection with at least two dwellings.
c.   Certain stairways in institutional buildings and places of assembly.
d.   Other stairways in institutional buildings, other residential and places of assembly together with offices, shops and storage and general buildings.

Purpose groups are as follows:

i   — Small residential
ii   — Institutional
iii — Other residential
iv   — Office
v   — Shop
vi   — Factory
vii — Other place of assembly
viii — Storage and general

Stairways

Flight – that part of a stairway or stepped ramp which consists of a step or consecutive steps.

Stairway – any part of a building which provides a route of travel and is formed by a single flight or by a combination of two or more flights and one or more intervening landings.

Parallel step – a step of which the nosing is parallel to the nosing of the step or landing immediately above it as shown in Sheet 69.1.

Tapered tread – a tread which has a greater width at one side than at the other and a going which changes at a constant rate throughout its length as shown in Sheet 69.2.

Pitch line – a notional or imaginary line drawn to connect all the nosings of the treads in a flight of stairs shown in Sheet 69.3.

Going – The going of a step is measured on plan between the nosing of its tread and the nosing of the next tread or landing above. The methods of measurement of the rise and going of a step are shown in Sheet 69.4.

Width of stairway – This is the unobstructed width measured clear of handrails and other projections as shown in Sheet 69.5

Deemed length – Sheet 69.6 shows a stairway having several consecutive tapered steps of differing lengths. In a stairway of this type, each such tread shall be deemed to have a length equal to the length of the shortest of those treads.

General Requirements for Stairways

The rise must be the same for each step in a flight between floors as shown in Sheet 69.7.

The going must be the same for each parallel step in a flight between consecutive floors.

The headroom must not be less than 2.000m measured vertically above the pitch line as shown in Sheet 69.7.

In an open stair with no risers, the nosing of the tread of any step or landing must overlap the back edge of the tread of the step below it by at least 15 mm as shown in Sheet 69.8. If the flight forms part of a building of purpose groups i, ii or iii used by persons under the age of five years the distance between the treads must not permit the passage of a 100 mm sphere.

For any parallel step in a building in groups i or iii the sum of its going plus twice its rise must not be less than 550 mm and not more than 700 mm. The rise of any step must not be more than 200 mm and the going not less than 220 mm with the maximum angle of pitch 42 to the horizontal as shown in Sheet 69.8.

Any tapered steps must comply to the regulations (see tapered steps).

Sheet 69.9 shows the graph from which the maximum and minimum permitted going and rise can be obtained within the shaded area for stairways in buildings in groups i or iii. There must not be fewer than 2 nor more than 16 risers in any flight in these groups.

The rise of any step in stairways in groups ii or vii must not be more than 180 mm and the going not less than 280 mm. The rise of stairways in all other groups must not be more than 190 mm and the going not less than 250 mm as shown in Sheet 70.10 with not fewer than 3 nor more than 16 risers per flight.

Sheet 70.11 shows the graph from which the maximum and minimum permitted going and rise can be obtained within the shaded area for stairways in groups ii, iii, iv, v, vi, vii, viii.

REGULATIONS FOR STAIRWAYS

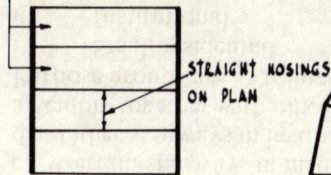

PARALLEL STEPS.
NOSINGS OF THE STEPS PARALLEL TO EACH OTHER ON PLAN

STRAIGHT NOSINGS ON PLAN

1

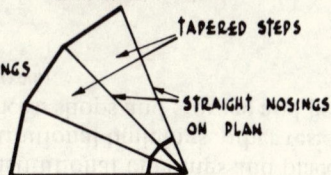

TAPERED STEPS.
NOSINGS NOT PARALLEL TO EACH OTHER ON PLAN

TAPERED STEPS

STRAIGHT NOSINGS ON PLAN

2

WIDTHS OF STAIRWAYS.

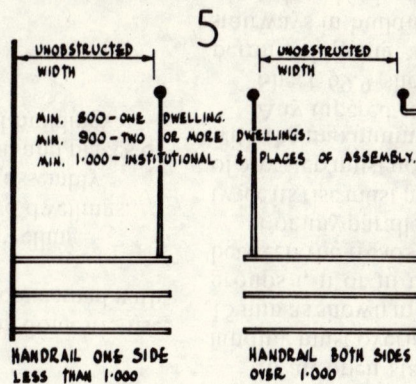

5

UNOBSTRUCTED WIDTH

MIN. 800 - ONE DWELLING.
MIN. 900 - TWO OR MORE DWELLINGS.
MIN. 1.000 - INSTITUTIONAL & PLACES OF ASSEMBLY.

HANDRAIL ONE SIDE LESS THAN 1.000

UNOBSTRUCTED WIDTH

HANDRAIL BOTH SIDES OVER 1.000

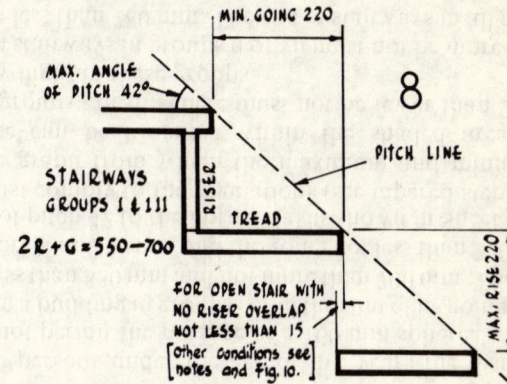

MIN. GOING 220

MAX. ANGLE OF PITCH 42°

PITCH LINE

RISER

TREAD

STAIRWAYS GROUPS I & III
2R+G=550-700

FOR OPEN STAIR WITH NO RISER OVERLAP NOT LESS THAN 15 [other conditions see notes and fig. 10.]

MAX. RISE 220

8

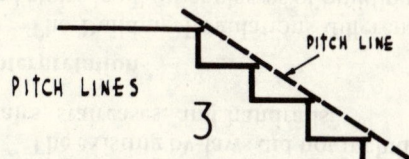

PITCH LINE

PITCH LINES

3

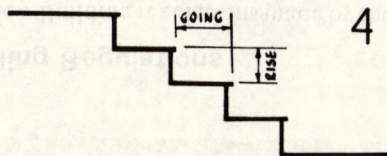

METHODS OF MEASURING RISE & GOING

GOING

RISE

4

HEADROOM

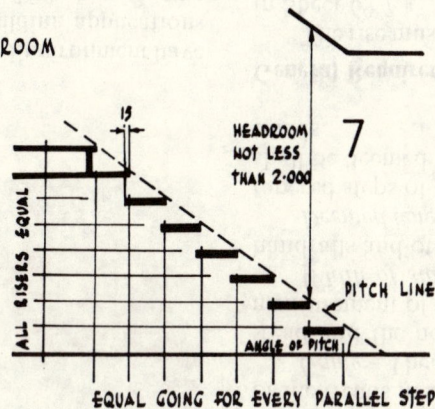

15

ALL RISERS EQUAL

HEADROOM NOT LESS THAN 2.000

PITCH LINE

ANGLE OF PITCH

EQUAL GOING FOR EVERY PARALLEL STEP

7

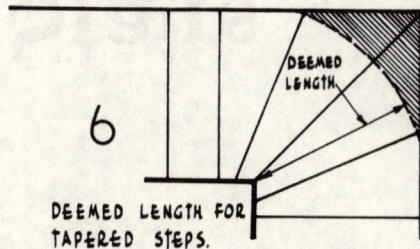

DEEMED LENGTH

DEEMED LENGTH FOR TAPERED STEPS.

6

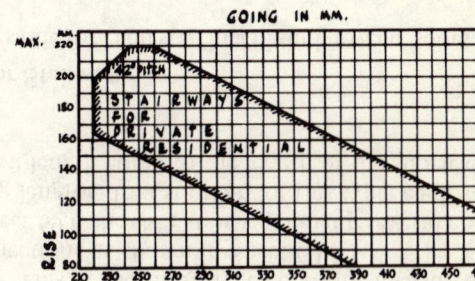

GOING IN MM.

STAIRWAYS FOR PRIVATE RESIDENTIAL

GRAPH SHOWING PERMITTED GOING AND RISE WITHIN SHADED AREA FOR PRIVATE RESIDENTIAL STAIRWAYS.

9

SHEET 69

REGULATIONS FOR STAIRWAYS

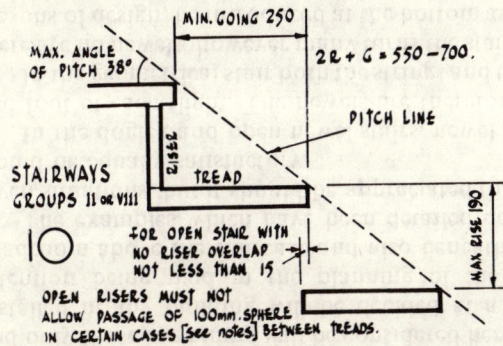

MAX. ANGLE OF PITCH 38°

MIN. GOING 250

$2R + G = 550 - 700$.

PITCH LINE

RISER

TREAD

STAIRWAYS GROUPS II or VIII

10

FOR OPEN STAIR WITH NO RISER OVERLAP NOT LESS THAN 15

OPEN RISERS MUST NOT ALLOW PASSAGE OF 100mm SPHERE IN CERTAIN CASES [SEE NOTES] BETWEEN TREADS.

MAX RISE 190

11 PERMITTED GOING AND RISE WITHIN SHADED AREAS FOR:

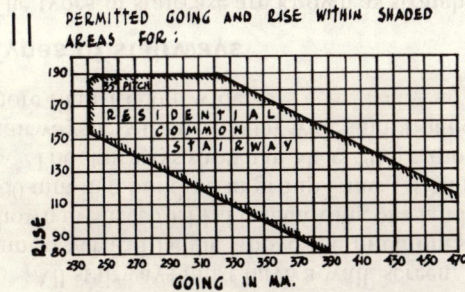

RESIDENTIAL COMMON STAIRWAY

RISE / GOING IN MM.

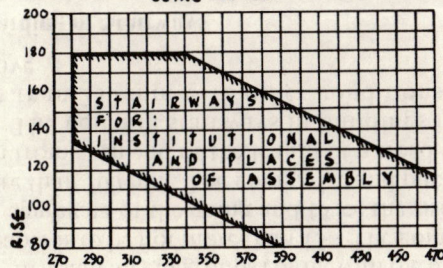

STAIRWAYS FOR: INSTITUTIONAL AND PLACES OF ASSEMBLY

RISE / GOING IN MM.

TAPERED STEPS STAIRWAYS OVER 1·000 WIDE.

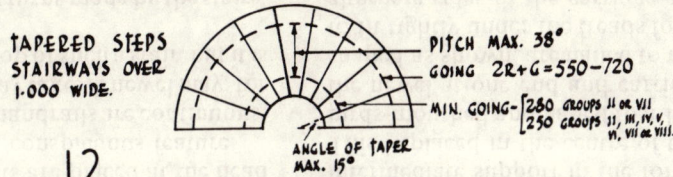

PITCH MAX. 38°
GOING $2R + G = 550 - 720$
MIN. GOING — 280 GROUPS II or VII
250 GROUPS II, III, IV, V, VI, VII or VIII

ANGLE OF TAPER MAX. 15°

12

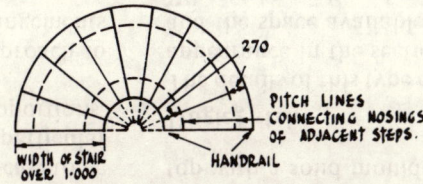

270

PITCH LINES CONNECTING NOSINGS OF ADJACENT STEPS.

WIDTH OF STAIR OVER 1·000

HANDRAIL

DEEMED WIDTH OF STAIRWAY OVER 1·000

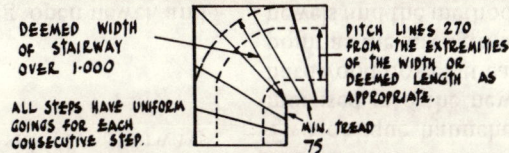

PITCH LINES 270 FROM THE EXTREMITIES OF THE WIDTH OR DEEMED LENGTH AS APPROPRIATE.

ALL STEPS HAVE UNIFORM GOINGS FOR EACH CONSECUTIVE STEP.

MIN. TREAD 75

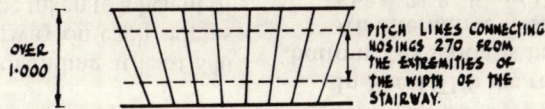

OVER 1·000

PITCH LINES CONNECTING NOSINGS 270 FROM THE EXTREMITIES OF THE WIDTH OF THE STAIRWAY.

13 TAPERED STEPS STAIRWAYS IN GROUP 1 UNDER 1·000 WIDTH.

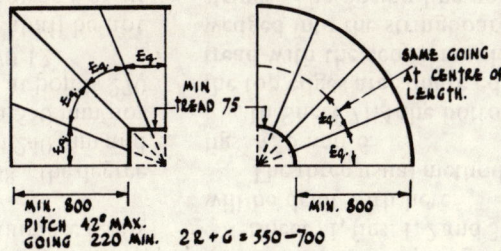

MIN TREAD 75

SAME GOING AT CENTRE OF LENGTH.

MIN. 800
PITCH 42° MAX.
GOING 220 MIN.

MIN. 800
$2R + G = 550 - 700$

GUARDING OF STAIRWAYS

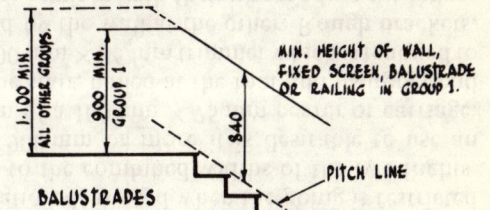

1·100 MIN. ALL OTHER GROUPS

900 MIN. GROUP 1

840

MIN. HEIGHT OF WALL, FIXED SCREEN, BALUSTRADE OR RAILING IN GROUP 1.

PITCH LINE

BALUSTRADES

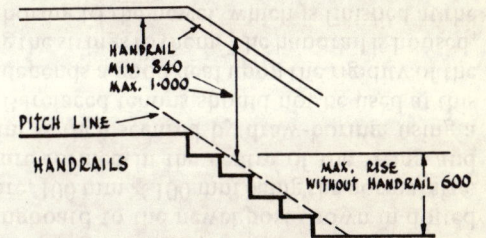

270

HANDRAIL MIN. 840 MAX. 1·000

PITCH LINE

PITCH LINE

MAX. RISE WITHOUT HANDRAIL 600

14

HANDRAILS

PERMITTED GOING AND RISE WITHIN SHADED AREAS FOR:

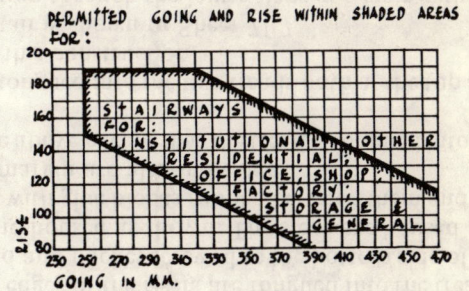

STAIRWAYS FOR: INSTITUTIONAL, OTHER RESIDENTIAL, OFFICE, SHOP, FACTORY, STORAGE & GENERAL.

RISE / GOING IN MM.

SHEET 70

Tapered Treads

The going of any part of a tread within the width of the stairway in all groups of buildings to be not less than 75 mm.

In a stairway in group i buildings the minimum going is 220 mm and the aggregate of the going and twice the rise to be not less than 550 mm nor more than 700 mm with the pitch not more than 42°. The going, rise and pitch is to be measured at the central point of the length of a tread if the staircase is less than 1.000m in width as shown in Sheet 70.13.

In a stairway in groups i or iii, with the pitch not more than 38°, the degree of taper is to be not more than 15°. The going to be not less than 240 mm and the aggregate of the going and twice the rise to be not less than 550 mm nor more than 700 mm. The going, rise and pitch is to be measured at points 270 mm from each end of the length of a tread as shown in Sheet 70.12.

The going in stairways in buildings in all other groups shall be not less than 280 mm with other conditions the same as in groups i or iii above.

Guarding of Stairways

All stairways must have a wall, screen, balustrade or railing at least 840 mm in vertical height above the pitch lines of the stairway on each side as shown in Sheet 70.14. The guarding of a landing should be taken to a height of 900 mm in a building in group i and 1.100m in buildings in all other groups.

The stairways shown in Sheet 70.5 require a handrail on one side only for stairways less than 1.000m wide and a handrail on both sides for stairways more than 1.000m wide.

Types of stairways

The types of staircase are known as straight flight, dogleg, open newel, and geometrical. The first type it is assumed will have been covered in earlier work and only those remaining will be considered here. The type of staircase to be installed in any building will be decided at a very early stage, particular attention being paid in the planning of the stair to step proportions, headroom above the staircase and also beneath the landings.

The examples which have been detailed demonstrate the approach to given situations, but it should be appreciated that many other arrangements would be equally satisfactory.

In the dogleg and open newel stairs, newel posts are placed at the head and foot of each flight. The newels are therefore a conspicuous feature.

In the geometrical stair both the strings and the handrails are continuous, there are no newels, however many turns the stair may take. A newel may, for reasons of design, be introduced at the bottom and top of such a stair, but it is not an essential part of the construction.

Stairs may be further described by the number of turns made by the stairs, by the type of strings, by the landings or whether tapered steps are used in the construction.

Quarter-turn stair

This type of stair changes its direction either by a quarter-space landing or by tapered steps.

Half-turn stair

The direction is reversed in this stair either by a half-space landing, or two quarter-space landings, or a quarter-space landing and a quarter space of tapered steps, or completely by tapered steps. Newel stairs are confined to rectangular plans but geometrical stairs may be arranged upon rectangles, circles, ellipses, or polygons.

Sheet 71, figs. 1, 2 and 3 are line diagrams of the three types of stairs which will be dealt with here.

The three usual methods of constructing the steps are shown in Sheet 71, figs. 4, 5 and 6.

In Sheet 71.4 the bottom edges of the risers are tongued into the treads, the top edges are square edged and pocket screwed to the bottom side of the tread with the scotia moulding housed as shown. The steps are housed and wedged into the stringboards with glue blocks rubbed between them and the strings. The nosing line and margin are also indicated.

Sheet 71.5 is similar to the above except that the tread is tongued into the riser.

In Sheet 71.6 the riser is tongued into the two treads both at the top and bottom. No scotia is used in this construction.

An alternative nosing detail is shown in Sheet 71.7.

Sheet 71.8 shows the section through the lower steps in a stair with the jointing of the handrail and stringboard to the newel post shown in dotted lines. The newels are usually square, 100 mm × 100 mm being the normal size. Two oblique haunched tenons are formed in the centre of the string and mortised into the newel, the joint is then secured by draw-boring, using a hardwood dowel at each tenon. Barefaced tenons should not be used at this point as the strength of the stair depends a good deal upon the rigidity of the newels and the method of jointing the strings to them. The handrail is housed, tenoned, and dowelled by draw-boring to the newel. which is finished at the top with a solid moulded cap.

Dogleg stair

The details of this type of stair are shown in Sheet 72.9, so called because of its appearance in the sectional elevation. It is used when the going is restricted and the space available is equal to the combined widths of the two flights. Where the width of a flight is 900 mm or more it is desirable to use an intermediate support in the form of a 100 mm × 75 mm bearer or carriage. This is placed in the centre of the stair, nailed at the foot and its upper end birds-mouthed and nailed to a 100 mm × 75 mm trimmer which is tenoned to the newel at one end and carried by the wall at the other. Rough brackets, shaped as shown, are nailed to the carriage with their upper edges cut square to fit tightly under the treads for further support; these brackets are fixed on alternate sides of the carriage and are usually short ends of floorboard.

The two outer strings are secured to 100 mm × 100 mm newels placed at the foot and the head of each flight, with the centre newel fixed to the landing trimmer and continued to the floor. The top newel is notched and fixed to the trimmer on the upper floor.

The half-space landing is constructed of 100 mm × 50 mm joists housed at one end to a 100 mm × 75 mm trimmer and supported by the wall at the

TYPES OF STAIRS WITH STEP DETAILS

TYPES OF STAIR

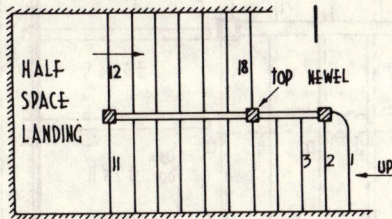

HALF SPACE LANDING

12   18   TOP NEWEL

11   3 2 1   ← UP

a) PLAN OF DOGLEG STAIR

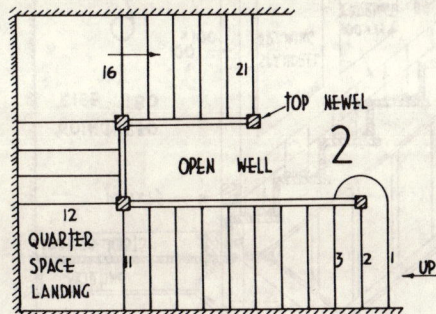

16   21   TOP NEWEL

OPEN WELL

12   11   3 2 1   UP

QUARTER SPACE LANDING

b) PLAN OF OPEN NEWEL STAIR

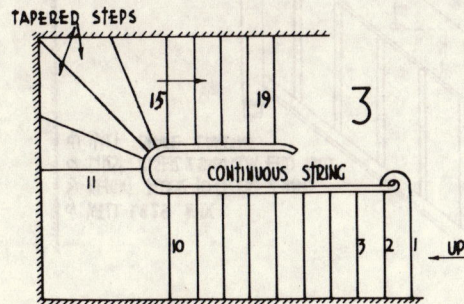

TAPERED STEPS

15   19

CONTINUOUS STRING

11   10   3 2 1   UP

c) PLAN OF GEOMETRICAL STAIR

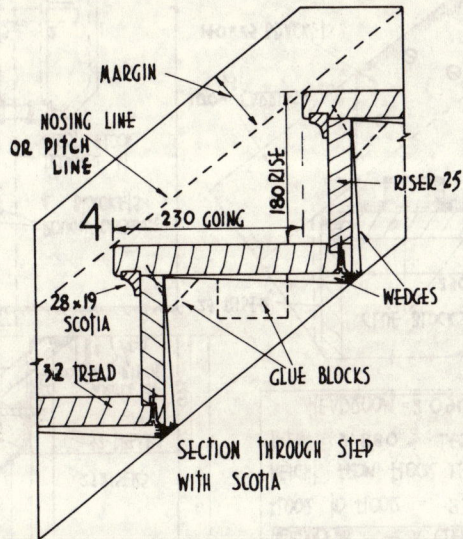

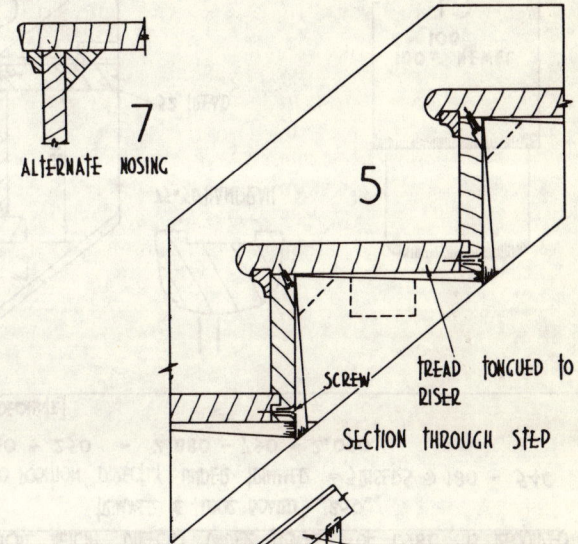

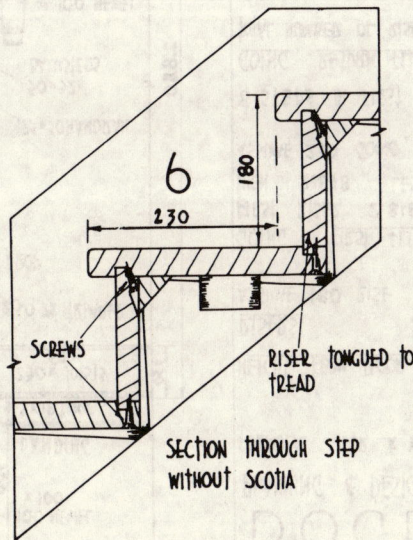

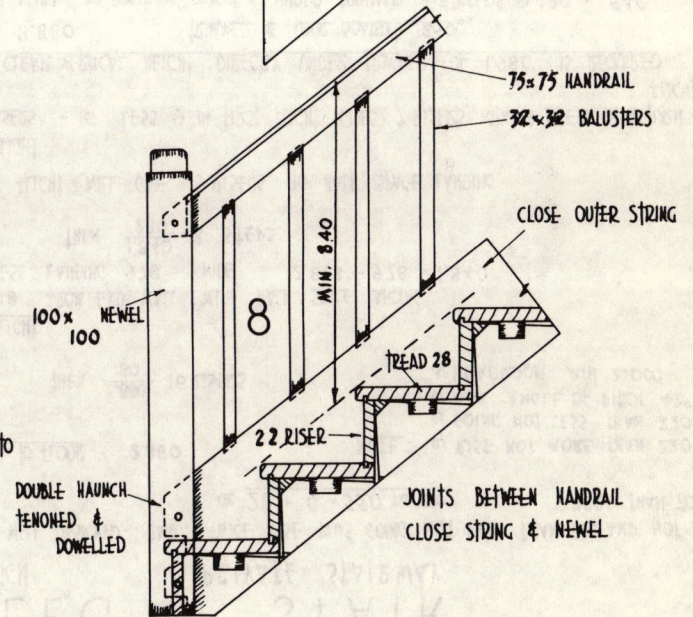

MARGIN

NOSING LINE OR PITCH LINE

180 RISE

230 GOING

RISER 25

28 x 19 SCOTIA

32 TREAD

WEDGES

GLUE BLOCKS

SECTION THROUGH STEP WITH SCOTIA

180

230

SCREWS

RISER TONGUED TO TREAD

SECTION THROUGH STEP WITHOUT SCOTIA

7

ALTERNATE NOSING

5

SCREW

TREAD TONGUED TO RISER

SECTION THROUGH STEP

75 x 75 HANDRAIL

32 x 32 BALUSTERS

CLOSE OUTER STRING

100 x 100 NEWEL

MIN. 840

TREAD 28

22 RISER

DOUBLE HAUNCH TENONED & DOWELLED

JOINTS BETWEEN HANDRAIL CLOSE STRING & NEWEL

SHEET 71

217

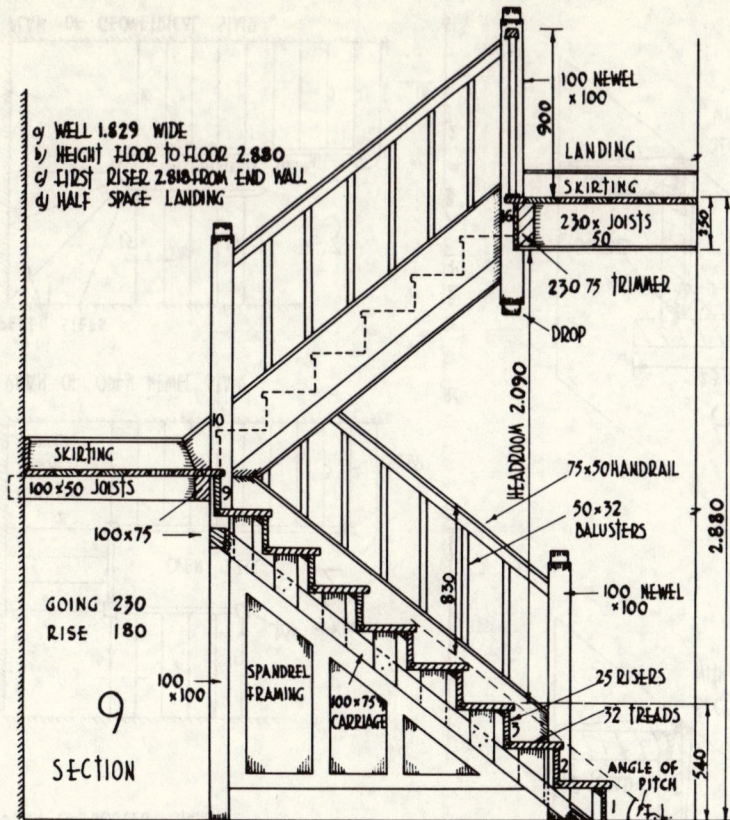

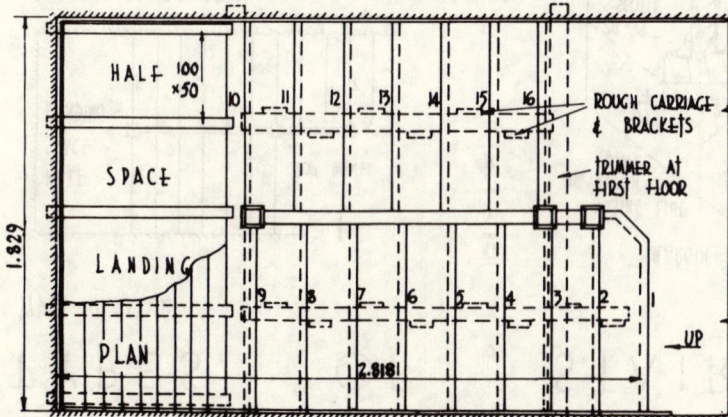

DOGLEG STAIR
PLANNING & DESIGN   PRIVATE STAIRWAY

FORMULAE FOR A WELL DESIGNED STAIR :- TWICE RISE PLUS GOING NOT LESS THAN 550 AND NOT
OR $2R + G = 550$ to 700   MORE THAN 700

HEIGHT FROM FLOOR TO FLOOR - 2.880

RISERS

ASSUME 180 RISE   THEN $\frac{2.880}{180} = 16$ RISERS

NOTE : a) RISE NOT MORE THAN 220
b) GOING NOT LESS THAN 220
c) MAX ANGLE OF PITCH 42°
d) HEADROOM MIN. 2.000

GOING FIRST FLIGHT
FIRST RISER 2.818 FROM END WALL WITH HALF SPACE LANDING
THEN 2.818 LESS LANDING 978 WIDE = 2.818 - 978 = 1.840
ASSUME 230 GOING, THEN $\frac{1.840}{230} = 8$ STEPS

8 STEPS IN FIRST FLIGHT WILL GIVE 9 RISERS TO HALF SPACE LANDING.

GOING RETURN FLIGHT
TOTAL NUMBER OF RISERS = 16 LESS 9 IN FIRST FLIGHT GIVES 7 RISERS WITH 6 STEPS IN RETURN FLIGHT.

HEADROOM — A CLEAR VERTICAL HEIGHT DIRECTLY UNDER TRIMMER OF 1.980 IS REQUIRED
FLOOR TO FLOOR = 2.880   TRIMMER & FLOOR BOARD = 250
HEIGHT FROM FLOOR LEVEL TO POSITION DIRECTLY UNDER TRIMMER = 3 RISERS @ 180 = 540
THEN 2.880 LESS 540 + 250 = 2.880 - 790 = 2.090

HEADROOM = 2.090 [ADEQUATE]

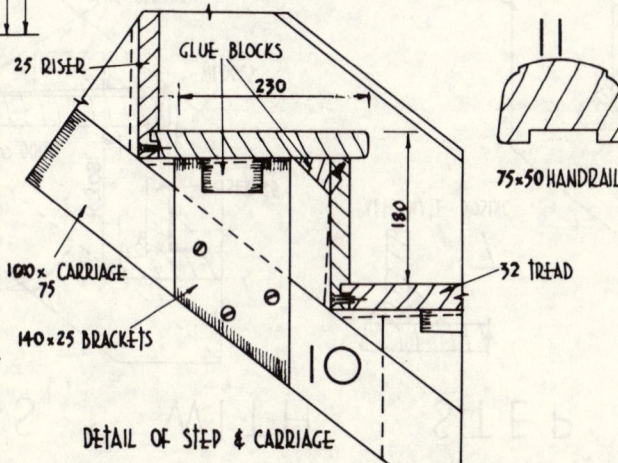

a) WELL 1.829 WIDE
b) HEIGHT FLOOR TO FLOOR 2.880
c) FIRST RISER 2.818 FROM END WALL
d) HALF SPACE LANDING

100 NEWEL × 100

900

LANDING

SKIRTING

230 × JOISTS 50

230 75 TRIMMER

DROP

HEADROOM 2.090

2.880

SKIRTING

100 × 50 JOISTS

100 × 75

GOING 230
RISE 180

100 × 100

9

SECTION

SPANDREL FRAMING

100 × 75 CARRIAGE

75 × 50 HANDRAIL

50 × 32 BALUSTERS

830

100 NEWEL × 100

25 RISERS

32 TREADS

ANGLE OF PITCH

540

HALF 100 × 50

SPACE

LANDING

1.829

PLAN

10   11   12   13   14   15   16

9   8   7   6   5   4   3   2   1

ROUGH CARRIAGE & BRACKETS

TRIMMER AT FIRST FLOOR

2.818

UP

25 RISER

GLUE BLOCKS

230

100 × CARRIAGE 75

140 × 25 BRACKETS

180

32 TREAD

10

DETAIL OF STEP & CARRIAGE

11

75 × 50 HANDRAIL

100 × NEWEL 100

12

SHEET 72

OPEN NEWEL STAIR
COMMON STAIRWAY

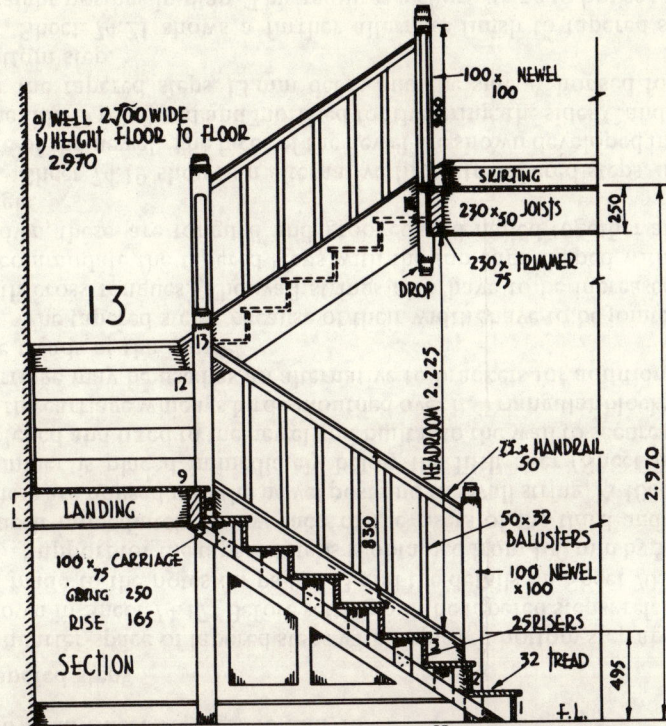

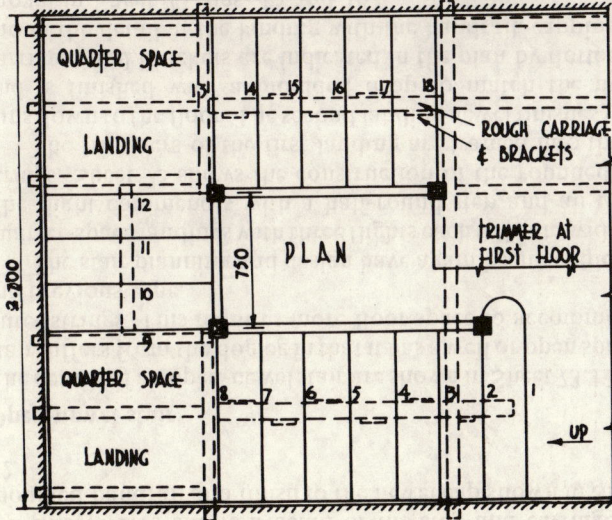

PLANNING & DESIGN

FORMULAE – 2R + G = 550 – 700

HEIGHT FROM FLOOR TO FLOOR – 2.970

RISERS

ASSUME 175 RISE   THEN $\frac{2.970}{175} = 16\frac{34}{35}$ RISERS.   TRY 165 RISE THEN $\frac{2.970}{165} = 18$ RISERS

18 RISERS AT 165

GOING

WELL 750 WIDE
ASSUME 250 GOING   THEN $\frac{750}{250} = 3$ STEPS IN WELL

NUMBER OF STEPS IN WELL = 3 WHICH GIVES 4 RISERS   ∴ 18 RISERS – 4 = 14 RISERS IN 2 FLIGHTS

ASSUME 8 RISERS IN FIRST FLIGHT WITH 4 RISERS IN WELL & 6 RISERS IN SECOND FLIGHT

HEADROOM   FLOOR TO FLOOR = 2.970   TRIMMER & FLOOR BOARDS = 250

HEIGHT FROM FLOOR LEVEL TO POSITION DIRECTLY UNDER TRIMMER = 3 RISERS @ 165 = 495

THEN 2.970 LESS 495 + 250 = 2.970 – 745 = 2.225

HEADROOM = 2.225 [ADEQUATE]

NOTE: a) RISE NOT MORE THAN 190
b) GOING NOT LESS THAN 230
c) MAX. ANGLE OF PITCH 38°
d) HEADROOM MIN. 2.000

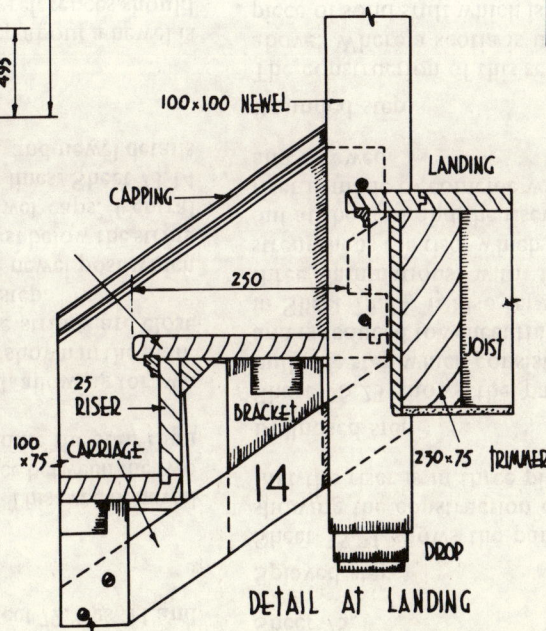

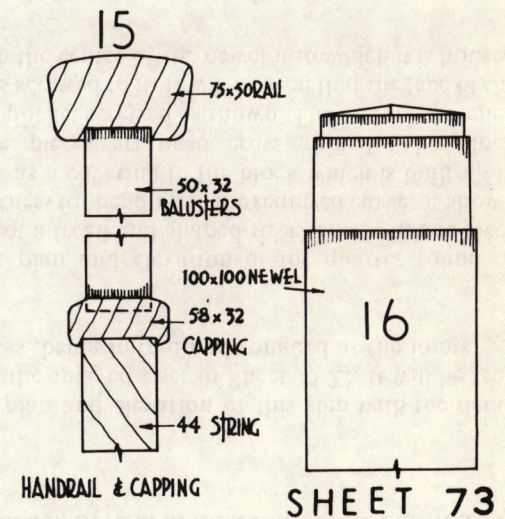

SECTION (13)

a) WELL 2,700 WIDE
b) HEIGHT FLOOR TO FLOOR 2.970

100 x 100 NEWEL
SKIRTING
230 x 50 JOISTS
230 x 75 TRIMMER
DROP
75 x 50 HANDRAIL
50 x 32 BALUSTERS
100 x 100 NEWEL
25 RISERS
32 TREAD
100 x 75 CARRIAGE
GOING 250
RISE 165
LANDING
HEADROOM 2.225
2.970
830
495
F.L.

PLAN

QUARTER SPACE
LANDING
ROUGH CARRIAGE & BRACKETS
TRIMMER AT FIRST FLOOR
QUARTER SPACE
LANDING
750
2.100
UP

DETAIL AT LANDING (14)

100 x 100 NEWEL
CAPPING
LANDING
250
JOIST
25 RISER
BRACKET
100 x 75 CARRIAGE
230 x 75 TRIMMER
DROP

15
75 x 50 RAIL
50 x 32 BALUSTERS
100 x 100 NEWEL
58 x 32 CAPPING
44 STRING
HANDRAIL & CAPPING

16

SHEET 73

219

other. The trimmer which spans the opening is carried by the outer walls. All the strings are close strings and the flight commences with a splayed step. The construction and details of this step are shown on Sheet 75.

Sheet 72.10 shows a detail of the step and carriage supports with the moulded handrail and finish to the newel cap shown in Sheet 72, figs. 11 and 12.

Open-newel stair

The details of an open-newel stair are shown in Sheet 73.13. This type of newel stair differs from the dogleg in that it has a well or open space between the two outer strings. This requires more floor space to accommodate the stair than the previous type.

The stair planning and design have again been detailed, allowing for two quarter-space landings with three flights each 975 mm wide shown in the plan. The flight commences with a half-round step and all the strings are close strings. Sheet 75 shows the construction of the rounded step.

The trimmers of the first landing are framed into the newel post which runs down to the floor. The second landing newel finishes just below the string and is finished with a moulded drop to match the newel caps. Central carriages and brackets are indicated in the plan by dotted lines. Sheet 73.14 shows the detail at the landing with the handrail, capping, and newel details shown in Sheet 73, figs. 15 and 16.

Tapered steps

A quarter-space of tapered steps with a splayed bottom step about a newel is shown in Sheet 74.17. Before setting out the tapered steps, references should be made to the notes on page 213 and the details on Sheet 70.12 and 70.13.

Support for the tapered steps is obtained from 100 mm by 50 mm bearers placed immediately below each of the risers of the third and fourth steps. These are housed into the newel post and the wall string. A 100 mm × 75 mm trimmer is placed immediately below the fifth riser (Sheet 74.18). This is notched and fixed to the newel and built into the wall to secure the lower end of the carriage which is birds-mouthed over it. Triangular blocks nailed to the carriage may be used as an alternative to brackets for additional support to the treads at the centre.

The tapered steps, because of their width, have to be jointed, preferably with cross tongues. The wall strings also have to be increased in width to accommodate the tapered steps with the top edge shaped with an easing as shown, these are tongued and grooved and nailed together at the internal angle.

Sheet 74.19 shows an alternative finish to tapered steps, increasing the size of the newel. The faces of the newel are shown developed in Sheet 74.20. The side D is housed and mortised for the string; the sides C and B are housed for the tapered steps 13 mm deep; and the side A housed for the shaped bottom step.

Sheet 74.21 shows a further alternate finish to tapered steps keeping straight nosings in plan. This requires a short string to house the ends of the steps which fall outside the newel when meeting the conditions of the regulations on angle of taper and minimum going at the narrowest point of the tapered step.

Step construction

The bottom step of the lowest flight in a staircase is often specially shaped and made to project beyond the newel. Several of these finishes are illustrated in Sheet 75.

Splayed step

Sheet 75.21 shows the part plan and elevation of this step with the detail showing the construction of the splayed riser in Sheet 75.22. It will be seen that the riser is in three pieces, being mitred and tongued at the joints.

Bullnosed step

Sheet 75.23 shows the part plan and elevation of the quarter round or bullnose step which consists of a tread and shaped riser. This is in one piece and is reduced to veneer thickness for bending to the required curve, as shown in Sheet 75.24. It also serves as a covering to the block which is built up in three laminations with the pieces arranged cross-grained. The block strengthens the riser which is pocket screwed as shown. The newel is notched out at the foot and the riser is screwed to it. It will be seen that the face of the riser is made to coincide with the centre of the newel into which it is housed and screwed.

Rounded step

The construction of this semicircular-ended step is similar to that described above. Where a scotia is used the curve of the scotia mould is worked on a piece of solid stuff which is fixed between the tread and the riser. Sheet 75.25 should make the construction clear. The newel prepared to receive the rounded step is also shown.

Geometrical stairs

Sheet 76.26 shows the plan and sectional elevation of a geometrical stair having a cut and bracketed string with a half-space landing.

Geometrical stairs may be constructed with continuous cut and mitred strings or cut and bracketed strings. Both types are shown and will be described.

A cut and mitred string

A cut and mitred string suitable for the above stairs is shown in different stages of construction in Sheet 76.27. At the upper end one tread and riser is shown finished with the balusters and return nosing in position. The two other steps show the preparation of the string to receive the riser. Here, the return nosing is removed to show the fixing of the balusters which are dovetailed and screwed to the tread. The return nosing and scotia similar to Sheet 76.28 are cut to mitre with the tread nosing in front and mitred to a short return piece of itself at the back. The nosing pieces are fixed either by slot-screwing as shown, or by tongue and groove.

A cut and bracketed string

This is shown in the sketch (Sheet 76.29) and a plan and section in Sheet 76.30. These show the difference in treatment required when ornamental brackets

QUARTER TURN OF TAPERED STEPS

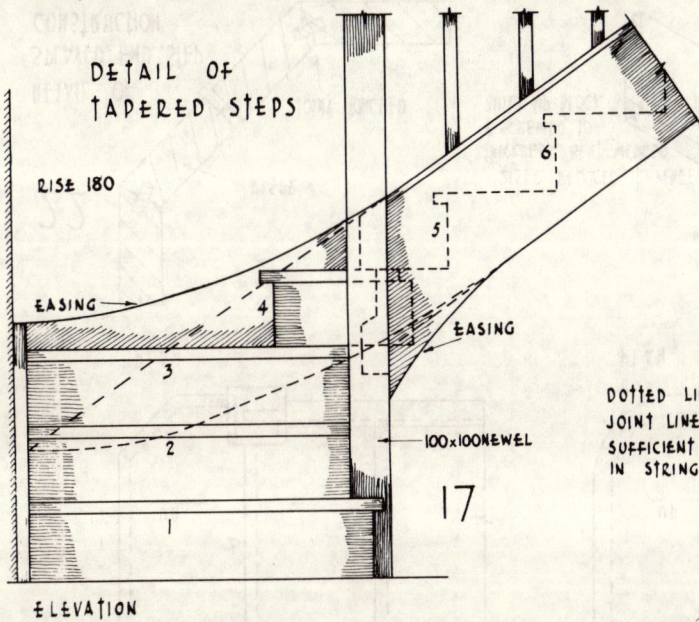

DETAIL OF
TAPERED STEPS

RISE 180

EASING

100 x 100 NEWEL

EASING

4

3

2

1

5

6

ELEVATION

17

SKETCH OF TAPERED
STEPS FOR
QUARTER TURN

WALL STRING

18

EASING

EASING

DOTTED LINES INDICATE
JOINT LINES PROVIDING
SUFFICIENT WIDTH OF MATERIAL
IN STRINGS & TREADS

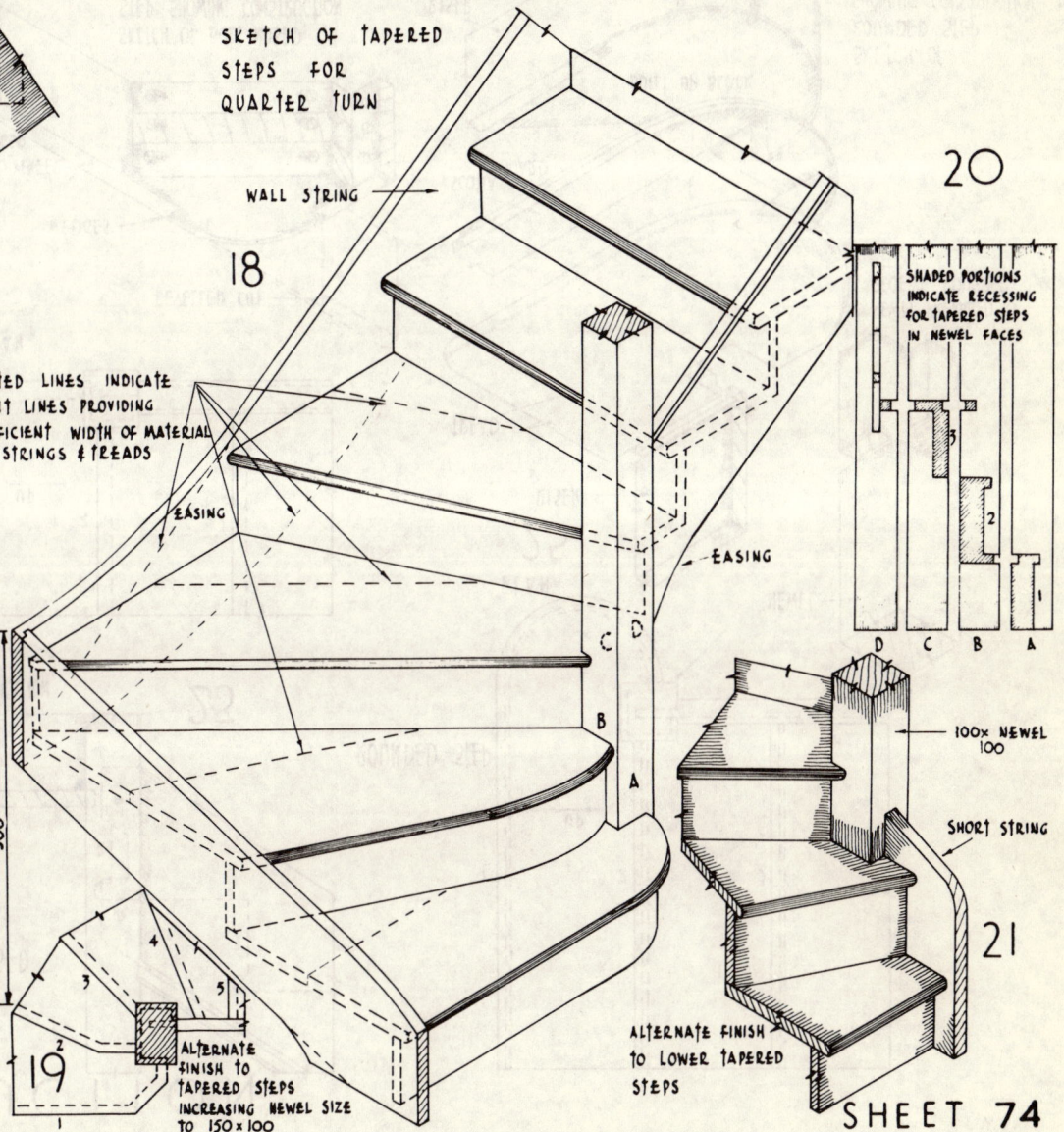

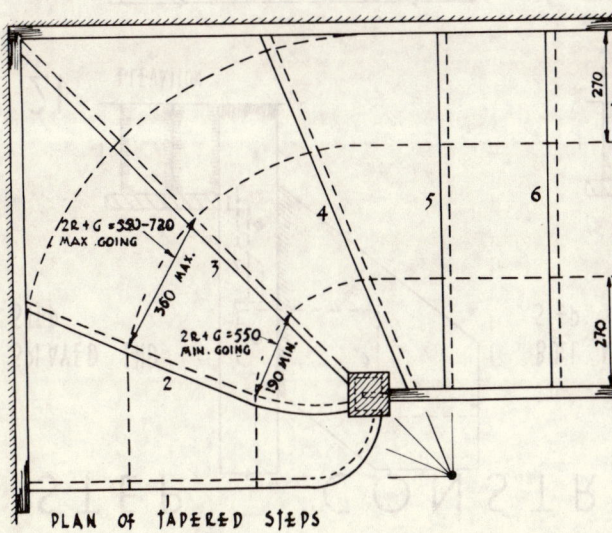

A

B

C

D

20

SHADED PORTIONS
INDICATE RECESSING
FOR TAPERED STEPS
IN NEWEL FACES

D   C   B   A

PLAN OF TAPERED STEPS

270

906

270

2R + G = 590-720
MAX. GOING

360 MAX.

2R + G = 550
MIN. GOING

190 MIN.

2

3

4

5

6

19

ALTERNATE
FINISH TO
TAPERED STEPS
INCREASING NEWEL SIZE
TO 150 x 100

2

3

4

5

1

100 x
100 NEWEL

SHORT STRING

21

ALTERNATE FINISH
TO LOWER TAPERED
STEPS

SHEET 74

221

STEP CONSTRUCTION

SPLAYED END STEP

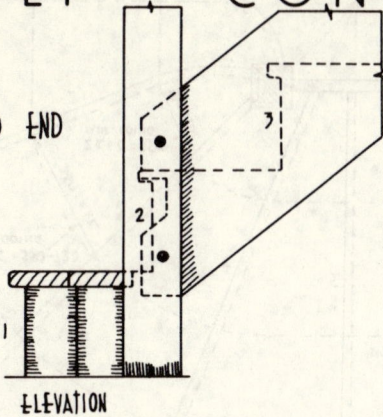

21

ELEVATION

BULL NOSED STEP

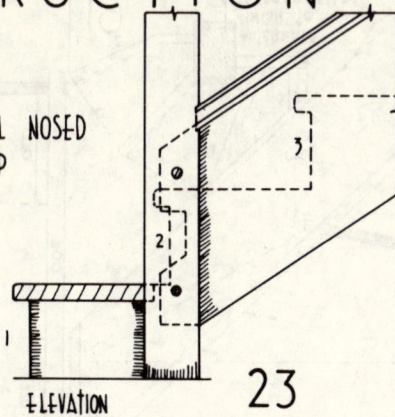

23

ELEVATION

ROUNDED STEP

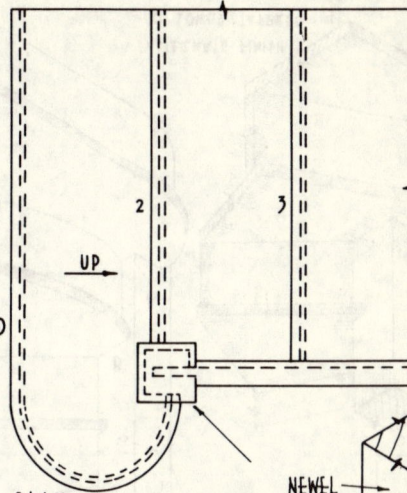

UP

PLAN

25

RISER

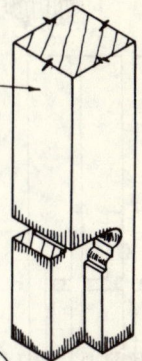

NEWEL

NEWEL PREPARED TO RECEIVE ROUNDED STEP

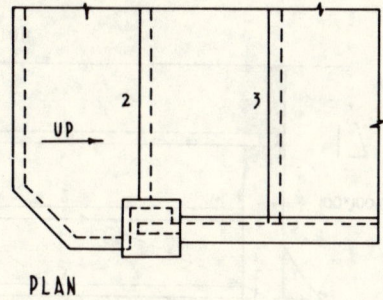

1   UP   2   3

PLAN

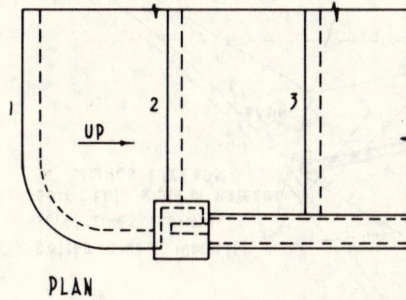

1   UP   2   3

PLAN

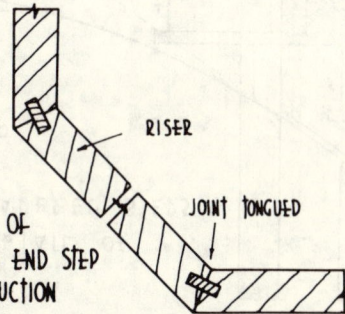

22

RISER

JOINT TONGUED

DETAIL OF SPLAYED END STEP CONSTRUCTION

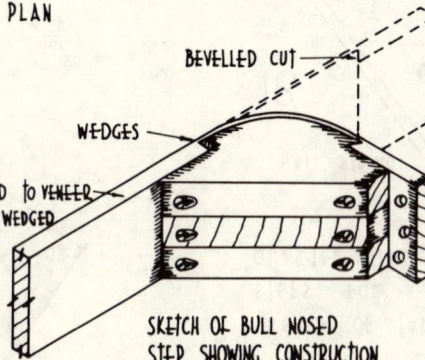

BEVELLED CUT

WEDGES

RISER REDUCED TO VENEER THICKNESS, BENT, WEDGED & SCREWED TO BUILT UP BLOCK

SKETCH OF BULL NOSED STEP SHOWING CONSTRUCTION

24

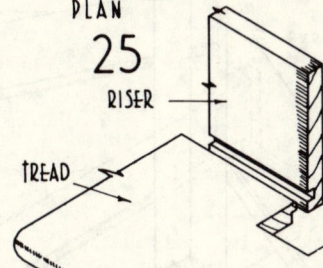

TREAD

SCOTIA BOARD

BUILT UP BLOCK

RISER

SKETCH OF ROUNDED STEP SHOWING CONSTRUCTION

SHEET 75

GEOMETRICAL STAIR
COMMON STAIRWAY

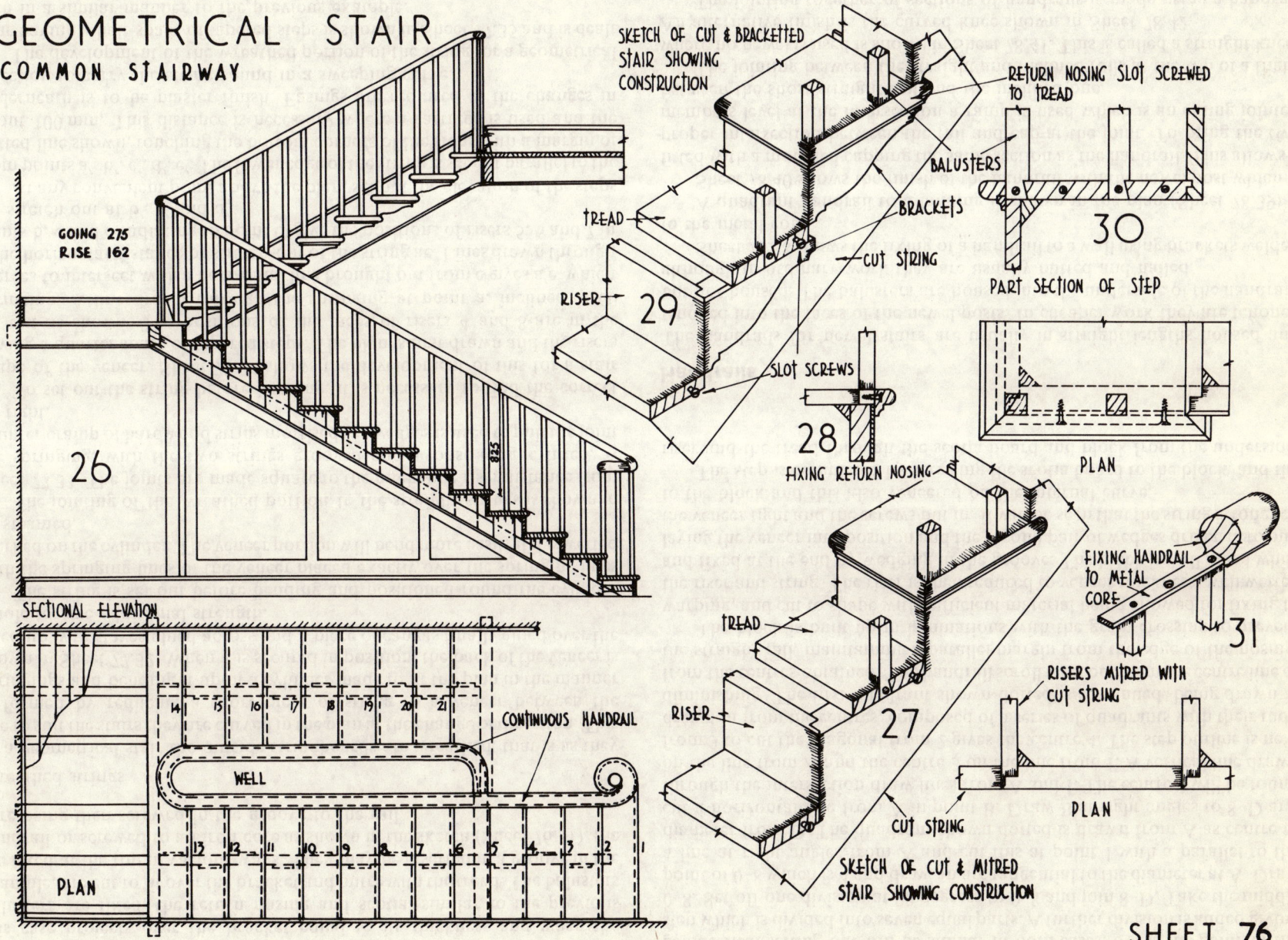

GOING 275
RISE 162

26

825

SECTIONAL ELEVATION

PLAN

14 15 16 17 18 19 20 21

CONTINUOUS HANDRAIL

WELL

13 12 11 10 9 8 7 6 5 4 3 2 1

SKETCH OF CUT & BRACKETTED
STAIR SHOWING
CONSTRUCTION

RETURN NOSING SLOT SCREWED
TO TREAD

BALUSTERS

BRACKETS

TREAD

30

CUT STRING

RISER

29

PART SECTION OF STEP

SLOT SCREWS

28

FIXING RETURN NOSING

PLAN

FIXING HANDRAIL
TO METAL
CORE

TREAD

31

RISERS MITRED WITH
CUT STRING

RISER

27

CUT STRING

PLAN

SKETCH OF CUT & MITRED
STAIR SHOWING CONSTRUCTION

SHEET 76

are planted on the face of the string under the return ends of the treads. The bracket, cut from plywood, is mitred with the end of the riser which runs over the face of the string as shown, and the string is rebated to receive the riser to which it is nailed before the brackets are fixed. It will be seen that the tread in this case projects over the bracket equal to its thickness and when the balusters are fixed, the return nosing and scotia, similar to the previous example, are cut to fit over the bracket and mitre with the tread. The balusters are fixed at the top in various ways. They may be inserted in a groove in the handrail or screwed to an iron core as shown in the sketch (Sheet 76.31), the core being then screwed in the groove to the rail.

Wreathed strings

In a geometrical stair the outside stringboards are wreathed, that is as they rise to suit the stairs they are curved in the plan at the change of direction. This is formed by reducing a short piece of string to a veneer between the springings and bending it upon a cylinder made to fit the plan in the manner shown in Sheet 77.32. When it is secured in position, the back of the veneer is filled up with staves glued across and a piece of canvas finally glued over the whole to give additional strength.

The string is set out before bending and positioned round the cylinder with the springing lines of the veneer placed exactly over the springing lines marked on the cylinder. The veneer portion will bend more easily if it is wetted or steamed.

The jointing of the wreathed portion to the straight strings is shown in Sheet 77.33. The joints are made square to the pitch, one riser distance from the springing, with the two strings grooved and a loose tongue fitted. A counter cramp of hardwood strips mortised for wedges is used to pull the joint up tight.

To set out the string before bending, it is necessary to find the correct shape of the veneer. Sheet 77.34 shows the development of this for a stair having a quarter space of tapered steps. The plan is first drawn and the risers set out as shown. The positions of the faces of risers 4 and 8 are in the springing. A line drawn through the springing at point a, inclined at 60 degrees to intersect with a horizontal line brought out from e gives a'e' which is the horizontal distance or stretch-out of the string ae. Lines drawn through points b, c and d radiating to point O give the positions of risers 5, 6 and 7 in the stretch out at b', c' and d'.

At any convenient point above the plan, set out the elevation of the steps from points a', b', c', d', e'. The lower edge of the string is set out parallel to the dotted line shown, touching the bottom corners of the steps with a margin of about 100 mm. This distance is necessary where a carriage is used and the underneath is to be plaster finish. Easings are required at the changes in direction to carry the string round in a sweeping curve.

The development of the wreathed portion of the string for a geometrical stair having a half-space of tapered steps is shown in Sheet 77.35 and is dealt with in a similar manner to the previous example.

Sheet 78.36 shows a sketch of the drum with the wreathed portion of the string previously developed in Sheet 77.34, prepared ready for jointing to the straight strings.

Curtail step

The details of the shaped end of the step are shown in Sheet 78.37. As the outline of the step follows the outline of the handrail immediately above it, the geometrical setting-out will be similar in both cases. 0–7 is the going of the step which is divided into seven equal parts. A further division is added giving 0–8. Set off one division at 90 degrees to 0–8 and join 8–D. Take the middle point of 0–8 which is 4 and draw an arc tangential to the diameter at A. Draw a line at right angles from A and cut this at point 1 with a parallel to the diameter from D. The quadrant shown dotted is drawn from A as centre to cut a horizontal line from A in point B. Draw B at right angles to 8–D and through the intersection draw lines from A and 1. The centre 2 will be found on the line from A and the centre 3 on the line from 1. A vertical line drawn from 3 to cut the diagonal from 2 gives the centre 4. The step outline is next drawn in from the centres, composed of a series of quadrants with their radii diminishing. The first quadrant shown dotted is continued, being drawn in from the centres obtained. The handrail scroll is set out from the centre line of the straight rail, maintaining a parallel margin from the edge of the nosing.

The block is built up in laminations with the grain crossing to prevent warping, and cut to shape with sufficient material being allowed for fixing to the riser and string. The riser is then reduced to veneer thickness, well wetted and fixed at the end by wedging in the groove. The block is well glued while laying the veneer into position and the second pair of wedges driven in to pull the veneer tight and the screws put in. It will be seen that the string is tongued to the block and this also veneered on the internal curve.

The step is completed by screwing the scotia board to the block, and the riser and the tread through the scotia board and block from the underside.

Handrails

The handrails for newel stairs are usually in straight lengths housed and tenoned into the faces of the newel posts. In cheaper work they are tenoned but not housed. The balusters are housed into the underside of the handrail, although in ordinary work they are usually butted and nailed.

Sheet 78.38 shows the fixing of a handrail to a wall using brackets welded to the metal core.

A quadrant handrail to a landing is shown in the plan (Sheet 78.39).

Sheet 78.40 shows the finish of the handrail with the newel post which is fitted with a moulded capping the same section as the handrail. This allows a proper intersection between the rail and cap at the joint. To bring the two members level at the intersection a ramp is used which is an easing jointed between the short straight rail and the inclined one.

The jointing between the straight and inclined rails at the top of a flight where no newel is used is shown in Sheet 78.41. This is called a straight knee. An alternative finish is the curved knee shown in Sheet 78.42.

The jointing together of sections of handrails is made using a handrail bolt and dowels shown in Sheet 78.43.

WREATHED STRINGS

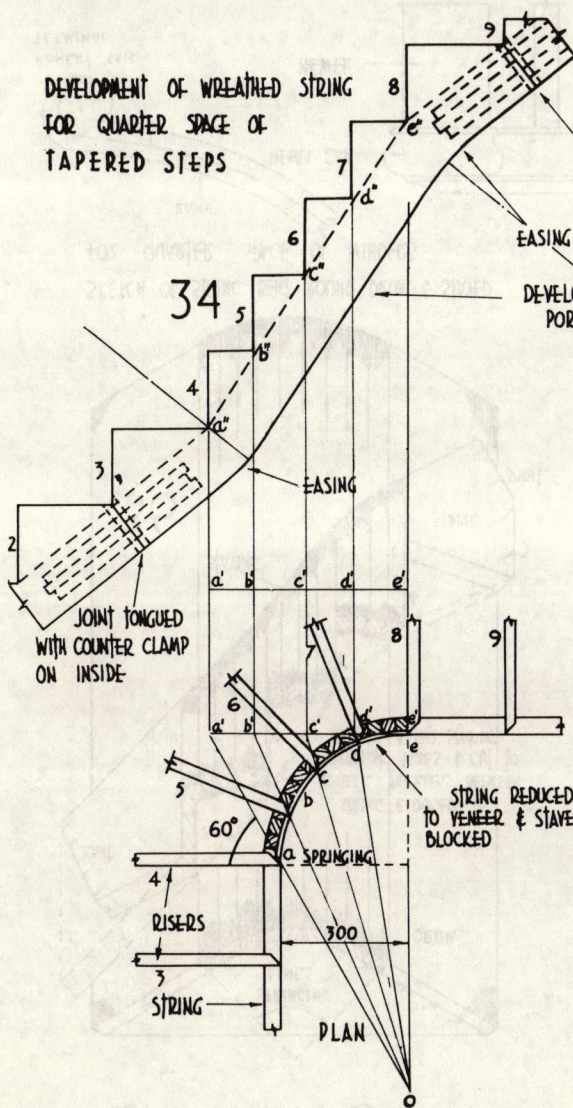

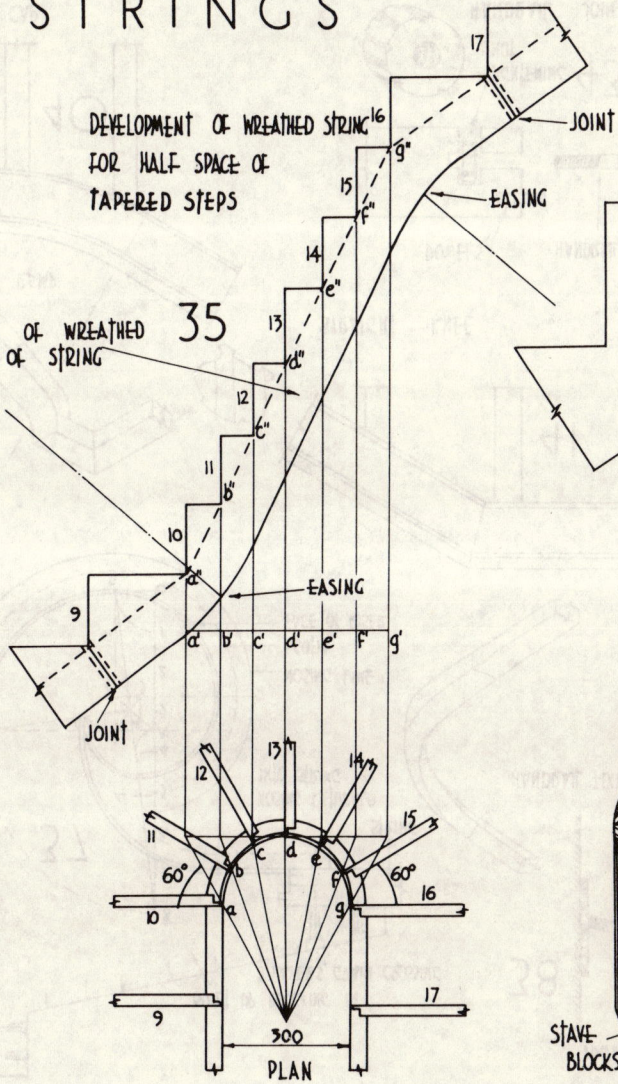

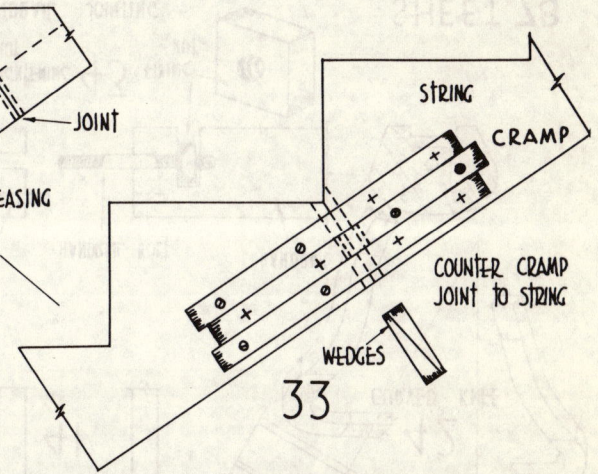

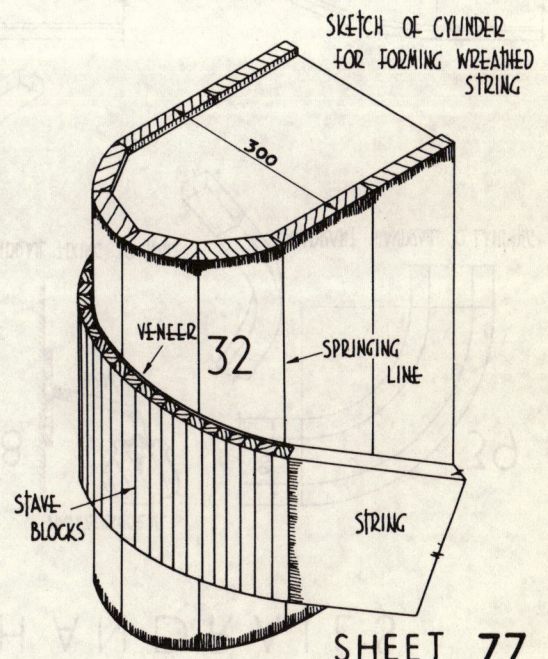

34

DEVELOPMENT OF WREATHED STRING FOR QUARTER SPACE OF TAPERED STEPS

JOINT

EASING

DEVELOPMENT OF WREATHED PORTION OF STRING

EASING

JOINT TONGUED WITH COUNTER CLAMP ON INSIDE

STRING REDUCED TO VENEER & STAVE BLOCKED

SPRINGING

RISERS

STRING

300

60°

PLAN

35

DEVELOPMENT OF WREATHED STRING FOR HALF SPACE OF TAPERED STEPS

JOINT

EASING

EASING

JOINT

300

60°   60°

PLAN

33

STRING

CRAMP

COUNTER CRAMP JOINT TO STRING

WEDGES

SKETCH OF CYLINDER FOR FORMING WREATHED STRING

32

300

VENEER

SPRINGING LINE

STAVE BLOCKS

STRING

SHEET **77**

225

STRINGS   STEPS   &   HANDRAILS

SCROLL END STEP

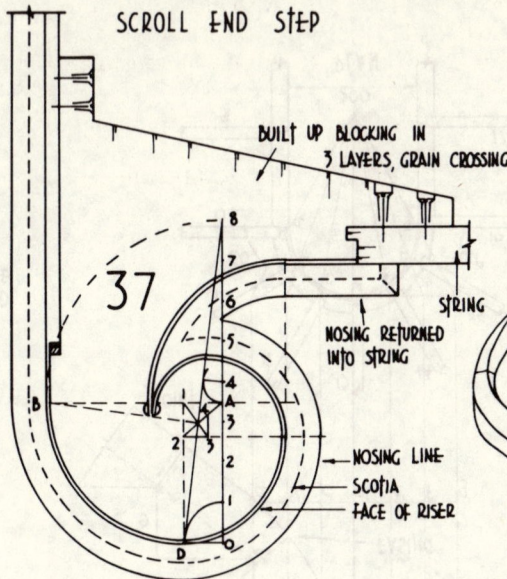

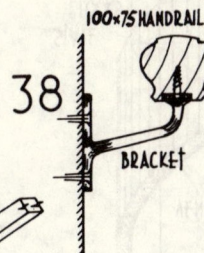

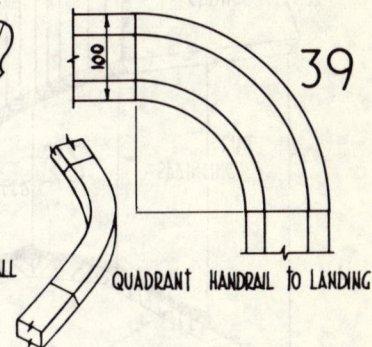

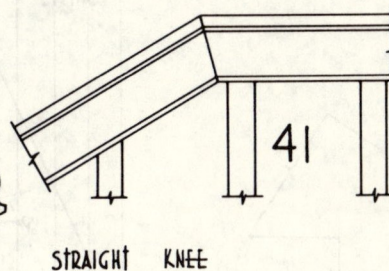

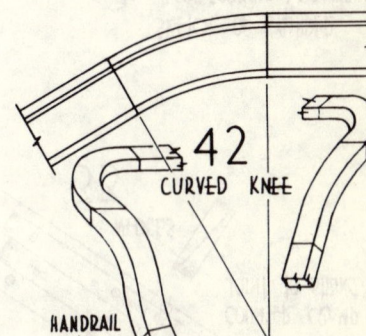

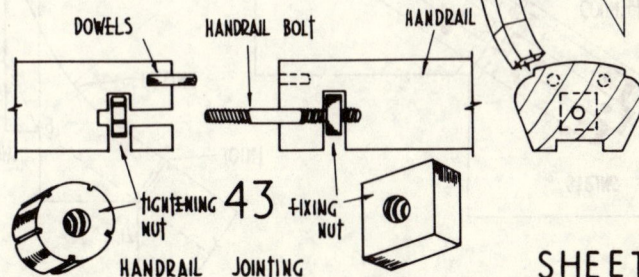

SPRINGING LINES

DRUM

JOINT

STRING REDUCED TO VENEER THICKNESS BETWEEN SPRINGING LINES & CUT TO PROFILE AFTER STAVING

STAVING

STRING

JOINT

36

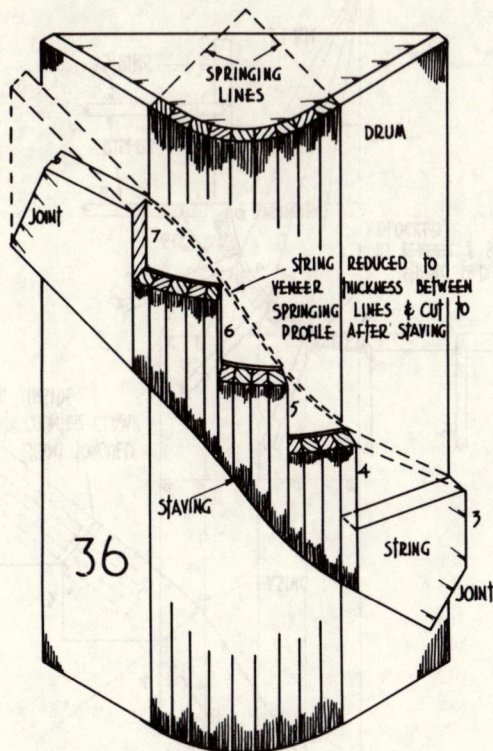

SKETCH OF STRING BENT ROUND DRUM & STAVED FOR QUARTER SPACE OF WINDERS

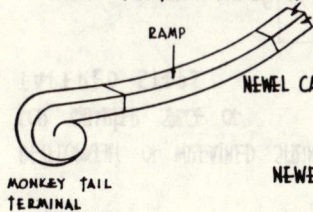

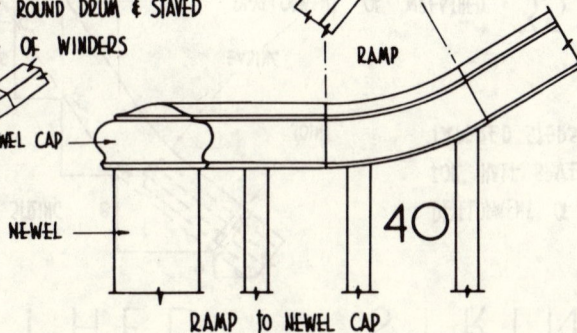

BUILT UP BLOCKING IN 3 LAYERS, GRAIN CROSSING

STRING

NOSING RETURNED INTO STRING

37

NOSING LINE
SCOTIA
FACE OF RISER

100×75 HANDRAIL

38

BRACKET

HANDRAIL FIXING TO WALL

100

39

QUADRANT HANDRAIL TO LANDING

41

STRAIGHT KNEE

42

CURVED KNEE

RAMP

RAMP

NEWEL CAP

NEWEL

40

RAMP TO NEWEL CAP

MONKEY TAIL TERMINAL

DOWELS

HANDRAIL BOLT

HANDRAIL

TIGHTENING NUT

43

FIXING NUT

HANDRAIL JOINTING

SHEET **78**

Scrolls

In geometrical stairs the handrail usually terminates in the form of a scroll. These scrolls are either vertical, when they are sometimes known as 'monkey's tails', or horizontal as shown at the foot of the stairs in Sheet 76.26. As has already been stated, scrolls are usually composed of a series of quadrants with their radii diminishing and to set these out, various methods are used.

In Sheet 79, figs. 44 and 45 the scrolls are set out in a similar way, these have been detailed for comparison of the finished outline. In Sheet 79.44 the overall width of the scroll is divided into nine equal parts, whereas in Sheet 79.45, eight divisions are used. The conclusion between the two examples is that if the angle 90b is altered by dividing 09 into more or less parts than eight, and the same setting out used, the volute approaches the centre at quicker or slower rate, according to the proportion of 9b to 09 in the figure. If 9b is less than one-eighth the diameter, the approach is slower, and vice versa.

Sheet 79.46 shows a further method of setting out a scroll where various proportions are adopted to determine the centres. C is the centre of the step, distances C–2 equal to 12 mm, 2–1 equal to 25 mm, and 1–5 equal to 12 mm. The centres 3 and 4 lie on diagonals drawn from 1 and 2 as shown.

Sheet 79.47 shows three sections of moulded handrails with a built-up handrail section shown in Sheet 79.48. A softwood core is faced with moulded top, sides, and bottom in hardwood tongued and grooved together as shown. As with the built-up newel (Sheet 79.49) this method conserves a certain amount of hardwood, and in work where newels larger than 100 mm square are required this method is the best on account of the possibility of splitting and warping in large solid members. Alternative face finishings and corner jointing to the newel are shown.

The jointing between the built-up newel and rail is by mortising the newel core to receive the tenon worked on the core of the handrail. Similar jointing is used between the string and the newel.

Open-tread stairs

In modern buildings such as shops, multi-stores, offices, and showrooms, stairs are much more contemporary in design than those which have been dealt with so far. The whole treatment is simple and effective, where elaborately moulded nosings, cappings, newels, and handrails are avoided. In many cases the closed strings, employed in newel stairs to receive the steps, give way to an open design of stair where risers, too, are often left out with the support provided in various other ways.

Examples of this modern treatment are to be found where a combination of materials is used, i.e. concrete stair with timber finishings; open timber treads with steel stringer supports; open timber treads with laminated beam supports, metal balusters, and either timber or extruded plastic handrails.

In most cases the work of the joiner is much simpler than hitherto, although the planning is the same for whatever form the stair may take. Various finishings to balustrades include timber panelling, open metal balusters, etched plate glass, wrought iron, aluminium alloy castings, and woven ribbon wirework.

The stairs in large stores and showrooms are usually wide to allow people to pass in both directions, these may be divided down the centre and up to 1·800 m or more in width, and for this reason the treads of hardwood are at least 38 mm thick. Iroko, agba, afrormosia, oak, teak, and meranti are suitable timbers for this type of work.

A suitable finish to a concrete stair is shown in Sheet 80.50. The outer faces of the stair are faced with a hardwood string cut to the shape of the steps and left projecting 19 mm. This is fixed by screwing through the face and pelleting. Non-slip stairtreads are fitted between the strings and screwed to the concrete treads, the riser and the space between the back of the stairtread and the riser filled in with linoleum or cork, using impact adhesive.

The stair shown in Sheet 80.51 has laminated stringers cut to receive the wide 38 mm thick treads and inclined risers which are fixed by screwing and pelleting to the stringers. It will be seen in the part elevation of the same figure that the risers finish flush to the stringer with the treads projecting some 100 mm with the corners rounded. The bases of the metal balusters are housed and fixed to the treads as shown.

Sheet 80.52 shows a portion of a stair having steel channel stringers with support for the treads obtained from 50 mm by 6 mm mild steel straps. These are welded to the channel and housed and fixed to the underside of the treads.

Sheet 80.53 shows a stair with a laminated stringer, the tread supports, which are from 200 mm × 50 mm hardwood, are formed with the lamination of the stringer.

Sheet 80.54 shows a part section of a stair with steel channel stringers and welded steel angle brackets, faced with timber treads and risers. The step detail (Sheet 80.55) shows the hardwood riser tongued into the treads on both edges with fixing by screwing through the angle bracket. Blocking pieces, coach screwed to the channel, provide for the fixing of the hardwood cut string which is screwed and pelleted from the face. The treads project over the strings with metal balusters fixed to them.

A further type of modern laminated stair is designed using shaped laminated beams, built up of 19 mm softwood lamina as central supports to laminated treads, fixed to mild steel brackets which are welded to a steel plate housed in the top edge of the central beams.

Sheet 81.1 and 81.2 shows the plan and elevation of a geometrical stair having curved strings. The setting out follows previous examples conforming to the regulations.

The finish at the lower steps shows curved risers finishing with a curtail step with a handrail scroll as shown in Sheet 81.3.

An alternate finish to the stair is shown in Sheet 81.4.

The development of the two curved strings to the stair is shown in Sheet 81.5 and 81.6.

There is a wide variety of designs of geometrical stairs and the development of a particular design depends upon the characteristics of the shapes chosen for the plan. These may be elliptical or part of an ellipse or circular in plan and are usually constructed without newel posts so that the wreathed handrail is continuous. Stairs of the more unusual design in the modern trend are now often called for. These include stairs with open treads and carriages, treads mounted on to a single carriage or string which may be either straight or curved.

SCROLLS, BUILT UP NEWELS & HANDRAILS

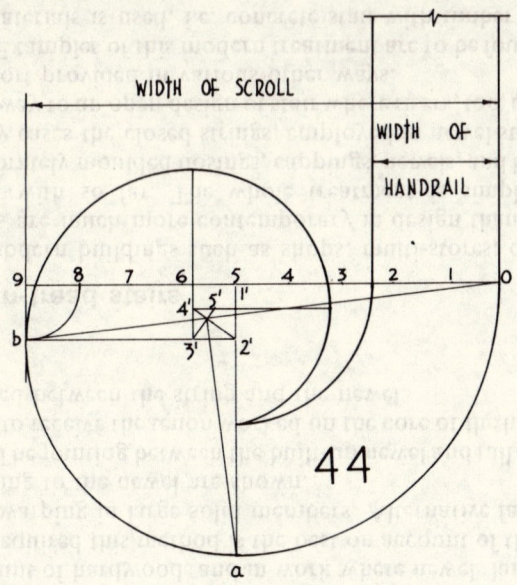

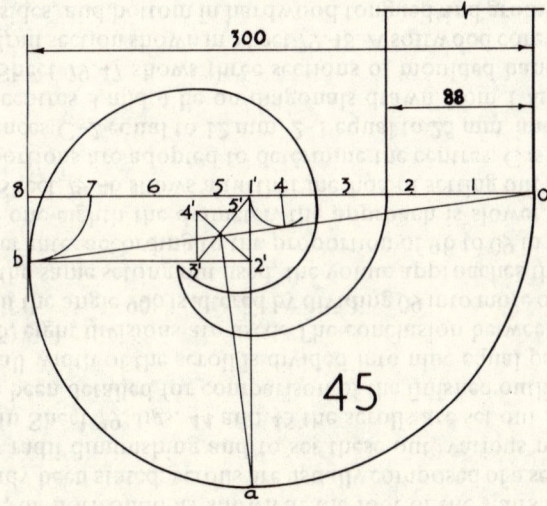

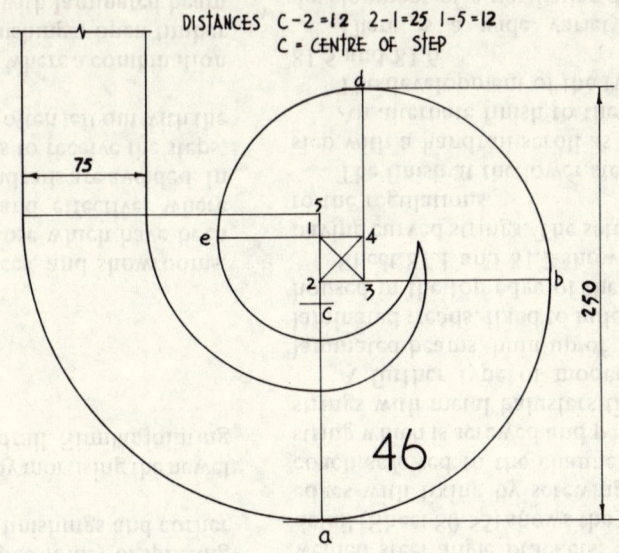

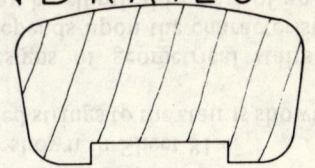

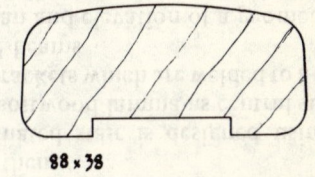

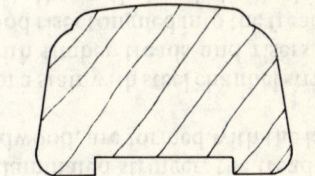

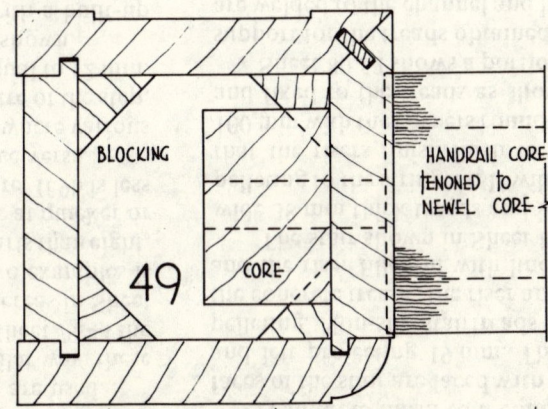

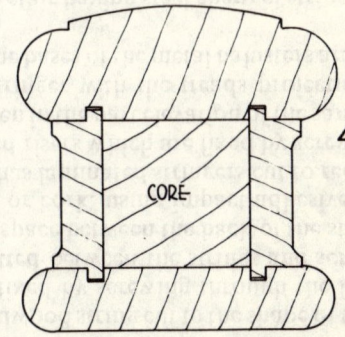

DISTANCES C-2 =12 2-1=25 1-5=12
C = CENTRE OF STEP

WIDTH OF SCROLL

WIDTH OF HANDRAIL

75

44

46

88 × 50

47

88 × 38

300

88

45

BLOCKING

CORE

49

HANDRAIL CORE TENONED TO NEWEL CORE

75 × 50

CORE

48

DETAILS OF BUILT UP NEWEL   &   HANDRAIL   SHEET 79

STAIRS

FINISH TO CONCRETE STAIR

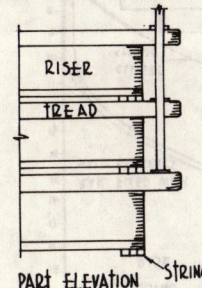

PART ELEVATION

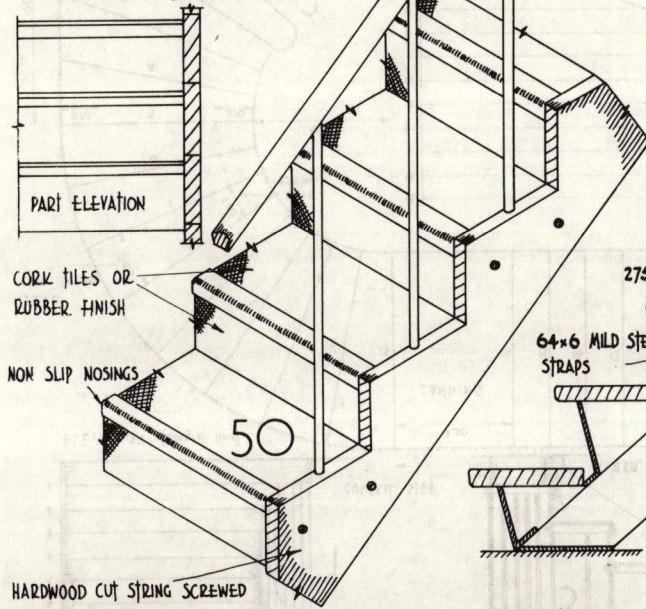

CORK TILES OR RUBBER FINISH

NON SLIP NOSINGS

50

HARDWOOD CUT STRING SCREWED & PELLETTED TO CONCRETE STAIR

RISER

TREAD

PART ELEVATION — STRING

51

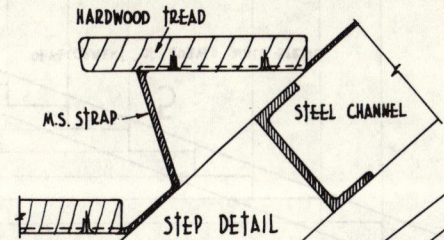

HARDWOOD TREAD

M.S. STRAP

STEEL CHANNEL

STEP DETAIL

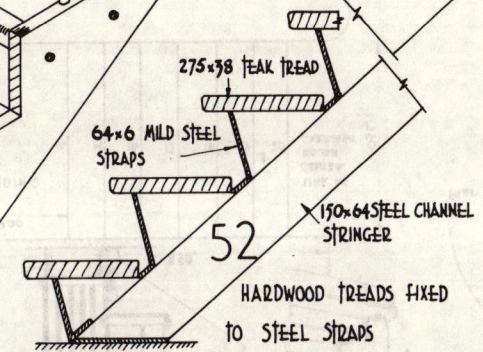

275×38 TEAK TREAD

64×6 MILD STEEL STRAPS

150×64 STEEL CHANNEL STRINGER

52

HARDWOOD TREADS FIXED TO STEEL STRAPS

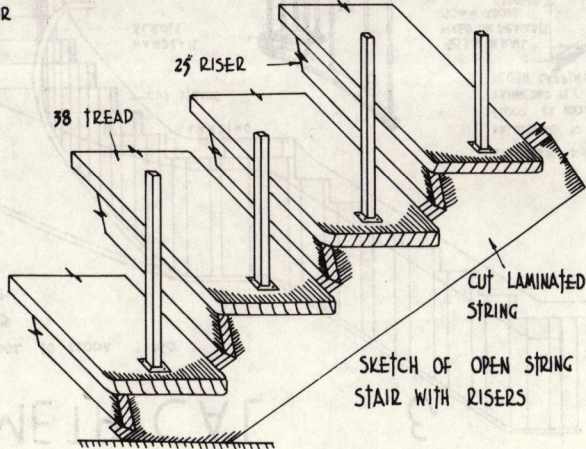

25 RISER

38 TREAD

CUT LAMINATED STRING

SKETCH OF OPEN STRING STAIR WITH RISERS

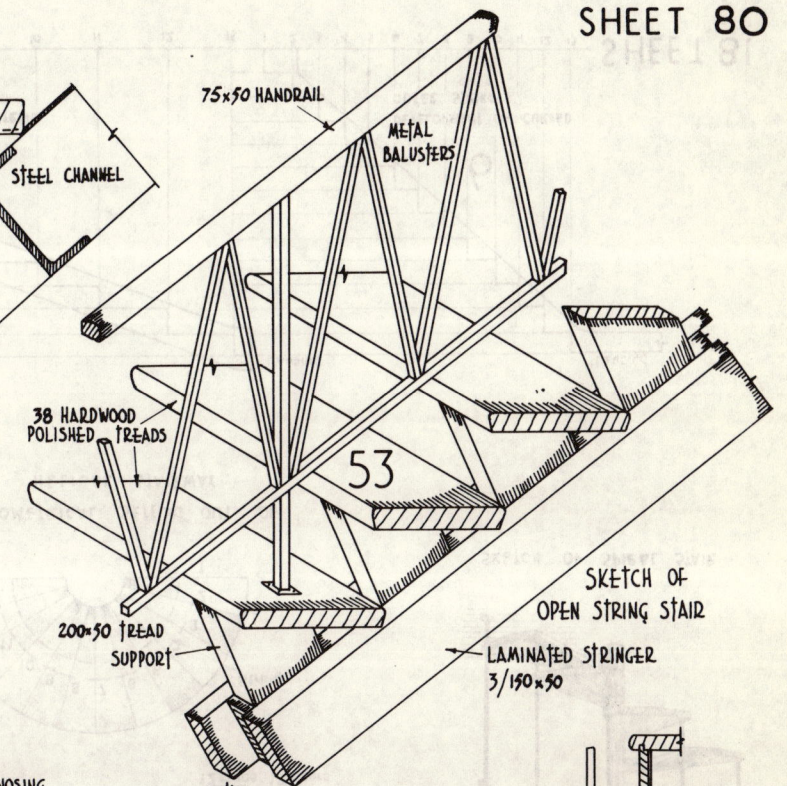

75×50 HANDRAIL

METAL BALUSTERS

38 HARDWOOD POLISHED TREADS

200×50 TREAD SUPPORT

53

SKETCH OF OPEN STRING STAIR

LAMINATED STRINGER 3/150×50

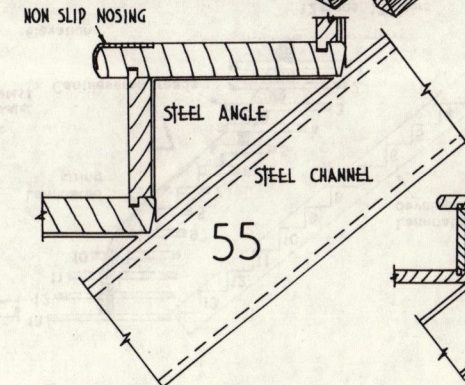

NON SLIP NOSING

STEEL ANGLE

STEEL CHANNEL

55

DETAIL OF OPEN STAIR WITH STEEL ANGLES & CHANNEL STRINGER

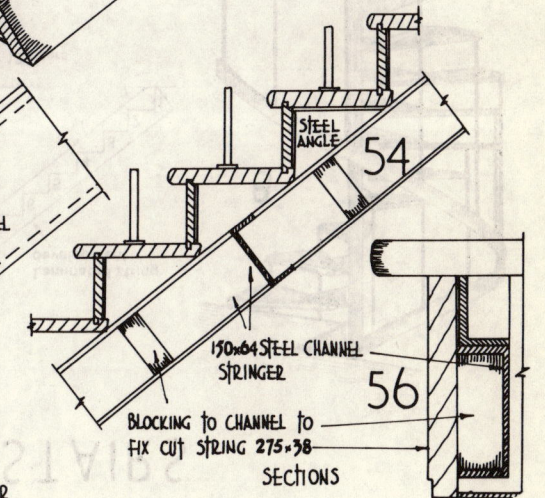

STEEL ANGLE

54

150×64 STEEL CHANNEL STRINGER

56

BLOCKING TO CHANNEL TO FIX CUT STRING 275×38

SECTIONS

GEOMETRICAL & HELICAL STAIRS

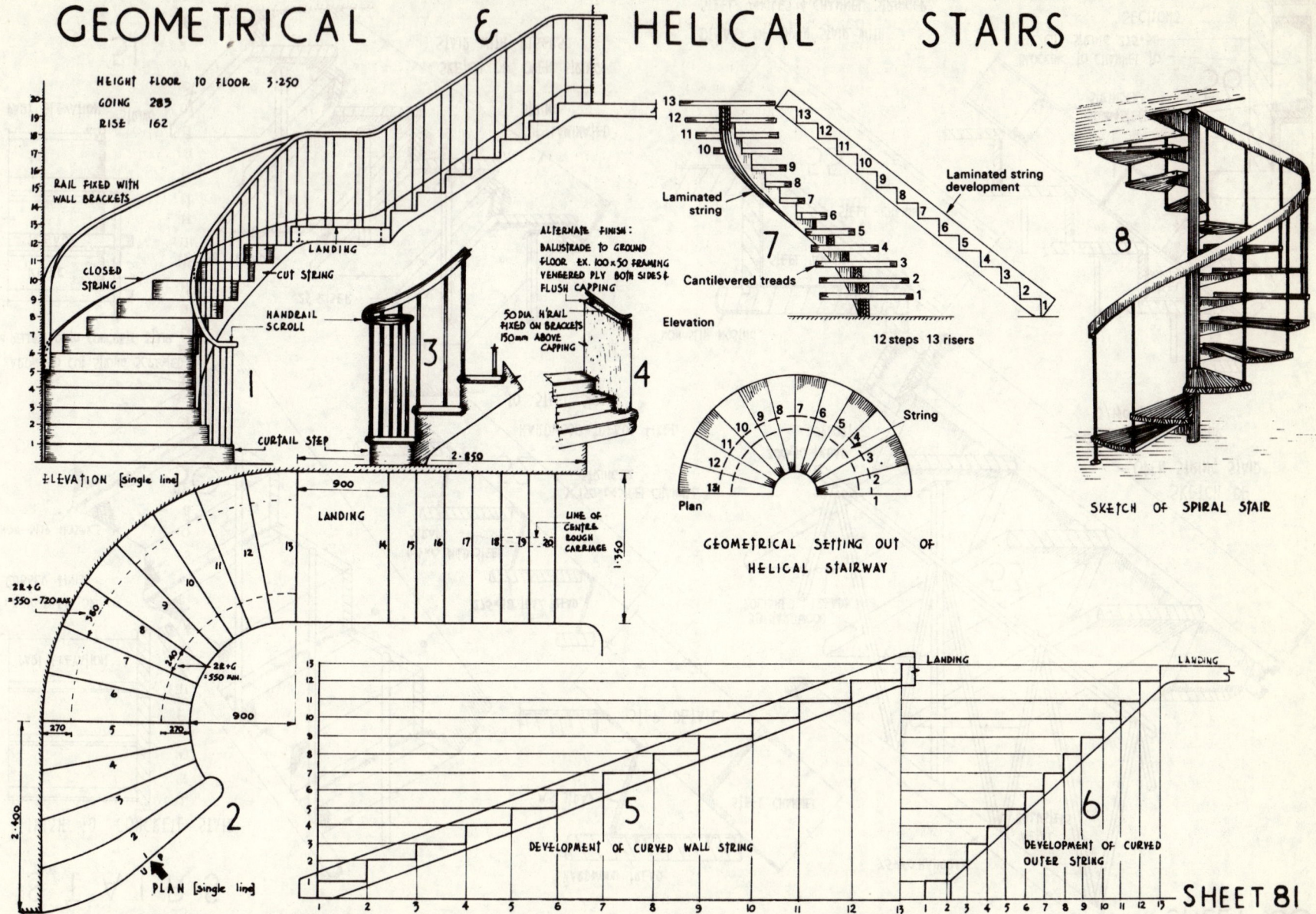

HEIGHT FLOOR TO FLOOR 3·250
GOING   285
RISE   162

RAIL FIXED WITH WALL BRACKETS

CLOSED STRING

CUT STRING

LANDING

HANDRAIL SCROLL

CURTAIL STEP

1

ELEVATION [single line]

3

4

ALTERNATE FINISH: BALUSTRADE TO GROUND FLOOR EX. 100×50 FRAMING VENEERED PLY BOTH SIDES & FLUSH CAPPING

50 DIA. H'RAIL FIXED ON BRACKETS 150mm ABOVE CAPPING

2·850

Laminated string

Cantilevered treads

7

Elevation

12 steps 13 risers

13 Laminated string development

String

8

SKETCH OF SPIRAL STAIR

Plan

GEOMETRICAL SETTING OUT OF HELICAL STAIRWAY

2R+G =550–720 MM.

2R+G =550 MM.

900

900

270 270

LANDING

LINE OF CENTRE ROUGH CARRIAGE

1·550

2·400

PLAN [single line]

2

5

DEVELOPMENT OF CURVED WALL STRING

LANDING

LANDING

6

DEVELOPMENT OF CURVED OUTER STRING

SHEET 81

An illustration of a helical stairway with treads cantilevered from a centre laminated post or column is shown in Sheet 81.8.

Sheet 81.7 shows the geometrical setting out for a helical stairway having twelve steps. The method of projecting the elevation and determining the development of the veneer for the laminated string is as described in Fig. 124.

Handrail-wreaths

Geometrical or continuous stairs are so called because the setting out of the strings and handrails is based upon geometrical principles. The handrails and one or both strings run continuously from top to bottom of the successive flights.

In the making of a continuous handrail the chief difficulty is in the wreaths, where the rail is of double curvature. This requires the joiner to produce a templet, or face mould, along with bevels which may be applied to the plank so that the wreath may be cut out, squared, and jointed correctly with straight rails. Many arrangements embracing handrails of double curvature are met, a selection of the more common forms only are dealt with here.

There are several systems of handrailing, all giving satisfactory results, but each requires a knowledge of solid geometry. The system described in this chapter is that known as the square cut tangent system. In this method the rail may be treated as a part of a hollow cylinder whose base is the plan of the rail, the edges of the squared wreath forming parts of the vertical surfaces of the cylinder.

The plan of the centre line of the curved portion of the rail starts at one corner of the plan of a four-sided prism and finishes at the opposite corner, as shown in Sheet 82.1. The inclined plane, which is assumed to cut the cylinder, is determined from the plan and vertical stretch-out of the stair. The section of the cylinder made by the inclined plane is elliptical and forms the pattern for cutting out the wreathed portion of the rail. This pattern is called the face mould. In the sketch of the block, the plan of the centre line of the wreath and its intersection with the portions of straight rails is shown. The upper rail is level and the lower one inclined. The tangent planes are represented by the vertical surfaces AB and BD of the block.

Sheet 82.2 shows the plan of the wreath connecting a straight flight and a level landing with the top riser placed at the springing. The plan of the rail and prism A–B–D–C is drawn, the tangents AB and BD being the centre lines of the two rails produced. The prism represents the block shown in the sketch (Sheet 82.1). Next, project the riser lines upwards vertically and construct one or two steps to give the pitch of the stair. This is also the inclination of the top surface of the block or prism. Project the plan and centre line of the straight rail upwards to cut the pitch line in point E. The rectangular section of the rail is drawn from E with the lower surface horizontal. A line drawn parallel to the nosing line passing through point F is the top face of the plank. The bottom face of the plank is taken as the nosing line. The thickness of the plank is the distance between the two surfaces and the bevel is equal to the pitch of the straight flight.

Draw the VT and at points from where the lines projected from the plan cut this line, draw out at right angles. From A′ and B′ set off the distances AC and BD to give points C′D′ in the development. A′B′D′C′ is the development of the inclined surface of the block, C′D′ is the semi-major axis and C′A′ the semi-minor axis. The width of the face mould at D′ is obtained from the sides of the rectangular rail section projected on to the pitch line. The width of the face mould at A′ is equal to the width of the rail shown in plan.

Draw the curves of the face mould with the trammel, the shank may be made any convenient length.

The sketch of a prism (Sheet 82.3), cut by an oblique plane shows the bevels or angles which the cutting plane makes with the vertical faces of the prism. The upper pitch is produced to meet the ground line at K, through this point and B draw the horizontal trace to find the angle between the plane and this face. With A as centre, draw an arc tangential to the pitch line to cut the perpendicular AE, draw a line from the intersection to B to give the first bevel. The top face of the prism is contained in a second plane. Produce B1 to meet edge FE produced, then with E as centre, draw an arc tangential to B1 produced cutting the perpendicular AE. The dotted line drawn from this point to H gives the second required bevel.

Sheet 82.4 shows the above prism drawn in plan and elevation with the quadrant BD the centre line of the rail. AB and AD are the tangent lines. The bevels are obtained as already described for Sheet 82.2. The line X′–Y′ is drawn at right angles to the horizontal trace through the centre C so that a line which shows the true inclination will also contain the major axis of the elliptical section. Lines from A, C, and D are drawn parallel to HT, and on a line from D set up the height taken from elevation. Draw 1, 3′ which is the true inclination of the plane. From 1, 2′, C′, 3′ draw lines at right angles and step off distances 1B′ equal to 1B, 2′A′ equal to 2A, C′ minor to C minor, 3′D′ to 3D. Join A′B′C′D′, this is the section of the prism and to be correct A′B′ must equal B″L and A′D′ equal to HL. The section of the cylinder is drawn by continuing the centre line of the rail in the plan to meet X′Y′ at 4. Draw 4–4′ parallel to 3–3′ then from C′ to 4′ is the semi-major axis, the semi-minor being drawn, the semi-ellipse is completed to pass through B′ and D′.

Sheet 82, figs 5, 6, 7, and 8 show the various stages of marking and shaping the wreath.

The wreath is cut from a plank the required thickness to the outline of the face mould, with additional material left on at the edges if required. The tangent lines and springing lines are marked on both faces of the face mould. After truing up the faces of the wreath the face mould is applied and the tangent lines marked as shown in Sheet 82.5, along with the joint lines square to the tangents and to the face of the plank. The positions for the handrail bolts are next accurately marked.

Sheet 82.6 shows the marking of the wreath for bevelling, which requires the bevel to be marked at the end, and the face mould slid along so that the tangent lines tie exactly on the new tangent lines drawn from the twist bevel applied at the centre of the joint. The mould is then marked on the face of the plank and repeated on the lower face. The material may be cut away with a bow saw, gouge, and chisel and finished by spokeshave. The bevelling of the wreath may also be done on the bandsaw.

HANDRAILING

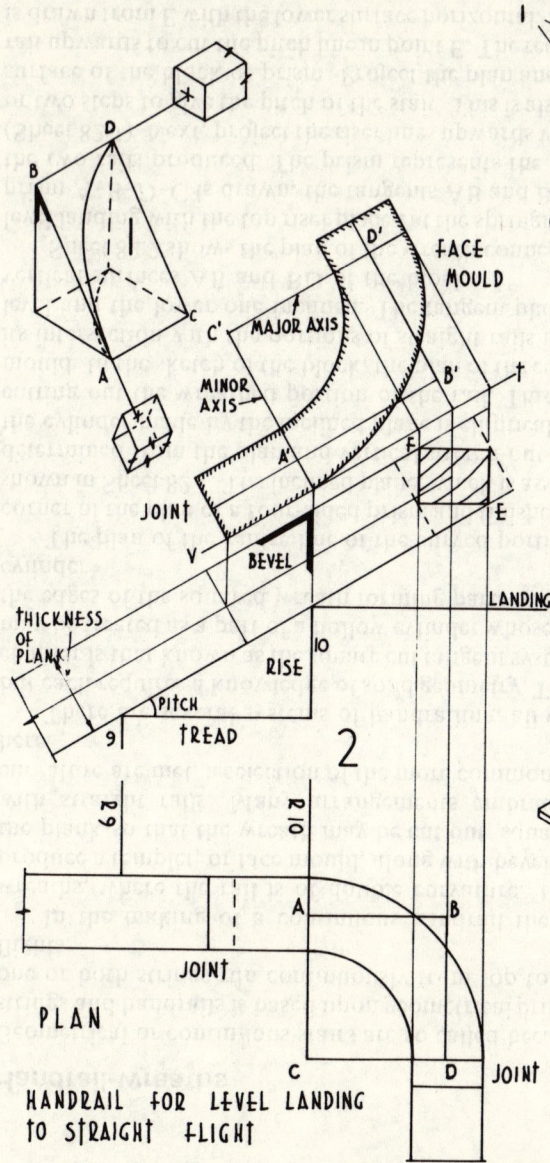

SKETCH OF PRISM

FACE MOULD

MAJOR AXIS

MINOR AXIS

JOINT

BEVEL

THICKNESS OF PLANK

RISE

PITCH

TREAD

2

9   10

PLAN

JOINT

JOINT

A   B

C

D

HANDRAIL FOR LEVEL LANDING TO STRAIGHT FLIGHT

LANDING

1

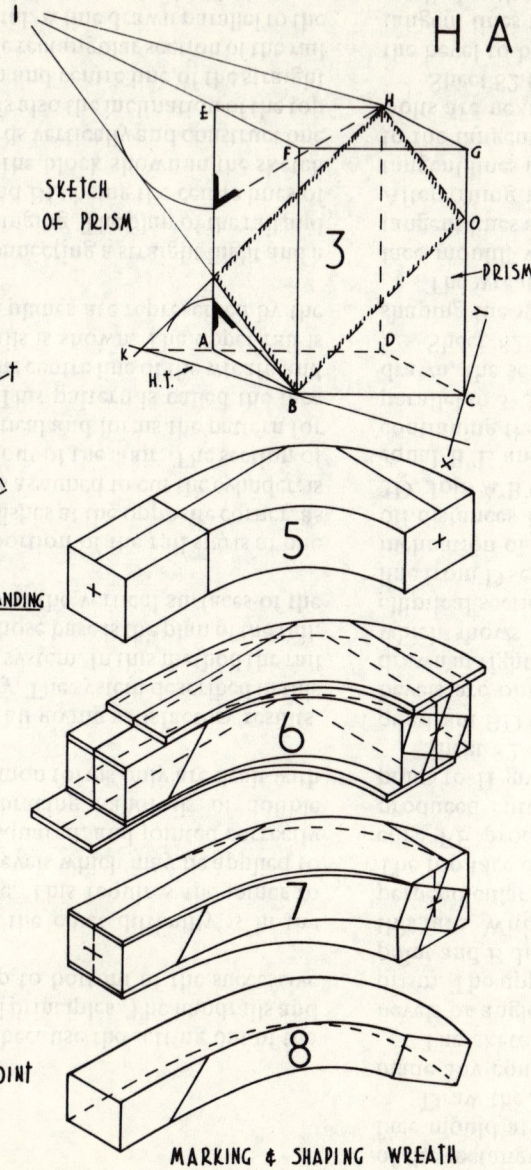

3

PRISM

H.T.

H

E   F   G

O

K   A   D   Y

B   C

X

5

6

7

8

MARKING & SHAPING WREATH

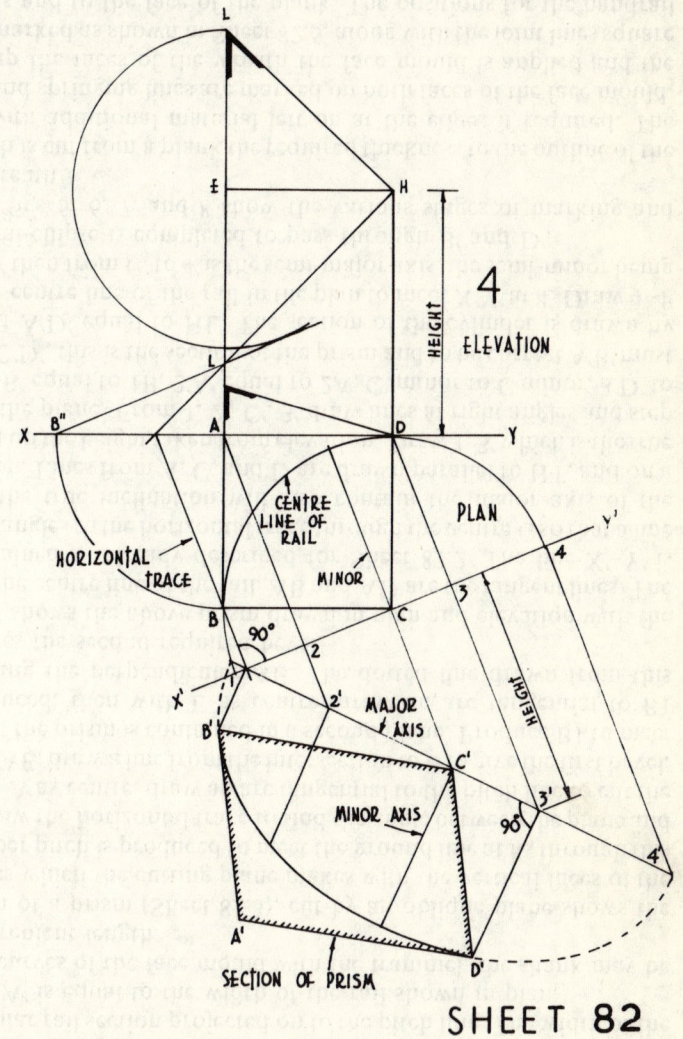

4

ELEVATION

HEIGHT

L

E   H

PLAN

CENTRE LINE OF RAIL

MINOR

HORIZONTAL TRACE

X   B"   A   C   D   Y

90°

2

3

2'

X'

MAJOR AXIS

MINOR AXIS

3'

90°

4'

Y'

HEIGHT

B'

C'

A'

D'

SECTION OF PRISM

SHEET 82

H A N D R A I L I N G

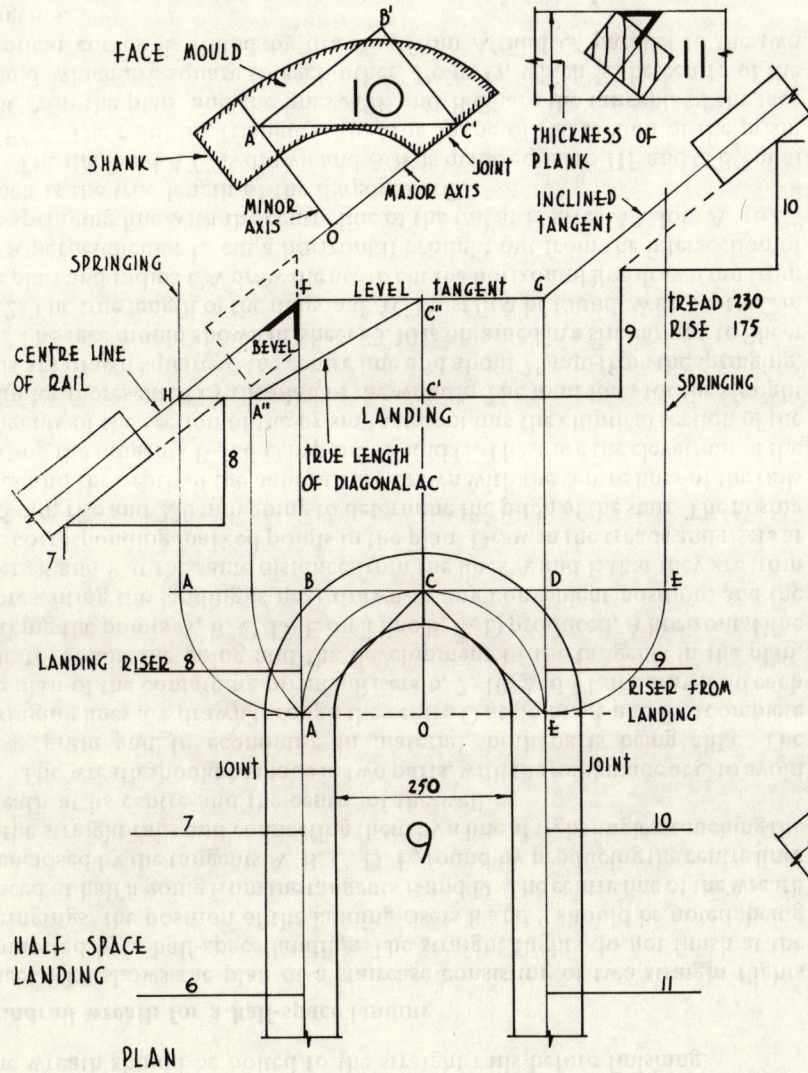

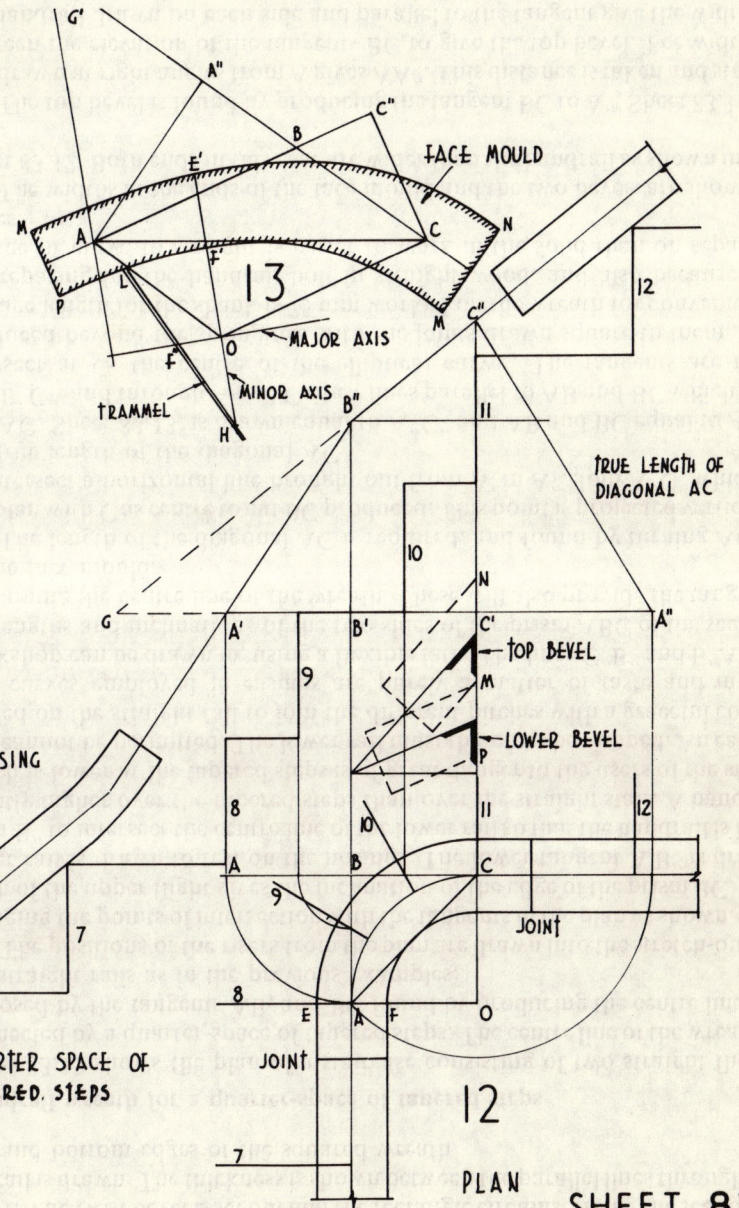

SHEET 83

The top and bottom surfaces are next prepared, being marked as in Sheet 82.7 with freehand falling lines on the edge of the material kept to the thickness of the handrail.

Either the top or bottom of the wreath may be finished first (Sheet 82.8), but where a metal core is fitted to the lower face, this must be finished first. The wreath should be bolted to the straight rails before finishing.

Handrail wreath for a half-space landing

Sheet 83.9 shows the plan of a staircase consisting of two straight flights connected by a half-space landing. The straight flights do not finish at the springings, the position of the landing risers 8 and 9 should be noted, being placed at half a going from the tangents B and D. The centre line of the wreath is enclosed by the tangents A, B, C, D, E, found by producing the centre lines of the straight rails and connecting them by a line at right angles, touching the wreath at its centre and the centre of the well.

The wreath should be made in two parts, with the joint made at C to avoid cross grain and to economise in material, both parts being alike. The springing lines are drawn through the centre O at points A and E to complete the plan of the containing prisms. Risers 6, 7, 10, and 11 are drawn on each side to obtain the going and the development of the tangents in the plan, making the points A, B, C, D, E on a line B, C, D produced. A horizontal line representing the landing is next drawn in any convenient position, and the risers 8 and 9 at the same distance from the lines A and E that they are from the corresponding marked points in the plan. Draw in the treads and risers at 175 mm rise and 230 mm going to determine the pitch of the stair. The nosing lines and the depth of the handrail are drawn with the centre lines of the rails cutting the tangents B and D in points F and G. These are the elevations of the tangents of the section of the prism that contains the elliptical section of the cylinder represented by the plan of the wreath. The joint lines for the straight rails are drawn square to the centre line and about 75 mm from the springing.

The face mould shown in Sheet 83.10 is obtained in a similar way to Sheet 82.2. The true length of the diagonal AC must first be found. With centre C in the plan and radius CA draw the arc to cut the horizontal line drawn out from C. A perpendicular to cut a horizontal brought out from the intersection of the springing line with the centre line of the rail at H gives A″. Join A″ to C″ which is the true length of the diagonal AC.

The diagonal A′C′ is drawn and A′B′ is made equal to HF and C′B′ equal to FC″. The resulting triangle is the true shape of the section of the prism ABCA in the plan, and the lines A′B′ and B′C′ are the tangents of the face mould which are square to each other. Point O, which is the centre of the elliptical curves, is found by drawing from A′ and C′ parallel to the two tangents.

The width of the rail in the plan is marked off equally on each side of the tangent at A′ in the face mould, and the shank is drawn with the joint made at right angles to the tangent. The width of the face mould at the other end is found by stepping off the width of the handrail on each side of the tangent HF. Lines drawn parallel to the tangent to cut the perpendicular from B give the width of the mould at C′. The semi-major and semi-minor axes of the curves being determined, the curves are now drawn in by trammel. As both

halves of the wreath are alike, only one face mould is required.

The twist bevel is given by the angle which the tangent HF makes with the vertical shown.

The thickness of the plank from which to cut the wreath is shown in Sheet 83.11. The twist bevel is set out and the rectangle circumscribing the section of the rail is drawn. The thickness is shown between the parallel lines through the top and bottom edges of the squared wreath.

Handrail wreath for a quarter-space of tapered steps

Sheet 83.12 shows the plan of a staircase consisting of two straight flights connected by a quarter-space of tapered steps. The centre line of the wreath is enclosed by the tangents AB, and BC found by producing the centre lines of the straight rails as in the previous examples.

The positions of the risers from the plan are drawn into the stretch-out by swinging the points of intersection with the tangents in the plan as shown. The pitch of the upper flight gives the inclination of the edge of the prism BC. The lower rail is drawn to rest on the nosings. The lower tangent A′B″ is drawn from B″ to intersect the centre line of the lower rail so that the handrail is kept slightly higher over the tapered steps than over the straight stair. A handrail which is lower at the tapered steps is of great danger to the users of the stairs and cannot be permitted. The lower rail must therefore be ramped. An easing is used on the straight rail to join the different pitches with a graceful curve. The curves employed in easings are purely a matter of taste and in the workshop can be drawn in, using a flexible lath. The lines C″B″ and B″A′ are the lengths and inclinations of the two sides of the prism ABC of the section containing the centre line of the wreath. These will also provide the tangents of the face mould.

The length of the diagonal AC is required, and found by turning AC in the plan with C as centre to cut BC produced. This point is projected vertically to intersect a horizontal line brought out from A′ in A″. Join A″C which is the true length of the diagonal AC.

AC, Sheet 83.13, is drawn equal to A″C″ and AB and BC equal to A′B″ and B″C″, and through A and C draw lines parallel to AB and BC which will intersect at O, the centre of the elliptical curves. The tangents are next produced beyond the springings with the joints drawn square to them. The average length for the shank is 75 mm worked on the wreath for convenience of preparing for the handrail bolt in straight wood, and also because the change of curve to straight is easier to work in the solid than on separate pieces.

The widths at the ends of the face mould and the two bevels are shown in Sheet 83.12. Both ends in this case are wider than the handrail as shown in the plan.

The top bevel is found by producing the tangent BC to A″, Sheet 83.13. A line drawn at right angles from A gives AA″. This distance is taken and struck between the elevation of the tangents BC to give the top bevel. The width of the handrail drawn on each side and parallel to the tangent give the width of the face mould NM. The lower bevel and width of mould is obtained in a similar way.

As the axes of the elliptical curves are not in line with the springing lines

HANDRAILING

OBTUSE ANGLED PLAN

ACUTE ANGLED PLAN

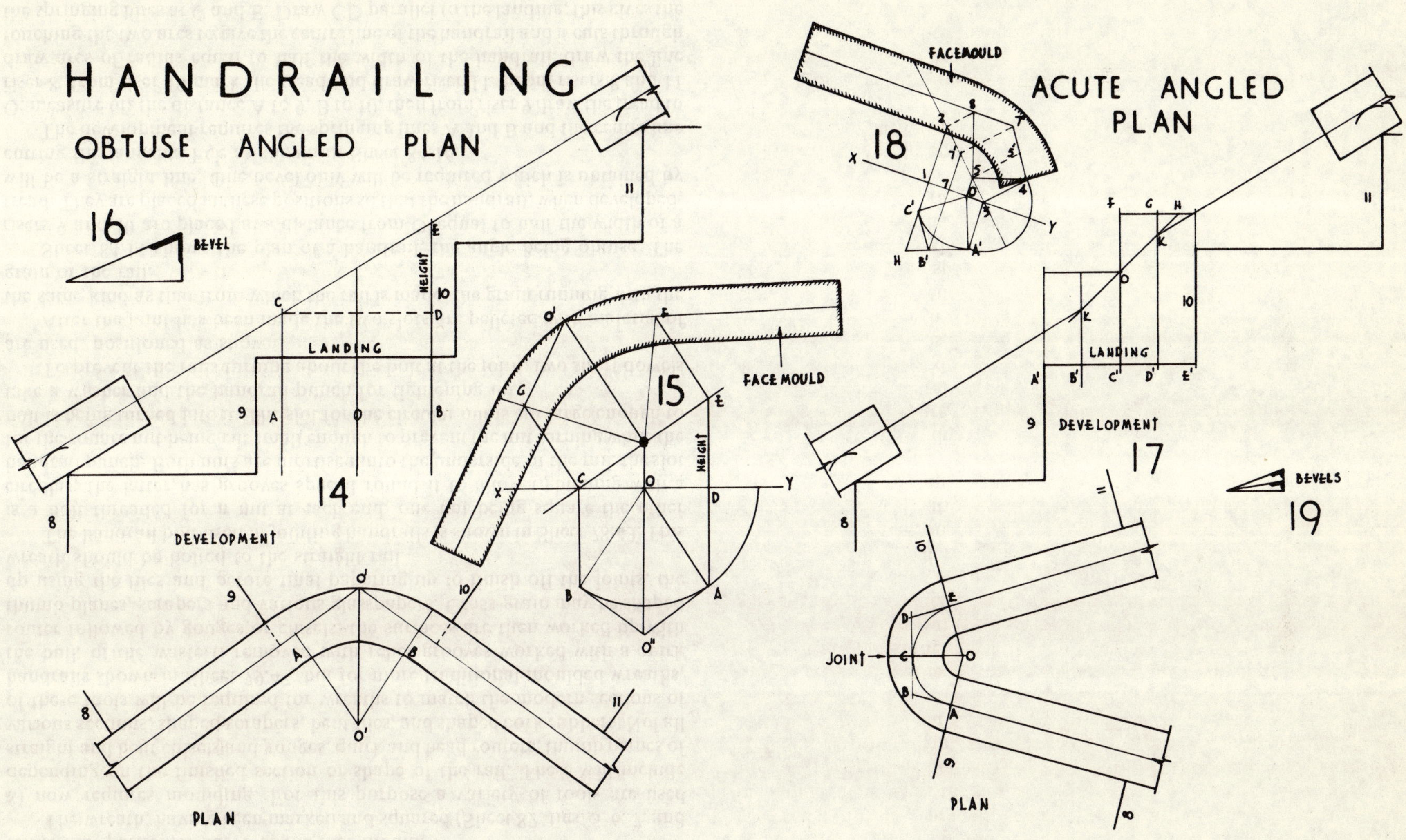

FACEMOULD

16

BEVEL

14

11

E

HEIGHT

10

C   D

LANDING

9   A   B

8

DEVELOPMENT

15

FACEMOULD

d

G

E

HEIGHT

x   C   O   D   Y

B   A

O"

d

9   10

A   B

11

O'

8

PLAN

18

x

1   7

c'

H   B'   A'

f   G   H

k

O

k   10

LANDING

A'   B'   C"   D'   E'

9   DEVELOPMENT

11

8

17

BEVELS

19

JOINT

E

D   O

C

B

A

9

PLAN

8

SHEET 84

due to section having two pitches, the direction of these is to be found. The tangent B″C″ (Sheet 83.12) is produced to a horizontal taken out from A′ in point G. The length B″G is added to the tangent BC in Sheet 83.13, to give point G′. A line from G′ through A is the horizontal trace, and a line drawn parallel to this from O will be the direction of the minor axis with the major axis drawn at right angles to it. The length of the minor axis OE′ is taken from the centre O in plan to the inside and outside faces of the rail, OF and OE. The trammel rod is laid across the axis from one of the given points with the point F″ resting in the direction of the major axis; then a mark made at point H where the rod passes the minor axis will be the length of the major axis to strike that particular curve of the face mould.

The wreath, having been marked and squared (Sheet 82, figs. 5, 6, 7, and 8) now requires moulding. For this purpose a variety of tools are used depending on the finished section or shape of the rail. These will include straight and bent chisels and gouges, quirk and bead routers, thumb planes of various sections, shaped scrapers, bent files, and shaped cork rubbers. Not all of these tools will be required for wreaths to match the modern sections of handrails shown in Sheet 79.47, but for more traditional moulded wreaths, the bulk of the waste is removed with relief grooves worked with a quirk router followed by gouges or chisels, the surfaces are then worked up with thumb planes, scrapers and various glasspapers. Cross-grain may be shaped up using the files and before final papering up to finish off the joints, the wreath should be bolted to the straight rail.

The handrail bolt used in jointing handrails is shown in Sheet 78.43. This is a bolt threaded for a nut at each end, one nut being square the other circular; the latter has grooves spaced round it to allow tightening with a handrail punch. Both nuts are mortised into the underside of the rail, the slot for the square nut being cut small enough to prevent the nut turning while the bolt is being turned into it. The slot for the circular nut is cut large enough to take a washer and the handrail punch for tightening up.

To prevent the rails turning about the bolt at the joint, two short dowels are used, positioned as shown.

After the joint has been made the two slots are pelleted with material of the same kind as that from which the rail is made, the grain running with the grain of the rail.

Sheet 84.14 shows the plan of a handrail, the angle being obtuse. The risers 9 and 10 are placed at a distance from O′ equal to half the width of a tread. They are placed in these positions so that the handrail, when developed, will be a straight line. One bevel only will be required which is obtained by cutting the tangents FG, as shown in Sheet 84.16.

The development requires the springing lines A and B and the centre line O; measure off the distance A to 9, B to 10, then from riser 9 draw the tread to riser 8; from riser 10 mark the tread and draw riser 11. From risers 8 and 11 draw arcs of radius equal to half the width of the handrail; draw the line touching the two arcs to give the centre line of the handrail and it cuts through the springing lines at C and E. Draw CD parallel to the landing; this gives the height the rail rises from A to B in plan.

To draw the face mould Sheet 84.15 draw AO″, BO″, equal to AO′ and BO′. Draw the ordinate through O″O and one from A and B parallel to O″O. Next draw the XY line and extend AD to E making DE equal to DE in

elevation, draw the pitch CE. At right angles draw OO′ equal to OO″; EF equal to AD; CG equal to CB. Draw the springing lines from O through FG. Draw the tangents and face mould as previously explained.

Sheet 84.17 shows the plan and development of a handrail the angle being acute in plan. The joint is shown at the centre of the well and four risers, two to each flight with risers 9 and 10 in the springing in plan.

The development is similar to previous work as is the face mould for the wreath shown in Sheet 84.18, with the necessary bevels shown in Sheet 84.19.

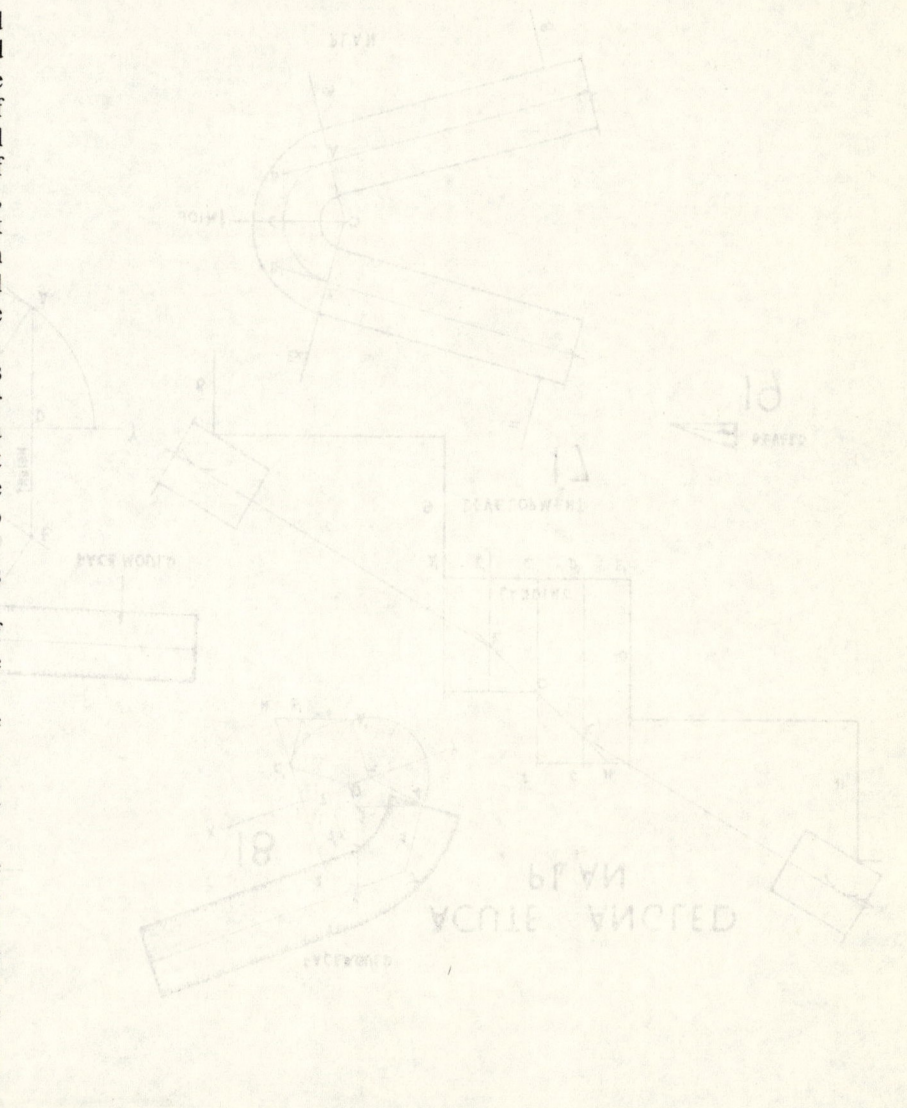

Chapter 19

Wall panelling

Wall panelling 1

The traditional type of panelling for the covering of walls in large houses and public buildings consists of mortised and tenoned framing prepared to receive well-figured and finely toned panels, combined with solid mouldings. The height varies considerably. Panelling which is about window bottom or chairback height is termed dado panelling. Panelling may be taken to the height of the doors in a room, or the whole of the walls may be covered to cornice height.

As this type of panelling still exists in older buildings, work on restoration, or in additions and extensions, requires to be done, and for this reason has been included here, followed by more modern designs and treatments.

Sheet 85.1 shows the part elevation and section of the lower portion of traditional wall panelling. The height of the moulded dado is 900 mm and represents the lower portion of the panelling in the key elevation (Sheet 85.2). This shows the arrangement of panelling suitable for a board-room where almost the full height of the wall is covered.

The framing of this panelling would be prepared in the shop and transported to the site for fixing. Special considerations should be given to work of this size from the point of view of handling, transporting and getting the framing upstairs and through doors. This particular design lends itself to the dividing up of the frames. It will be seen in the section that at dado height the framing has been split, the dado rail breaking the joint between the two sections. At this point, softwood lining-out pieces are shown where they are covered by the dado rail. This conserves a certain amount of hardwood and reduces the widths of the rails, thereby counteracting warping.

The construction of the panelling is similar to that for doors. Solid moulded framing is shown grooved for solid raised panels, the work is only finished on the face side, the back being left from the saw. Work of this type is secret-fixed wherever possible, but first it is necessary to fix a system of rough grounds. These are 19 mm thick, being the thickness of the plaster and placed behind the frame members so as to give a solid bearing to the work. Sheet 85.3 shows a typical arrangement of the groundwork where framed grounds are used at the doorway. These are fixed to plugs in the walls which must be plumbed and squared most carefully, particularly at openings and angles.

The fixing of the work to the grounds, whether by nailing or screwing, is done through the face where the fixing is covered by the skirting or dado. The fixing of the skirting is best done by first screwing a moulded fillet to the floor to receive the bottom edge, the top edge being pinned through the square or screwed through the face to skirting blocks or soldiers. Where screws cannot be hidden, the heads must be let into the face of the panelling and pelleted. Other forms of secret fixing will be dealt with later.

Sheet 85.4 shows the method of fixing panelled framing at the internal angles. The edge of one stile is tongued to fit a groove made in the stile of the other. The jointing of the skirting at all internal angles (Sheet 85.5) shows the skirtings stop grooved and tongued with the top mouldings scribed. All external angles are mitred.

Sheet 86.6 shows the finish at the cornice. This is the horizontal member fixed to the framing and is usually built up, with provision for ventilation at the back of the panelling as shown.

If concealed lighting is used at cornice level, metal brackets screwed to the moulded member and groundwork are used (Sheet 86.7). As a considerable amount of heat is given off by this type of lighting, care should be taken with the mitres and other jointing. Canvas glued on the back side of the members helps to strengthen the joints and prevents light from showing through in the event of shrinkage.

Two forms of pilasters used for joining up sections of wall panelling and as a decorative feature are shown in Sheet 86.8. The concealed fixing of the second form of pilaster is clearly shown in the sketch (Sheet 86.9) and requires no further explanation.

An alternative secret fixing is shown in Sheet 86.10, again illustrated from the rear and secret fixing, using slot screws shown in Sheet 86.11.

Wall panelling 2

The introduction of new materials and new applications has resulted in a new conception of joinery finishings. The modern trend is for solid mouldings with veneered plywood or laminboards, from 10 mm thick upwards. Sheet 87.12 shows an example of this type where veneered panels are used with bolection mouldings and finished with a moulded capping.

A section through the cornice is shown in Sheet 87.13, where it will be seen that the edges of the veneered ply panels are splayed and flush with the framing on the back side. This allows for a bolder bolection moulding to be used on the face. These are mitred at the corners and screwed through the panels from the rear.

Sheet 87.14 shows a section through a mullion and the jointing between the upper and lower sections of the panelling is shown in Sheet 87.15.

Another example of modern panelling suitable for a bank or board-room

WALL PANELLING

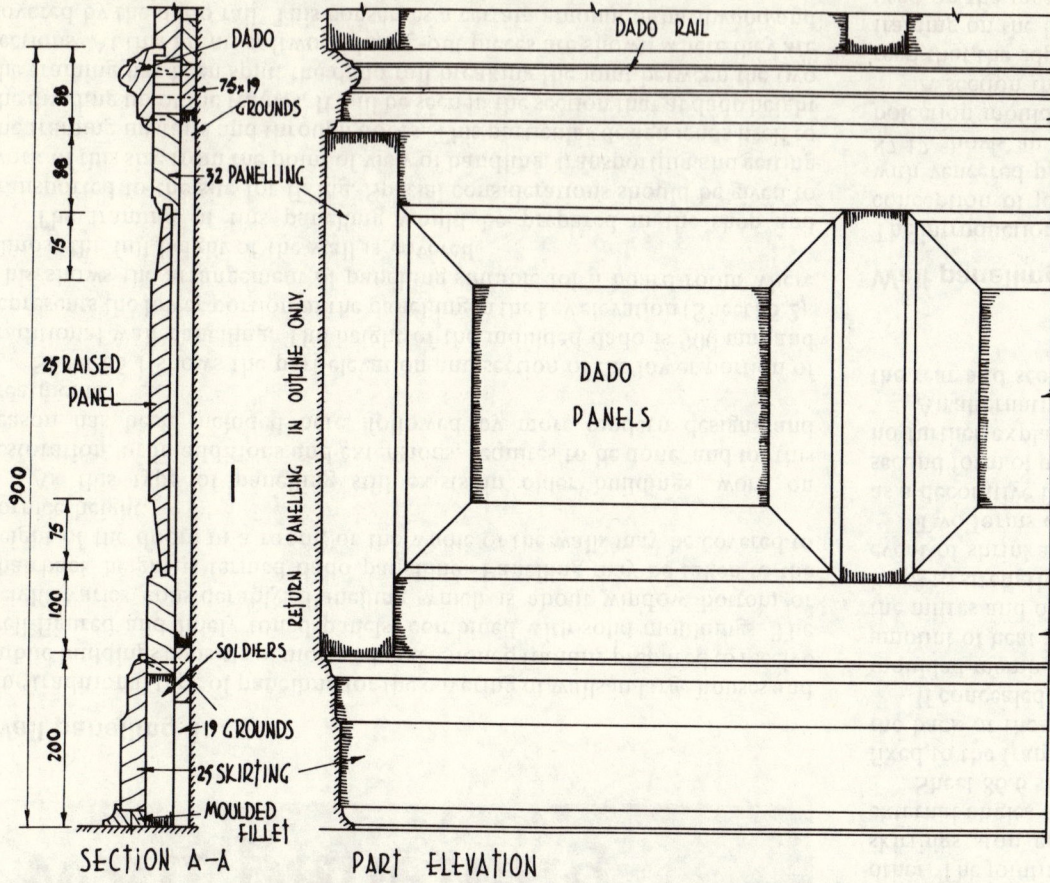

DADO

75 x 19 GROUNDS

32 PANELLING

25 RAISED PANEL

SOLDIERS

19 GROUNDS

25 SKIRTING

MOULDED FILLET

86
86
75
900
75
100
200

SECTION A-A

RETURN PANELLING IN OUTLINE ONLY.

DADO RAIL

DADO PANELS

PART ELEVATION

PORTION OF PANELLING FOR BOARD ROOM

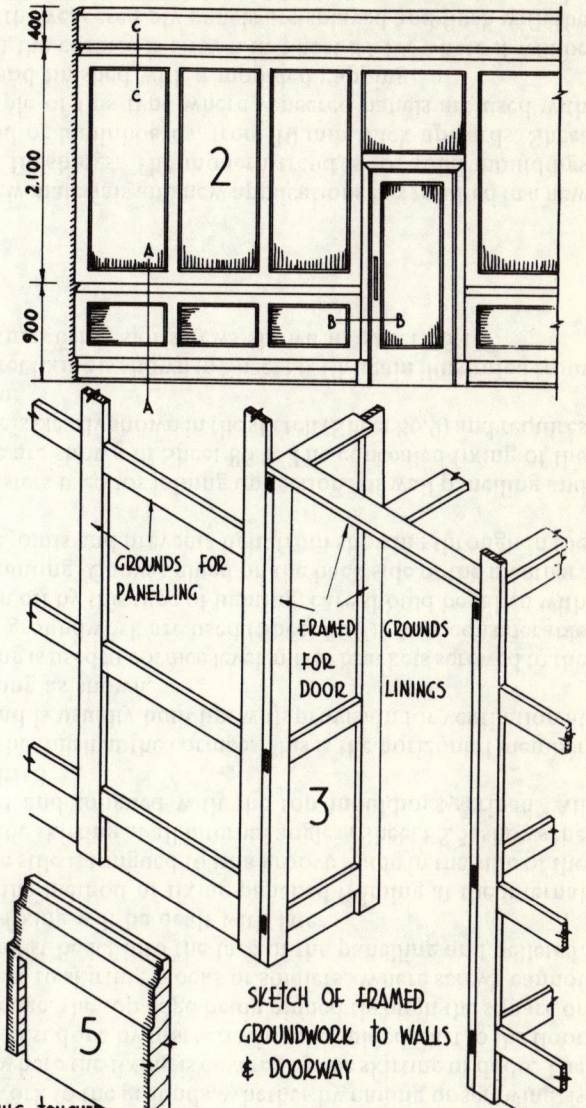

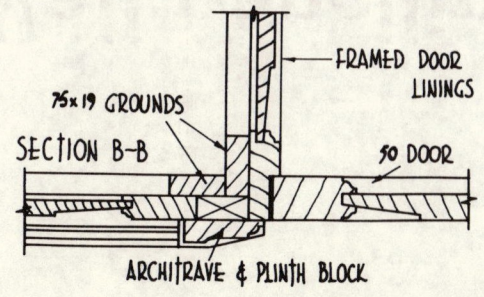

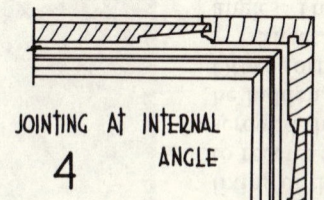

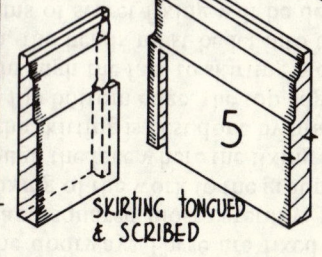

400
2.100
900

C
C
2
A
A
B
B
3

GROUNDS FOR PANELLING

FRAMED GROUNDS FOR DOOR LININGS

FRAMED DOOR LININGS

75 x 19 GROUNDS

SECTION B-B

50 DOOR

ARCHITRAVE & PLINTH BLOCK

JOINTING AT INTERNAL ANGLE

4

SKIRTING TONGUED & SCRIBED

5

SKETCH OF FRAMED GROUNDWORK TO WALLS & DOORWAY

SHEET 85

PANELLING   DETAILS

CORNICE

32 FRAMING

25 RAISED PANEL

PACKING SLIPS TO ALLOW FOR VENTILATION BEHIND PANELLING

75×19 GROUNDS

WALL LINE

6

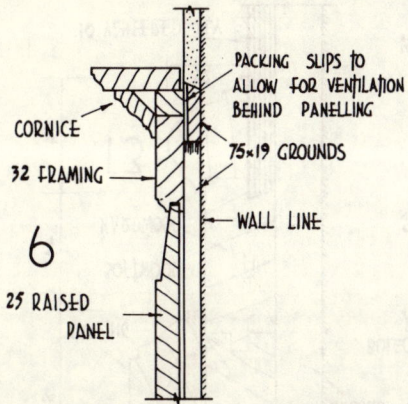

SECTION C-C
CORNICE DETAIL TO BOARD ROOM PANELLING

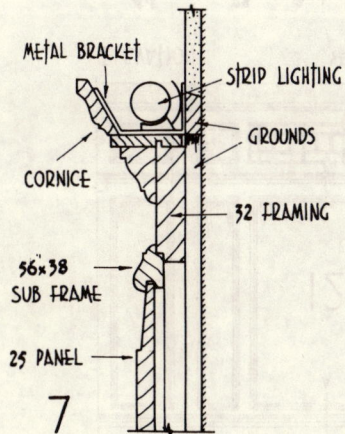

REEDED PILASTER

HORIZONTAL SECTION SHOWING PILASTER JOINING UP SECTIONS OF PANELLING

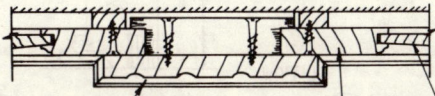

8

SECRET FIXING USING SLOTSCREWS

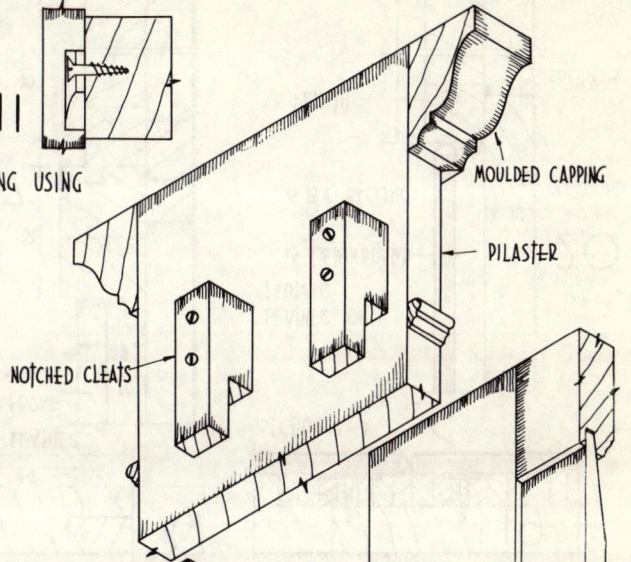

11

MOULDED CAPPING

PILASTER

NOTCHED CLEATS

METAL BRACKET

CORNICE

STRIP LIGHTING

GROUNDS

32 FRAMING

56×38 SUB FRAME

25 PANEL

7

ALTERNATE CORNICE DETAIL WITH CONCEALED LIGHTING

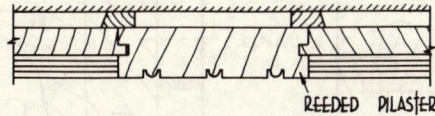

FLUTED PILASTER

BEARER FOR FIXING PILASTER, GLUED & SCREWED TO BACK

9

SKETCH OF CONCEALED FIXING OF PILASTER TO PANELLING

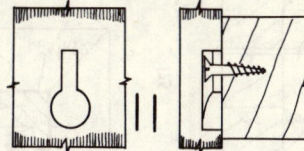

STILE

RAISED PANEL

STILE NOTCHED TO ALLOW BEARER TO PASS THROUGH WITH BEVELLED SEATING TO RECEIVE BEARER

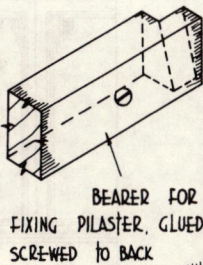

10

BLOCKS

SECTIONS OF PANELLING

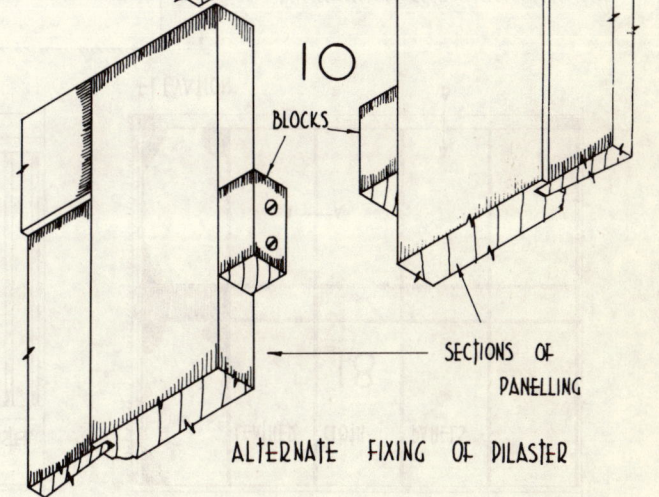

ALTERNATE FIXING OF PILASTER

SHEET 86

239

240

TYPES OF PANELLING

TRADITIONAL

12

A—A

B—B

C

ELEVATION

VENTILATOR

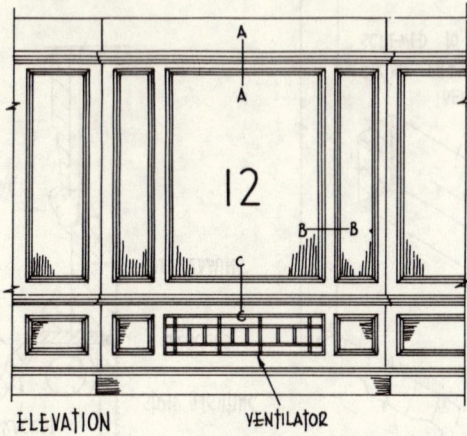

MOULDED STRIP WITH LEATHER

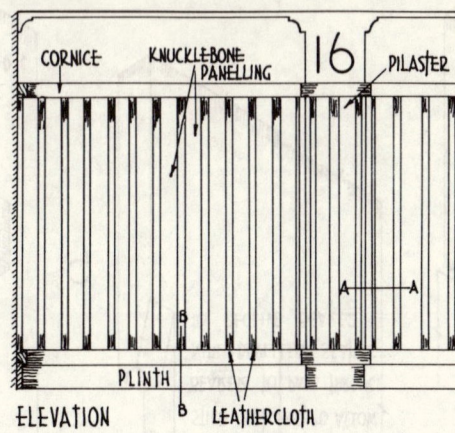

CORNICE

KNUCKLEBONE PANELLING

16

PILASTER

A—A

B

PLINTH

ELEVATION

B

LEATHERCLOTH

LEATHER COVERED

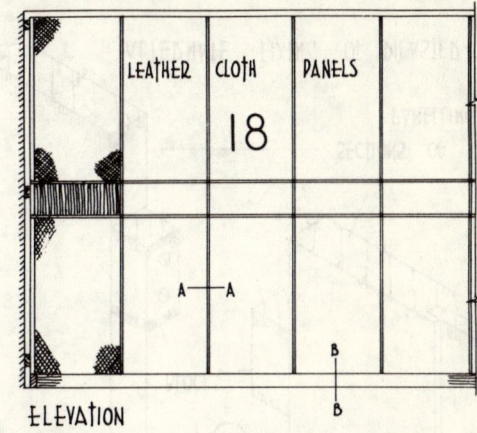

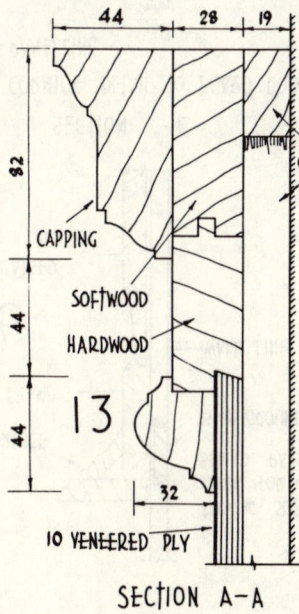

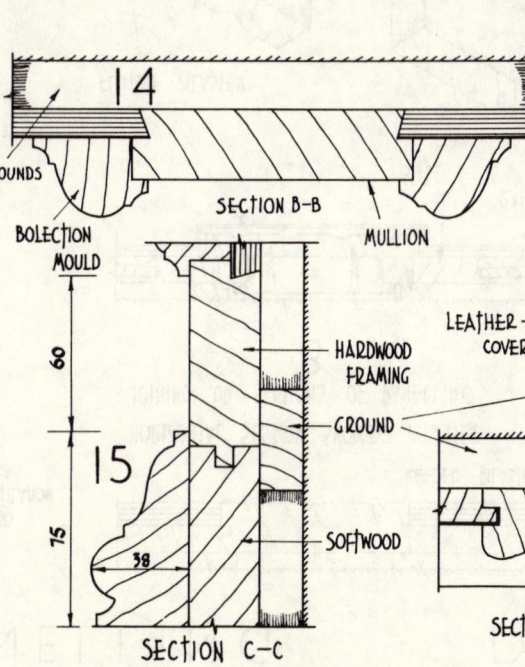

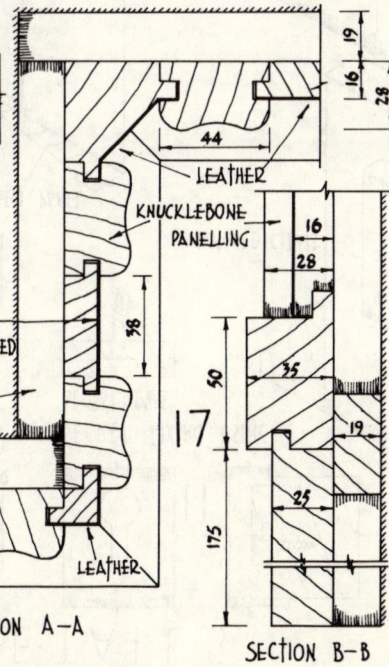

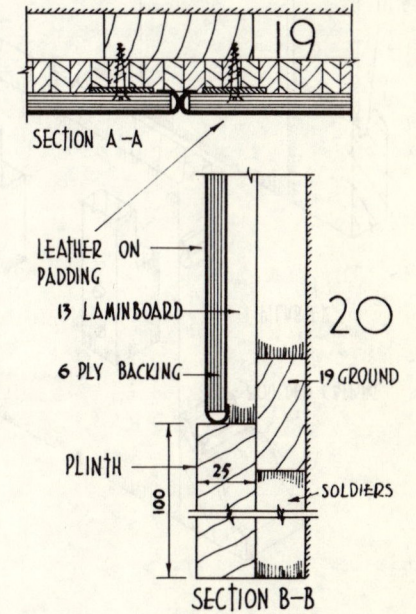

LEATHER CLOTH PANELS

18

A—A

B

ELEVATION

B

---

44 28 19

14

32

GROUNDS

CAPPING

SOFTWOOD

HARDWOOD

44

13

44

32

10 VENEERED PLY

SECTION A-A

SECTION B-B

MULLION

BOLECTION MOULD

60

HARDWOOD FRAMING

GROUND

15

75

38

SOFTWOOD

SECTION C-C

LEATHER

LEATHER COVERED

KNUCKLEBONE PANELLING

17

38

50

175

LEATHER

SECTION A-A

19
16
28

16
28

35

50

19

25

175

SECTION B-B

19

SECTION A-A

LEATHER ON PADDING

13 LAMINBOARD

6 PLY BACKING

20

19 GROUND

PLINTH

25

100

SOLDIERS

SECTION B-B

FLUSH DADO PANELLING

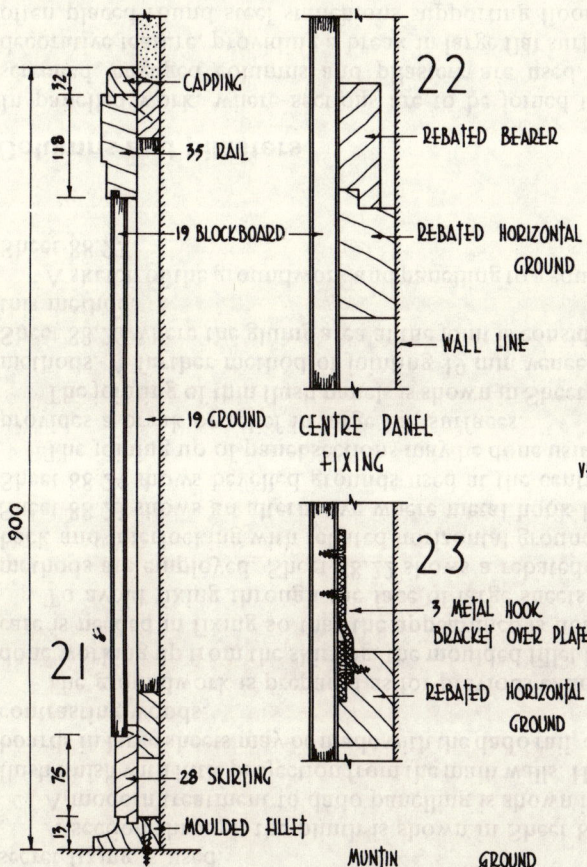

32
118

CAPPING

35 RAIL

19 BLOCKBOARD

19 GROUND

900

SECTION

75
13

28 SKIRTING

MOULDED FILLET

21

22 REBATED BEARER

REBATED HORIZONTAL GROUND

WALL LINE

CENTRE PANEL FIXING

23 3 METAL HOOK BRACKET OVER PLATE

REBATED HORIZONTAL GROUND

10 VENEERED PLY

FIXING THIN FLUSH PANELS

16 BLOCKBOARD

25

JOINTING FLUSH PANELLING

64×19 GROUND FIXED TO PANEL

64×19 GROUND FIXED TO WALL

VENEERED FLUSH PANEL

24

FIXING TO CENTRE GROUNDS

FIXED GROUND

CLEAT GLUED & SCREWED TO PANEL & GROUND

CLEAT GLUED & FIXED TO PANEL

26

JOINT & GLUE LINE

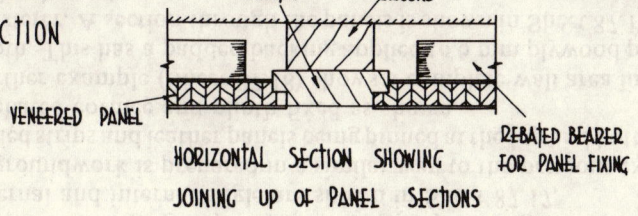

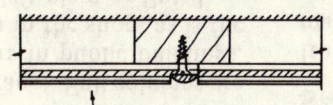

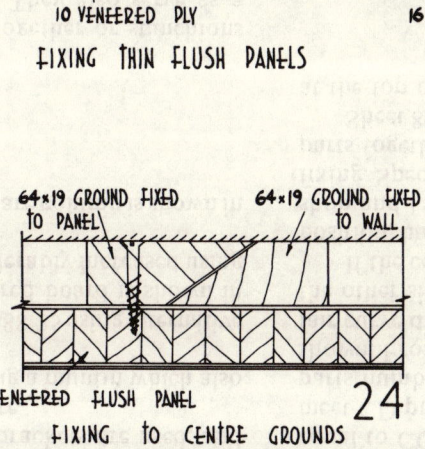

CONTINUOUS FLUSH PANELLING

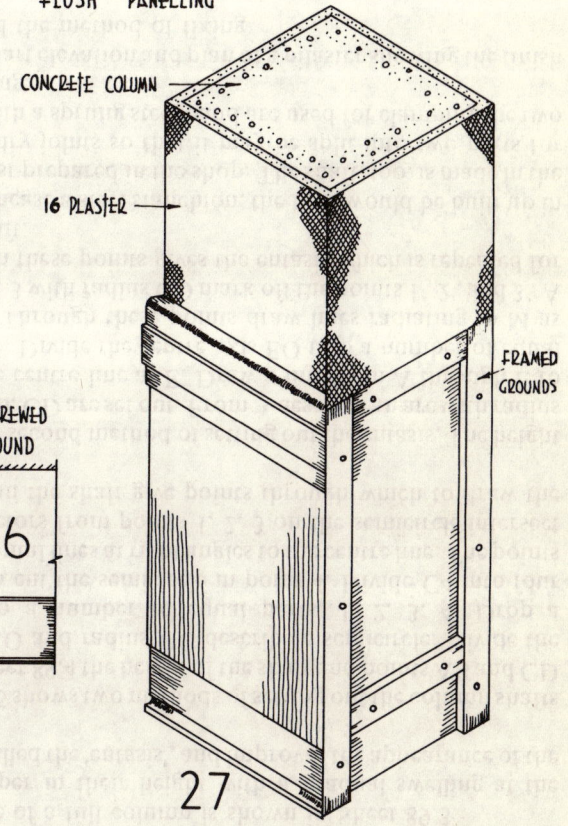

MUNTIN   GROUND

VENEERED PANEL

HORIZONTAL SECTION SHOWING JOINING UP OF PANEL SECTIONS

REBATED BEARER FOR PANEL FIXING

CONCRETE COLUMN

16 PLASTER

FRAMED GROUNDS

27

FLUSH PANELLING TO SQUARE COLUMN

SHEET **88**

241

is shown in Sheet 87.16. A combination of polished hardwood moulded strips alternating with leather-covered narrow panels is used. The hardwood strips are knucklebone in section, grooved in both edges to receive the contrasting leather panels, which are tongued.

A part section through a pilaster showing the panelling and the treatment at an external and internal angle are shown in Sheet 87.17.

The groundwork is prepared in a similar way to the previous examples, the moulded strips and leather panels being pinned at the top and bottom with a plain rebated cornice and plinth fixed as shown.

A further example (Sheet 87.18) shows a complete wall area lined with leathercloth. This has a padded backing applied to 6 mm plywood panels by the upholsterer. A section through the panels is shown in Sheet 87.19, where secret fixing is used.

A section through the plinth is shown in Sheet 87.20.

A modern treatment to dado panelling is shown in Sheet 88.21, giving a flush finish with little projection from the main walls. Here the use of veneered boards in large sheets may be made with the dado rail, capping and skirting of contrasting woods.

The groundwork is prepared as for previous examples, but the fixing is done working up from the skirting, the moulded fillet being fixed first. Again care is needed in fixing so that the appearance is not spoiled.

To avoid fixing through the face of large sheets at the centre, various methods are employed. Sheet 88.22 shows a rebated bearer screwed to the back and interlocking with rebated horizontal grounds plugged to the wall. Sheet 88.23 shows an alternative where metal hook brackets are used, and Sheet 88.24 shows bevelled grounds used at the centre.

The joining up of panel sections may be done using a muntin which also provides a break or relief in large flat surfaces.

The jointing of thin flush panels is shown in Sheet 88.25 using alternative methods. A further method of jointing 19 mm veneered board is shown in Sheet 88.26 where the gluing area at the joint is considerably increased using this method.

A sketch of the groundwork and panelling to a square column is shown in Sheet 88.27.

Columns and pilasters

In panelled work, where sections are to be joined together or stanchions screened, engaged columns and pilasters are used. They also serve as a decorative feature, providing a break in large flat surfaces. Full columns are often placed round steel stanchions supporting floors in public buildings. Both columns and pilasters are made and framed-up in the shop, as is the panelling, ready for fixing on the site. They may be left plain or fluted.

Sheet 89.1 shows a part elevation at the base of a half column, engaged with and joining up two sections of wall panelling. To the left of the centre line the finished work is shown, and to the right the section showing the built-up base and shaft.

The base (Sheet 89.2) is built up of four members arranged with cross or staggered joints shown in plan. The joints are made with loose tongues, stopped short of the face and glued. The plinth is formed by mitring at the angles, glue blocks being used on the inside, and screws used where they will not be seen on the face. The first moulded member is prepared and fitted on the plinth, being jointed on the centre line in two sections. The other members are prepared in the same way, glued and screwed as before. It will be seen that the top moulded member is rebated to receive the shaft. This is built up as shown to the left of the centre line in plan, the jointing between the staves being loose tongues, well glued with glue blocks behind. The staves should be tapered and fluted to suit the shape of the column before being glued.

The plan at the base of a full column is shown in Sheet 89.3.

Columns usually taper in their height with a gradual swelling at the centre. This swelling is called the 'entasis', and improves the appearance of the finished column.

Sheet 89, figs. 4 and 5 shows two methods of setting out the column shafts to give the entasis. In Sheet 89.4 the height of the shaft and points AB and CD are set out. With centre O and radius OC describe a semicircle. Divide the height of the shaft into a number of equal parts, 1, 2, 3, 4. Drop a perpendicular from A to cut the semicircle in point 4. Divide C4 into four equal parts. Draw horizontal lines at right angles to the centre line. The points where the vertical projectors from points 1, 2, 3 on the semicircle intersect horizontal lines 1, 2, 3 on the shaft give points through which to draw the curve of entasis.

Sheet 89.5 shows the second method of setting out the entasis. The height of the shaft and points AB, CD are set out. From A describe an arc with radius equal to CO, cutting the centre line at E. Draw a line from A through E to meet CD produced in M. Divide the centre axis EO into a number of equal parts numbered 1, 2, 3. Through these points draw lines radiating to M as shown. From points 1, 2, 3 with radius CO mark off the points 1', 2', and 3'. A fair curve drawn through these points gives the entasis which is repeated for the other side of the shaft.

If the column is to encase a steel stanchion, the base would be built up in position, having been first prepared in the shop. The shaft, too, is made in the shop and left with two dry joints so that it may be split into two parts for fixing. Special clamps with a sprung steel band are used for clamping the two parts together after gluing.

Sheet 89.6 shows a part elevation and plan of a pilaster showing the finish at the top or capital and the method of fixing.

Seating

It may be said that seating for use in a particular type of building is purpose made. The range in style, design, dimensions, and construction is wide and varied, with most of the work from the design stage to completion carried out by specialist manufacturers of furniture and upholstery.

It is the intention in this section to detail some of the forms of seating to be found in public buildings, to compare designs, particularly heights of seating, and detail those where any joiner's work is involved.

Seating suitable for reception areas, lounges, restaurants, hotels, and bars using metal and wood frames is shown on Sheet 90.1.

COLUMN CONSTRUCTION AND

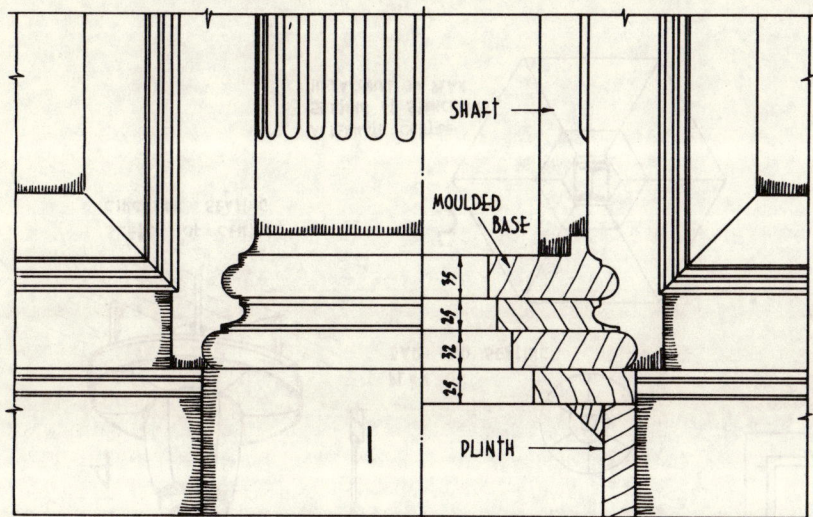

SETTING OUT

AB = DIAMETER AT CAPITAL
CD = DIAMETER AT BASE
AE = CO

SHAFT

MOULDED BASE

35 35
25 25
25 32 25
25

PLINTH

PART ELEVATION   PART SECTION

1

ENTASIS   SHAFT

4

5

LINES RADIATING TO POINT M

SETTING OUT OF COLUMNS

PANELLING

EX. 88×38

19×5 TONGUES

19×5 FLUTES

168

338

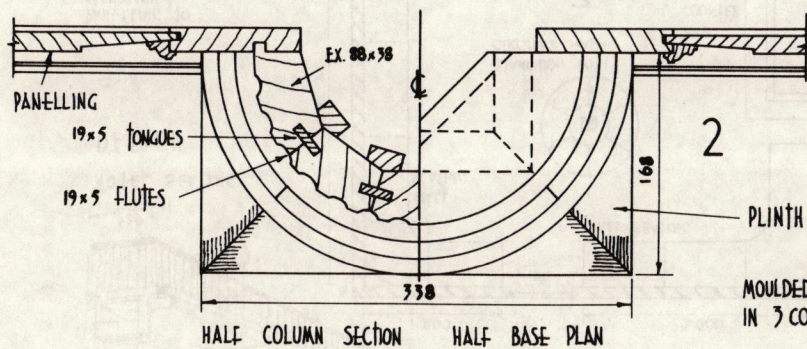

HALF COLUMN SECTION   HALF BASE PLAN

DETAILS OF ENGAGED COLUMN

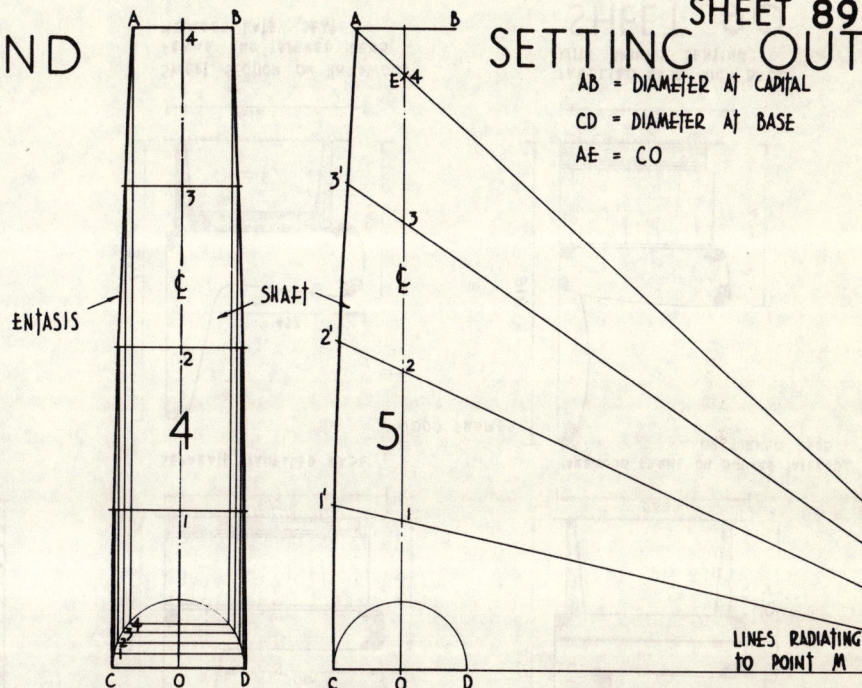

2

COLUMN

PLINTH

MOULDED BASE IN 3 COURSES

SHAFT

3

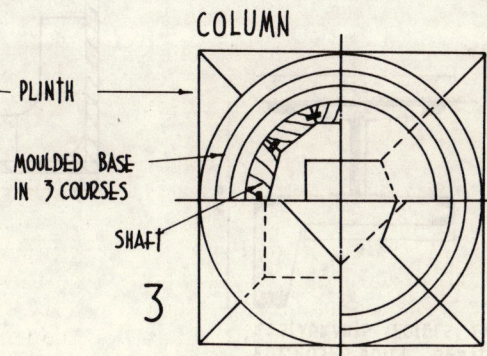

PLAN AT BASE OF COLUMN

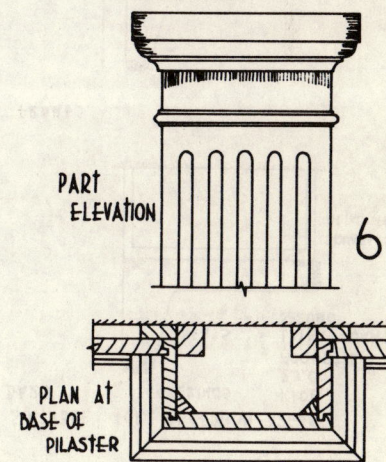

PILASTER

PART ELEVATION

6

PLAN AT BASE OF PILASTER

243

SEATING

HOTEL SEATING DESIGN

1·200
700
525

2

PANELLING to STANCHION

SKETCH OF CENTRE CIRCULAR SEATING

4

WALL SEATING

WALL SEATING

1·800
3·000

4·200

STANCHION WITH CIRCULAR SEATING

BAR COUNTER

PLAN OF BAR AND SEATING

3

ALTERNATE CENTRE SEATING AT STANCHIONS HEXAGONAL IN PLAN

5

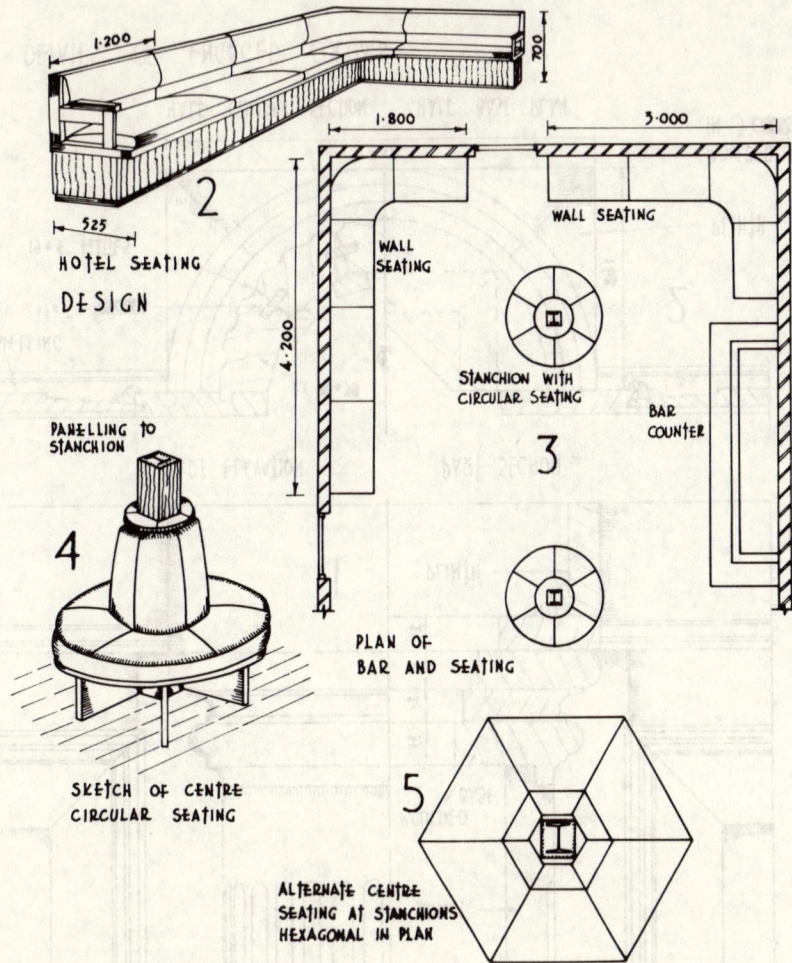

PURPOSE BUILT SEATING SUITABLE FOR RECEPTION AREAS, RESTAURANTS, HOTELS AND BARS. COVERINGS : HIDE
P.V.C
MOQUETTE
SEATS : LATEX, FOAM, SPRUNG

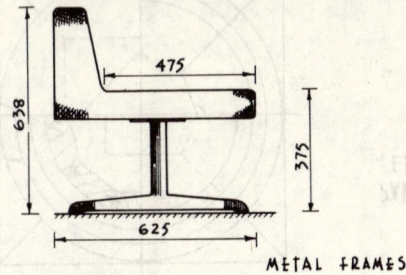

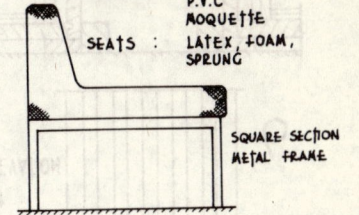

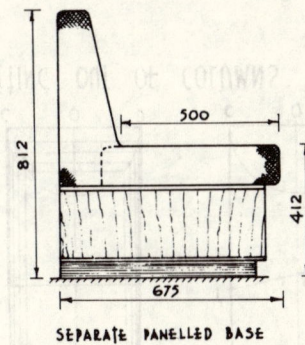

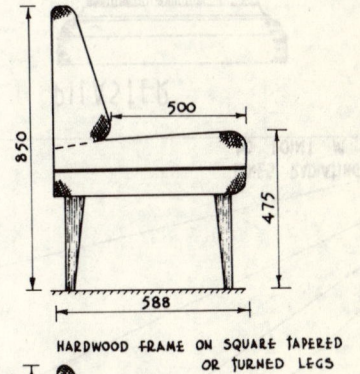

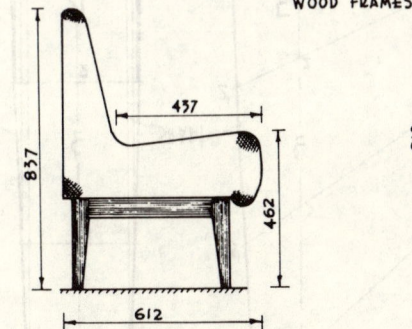

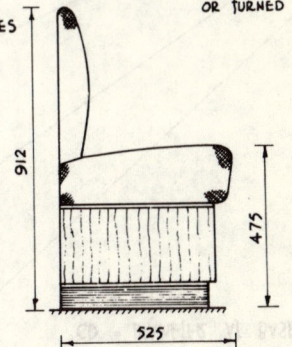

638
475
375
625

METAL FRAMES

SQUARE SECTION METAL FRAME

812
500
412
675

SEPARATE PANELLED BASE

850
500
475
588

HARDWOOD FRAME ON SQUARE TAPERED OR TURNED LEGS

WOOD FRAMES

837
437
462
612

SWEPT SECTION ON HARDWOOD FRAME AND TAPERED LEGS, MOULDED LATEX SEAT

912
475
525

PANELLED BASE, HIGH BACK WITH SPRUNG SEATING

SHEET 90

SEATING

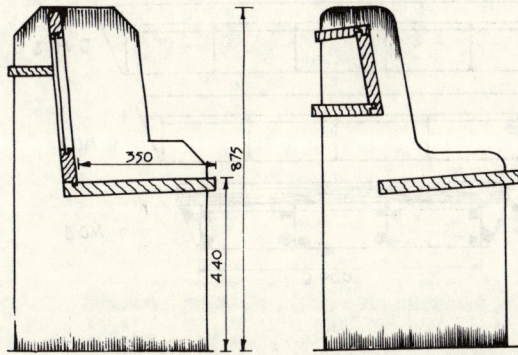

CHURCH SEATING WITH
SOLID ENDS.
3

350
875
440

SEATING FOR RECEPTION AREAS
LOUNGES, RESTAURANTS, HOTELS & BARS
1

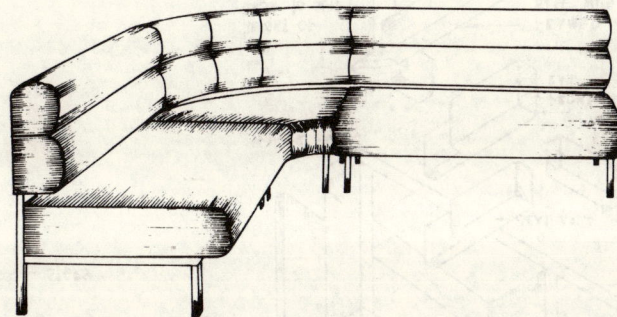

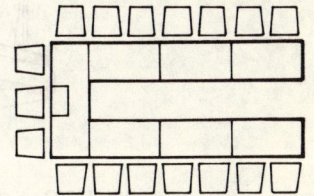

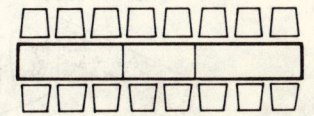

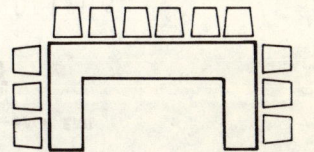

2

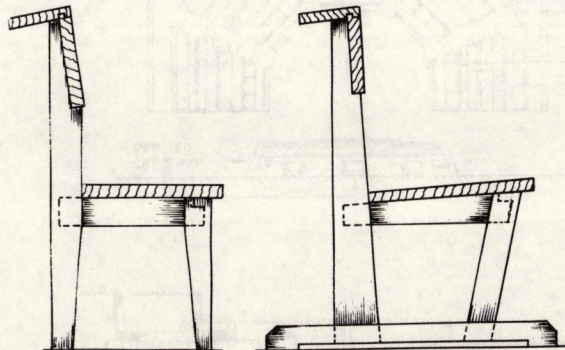

CHURCH SEATING - BENCH
TYPE WITH SUPPORTS

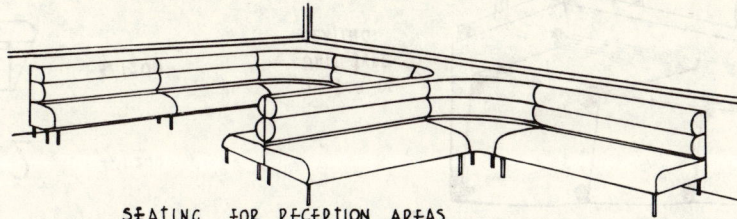

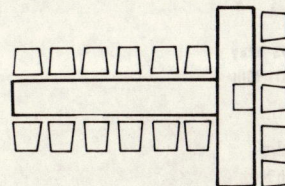

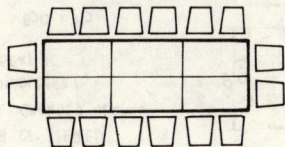

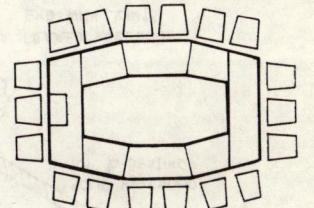

LAYOUT OF SEATINGS IN:
CONFERENCE ROOMS,
BOARD ROOMS,
COUNCIL CHAMBERS.

SHEET 91

SEATING

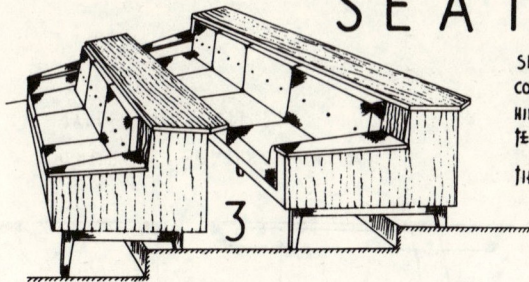

SKETCH OF TIERED
COUNCIL CHAMBER SEATING
HIDE UPHOLSTERY,
TEAK SEATING

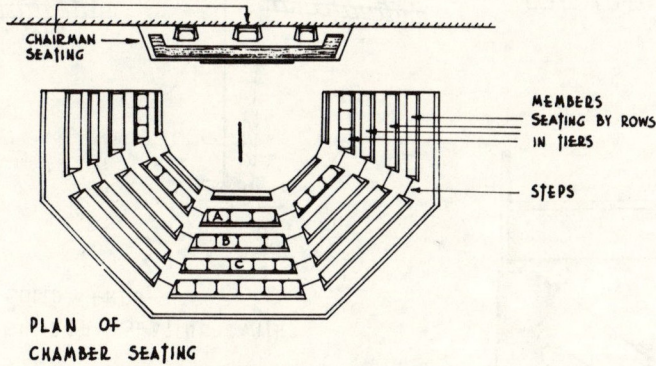

TIERS   850 x 115

3

CHAIRMAN
SEATING

MEMBERS
SEATING BY ROWS
IN TIERS

STEPS

1

A
B
C

PLAN OF
CHAMBER SEATING

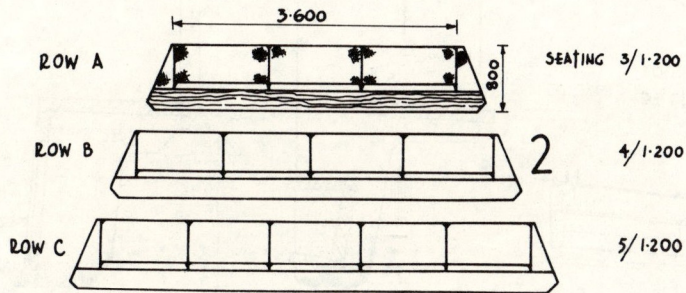

3·600

ROW A   800   SEATING 3/1·200

ROW B   4/1·200

2

ROW C   5/1·200

PLAN OF 3 CENTRE ROWS
OF SEATING

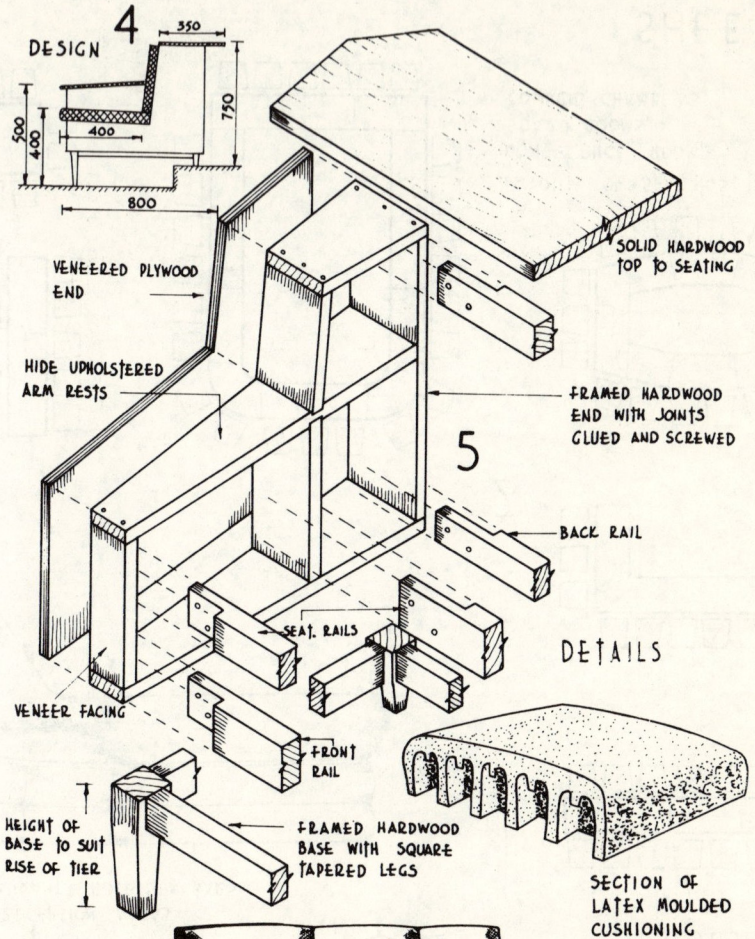

DESIGN 4

350

500   750
400   400

800

VENEERED PLYWOOD
END

HIDE UPHOLSTERED
ARM RESTS

SOLID HARDWOOD
TOP TO SEATING

FRAMED HARDWOOD
END WITH JOINTS
GLUED AND SCREWED

5

BACK RAIL

SEAT RAILS

DETAILS

VENEER FACING

FRONT
RAIL

HEIGHT OF
BASE TO SUIT
RISE OF TIER

FRAMED HARDWOOD
BASE WITH SQUARE
TAPERED LEGS

SECTION OF
LATEX MOULDED
CUSHIONING

SKETCH OF
UNIT TYPE
SEATING

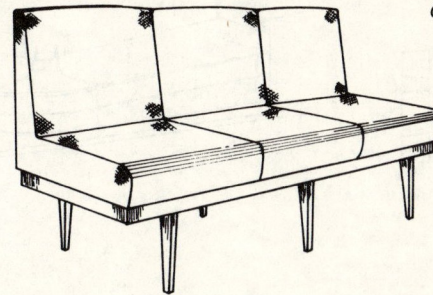

SHEET 92

PANELLING & SEATING

DESIGN

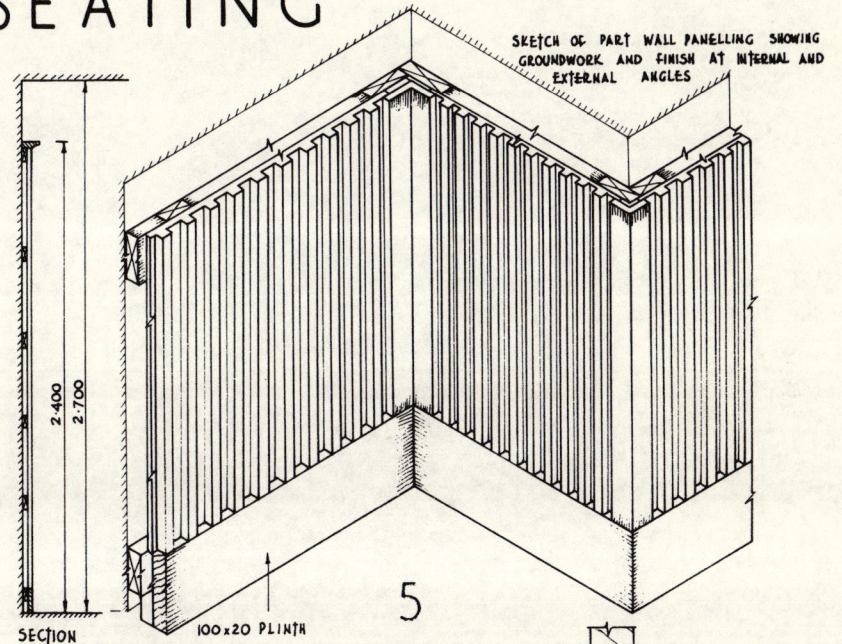

SKETCH OF PART WALL PANELLING SHOWING GROUNDWORK AND FINISH AT INTERNAL AND EXTERNAL ANGLES

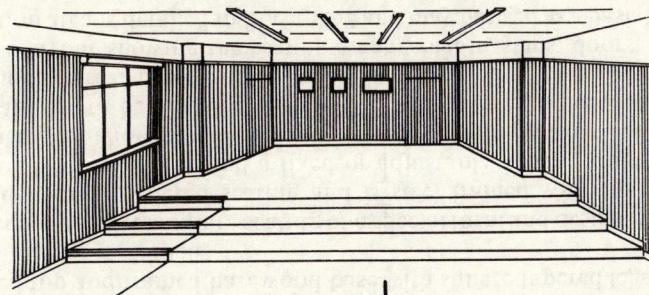

1

2.400
2.700

SECTION

100 x 20 PLINTH

5

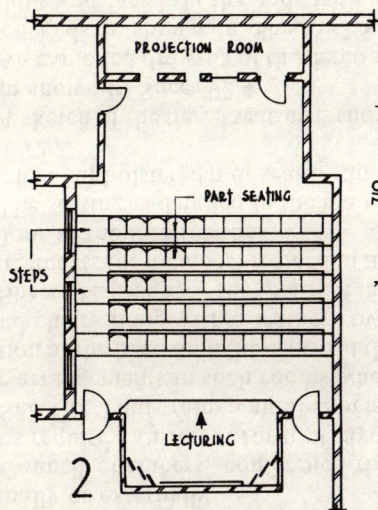

PROJECTION ROOM

PART SEATING

STEPS

LECTURING

2

PLAN OF LECTURE THEATRE SEATING

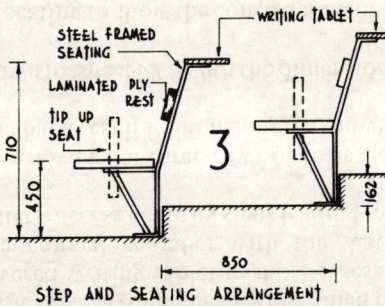

WRITING TABLET

STEEL FRAMED SEATING

LAMINATED PLY REST

TIP UP SEAT

3

710
450

162

850

STEP AND SEATING ARRANGEMENT

DETAILS

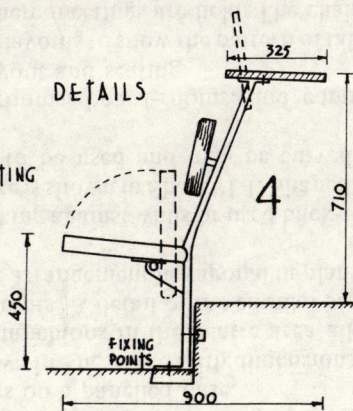

325

710

450

FIXING POINTS

900

4

DETAILS

20

12   9   9

PANELLING DETAIL WALNUT

25

EXTERNAL ANGLE MAPLE

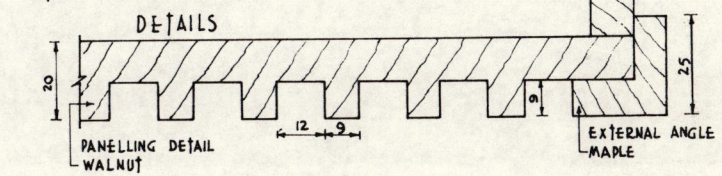

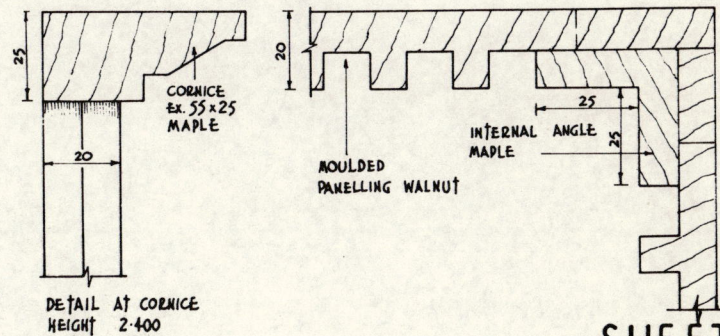

25

20

CORNICE EX. 55 x 25 MAPLE

DETAIL AT CORNICE HEIGHT 2.400

20

MOULDED PANELLING WALNUT

INTERNAL ANGLE MAPLE

25

25

SHEET 93

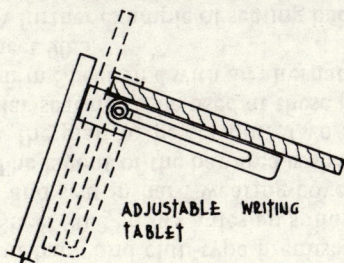

ADJUSTABLE WRITING TABLET

248

The coverings used may be hide, PVC, moquette, etc. with seats of latex, foam, or sprung. Designs include high backs, panelled bases, moulded seats, and metal or tapered wooden legs.

Wall seating is a form of seating usually associated with public houses, certain bars, and club-type premises.

Sheet 90.2 shows a design suitable for a bar area with upholstered high back and seat in hard-wearing covers on a panelled base.

The layout of the bar area is shown in Sheet 90.3 with dimensions which allow the plan to be set out. Two stanchions in the centre area allow for circular seating to be used at these points. A detail of the circular seating is shown in Sheet 90.4 with an alternate arrangement hexagonal in plan shown in Sheet 90.5.

A further example of seating backing against walls or used back-to-back in reception areas, lounges, and the like is shown in Sheet 91.1. Shaped corner units allow varying seating layouts to be used and may be curved either internally or externally.

Council chambers, conference rooms, board-rooms, and committee-rooms require a different form of layout and seating.

Sheet 91.2 illustrates a number of layouts to show the pattern of table and chair arrangements in such rooms where meetings are held. The chairman's position at the head or in the centre of the table layout is clearly identified by a special chair, along with a leather-covered writing tablet in most cases.

Sheet 92.1 shows the plan of a council chamber with the seating arrangements for members formed in tiers, access being by steps leading from the floor at the intersections.

The seating is detailed in the plan drawings of three rows of seats, each made up of upholstered hide seating and backs each 1·200 m long, as shown in Sheet 92.2.

A sketch of the tiered seating is shown in Sheet 92.3 with the dimensioned design shown in Sheet 92.4.

An exploded drawing of the tiered seating to show the construction of the framed ends is shown in Sheet 92.5. These are upholstered at the arm positions and faced on the ends with veneered plywood. A solid hardwood polished top and framed hardwood base with square tapered legs to suit the rise of the tier is shown.

Lecture theatres require seating of a special form and design. Usually this takes the form of tiered seating and is steel framed with tip-up seats, a laminated-ply back rest, and a fixed or adjustable writing tablet.

Sheet 93.1 shows a panelled lecture theatre with part of the floor tiered. The walls are to be panelled in moulded strip with entrance and exit doors alongside the rear projection-room.

The layout showing the seating arrangements, steps, doors, stage, and projection area is detailed in Sheet 93.2 and may be used as a basis for dealing with the various joinery items in the project.

Sheet 93.3 and 93.4 shows the details of the step and seating arrangements.

The wall-panelling details are shown in Sheet 93.5 where contrasting timbers are used. The moulded strip is walnut with the internal and external angles finished in maple.

Horizontal and vertical grounds are required as for traditional panelling.

Sheet 91.3 shows the sections of varying designs of pews or seating in church work. Two have pew ends into which the actual seat and back are housed. The lower two are bench type in design and have no ends, therefore the seats must have framed supports along their length.

Chapter 20

Fittings

The types of fittings needed for shops, libraries, banks, schools, and other public buildings are many and varied. Independent units or built-in fixtures include counters, cupboards, display units, benches, wall fittings, open shelves, and the like. Such fittings involves framing members together in three directions. For example, a bench may be made of a front frame and a back frame joined in the other direction with cross-rails as shown in Sheet 94.1.

The front and back frames consist of top and bottom rails dovetailed into end rails, with intermediate rails forming the cupboard areas housed into them. The cross-rails are tongued at each end into grooves on the inside edges of the front and back frames. The ends and vertical divisions to the cupboards are formed by panels of plywood where a flush finish is required or by framed ends. The cut-away section shows how the frames stand on a plinth or base which is used to form the bottoms to the cupboards. The solid jointed top is allowed to oversail the front and sides and is fixed with buttons. These are wooden buttons in this case, which are screwed into the top and engage in grooves or rebates in the carcass. This arrangement allows for any movement in the top. Slotted metal table plates (illustrated on page 278) are more often used as an alternative to wooden buttons for the fixing of unit tops.

A unit designed with a finely finished hardwood top, front, and ends would have the remaining carcassing in softwood.

Fittings which are too large to be transported and installed as a single unit have to be designed to be broken down into smaller sections and fitted together on site.

Sheet 94.2 shows the detail of the jointing between the members in the three directions with provision for wooden button fixing of the top.

Sheet 94.3 shows the framing of members to accommodate drawers in the end section, with a detail of the machine dovetailed jointing of a drawer.

Laboratory benches

Sheet 95.1 shows a pictorial view of a typical laboratory in an educational establishment. The layout is given in the detailed plan (Sheet 95.2) and includes wall benches with drawer and storage space under each unit running the length of the windowed exterior wall.

Three demonstration benches provided with under-bench units back-to-back are sited to allow part seating and part standing arrangements. Across the end wall are side-by-side fume cupboards to complete the fittings which are all provided with the usual services of a laboratory – gas, water, electricity, and waste disposal.

It is intended to use the laboratory situation in these examples as a student project, to deal in detail with each fitting in turn so that the student will appreciate the work involved in a given situation of this type involving purpose-made joinery.

The design aspect is determined and dealt with by the architect as expressed in the pictorial sketch, the plan is the most important detail which requires the dimensions to be checked on site before any actual setting out of the work can be started by the setter-out in the shop.

It will be seen that the benches are too large to handle and install in one unit and must therefore be made up of a series of complete carcasses which are placed against each other, or joined by rails, the top being fixed afterwards. The carcasses are machined and assembled in the shop, resulting in very little site work.

Sheet 95.3 shows a wall-bench system consisting of a series of framed leg supports to the hardwood work top. Cleating supports at the wall provide the necessary space behind the legs and under the top for the gas and water services to points where required. The leg supports are connected at the front and back by rails acting as bracing members as shown.

The sketch (Sheet 95.4) shows a run of drawer and storage cupboard under-bench units forming either wall benches or demonstration benches. The units are standard heights, widths, and lengths so that any pattern of units may be used singly or back to back, with the work top fixed over and to the units, using shrinkage plates.

The construction of each unit, whether drawer or cupboard design, follows the carcass construction in previous work except that veneered blockboard ends are used, as shown in Sheet 95.5. The rails are dovetailed into the ends and are covered by the drawer fronts.

Flush-fitting handles are used fixed to bevelled drawer fronts which allow finger room, as shown in Sheet 95.6.

Research workers, laboratory technicians, and the like using the laboratory continually require that benches be provided with storage space beneath them related to the needs of the user – everything from paperwork and filing to glassware and apparatus.

A work station requires a kneehole recess to provide a comfortable seated position for practical and written work. The kneehole can be fitted with a plain rail or a drawer for the storage of small items and can be positioned where required.

BASE UNITS & FUME CUPBOARDS

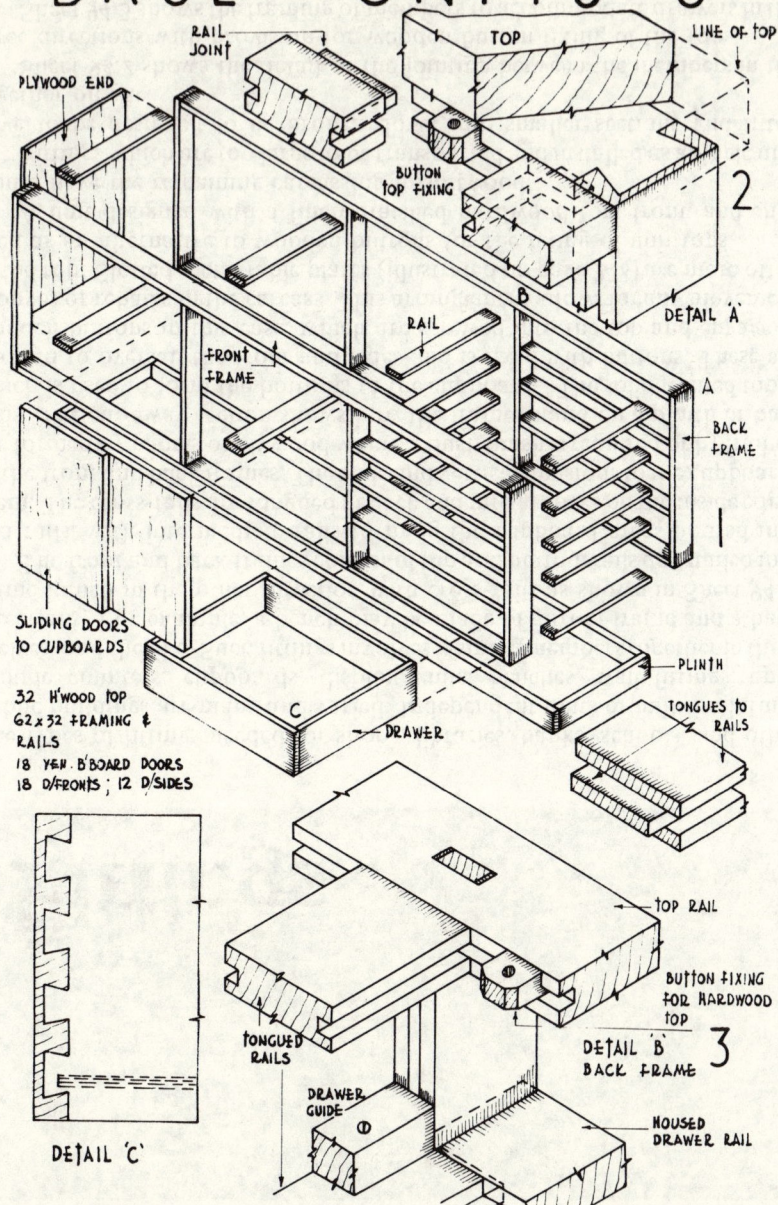

PLYWOOD END

RAIL JOINT

TOP

LINE OF TOP

BUTTON TOP FIXING

DETAIL 'A'

2

RAIL

FRONT FRAME

A

BACK FRAME

B

1

SLIDING DOORS TO CUPBOARDS

32 H'WOOD TOP
62 x 32 FRAMING &
RAILS
18 YEN. B'BOARD DOORS
18 D/FRONTS; 12 D/SIDES

PLINTH

TONGUES TO RAILS

C

DRAWER

TOP RAIL

BUTTON FIXING FOR HARDWOOD TOP

TONGUED RAILS

DRAWER GUIDE

HOUSED DRAWER RAIL

DETAIL 'B'
BACK FRAME

3

DETAIL 'C'

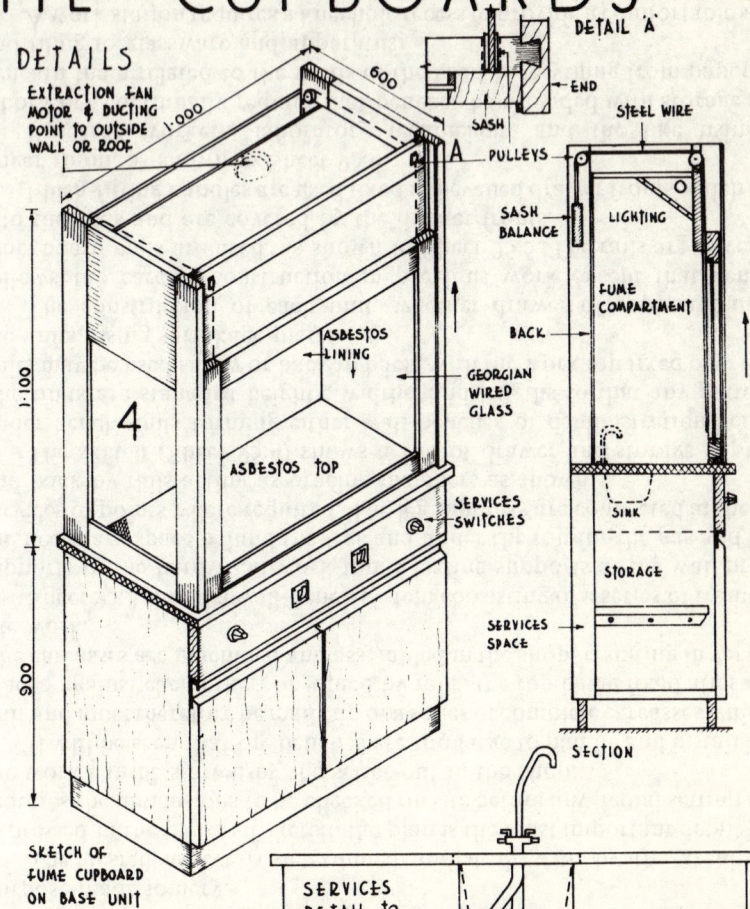

DETAILS

DETAIL 'A'

EXTRACTION FAN MOTOR & DUCTING POINT TO OUTSIDE WALL OR ROOF
1:000

600

SASH

END

STEEL WIRE

PULLEYS

SASH BALANCE

LIGHTING

1:100

4

ASBESTOS LINING

GEORGIAN WIRED GLASS

ASBESTOS TOP

FUME COMPARTMENT

BACK

SINK

SERVICES SWITCHES

STORAGE

SERVICES SPACE

900

SECTION

SKETCH OF FUME CUPBOARD ON BASE UNIT

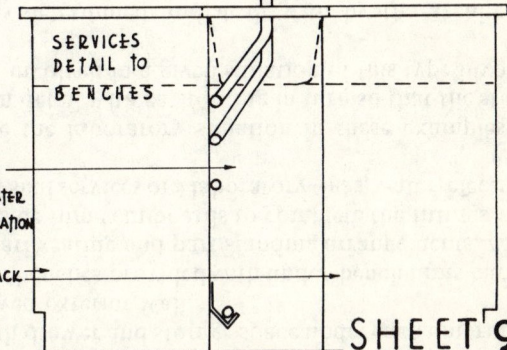

SERVICES DETAIL to BENCHES

5

PROVISION FOR SERVICES, GAS, WATER
ELEC. IN DEMONSTRATION BENCHES

UNITS BACK TO BACK

SECTION

SHEET 94

LABORATORY WORK

1 DESIGN

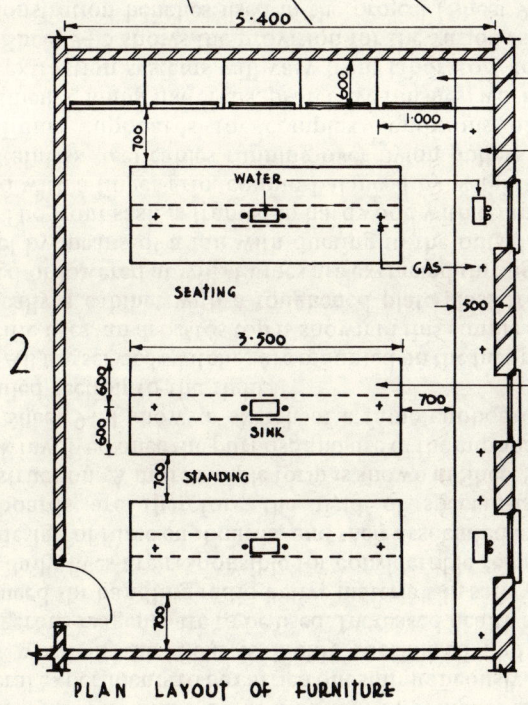

2 PLAN LAYOUT OF FURNITURE

- 5·400
- 1·000
- 700
- WATER
- GAS
- SEATING
- 3·500
- 600
- 600
- SINK
- 700
- STANDING
- 700
- 500

FUME CUPBOARDS SIDE BY SIDE WITH SERVICES – LIGHTING, GAS, WATER

WALL BENCH WITH DRAWER AND STORAGE UNDER BENCH UNITS WITH SERVICES – GAS AND WATER

DEMONSTRATION BENCHES WITH UNDER BENCH UNITS BACK TO BACK SERVICES AS SHOWN

FUME CUPBOARDS & LABORATORY BENCHES

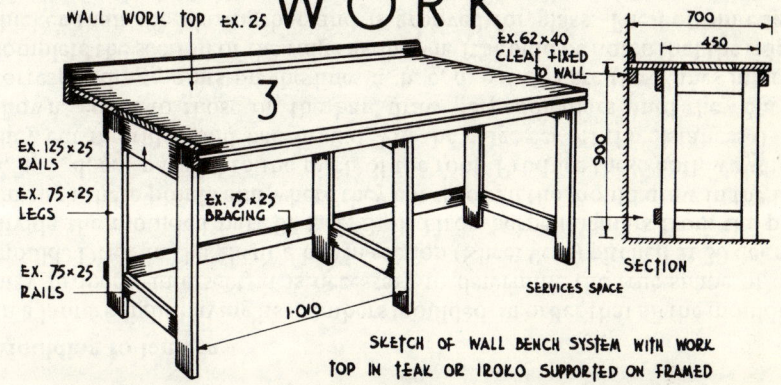

3

- WALL WORK TOP EX. 25
- EX. 62×40 CLEAT FIXED TO WALL
- EX. 125×25 RAILS
- EX. 75×25 LEGS
- EX. 75×25 BRACING
- EX. 75×25 RAILS
- 1·010
- 700
- 450
- 900
- SECTION
- SERVICES SPACE

SKETCH OF WALL BENCH SYSTEM WITH WORK TOP IN TEAK OR IROKO SUPPORTED ON FRAMED LEG SUPPORTS WITH SERVICES SPACE BEHIND

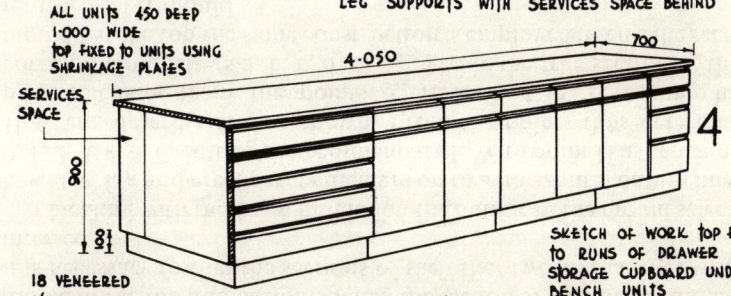

4

- ALL UNITS 450 DEEP 1·000 WIDE TOP FIXED TO UNITS USING SHRINKAGE PLATES
- SERVICES SPACE
- 4·050
- 700
- 900

SKETCH OF WORK TOP FIXED TO RUNS OF DRAWER AND STORAGE CUPBOARD UNDER-BENCH UNITS

5 PART PLAN AT FRONT CORNER

- 18 VENEERED B'BOARD END
- 150×15 DRAWER SIDE
- EX. 62×25 TOP RAIL

6 TOP AND DRAWER HANDLE DETAIL

- EX. 25 TEAK TOP
- EX. 62×25 TOP RAIL
- FLUSH FITTING BLACK PLASTIC HANDLE
- EX. 150×18 DRAWER FRONT

SHEET 95

Fume cupboards

The laboratory designed in Sheet 95.1 has side-by-side fume cupboards sited at the end wall.

Fume cupboards are used for the containment and removal of hazardous or toxic materials. In the majority of laboratories, total fume-cupboard capacity is best made up of a battery of compact fume cupboards enabling several experiments to be carried out simultaneously. It should be pointed out that work in this field is highly specialised and involved where highly dangerous reagents are to be used. Increased health and safety requirements, the need for handling radio-active materials in safety, the amount of noise in our daily lives are responsible for considerable research into air movement, the design of fume cupboards, and their associated extraction systems. Fume cupboards are therefore the field of specialists in their design and construction. A more simple form is shown in Sheet 94 to indicate the joiner's work involved once the purpose and use of the cupboard has been established.

Sheet 94.4 shows a sketch of a fume cupboard on a base unit with a detailed section to the right.

All the service switches are mounted on the face panel with a work top for specific uses, an asbestos top is shown in this simple system. The cupboard is basically a cabinet with a toughened plate-glass front panel which can be raised or lowered at will. Fumes are extracted through an aperture in the top panel by means of a fan with ducting to the outside.

The front sash is framed in hardwood with 6 mm toughened glass and is fitted with a finger grip, counterbalanced by weights at each side suspended on stainless steel cables running over nylon pulleys.

Fume cupboards for complex conditions of use require different treatments; lining materials, protective finishes, work-top materials, services, and extraction systems will vary from laboratory to laboratory.

Sheet 94.5 shows the provision for the various services used in island or demonstration benches used in the project (Sheet 95.1).

Raking mouldings

Before dealing with the problems on this subject, it is necessary to mention certain points. Moulds may be required to suit varying conditions as follows:

(a) Two level moulds and one raking over a square plan.
(b) Two raking moulds over a square plan.
(c) One level mould and one raking over an obtuse-angled plan.
(d) Two level moulds and one raking over obtuse- and acute-angled plans.
(e) Two raking moulds over an obtuse-angled plan.

Any mould in the above conditions may be taken as the given mould, from which the others may be developed.

The following points should be noted:

1. The true shape at the intersection of two moulds with a common intersection on a vertical mitre is the same for both moulds.
2. The section of one mould must always be known or given.
3. The inclination of the raking moulds and the plan angles will be known.

4. The thickness of raking moulds with common intersections is the same for each pair of moulds.
5. The widths of raking moulds differ in accordance with the pitch of the moulds.

Sheet 96.1 shows the key plan for condition (c) above over an obtuse-angled plan, Sheet 96.2 the geometrical construction with the level mould being the given mould. The plan is first drawn and the section of the given mould set out as shown, giving the points 0, 1, 2, 3, 4, 5, 6, 7, 8. Ordinates projected from the section of this mould on to the mitre line in plan are projected up into elevation to intersect with the given mould ordinates to determine the elevation of the mitre. The elevation of the mitre is reversed and the ordinates projected on to this. These points are drawn to the pitch of the raking mould to determine the required section as shown.

Sheet 96.3 shows the method of determining the bevels for the moulds. The bevel for the inclined mitre is shown at AB″ while the bevel for the level mould is shown in plan AB. The bevel at A′ in the elevation is the side bevel for the back of the moulding.

Mouldings to pediments

Sheet 96.4 shows the elevation of an inclined mould with a returned level mould at the top and bottom. When the section of the bottom mould is given, it is necessary to find the sections of the other two so that they will properly intersect.

Dealing with the given mould: on the outline of the mould select points a, b, c, d, e, f, g and erect perpendiculars on to a horizontal datum line in points 0, 1, 2, 3, 4, 5, 6, cutting the pediment at H. With point H as centre and radius H–6, etc. describe arcs as shown. From the points thus determined, drop perpendiculars from the points 0, 1, 2, 3, 4, 5, 6, cutting lines from corresponding letters a, b, c, d, e, f, g, giving the true section of the inclined mould. At the top the same construction is employed to give the section of the returned level mould.

Moulding to lantern

In a lantern light having its members moulded, in order that all the mouldings may properly intersect, it is necessary to determine the true shape of each mould. Draw in the glazing bar in section (Sheet 96.5) pitched at 30 degrees, divide the moulded part 0, 1, 2, 3, 4. Drop perpendiculars from the pitch through these points and where they cut through the mould draw in the lines a, b, c, d, e, f parallel to the pitch of the roof. Produce these both ways until they cut the ridge and bed mould. On the ridge set off the distances 0–4 as shown, equal to those on the bar, draw perpendiculars until they cut the corresponding points on the lines a, b, c, d, e, f. Where these lines intersect complete the section of the ridge as shown. It should be noted that the ridge is thicker than the glazing bar and is grooved for glass. The bottom edge is usually finished as shown.

The true shape of the bed mould is found in the same way as for the ridge.

To find the true shape of the hip rafter the dihedral angle is required. The distances a, b, c, d, e, f, g, h are taken from the glazing bar section and the geometrical construction to determine the true section, the same as for the ridge.

RAKING MOULDINGS

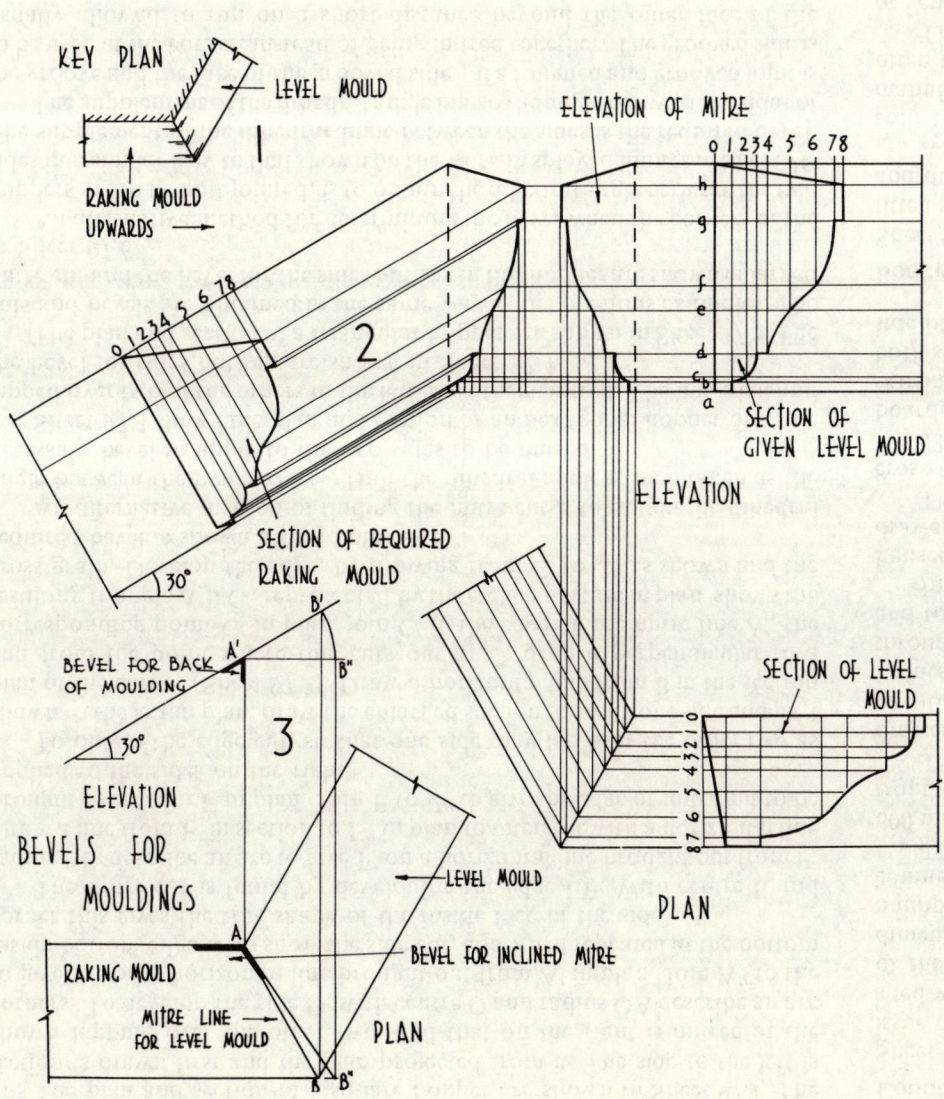

KEY PLAN

LEVEL MOULD

1

RAKING MOULD
UPWARDS →

ELEVATION OF MITRE

0 1 2 3 4 5 6 7 8

2

h
g
f
e
d
c
b
a

SECTION OF
GIVEN LEVEL MOULD

0 1 2 3 4 5 6 7 8

SECTION OF REQUIRED
RAKING MOULD

30°

ELEVATION

MOULDINGS TO PEDIMENTS

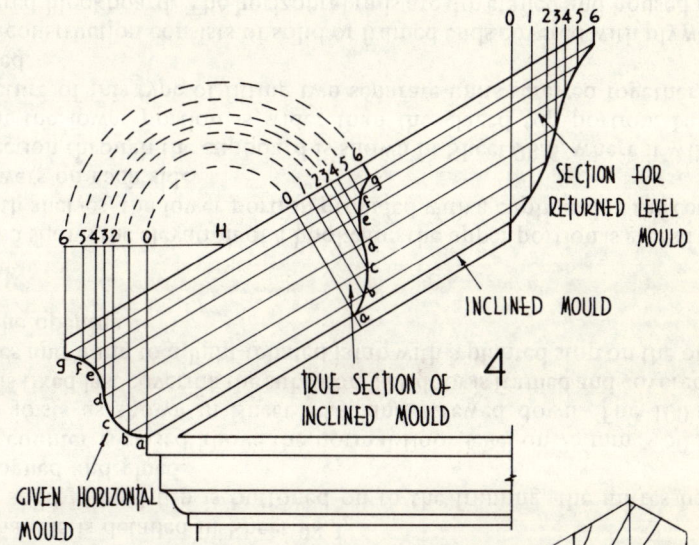

0 1 2 3 4 5 6

0 1 2 3 4 5 6
H

SECTION FOR
RETURNED LEVEL
MOULD

INCLINED MOULD

4

TRUE SECTION OF
INCLINED MOULD

GIVEN HORIZONTAL
MOULD

BEVEL FOR BACK
OF MOULDING

A'
B'
B"

30°

3

ELEVATION

BEVELS FOR
MOULDINGS

A

RAKING MOULD

B"

MITRE LINE
FOR LEVEL MOULD

PLAN

LEVEL MOULD

BEVEL FOR INCLINED MITRE

SECTION OF LEVEL
MOULD

8 7 6 5 4 3 2 1 0

PLAN

MOULDINGS TO LANTERN

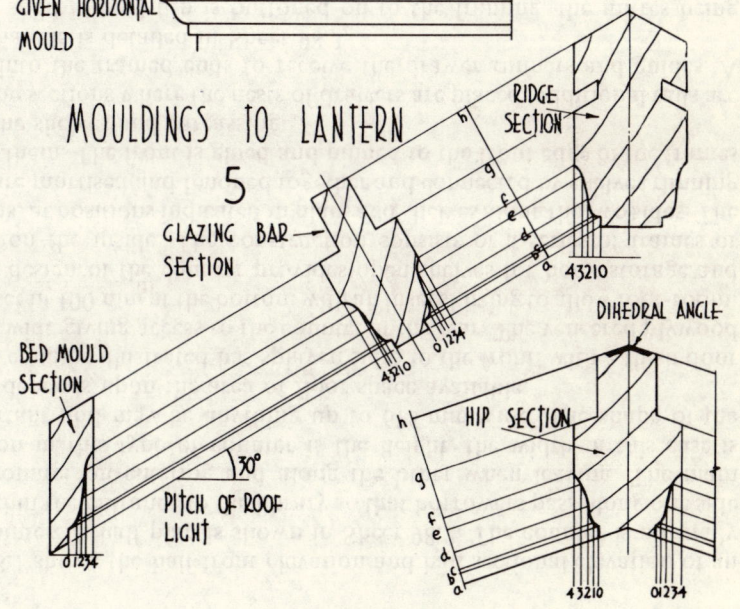

RIDGE SECTION

5

GLAZING BAR
SECTION

h
g
f
e
d
c
b
a

4 3 2 1 0

0 1 2 3 4

4 3 2 1 0

DIHEDRAL ANGLE

BED MOULD
SECTION

HIP SECTION

h

g

f
e
d
c
b
a

30°

PITCH OF ROOF
LIGHT

0 1 2 3 4

4 3 2 1 0   0 1 2 3 4

Hoppers

A hopper is an open box having inclined sides with the top edges made parallel to the horizontal or left square.

The plan and section of a square hopper are shown in Sheet 97.1. The section is drawn first and the plan projected from it. The side to the left is shown lapping over the other two, and that on the right is mitred at the corners. To develop the side D, with centre O and radius OA describe an arc to give A″ on a horizontal line brought out from A′ in plan. Join A″ to the inside bottom edge O′ to show the side cut, and when repeated at the bottom corner this gives the true shape of the inside face of the side.

The mitre cut is found by developing the edge EF. With centre E and radius EF, describe an arc to give F′ on a horizontal line brought out from E. Draw a line from F′ in section to F″ in plan to intersect with a horizontal line brought over from F in plan. Join E to F″ to give the edge or mitre cut to be applied to the sides on the right.

To obtain the edge cut, so that one side may lap over the other two as shown to the left in plan, draw the enlarged section through one side and part plan of the corner (Sheet 97.2). Draw a horizontal line from B in the section and from the point where this cuts out at C, drop a perpendicular to a corresponding point C′ in plan; join C′ to the edge on the mitre line A′, the resulting triangle A′B′C′, represented by the etched portion in plan, shows the cross grain. Develop the edge, by drawing from C′ to C″ as shown and the required bevel is shown after joining A′C″.

An alternative method of finding the mitre cut is to obtain the dihedral angle between the two surfaces. Half the dihedral angle is then taken as the necessary bevel to apply to the two sides to be mitred.

Sheet 97.3 shows the plan and section of an hexagonal hopper one side lapped over two of the others to the left, the others mitred. The mitre cut and the bevel for sides butting are shown in Sheet 97.4.

The plan and section of a triangular hopper are shown in Sheet 97.5. The cuts and bevels are obtained in the same way as the previous examples. The mitre cut and the bevel for the sides butting in this hopper are shown enlarged in Sheet 97.6.

An alternative method for determining the bevel when the corners of the hoppers are to be butt jointed, is to obtain the dihedral angle between the two sides in a similar way to that shown on the work in splayed linings in Sheet 46. The supplement of the dihedral angle between the sides is the required bevel.

The supplement of the dihedral angle is also the angle between the sides of the groove and the face of the grooved side – if a tongued and grooved joint is to be used at the corner, instead of being mitred together. The grooved side is usually allowed to run on a short distance beyond the outer face of the tongued side to give additional support to the tongue when this particular jointing is used.

Library furniture

This comprises issue counters, bookcases, wall fitments, shelves, card index cabinets, magazine racks, tables, and newspaper stands. Again this type of work provides the highest class of joiner's work, done in hardwood and polished, all to match. The following illustrations are typical of those to be found in most parts of this country.

Counters

Sheet 98.1 shows the half front elevation and half sectional elevation of an issue counter. A half plan is shown in Sheet 98.2. The counter is normally sited within the entrance to the library so that borrowers pass along one side of the counter on entering and along the other when leaving. The main dimension in this type of counter is the height, the width in this case is unimportant and may be anything up to 675 mm wide. The shape of the counter depends upon the area of floor space available.

The counter illustrated has splayed sides to the front, with a flush door 600 mm wide giving access to the counter at the rear. The veneered plywood front is set in 100 mm at the bottom with an inset skirting to allow foot-room.

The design of the counter provides open shelves for book storage and drawers on the inside. The construction consists of a series of frames or partitions, at positions indicated in plan, and shelves along the two sides. The frames are mortised and tenoned together and connected by shelves running through them. The front is glued and pinned to the front edge of the frames and to the shelves and carcassing.

In the sections where the nests of drawers are placed, additional rails are housed into the framed ends to receive the drawer runners and guides. A drawer handle is detailed in Sheet 98.3.

The solid jointed top is buttoned on to the framing, the mitres being cross-tongued and glued.

The counter is raised above the normal floor level on 75 mm × 50 mm boarded joists as shown in Sheet 98.4, and screwed down. The 100 mm skirting is fixed last covering the subfloor. The door is framed and covered on both sides and hung to a light rebated jamb with a planted stop on the other side of the opening.

Bookcase

Sheet 98.5 shows the elevation of a bookcase, the upper portion is glazed and fitted with shelves, the lower portion provided with a cupboard in the centre and drawers on each side.

A section through the cupboard is shown in Sheet 98.6, where it will be seen that the lower portion is wider than the glazed top portion. In the manufacture of this type of fitting two separate units screwed together are often used.

The construction consists of solid or framed ends covered with plywood or veneered blockboard. The horizontal rails are dovetailed and housed into the ends, with the vertical members housed into the rails. The tops of both units are slot-screwed to the rails. Both units may have independent plywood backs nailed on, or a single back rebated in as shown.

Veneered blockboard doors with all the edges lipped and rebated at the centre are fitted to the lower cupboard, with the plywood bottom forming the rebate and a planted stop at the top.

The shelves in the top portion are loose and are adjustable on patent strip fittings shown in Sheet 98.8 for varying sizes of books. The strip fittings are

OBLIQUE WORK - HOPPERS

a) SQUARE
b) HEXAGONAL
c) TRIANGULAR

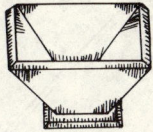

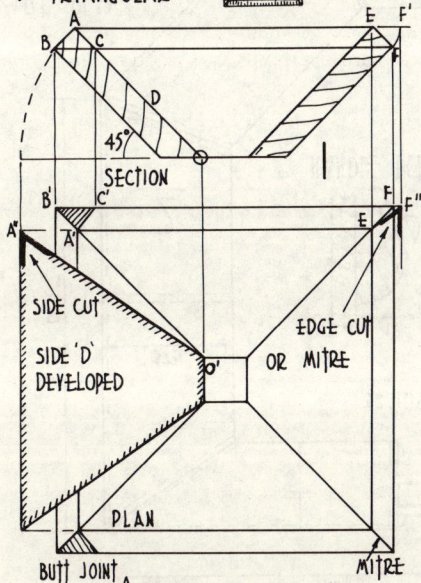

A B C E F'
SECTION 45°
D
1

SIDE CUT
SIDE 'D' DEVELOPED
EDGE CUT OR MITRE
B' C' F F''
A' A' E
O'
PLAN
BUTT JOINT   MITRE

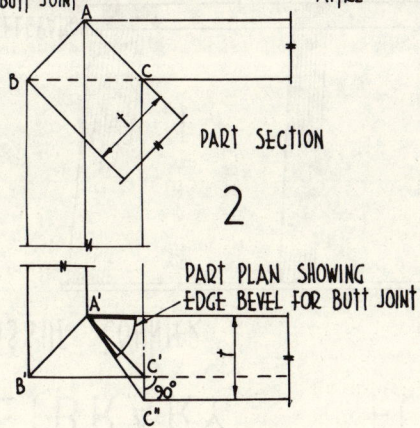

A
B   C
PART SECTION
2
N
A'
B'   C'
90°
C''
PART PLAN SHOWING EDGE BEVEL FOR BUTT JOINT

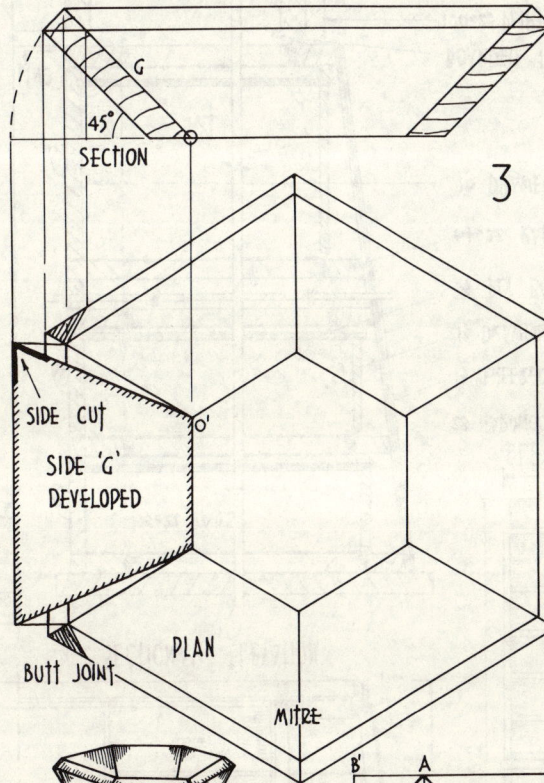

G
SECTION 45°
3
SIDE CUT
SIDE 'G' DEVELOPED
O'
PLAN
BUTT JOINT
MITRE

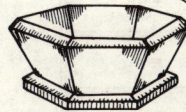

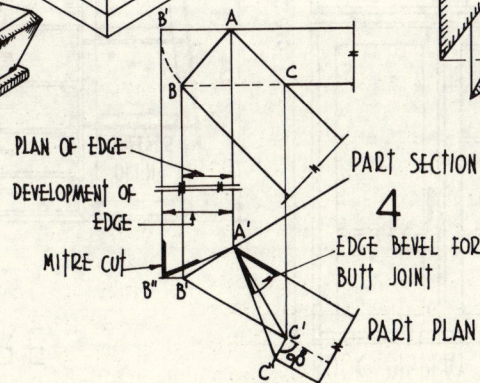

B'   A
B   C
PART SECTION
PLAN OF EDGE
DEVELOPMENT OF EDGE
4
A'
MITRE CUT
B'' B'
EDGE BEVEL FOR BUTT JOINT
C'
PART PLAN
90°
C''

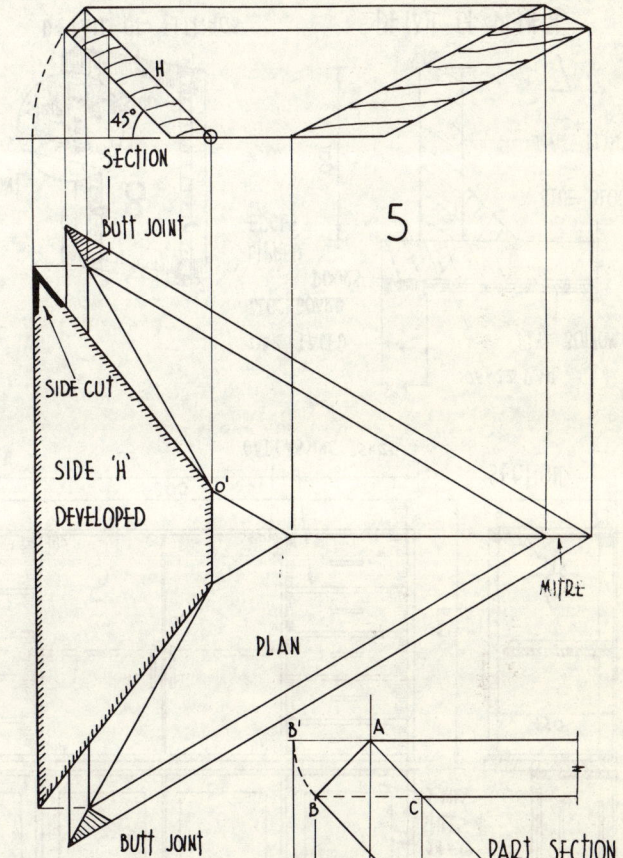

H
SECTION 45°
BUTT JOINT
5
SIDE CUT
SIDE 'H' DEVELOPED
O'
PLAN
BUTT JOINT
MITRE

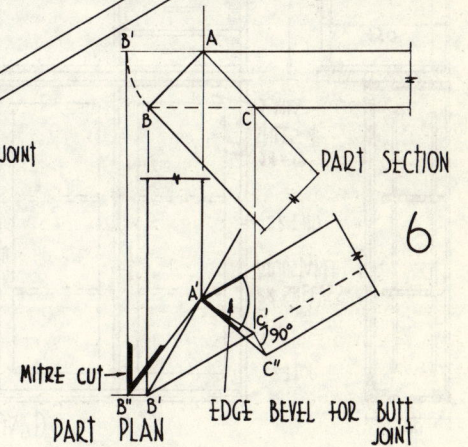

B'   A
B   C
PART SECTION
6
A'
C'
90°
MITRE CUT
B'' B'
PART PLAN
C''
EDGE BEVEL FOR BUTT JOINT

LIBRARY FURNITURE

ISSUE COUNTER

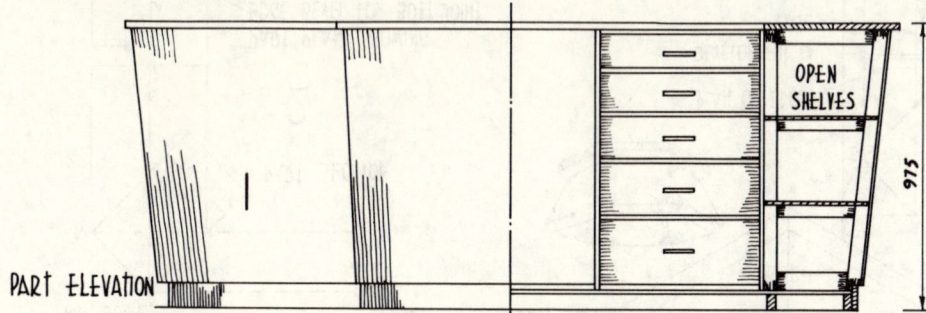

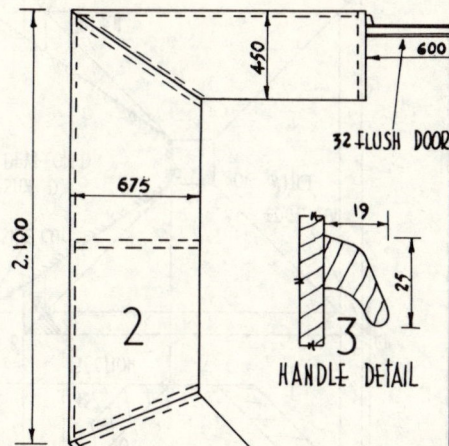

PART ELEVATION

975

OPEN SHELVES

1

SECTIONAL ELEVATION

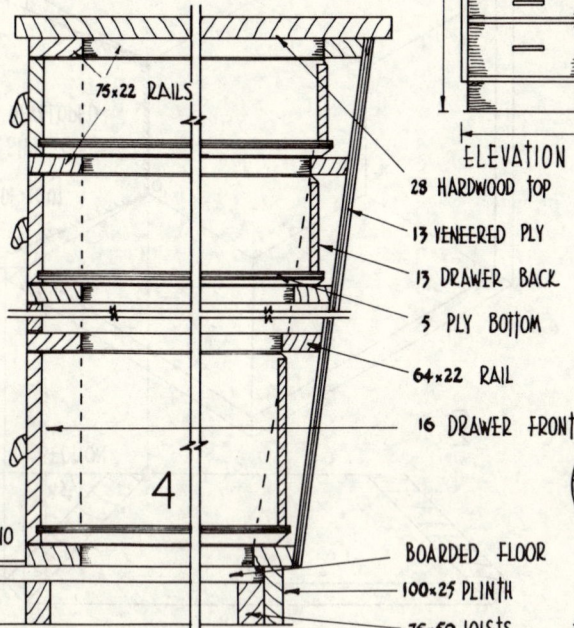

PART PLAN

2.100

675

450

600

32 FLUSH DOOR

2

19

25

3

HANDLE DETAIL

3.600

75×22 RAILS

28 HARDWOOD top

13 VENEERED PLY

13 DRAWER BACK

5 PLY BOTTOM

64×22 RAIL

16 DRAWER FRONT

4

LINO

BOARDED FLOOR

100×25 PLINTH

75×50 JOISTS

DETAILS OF COUNTER

BOOK DISPLAY & CUPBOARD

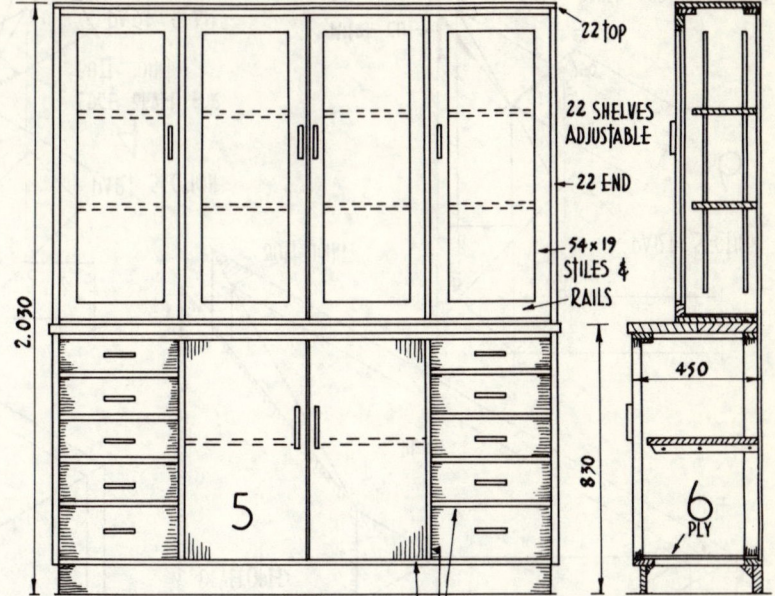

22 top

22 SHELVES ADJUSTABLE

22 END

54×19 STILES & RAILS

2.030

5

1.800

ELEVATION

CARCASSING 75×25

300

6 PLY

SECTION

450

830

OAK FACED BLOCKBOARD DOORS LIPPED ALL EDGES

8

STRIP

Stud

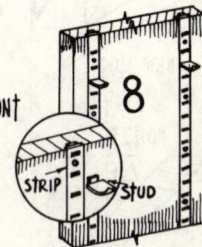

DETAIL OF FITTINGS FOR SHELF ADJUSTMENT

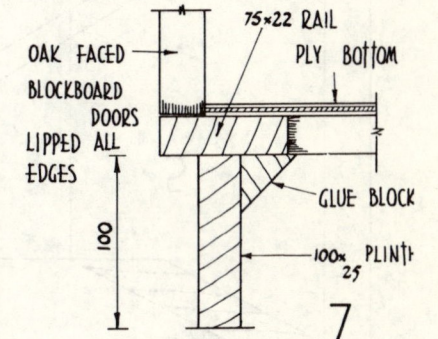

75×22 RAIL

PLY BOTTOM

GLUE BLOCK

100×25 PLINTH

100

7

DETAIL OF PLINTH

SHEET 98

LIBRARY FURNITURE

WALL UNITS FOR BOOKS

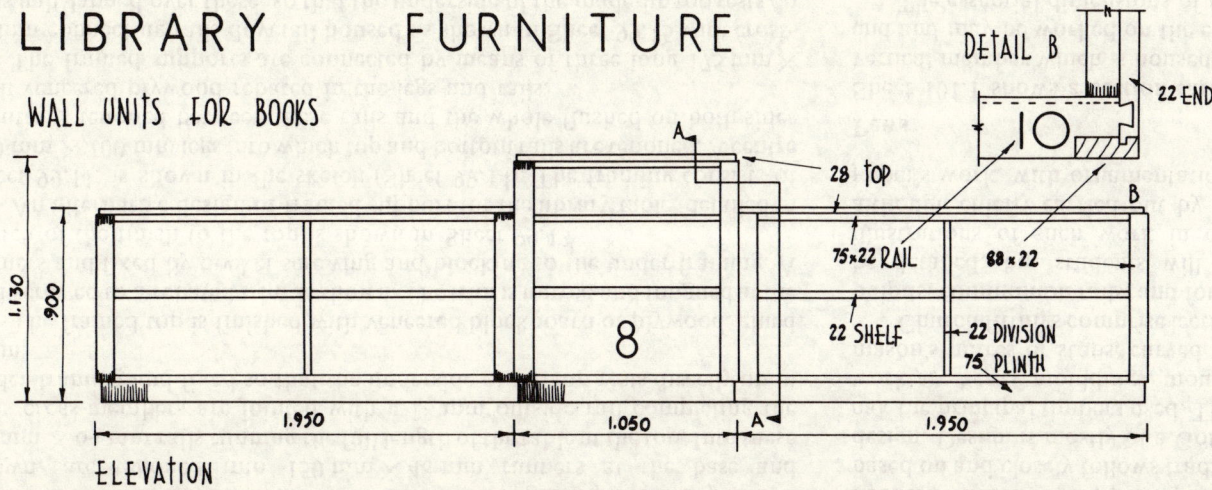

28 TOP
75×22 RAIL
88×22
22 SHELF
22 DIVISION
75 PLINTH

8

1.130
900

1.950   1.050   1.950

ELEVATION

A
A

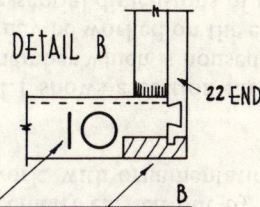

DETAIL B

10

22 END

B

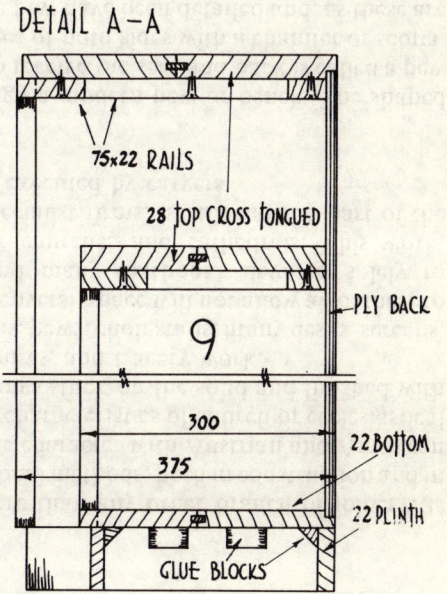

DETAIL A-A

75×22 RAILS
28 TOP CROSS TONGUED
PLY BACK

9

300
375

22 BOTTOM
22 PLINTH
GLUE BLOCKS

75

TABLE

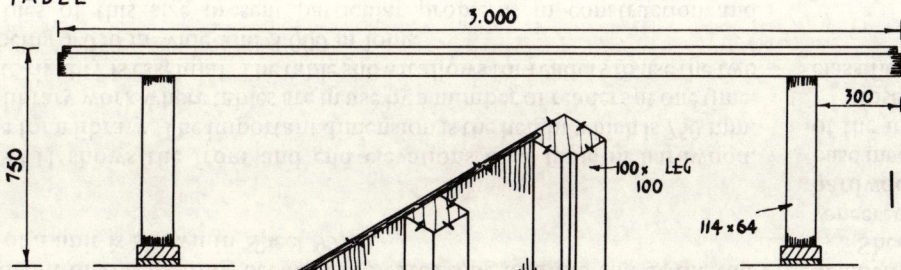

3.000

750

100× LEG
100

300
114×64

ELEVATION

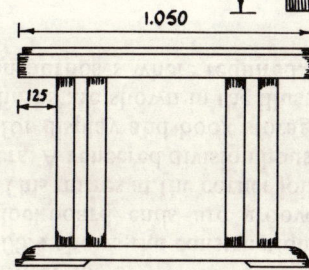

1.050

125

11

END ELEVATION

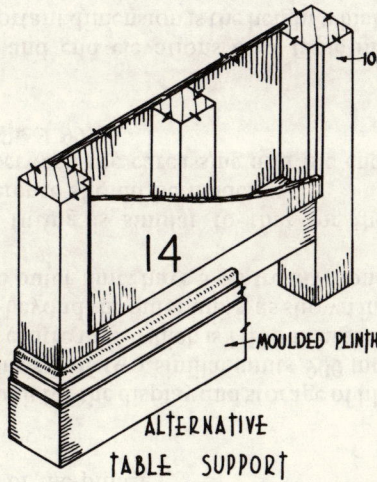

14

ALTERNATIVE
TABLE SUPPORT

MOULDED PLINTH

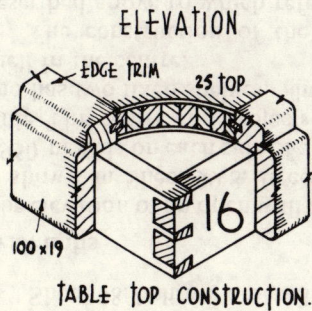

EDGE TRIM   25 TOP

16

100×19

TABLE TOP CONSTRUCTION.

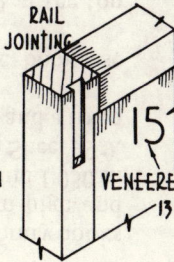

RAIL
JOINTING

15

VENEERED
13 PLY

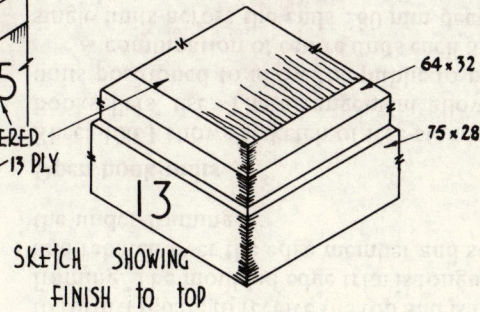

64×32 TRIM

75×28

13

SKETCH SHOWING
FINISH TO TOP

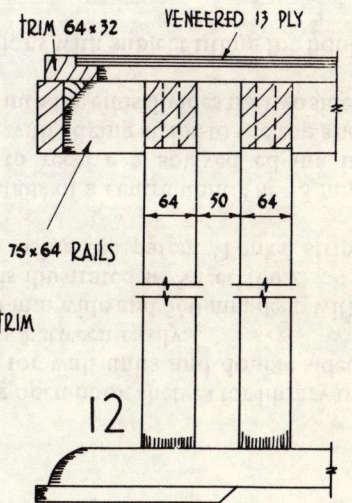

TRIM 64×32   VENEERED 13 PLY

75×64 RAILS

64   50   64

12

SECTION   SHEET 99

housed and screwed into the ends and divisions, and for this reason these members should not be framed, but either solid timber or blockboard used. Cleats screwed to the lower division members support the shelf as shown.

The upper doors are glazed with beads on the inside. The centre pair are rebated at the centre with planted stops.

Sheet 98.7 shows a detail of the plinth.

Wall units

The elevation of an open wall unit for the display and storage of library books is shown in Sheet 99.8. It consists of two similar units 900 mm high and 1·950 m wide on each side of a centre unit, which is 1·130 m high and 1·050 m wide. This centre unit projects beyond the outer units, as shown in Sheet 99.9, and has two fixed shelves. The outer units have a centre division and a fixed shelf in the centre.

The construction of the fitting is similar to that for the bookcase described above to which reference should be made.

A detail of the jointing between the carcassing and the end at the top corner of a unit is shown in Sheet 99.10.

Tables

Sheet 99.11 shows the front and end elevations of a table in hardwood, suitable for a library. The important dimension is the height which is 750 mm.

In library work where tables are in use by a number of readers at one time, absolute rigidity is essential. The table shown allows for readers to use the two sides, being 1·050 m wide and 3·000 m long.

Tables of this size present particular problems in construction and technique, as the construction must be such as to prevent movement in any members which causes loss of rigidity. The construction of the framing is shown in Sheet 99.12, 114 mm × 64 mm legs, arranged in pairs and spaced as shown, are tenoned into 150 mm × 48 mm runners at the base and 75 mm × 64 mm rails running the full length of the table at the top. Into these rails cross-members are jointed with a 28 mm outside rail completing the under-framing and fixed so that the underside of the top rests directly upon them.

The framed top is finished with veneered blockboard or plywood, glued and screwed to a rebated trim as shown. The trim is mitred and tongued at the corners and fixed by pocket screwing and blocking to the under framing. A sketch of the finish to the top is shown in Sheet 99.13.

An alternative design of framed support for the library table, detailed in Sheet 99.11, is shown in the sketch (Sheet 99.14). The framing consists of 100 mm × 100 mm legs into which top and bottom rails are tenoned. A centre muntin is tenoned between these rails and the whole flushed on both sides with veneered plywood rebated in the legs and rails.

The framed supports are connected by means of three long 175 mm × 64 mm connecting rails dovetail housed as shown in Sheet 99.15 with cross-rails half-lapped over these, so that the underside of the made-up top rests on the tops of the connecting rails and the framed ends.

A moulded plinth is mitred round the supports as shown, which may be protected by metal kicking strips. These are often used round the bases of furniture and fittings in public buildings, not only to protect the woodwork from damage by kicking, but from staining by floor cleaning appliances and materials.

The construction of the top is shown in Sheet 99.16. The under framing is through dovetailed at the corners with a moulded edge member mitred round to form a rebate to receive the top and is fixed by screwing through the under framing. The moulded edge trim is tongued to the veneered laminboard top, and rebated over the edge member and secured by pocket screwing through the under-framing.

Open book units

Sheet 100.1 shows a sketch of free-standing open book shelves for library or booksellers' use. The arrangement allows for wall units and double-sided units positioned to allow the public to pass between easily.

A combination of centre units each 850 mm wide and 560 mm deep with single units across the ends 280 mm deep is illustrated in Sheet 100.2.

All the units have adjustable shelves using the patent 'Tonks' strip, detailed in Sheet 100.3.

Sheet 100.4 shows the construction details of a centre unit. The 18 mm veneered blockboard ends are grooved to receive a splayed edging in hardwood. This mitres at the corner joints with edging strips to the top and base members. A veneered division housed into the ends divides the two sides of the unit for display and book storage.

Inset plinths are shown in the illustrations with subject titling for book classification purposes where required.

Church work

Modern church work, perhaps more than any other branch of joinery, is based on and closely follows traditional methods, both in construction and in design. Design is mostly of a Gothic character with Austrian and American oak the principal timbers used. The characteristics of church or ecclesiastical work are heavy mouldings, mouldings stuck on the solid and finished with mason's mitres or stops, carved figures, and tracery work.

Church fittings comprise lecterns, pews, choir stalls, litany desks, screens, pulpits, communion rails, and font covers. Space will not allow all of these to be detailed, but students will find many textbooks devoted solely to illustrations of such work in our churches and cathedrals. This work, although chiefly carried out by specialist firms, is still much a part of the joiner's work, with ornamentation executed by carvers.

Pews

Sheet 101.1 shows a section through a modern pew or bench. The shaped vertical member which is housed to receive the seat and back is called a pew end and may be worked on the edges of both sides with a chamfer or scotia.

The essential dimensions of the pew have been detailed and, as these are most important, they should be remembered. Some further points concerning design are that seats should not be less than 350 mm wide, the back may slope between 25 mm and 100 mm, and 500 mm allowed for the seating of each person.

LIBRARY FURNITURE

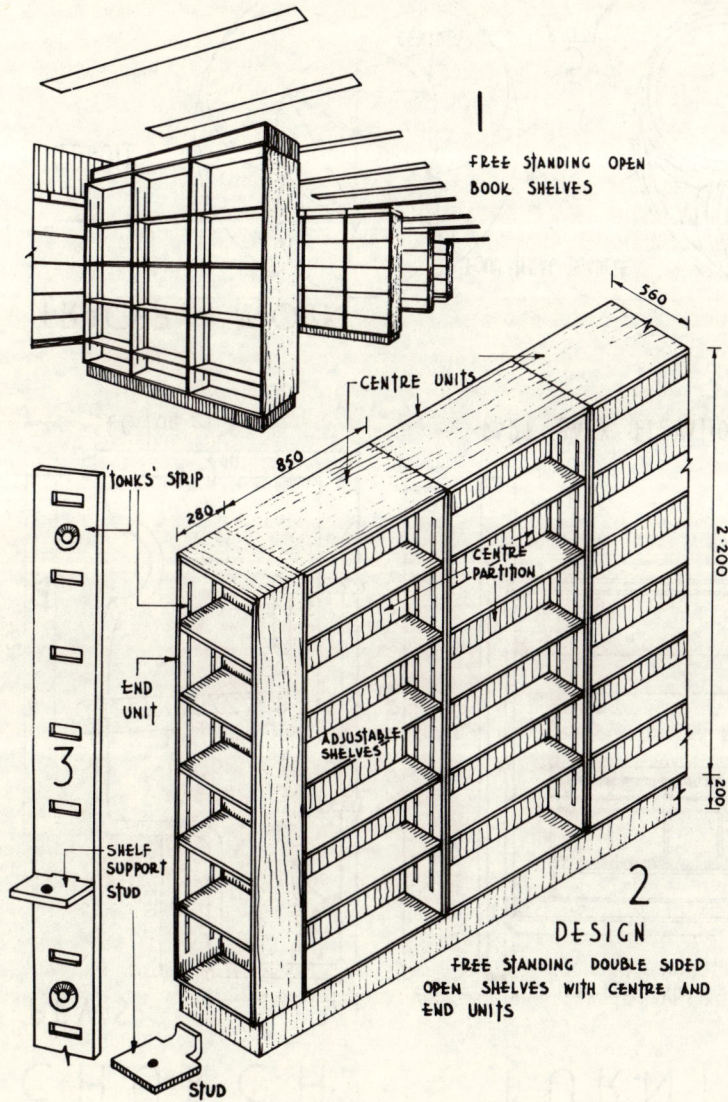

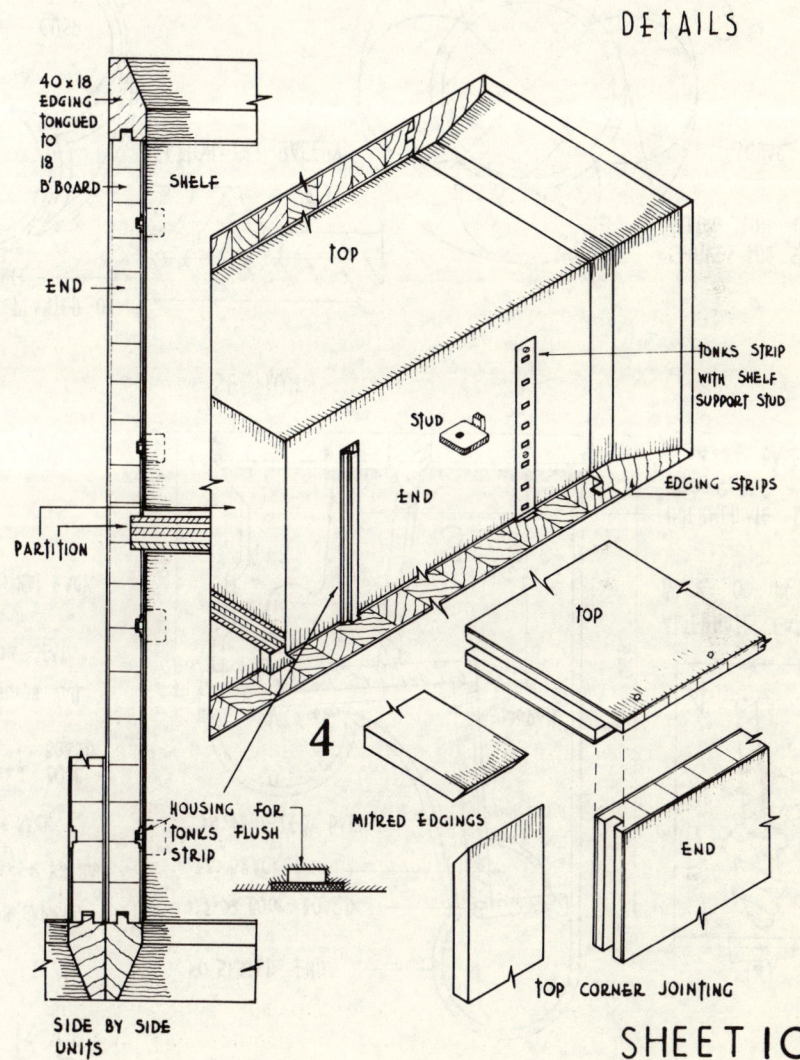

1

FREE STANDING OPEN
BOOK SHELVES

CENTRE UNITS

560

850

280

TONKS' STRIP

CENTRE
PARTITION

2·200

END
UNIT

ADJUSTABLE
SHELVES

200

SHELF
SUPPORT
STUD

3

STUD

2

DESIGN

FREE STANDING DOUBLE SIDED
OPEN SHELVES WITH CENTRE AND
END UNITS

DETAILS

40 × 18
EDGING
TONGUED
TO
18
B'BOARD

SHELF

END

PARTITION

HOUSING FOR
TONKS FLUSH
STRIP

SIDE BY SIDE
UNITS

4

TOP

TONKS STRIP
WITH SHELF
SUPPORT STUD

STUD

END

EDGING STRIPS

TOP

MITRED EDGINGS

END

TOP CORNER JOINTING

SHEET 100

259

CHURCH FURNITURE

PEWS

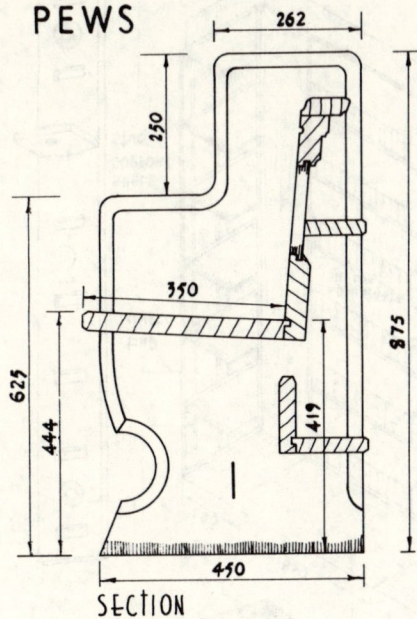

262

250

350

625

444

419

875

450

SECTION

1

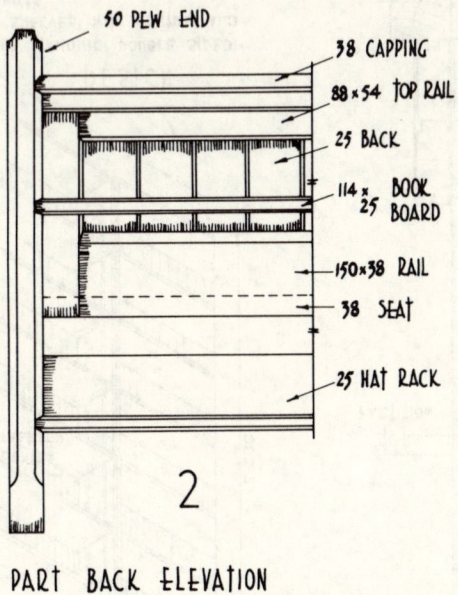

50 PEW END

38 CAPPING

88 x 54 TOP RAIL

25 BACK

114 x 25 BOOK BOARD

150 x 38 RAIL

38 SEAT

25 HAT RACK

PART BACK ELEVATION

2

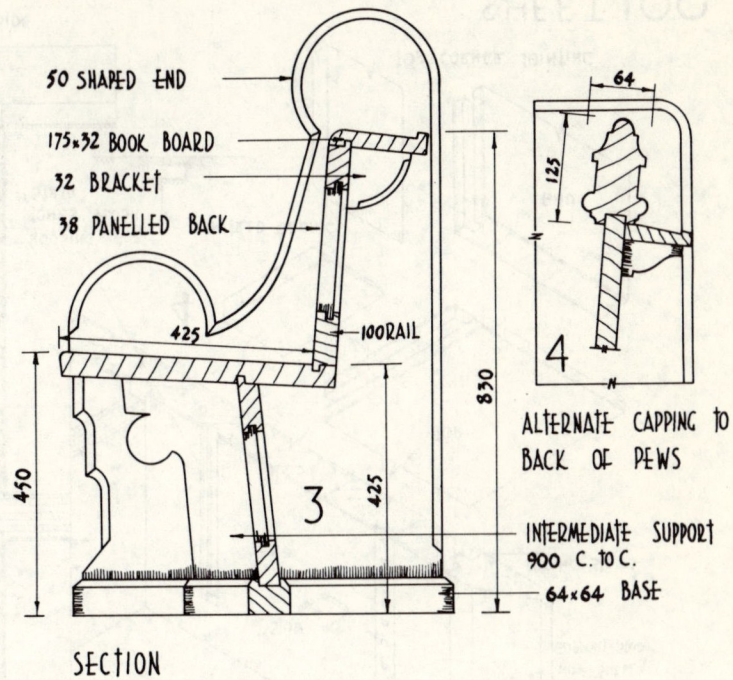

50 SHAPED END

175 x 32 BOOK BOARD

32 BRACKET

38 PANELLED BACK

425

100 RAIL

450

830

425

3

SECTION

INTERMEDIATE SUPPORT 900 C. to C.

64 x 64 BASE

64

125

4

ALTERNATE CAPPING to BACK OF PEWS

TRACERY WORK

7

TREFOIL

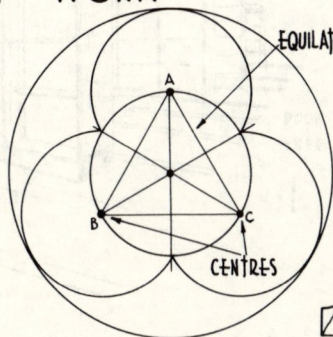

EQUILATERAL TRIANGLE

A

B

C

CENTRES

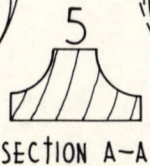

5

SECTION A-A

TRACERY PLANTED ON SOLID PANEL

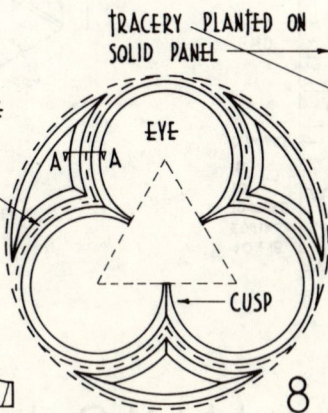

FOIL

A   A

EYE

CUSP

8

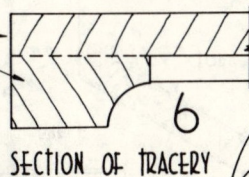

SECTION OF TRACERY

6

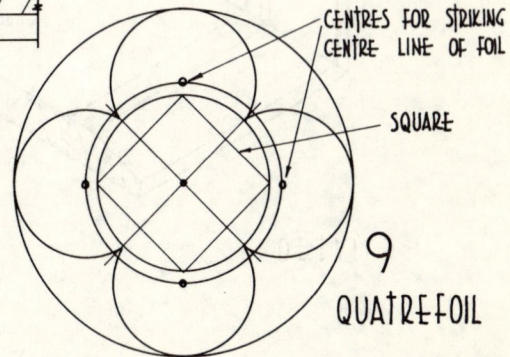

CENTRES FOR STRIKING CENTRE LINE OF FOIL

SQUARE

9

QUATREFOIL

SHEET 101

The pew end may be carved or finished with some form of decorative treatment on one or both sides, although on modern work much of this decoration has been dispensed with.

Sheet 101.2 shows a part back elevation of the pew. The back is framed and filled between muntins with tongued and grooved boards, tongued top and bottom to the rails and finished flush on the face side. The top rail is moulded and finished on the top edge with a moulded capping slot screwed or screwed through the top and pelleted. The capping serves as a handrail and may be used to aid rising, after kneeling, or from the seat.

A book-board is shown and a hat-rack in the form of a shelf, which may also store kneeling cushions below the seat.

The seat, which is tongued to the back is pocket-screwed from below and through the bottom rail and slopes 25 mm from front to back.

Sheet 101.3 shows a further example of an older type of pew having a wider seat. This has a framing acting as additional support below the seat, along with intermediate supports as shown. It will be seen that the wider book-board is tongued to the top rail of the back and supported with moulded brackets in this design.

The shaped pew end is tenoned and mortised to a moulded base, which is housed to receive the end, and also the lower framework.

An alternative capping finish is detailed in Sheet 101.4.

Tracery work

This was referred to earlier as a characteristic of church work and may be defined as a series of curved and straight mouldings which require to be set out geometrically. The basis of the designs is the trefoil and quatrefoil.

Tracery work may be cut from the solid or pierced and moulded on both sides, giving a section as shown in Sheet 101.5, or it may be planted on a solid panel (Sheet 101.6). The moulding used may be a chamfer or a scotia.

Sheet 101.7 shows the method of setting out a trefoil. The term refers to the number of part circles enclosed in the larger circle.

The setting out is constructed upon an equilateral triangle ABC. The angles of the triangle are bisected to give the centre of the circle enclosing the triangle, and the large circle. Points A, B, and C are the centres for striking the centre lines of the foil.

Sheet 101.8 shows the complete tracery based on the above construction. The leaf shape between each foil is called a 'cusp' and the piercing in the tracery the 'eye'.

The setting out of a quatrefoil, which has four part circles contained within a larger circle, is shown in Sheet 101.9. It is constructed upon a square with the setting-out similar to that for the trefoil.

Book-rest

Sheet 102.10 shows a part front and the end elevation of a book-rest to a choir stall. The choir stalls are usually placed on each side of the church chancel with a book-rest along the front of each side. A fitting of similar design placed along the front of the first row of pews in the body of the church is often used.

The seating for the choir has not been shown and although the construction follows closely that for pews, their dimensions differ. The book-board is much higher, and so are the ends, to allow music for the choir to be read while standing.

The book-rest illustrated has framed ends with tracery work prepared separately and planted on the back panel, which is housed into the framing. The rails are stub-tenoned into the stiles and pinned from the back without coming through on to the face. The front is framed in a similar way, but the top rail is allowed to run through to the stile. This is moulded and prepared to receive the top tracery panel as shown. This panel is of quatrefoil design divided at intervals along its length by moulded mullions.

The lower panel mouldings form part of the tracery work, and are scribed on to the chamfer of the bottom rail. A moulded capping is fixed by slot-screwing to the top rail and framed ends, being mitred at the corners. Shaped buttresses, moulded from the solid, are slot-screwed to the framed ends on the face providing additional decoration.

Sheet 102.11 shows a part back elevation of the choir stall with the book-rest housed into the framed end. A detail of the top rail and capping is shown in Sheet 102.12, and the bottom rail in Sheet 102.13.

Communion rail

The fitting is placed in front of the altar table and usually spans the full width of the chancel between walls. Sheet 102.14 shows a part elevation of the rail which is normally heavily moulded. As the congregation kneel at the communion rail, the height is most important, usually between 600 mm and 675 mm. Fixing is done at each end to posts plugged to the walls and through the bottom rail into the floor. The centre bay is hinged to fold to each side, giving access to the altar as it is only in position during the Communion Service.

The tracery work is pierced or open tracery, moulded on both sides and housed in the framework as shown in the detail (Sheet 102.15). A detail of the top rail with wide capping and tongued bed moulds is shown in Sheet 102.16. The built-up base detail is shown in Sheet 102.17.

In some churches a heavily moulded rail, usually about 125 mm × 75 mm in section, is supported on pillars carrying ornamental wrought-iron brackets. The pillars are generally octagonal and spaced about 900 mm apart. The centre section directly in front of the altar table is either hinged, so that it lies on the fixed rail when it is open, or made to be removed altogether with pins at each end to drop in slots cut in the fixed rails.

Litany desk

There are many variations in design of this small desk, which is used by the reader in a kneeling position during a church service. Sheet 103.18 shows the elevation and section of such a desk. Again, before attempting to design a litany desk, the dimensions, of which the height is the most important, should be noted. This church fitting consists of a framed base with two ends or uprights containing a decorative front panel or tracery, a book-rest, and kneeling board.

The framing for the panel is housed into the two shaped ends and the base and housed into the book-rest as shown. The panel is housed into the framing and placed in position when wedging up the framing.

The shaped ends are stub-tenoned and pegged into the base which

CHOIR STALL

PART ELEVATION

END ELEVATION

PART BACK ELEVATION

10

11

1,050

850

50

250

88 100

CAPPING

TOP RAIL

BOOK REST

TRACERY

200

19

PANEL

TRACERY

19 32

SECTION A-A

12

13

32

162

50

SECTION B-B

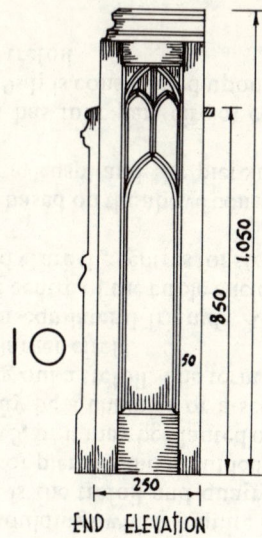

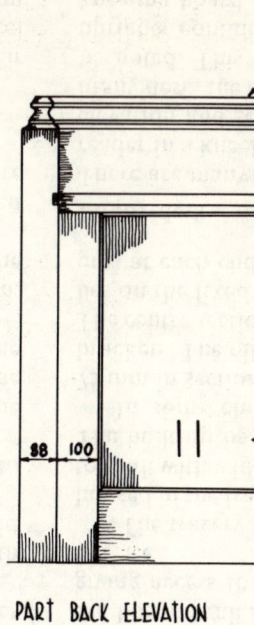

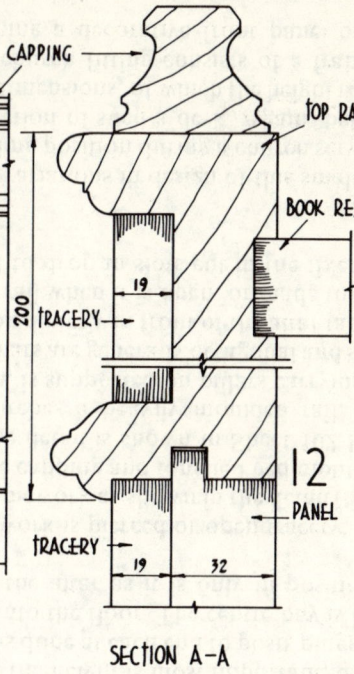

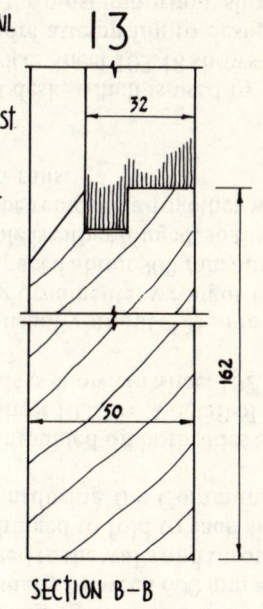

COMMUNION RAIL

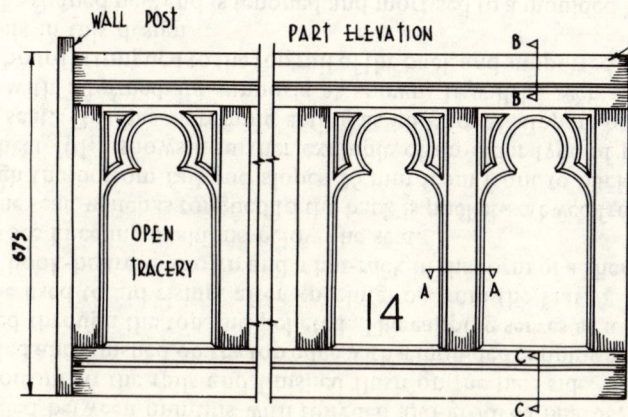

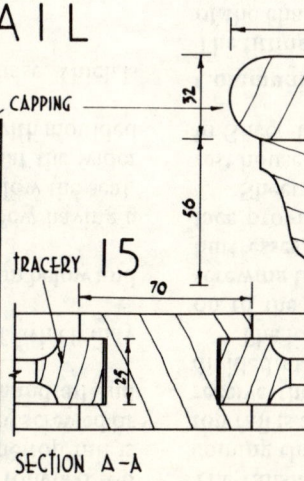

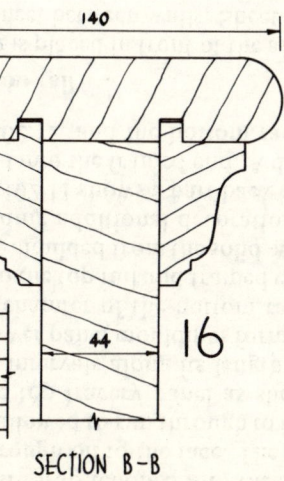

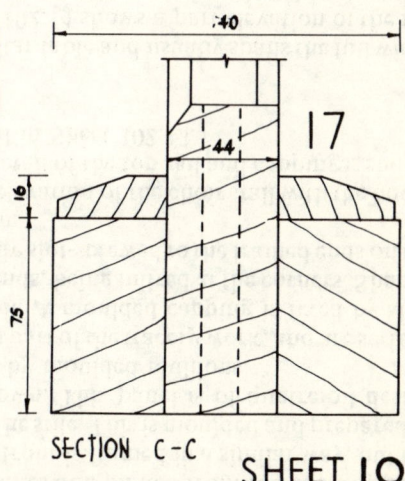

WALL POST

PART ELEVATION

CAPPING

OPEN TRACERY

675

14

A A

B B

C C

TRACERY

15

SECTION A-A

70

25

32

56

44

16

SECTION B-B

140

140

44

17

16

75

SECTION C-C

SHEET 102

262

LITANY DESK AND LECTERN

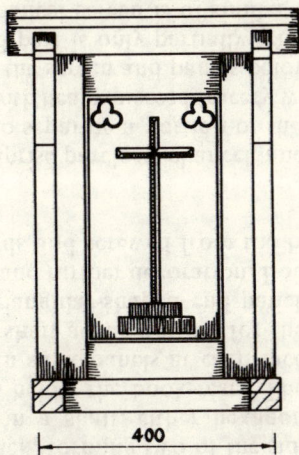

18

ELEVATION

BOOK REST 32
SLOPE 38
END 250×32
FRAMING 32
PANEL 25
BASE 64×64
PADDED KNEELING BOARD
230×32

686

700

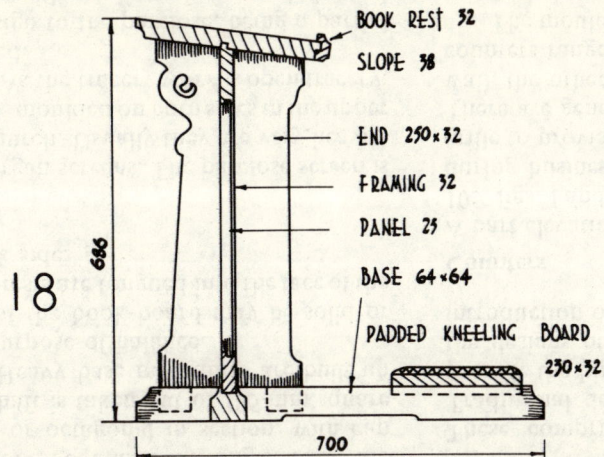

SECTION

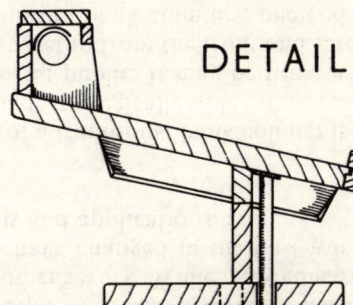

DETAIL

20

MAX. 1.200

1.000

550

19

ELEVATION

EX.125×50 SIDE BUTRESS
TENONED TO BASE

EX 175×50 CENTRE
BUTRESS
TENONED TO
BASE

175×32

125×64 BASE

762

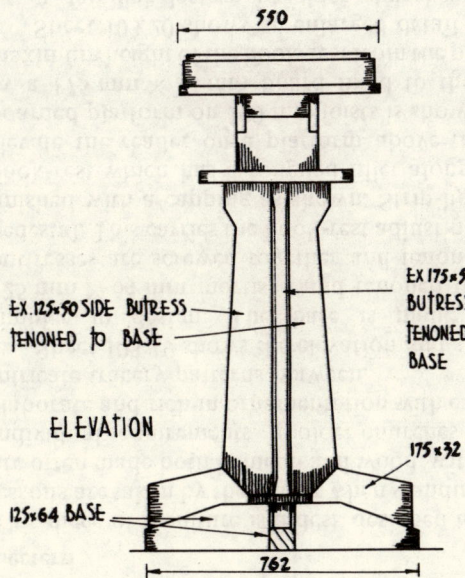

88

STRIP LIGHTING
BOOK REST 550×400
FILLET 44×19
CAPPING 25
BOOK REST ADJUSTMENT
TO MAX.1.200 HIGH FROM
RAISED PLATFORM

1.016

100× JOISTS
50

75×64 BASE

125

300

750

SECTION

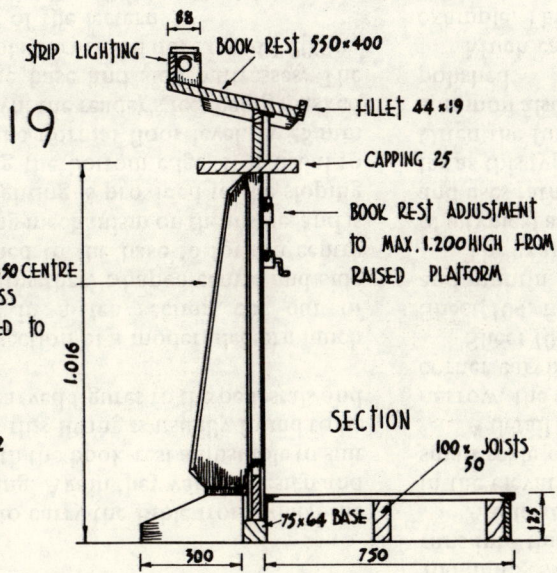

SHEET **103**

extends to receive a padded kneeling-board housed between. The book-rest, moulded on the front projecting edge, has a moulded fillet tongued on the bottom edge to rest the books against as the board slopes 38 mm.

Lectern

This piece of furniture is a desk designed to carry the Bible from which the lessons are taken by the reader when standing. Again they vary in design and are often made both in metal and wood, with the book-rest adjustable to suit individual requirements. In older churches this fitting is usually found to be elaborate and rich in ornamentation with carved figures to the pedestals and intricate tracery patterns between.

Sheet 103.19 shows the elevation and section of a modern lectern much simpler in design. The base is made in a tee section cut out of 125 mm × 64 mm mortised and tenoned together. Shaped centre and side buttresses are screwed together and tenoned to the base to form a centre pedestal. This carries the book-rest adjusting mechanism on the inside, and is finished with a capping as shown. Strip-lighting is provided to the sloping book-rest which has a tongued fillet along the bottom edge. It is usual to elevate the reader on a platform above the normal floor level. A 25 mm boarded platform on 100 mm joists is shown, the reader's feet being masked by a 175 mm × 32 mm board fixed to the base and side buttresses. The maximum height of the book-rest from the platform need not exceed 1·200 m.

Sheet 103.20 shows an enlarged detail of the lectern.

A familiar lectern in older churches is the 'Golden Eagle' in brass mounted on a heavily moulded shaft octagonal in section and carrying the Bible on its extended wings. The base of the shaft is also heavily moulded to the same shape to give the lectern balance.

Lecterns may have double-sided book-boards, that is, two sloping boards back-to-back, forming two of the sides of an equilateral triangle. These are mounted on a shaft, either hexagonal or octagonal in section, with cap mouldings below the book-rests. The shaft is taken out of 125 mm square timber with sunk panels at each face. Heavy base mouldings are built up round the shaft at the bottom for the purpose of balance.

The triangular-shaped end panels of the book-board may be solid or pierced to add further decoration. Book-rests are tongued into the face of the book-boards and screwed from the back side.

Screens

These comprise parclose, chancel, and organ screens. The parclose screen is provided to separate a portion of the church. Usually they are very heavily moulded with heavy pierced tracery work moulded on both sides in the upper portion of the screen and panels below. As the tracery work is open tracery, the divided part is only partially screened.

The chancel screen is of similar design to the parclose, being a partial screen, tracery work in the upper portion with solid panels or panels formed of tongued and grooved boards below. The priests' stalls are often formed against the lower portion in chancel screens, facing the altar.

Sheet 104.21 shows the elevation of a screen 1·800 m high suitable for screening an organ or the entrance to a vestry. It consists of solid panels in a number of bays tongued into 32 mm framing. This is chamfered and prepared to receive the tracery work in the upper portion as shown.

At the open end of the screen the framed panelling is tongued into a 100 mm × 70 mm end post which is stop-chamfered. The moulded plinth runs into this with the moulded capping mitred round it to finish the screen.

A quatrefoil design has been used for the tracery work to the screen, but in the elevation the details of the tracery have been omitted because of the small scale of the drawing.

A detail of part of this panel is shown in Sheet 104.22. As the panel is only narrow, the tracery is cut from the solid and moulded on one side with the corner carvings completing the decoration.

Sheet 104.23 shows a section through the moulded cappings and top rail, Sheet 104, figs. 24 and 25, sections through the panelled framing, end post, and muntin. Sheet 104.26 shows a part section through the plinth.

The examples of church fittings which have been covered in this book are of a typical and simple character, the object being to explain the construction and uses rather than depict elaborate designs. It should be mentioned that, so far as this type of work is concerned, the finish varies from church to church. Often the furniture is left in its natural state to mellow over the years. It is common also to treat the surface with wax or to have it stained or fumed and polished.

Much carved work is to be found in many churches. The reredos is an example. This is positioned as a background to the altar and is elaborate and rich in ornamentation with carved figures the highlight. This type of work, of course, is the work of specialists who are carvers or sculptors, not joiners.

Bank fittings

These comprise counters for cashiers and clerks, desks, and screens. Traditional designs in this type of work, invariably done in hardwood, provide the highest class of joiner's work. While basic constructions remain, the designs of these fittings have changed in the post-war years with the introduction of new materials and applications.

Counters

A part elevation and section of a traditional bank counter is shown in Sheet 105, figs. 1 and 2. The top is made strategically wide for the cashiers to work at during business with the general public. It may be fitted with a protective grille to provide extra security, set in from the front and screwed to the top. There are generally no openings in bank counters, one end abuts against a wall, the other framed into a return counter or screen. The height of bank counters ranges from 900 mm to 1·050 mm.

The moulded counter top overhangs the front framing to allow the public to stand close without kicking the skirting. Supporting the overhanging top are shaped trusses, 64 mm thick, reeded on the front edge to add further decoration, and fixed by screwing through the front framing from the rear.

The front panelling is made up in a series of frames with joints hidden by the moulded trusses. Bed moulds and moulded skirtings are cut between the

S C R E E N

100×38 MOULDED CAPPING

102

1,700

21

83×32
RAILS &
MUNTINS

19 PANEL

100×32 STILE

100×
70 POST

A

A

B

B

C   C

D
D

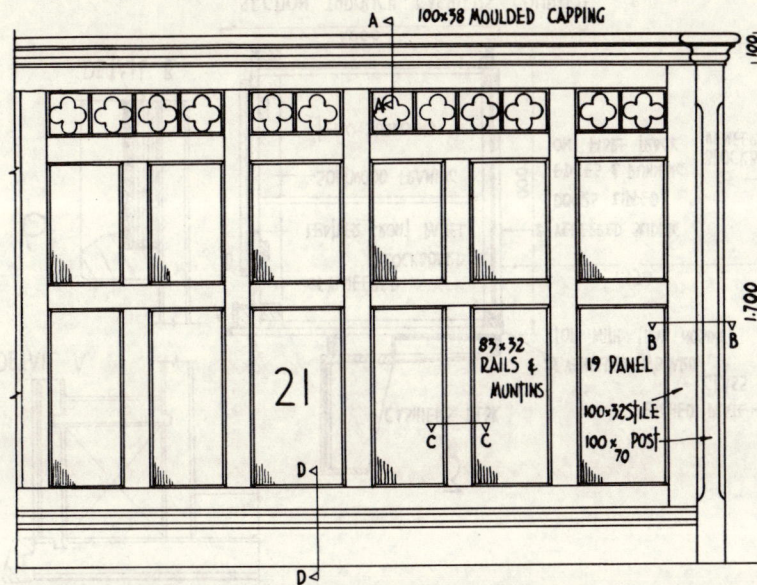

PART ELEVATION

TRACERY DETAIL

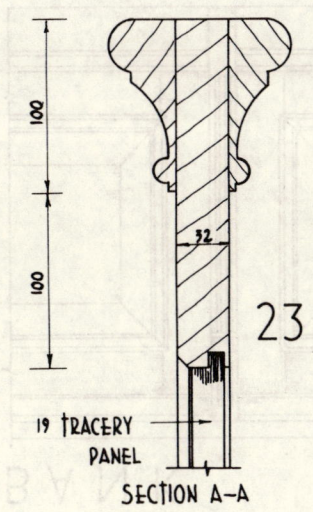

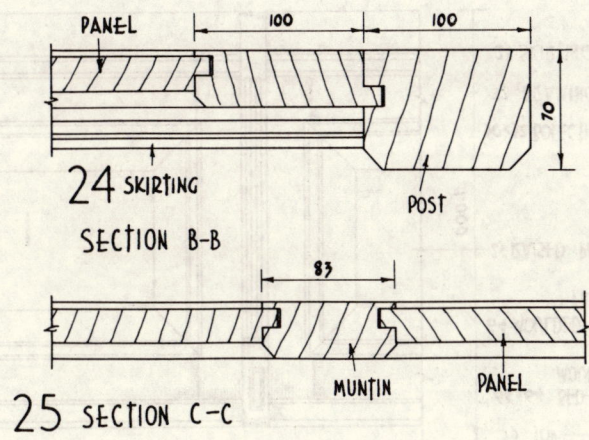

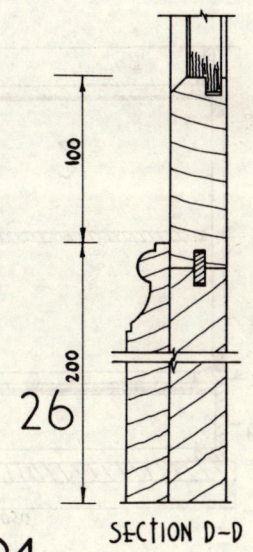

22

44

6   10

10

156

100

100

32

23

19 TRACERY
PANEL

SECTION A-A

PANEL

100   100

70

24 SKIRTING

POST

SECTION B-B

83

25 SECTION C-C

MUNTIN   PANEL

100

200

26

SECTION D-D

SHEET 104

265

BANK COUNTERS

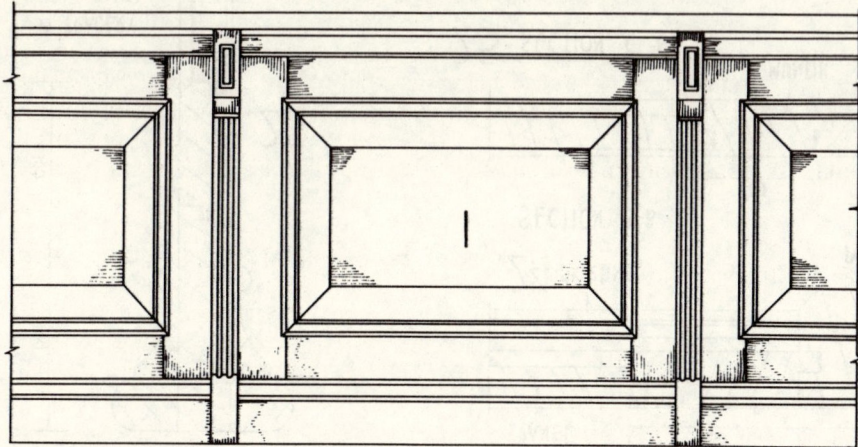

PART ELEVATION

1

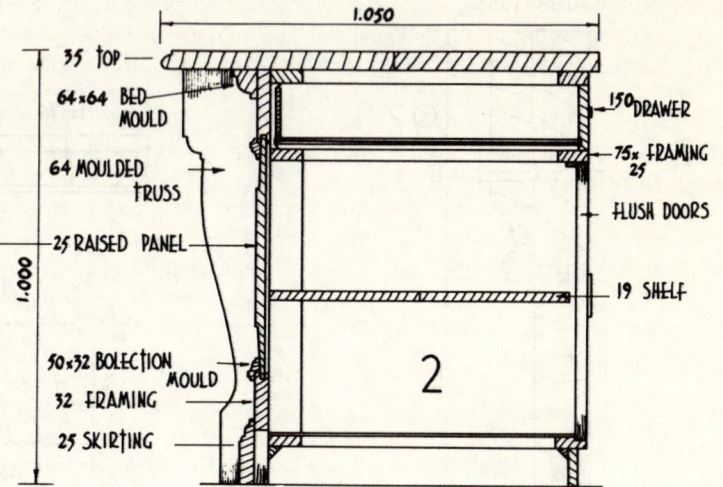

SECTION

35 TOP
64×64 BED MOULD
64 MOULDED TRUSS
25 RAISED PANEL
50×32 BOLECTION MOULD
32 FRAMING
25 SKIRTING
1.050
150 DRAWER
75× FRAMING 25
FLUSH DOORS
19 SHELF
2
1.000

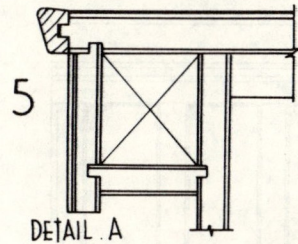

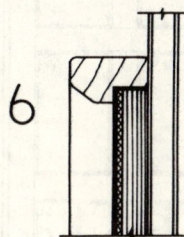

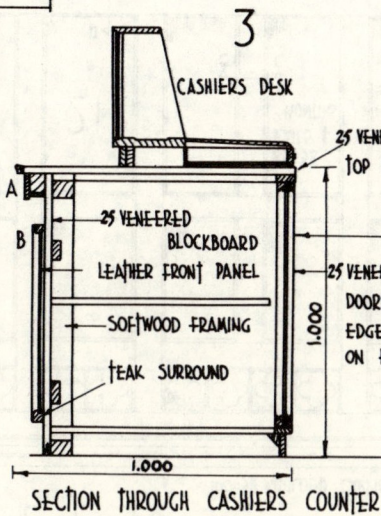

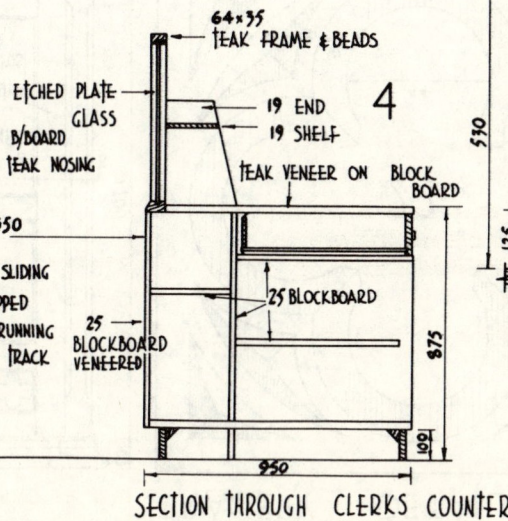

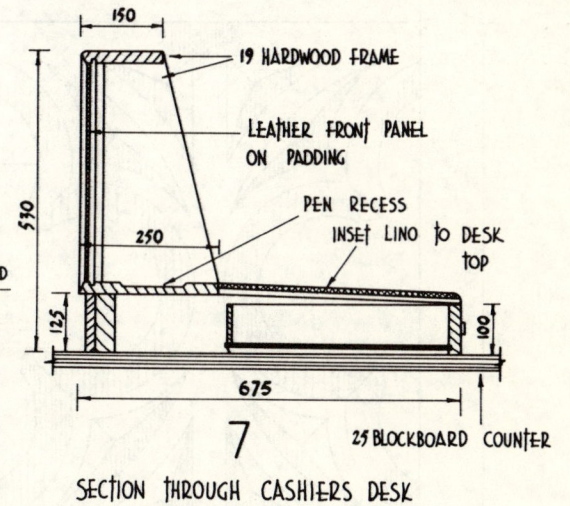

5

DETAIL A

6

DETAIL B

3

CASHIERS DESK

25 VENEERED B/BOARD TOP WITH TEAK NOSING
25 VENEERED BLOCKBOARD
LEATHER FRONT PANEL
SOFTWOOD FRAMING
TEAK SURROUND
25 VENEERED SLIDING DOORS LIPPED EDGES & RUNNING ON FIBRE TRACK
1.350
1.000
1.000

SECTION THROUGH CASHIERS COUNTER

4

64×35 TEAK FRAME & BEADS
ETCHED PLATE GLASS
19 END
19 SHELF
TEAK VENEER ON BLOCK BOARD
25 BLOCKBOARD
25 BLOCKBOARD VENEERED
950
875
190

SECTION THROUGH CLERKS COUNTER

7

150
19 HARDWOOD FRAME
LEATHER FRONT PANEL ON PADDING
PEN RECESS
INSET LINO TO DESK TOP
530
250
125
100
675
25 BLOCKBOARD COUNTER

SECTION THROUGH CASHIERS DESK

SHEET 105

trusses and fixed by screwing through the framing.

In the underfitting, cash drawers are fitted immediately below the top and below these a series of cupboards and shelves, with flush doors, either side-hung or sliding. A feature of the underfitting is the construction of the cash drawers. Since the drawers may have to hold considerable weights in cash, they must be robust and run freely. They are made of hardwood with 100 mm diameter removable till bowls and note divisions. Drawer sides 22 mm thick are used with patent ball-bearing runners screwed to them, the guides being fixed to the sides of the pedestals.

A section through a modern cashiers' counter and desk is shown in Sheet 105.3. The clerks' counter shown in Sheet 105.4 is placed 1·350 m clear on the staff side of the banking counter.

Blockboard, veneered, with all exposed edges lipped in hardwood, is used for the construction. The basic framework is built up out of 75 mm × 50 mm softwood for the front with 75 mm × 25 mm hardwood carcass rails dovetailed into the ends. The pedestal ends, shelves, bottom, and front are of blockboard with the horizontal members housed and glued. The top is of veneered blockboard or finished with plastic, with the nosing and lipped edges in teak and fixed by gluing and screwing to the carcassing.

A detail of the nosing and finish to the front edge of the counter top is shown in Sheet 105.5. The finish to the front of the counter shows leather, padded on a plywood backing, with a mitred teak surround glued and screwed through the blockboard, as shown in Sheet 105.6. Veneered blockboard sliding doors are fitted with lipped edges in hardwood and running on fibre track.

The clerks' counter (Sheet 105.4) is constructed in a similar manner. A plate-glass screen with teak surround and shelf behind is fixed to the counter top.

Cash drawers are similar to those already described for traditional bank counters.

All heading joints in the counter top are made with a counter cramp. The tops are grooved and cross-tongued and are pulled together and held firmly in place by the cramps. These are screwed across the joints on the underside of the top with folding wedges driven to pull the joint up tight.

The detail of the cashiers' desk is shown in Sheet 105.7.

Bank screens

In the introduction to work on bank fittings a classification of the joinery items is made and includes screens.

The security aspect in bank work is now of prime importance and staff-protection screens have been the subject of much change in design and construction of late. They are installed at the counter or wherever customer contact is made and automatically become the focal point of attraction. Anti-bandit glass is used in units of varying design which are secured between the counter top and room ceiling.

Sheet 106 shows two forms of staff-protection screens used in bank work. The units are formed in stainless steel with anti-bandit glass as shown in the plan (Sheet 106.1). The counter provides for service positions 800 mm wide and customer writing positions sited at the computer points, 600 mm wide.

The screen allows a 75 mm clearance for service above the counter top with enquiry door units sited at intervals where required.

Sheet 106.2 shows an alternate design of glass screen provided with louvres at the customer positions; partitioning between customers is by stabilising glazed units having the lower portion panelled to match the counter.

Self-locking, lift-up flaps or vertical sliding sashes may be used in screen designs as access doors for large parcels.

Screens and partitions

Many types of buildings require screens and partitions in one form or another. Hotels, banks, schools, offices, workshops, assembly halls, and places of entertainment use units which may be glazed or panelled for dividing floor or work areas.

Partitions are usually made the height of the room as part of the structure, with the required door and window openings formed in them. They may be less than the ceiling height and not a part of the structure. Partitions may also be folding and work on the same principle as folding doors.

Screens are usually designed for a particular purpose and situation. A dwarf screen used in the project on page 272 is an example where a flush panelled unit or part glazed and panelled units are used. Light, movable screens finished in materials are used principally in office situations, where a number may be used together in line, or staggered, forming screens to desks. Sheet 106.3 shows a typical office-type movable screen. The frame and glazing beads are polished hardwood with the lower portion covered on both sides in leathercloth on a softwood core. Shaped feet in either chromed metal or wood are shown.

In situations where this form of screen is used for visual screening they are non-acoustic. Acoustic treatment may be applied to screens to increase the sound-absorption qualities.

Sheet 109.7 shows the elevation of a glazed and panelled dwarf screen used in the counter-screen project for screens numbers 1, 2, and 3 with screen number 4 shown in Sheet 106.4.

The finish is polished hardwood to match the counter work and wall panelling décor to the office, and the screens are so arranged to give privacy to the office staff and areas for interviewing purposes.

A screen which may be used in a hallway, corridor, or entrance situation is shown in Sheet 107.1. Provision for access incorporating full-glazed, double-action swing doors is made, with glazed and panelled sections as shown.

The screen may be taken to the ceiling height or kept below, depending on the situation. Should a screen of some length be required some consideration must be given to the size and construction from a handling and transporting point of view, along with the need to get the units through doors and up stairs. Assembly may have to be carried out on site in difficult circumstances.

Jointing of such units in a lengthy screen would be made by reducing the two joining stiles in thickness and fixing through the rebates by screwing and pelleting.

SCREENS

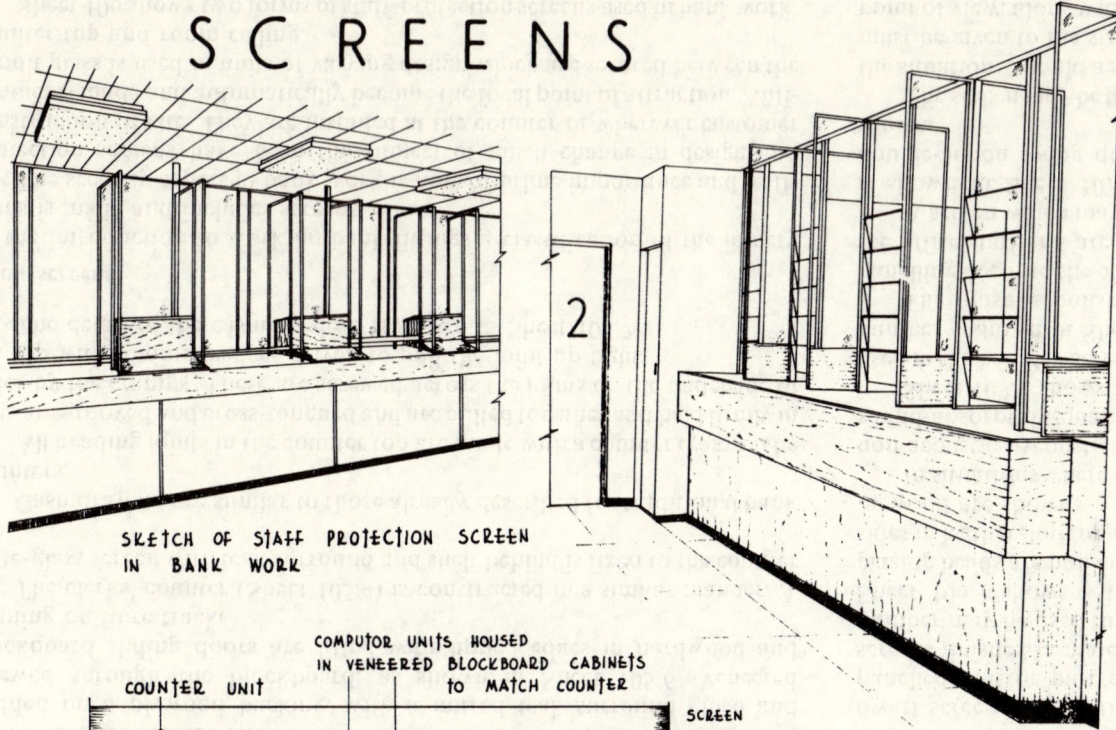

SKETCH OF STAFF PROTECTION SCREEN
IN BANK WORK

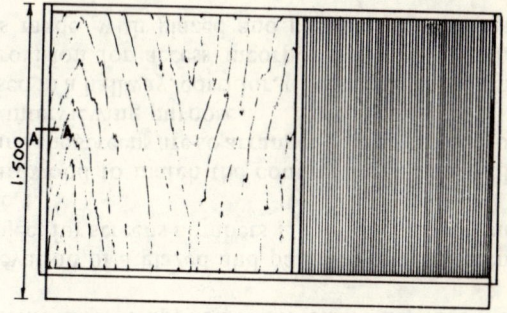

ELEVATION

SCREEN N° 4
COUNTER PROJECT SHEET

4

SECTION A-A

SECTION B-B

3

SKETCH OF
LIGHT MOVEABLE
SCREEN ON CHROME FEET

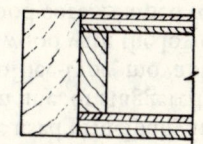

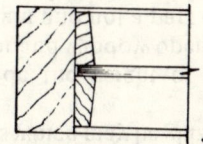

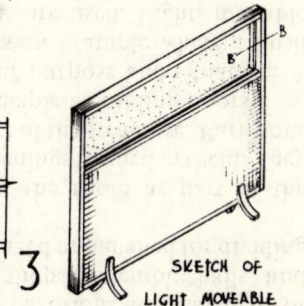

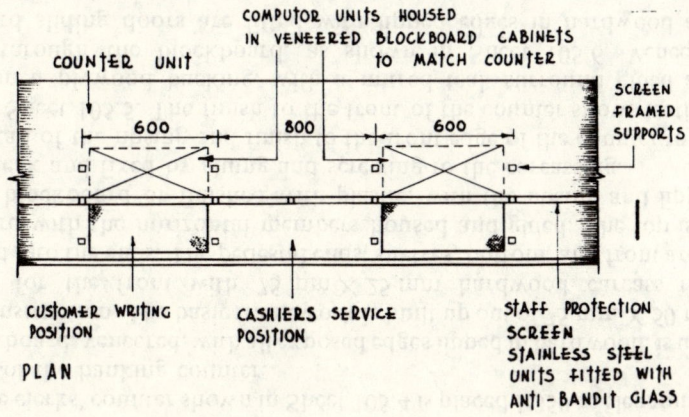

COMPUTOR UNITS HOUSED
IN VENEERED BLOCKBOARD CABINETS
TO MATCH COUNTER

COUNTER UNIT

SCREEN
FRAMED
SUPPORTS

600   800   600

1

CUSTOMER WRITING
POSITION

CASHIERS SERVICE
POSITION

STAFF PROTECTION
SCREEN
STAINLESS STEEL
UNITS FITTED WITH
ANTI-BANDIT GLASS

PLAN

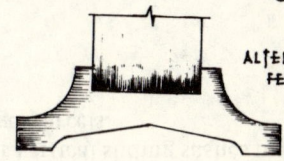

ALTERNATE SHAPED
FEET TO SCREEN

SHEET 106

SCREENS & PARTITIONS

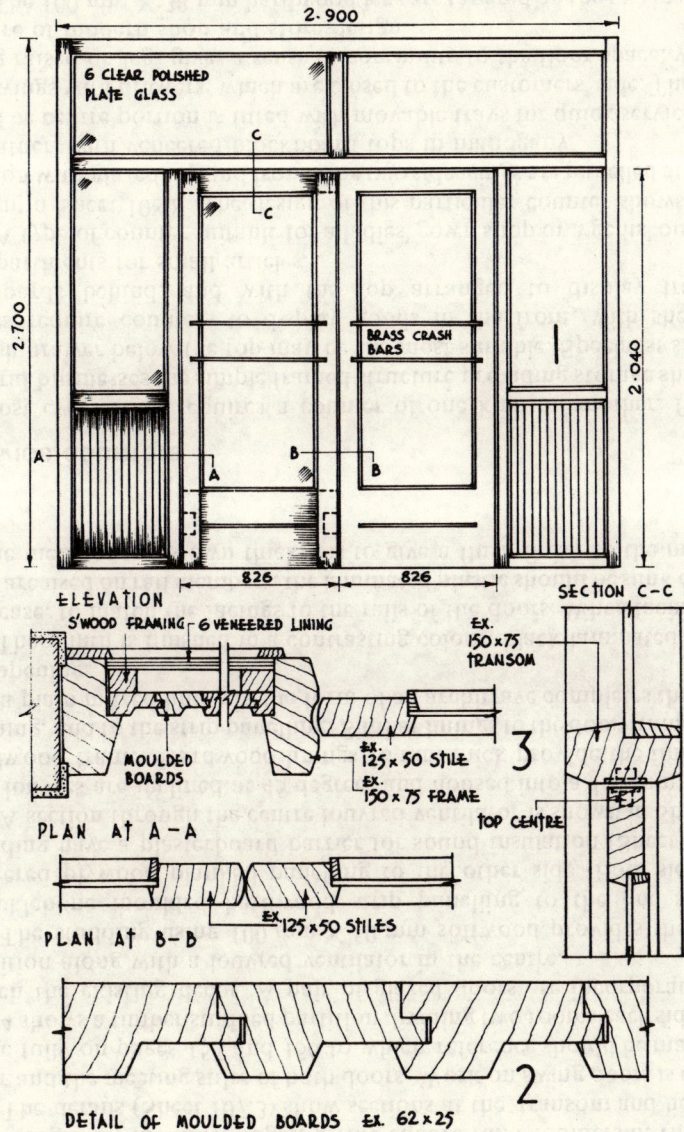

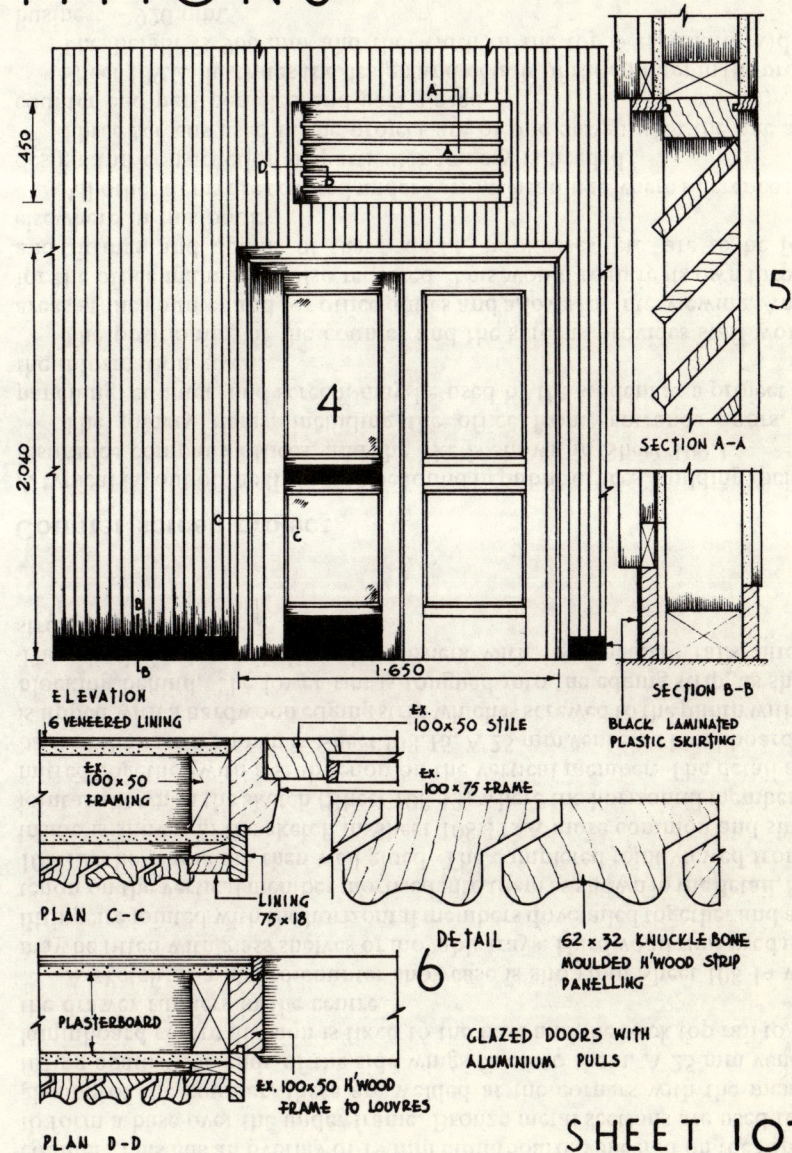

2·900

6 CLEAR POLISHED
PLATE GLASS

C

C

BRASS CRASH
BARS

1

2·700

2·040

A

A

B

B

826

826

ELEVATION

SECTION C-C

S'WOOD FRAMING   6 VENEERED LINING

EX.
150 x 75
TRANSOM

MOULDED
BOARDS

EX.
125 x 50 STILE

3

EX.
150 x 75 FRAME

TOP CENTRE

PLAN AT A-A

EX. 125 x 50 STILES

PLAN AT B-B

2

DETAIL OF MOULDED BOARDS EX. 62 x 25

A

D

b

5

SECTION A-A

450

4

2·040

C

C

B

B

1·650

ELEVATION

SECTION B-B

6 VENEERED LINING

EX.
100 x 50 STILE

BLACK LAMINATED
PLASTIC SKIRTING

EX.
100 x 50
FRAMING

EX.
100 x 75 FRAME

PLAN C-C

LINING
75 x 18

DETAIL

6

62 x 32 KNUCKLE BONE
MOULDED H'WOOD STRIP
PANELLING

PLASTERBOARD

GLAZED DOORS WITH
ALUMINIUM PULLS

EX. 100 x 50 H'WOOD
FRAME TO LOUVRES

PLAN D-D

SHEET 107

Moulded tongue and groove boards detailed in Sheet 107.2 are used in the lower portions of the sidelights with a softwood framing for fixing the flush veneered panelling to the inside. It will be seen that the moulded boards and the glass rebates are in line, allowing the stiles and mullions to be rebated in their length on the inner edges, giving square rail shoulders at the joints.

The details (Sheet 107.3) show sections at the transom and head of the door and the meeting stiles of both doors. Work on swing doors is dealt with more fully on pages 155 and 156 to which reference should be made. Sheet 107.4 shows a timber studded partition dividing two rooms each side faced to match the existing décor. A pair of glazed doors are incorporated in the partition along with a louvred ventilator in the centre.

The studding using 100 mm × 50 mm softwood provides the core for knucklebone moulded hardwood strip panelling to the one side, with veneered plywood planked panelling to the other side. Both sides of the studding have a plasterboard barrier for sound insulation (Sheet 107.6).

A section through the centre louvred ventilator is shown in Sheet 107.5. The louvres are inclined at 45 degrees and housed into a 100 mm × 50 mm hardwood frame. Hardwood linings 18 mm thick provide the finish to the opening, and to the strip panelling. Similar linings to the door-frame are used and a plain mitred lining in the form of an architrave completes the finish at the opening.

The plinth is finished in a contrasting colour, black laminated plastic in this case, to match the facings to the rails of the doors. When facings of this type are used on rail members, the laminated plastic should be sunk or housed in the member to its own thickness to give a flush finish to the member.

Service counters

Almost every shop requires a counter of one kind or another. For small general businesses the simple framed structure providing storage shelves and a cash drawer below the top may be the most suitable. Specialist shops and stores require counters to display goods in the front, with shelves and cupboards behind, and with the top arranged to display trays with compartments for small articles.

A type of counter suitable for a ladies' gown shop or a gents' outfitters is shown in Sheet 108.8. The design of this particular counter shows a centre portion with glazed top and front. The two side wings are panelled at the front in leather, with veneered blockboard tops in mahogany.

The centre portion is fitted with movable trays for quick service and the side wings with drawers, which are closed to the customers' side. The counter being raised on legs gives a sense of continuity to the floor space, which is a feature of modern shop and store design.

The 100 mm × 38 mm hardwood legs are tapered on the two inside edges and have 64 mm × 38 mm rails tenoned into them to form the underframe. The carcassing for the closed ends consists of 50 mm × 38 mm framing mortised and tenoned together and covered on the face side and ends with 25 mm blockboard. Sheet 108.9 shows the section through this portion of the counter with the drawers of varying depths, in position. It will be noticed that the fronts of the drawers are kept narrower than the sides and require no

handles. These run on hardwood runners with guides, and are easily taken out for quick service. The fixing of the counter to the base is by screwing through the framing into the rails of the base.

Sheet 108.10 shows a section through the glazed centre portion of the counter. This has an overlay of 19 mm laminboard, veneered on the top side, to form a base over the under frame. Bronze metal sections are used for the glazing of the counter, these are welded at the corners with the members fitting against the ends of the side wings fixed to them. A 25 mm veneered laminboard centre division is fixed to the base and the back top rail to carry the drawer runners in the centre.

A sketch of a glazed counter showcase is shown in Sheet 108.14 which may be fitted with glass shelves or movable trays. Hardwood rims used in this fitting are jointed with the horizontal members dovetailed together and a stub tenon on the vertical member mortised into them as shown in the detail, Sheet 108.12, the whole are then well glued. The completed joint viewed from the inside is shown in the sketch in Sheet 108.11. A more common and simpler joint is shown in the sketch (Sheet 108.13) where the horizontal members are mitred together with a stub tenon on the vertical member. The detail at the base of the case is shown in Sheet 108.15. A 25 mm veneered blockboard base is lipped with a hardwood edging strip which is screwed to the plinth with glue blocking behind. The lower rim is tongued into the edging strip, as shown. The plinth is dovetailed at the corners with intermediate rails fitted to strengthen the base.

Counter screen project

A typical layout of the fittings to be found in public offices, building societies, insurance company offices, and the like is shown in Sheet 109.1.

The joinery items, including the office front, entrance doors, wall panelling, counter, and screen, may be used by the student as a project from the information given.

The positioning of the counter and the screens provides staff working areas at the counter and for office duties and allows for interviewing. An area for the office manager is also required. This would require its own furniture and fittings and details of these, desks, bookcases, etc. are to be found elsewhere in this book.

All other items are covered under various headings where reference to the design and construction of particular joinery is needed.

Three screens used in the project are of one design with the one at the counter end part panelled and part glazed.

Sheet 109.2 illustrates the design and details of the counter in the project.

The height is 900 mm and the width of the top – especially wide for business – 920 mm.

The counter, it will be seen, starts at the window back and finishes against the teak screen numbered in the plan layout in Sheet 109.6 as number 4. A combination of materials in the finish includes a leather-covered top with quilted hide panels to the front buttoned with aluminium buttons. Reference to similar work on leather-covered wall panelling in Sheet 87.18 and page 242 should be made. Intermediate vertical panels in teak veneer with a banding of

SERVICE COUNTERS

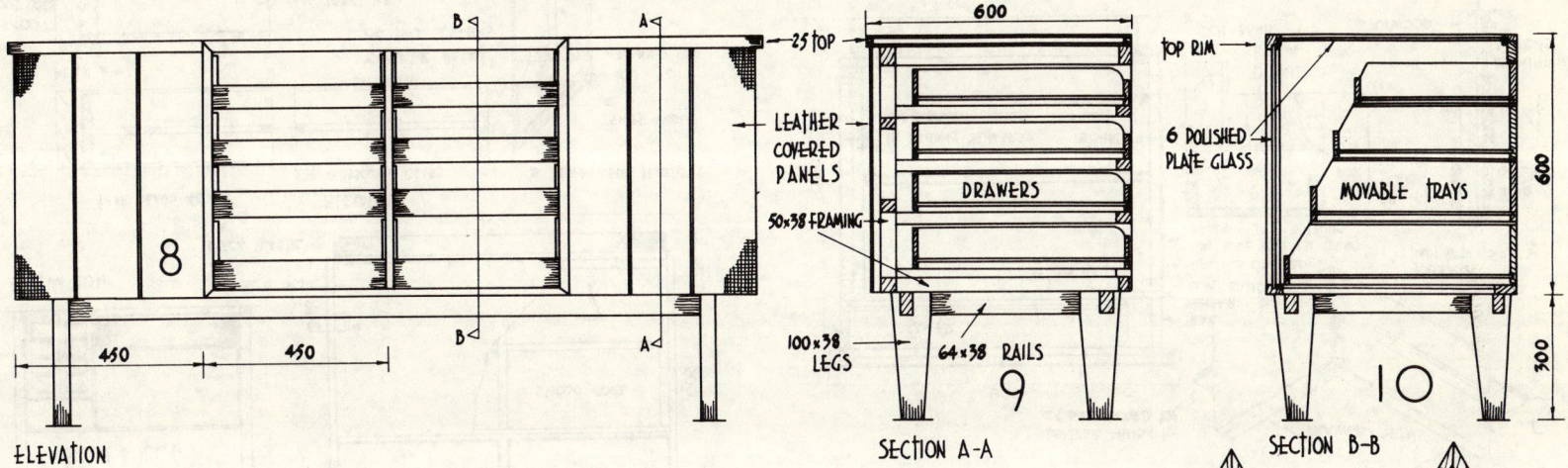

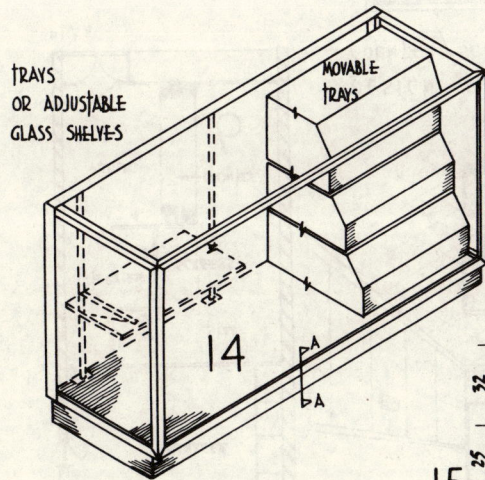

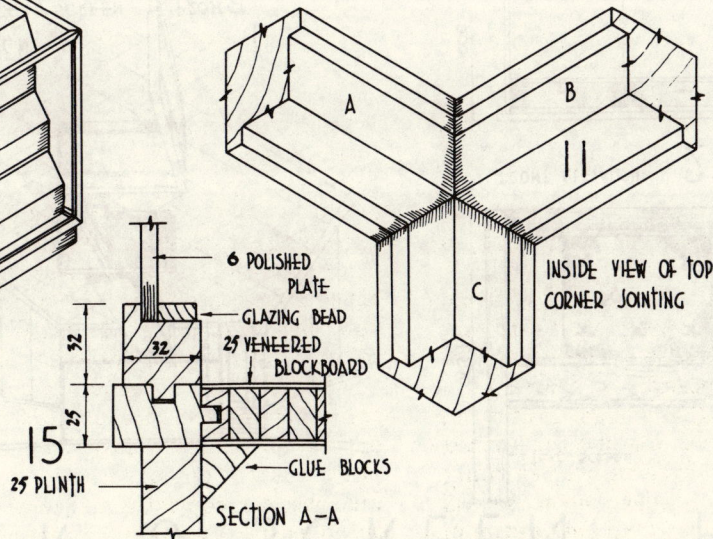

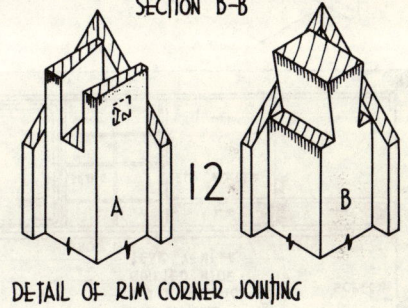

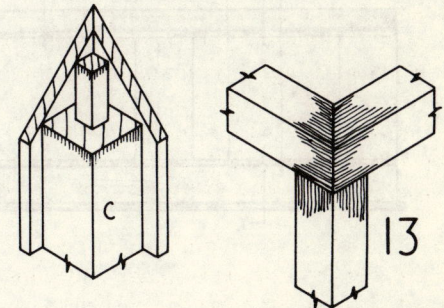

ELEVATION

8

450   450

B   A

25 TOP

LEATHER
COVERED
PANELS

50x38 FRAMING

DRAWERS

100x38
LEGS

64x38 RAILS

SECTION A-A

9

600

TOP RIM

6 POLISHED
PLATE GLASS

MOVABLE TRAYS

SECTION B-B

10

600

300

SKETCH OF COUNTER SHOWCASE

TRAYS
OR ADJUSTABLE
GLASS SHELVES

MOVABLE
TRAYS

14

6 POLISHED
PLATE
GLAZING BEAD

32   32

25 VENEERED
BLOCKBOARD

25
25

GLUE BLOCKS

15

25 PLINTH

SECTION A-A

A   B

11

C

**INSIDE VIEW OF TOP
CORNER JOINTING**

A   12   B

DETAIL OF RIM CORNER JOINTING

C   13

SHEET 108

271

COUNTER & SCREEN PROJECT

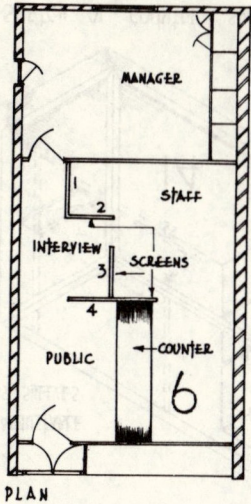

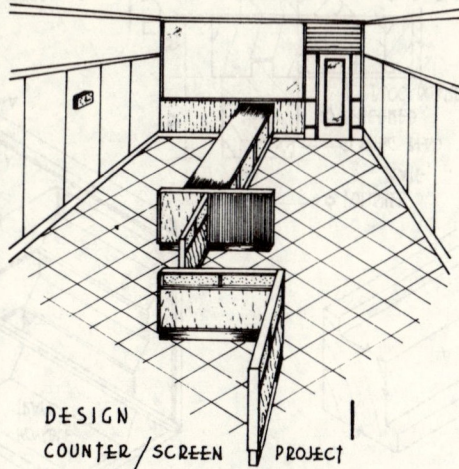

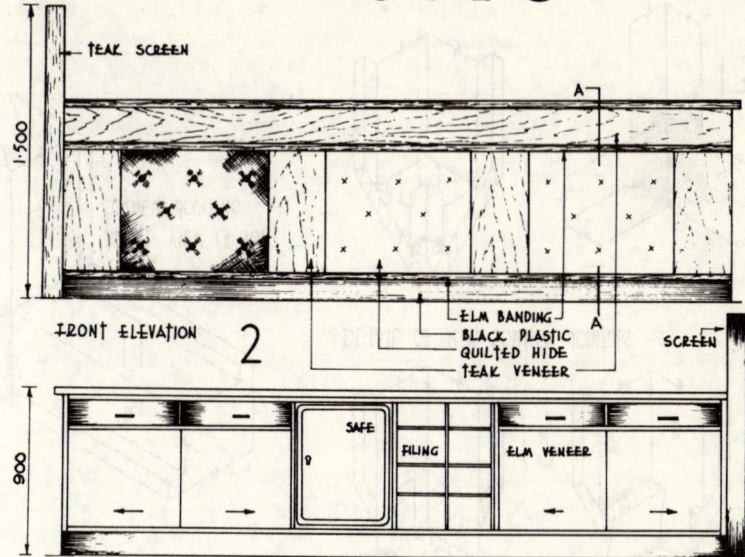

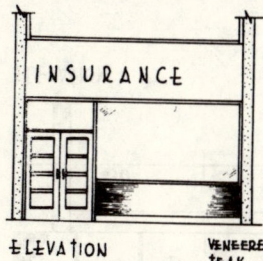

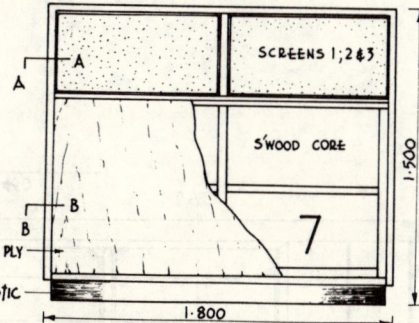

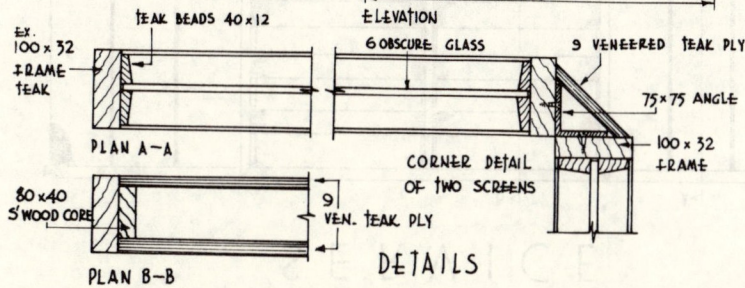

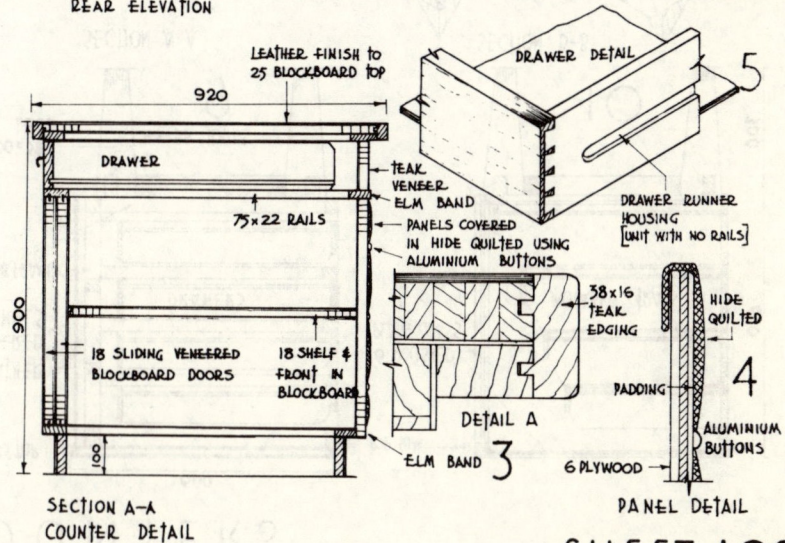

6 PLAN

MANAGER
STAFF
INTERVIEW
SCREENS
PUBLIC
COUNTER

1 DESIGN COUNTER/SCREEN PROJECT

INSURANCE

ELEVATION

7 SCREENS 1,2&3
S'WOOD CORE
VENEERED PLY TEAK
BLACK PLASTIC
ELEVATION
1·500
1·800

EX. 100 x 32 FRAME TEAK
TEAK BEADS 40 x 12
PLAN A-A
80 x 40 S'WOOD CORE
9 VEN. TEAK PLY
PLAN B-B

6 OBSCURE GLASS
9 VENEERED TEAK PLY
75 x 75 ANGLE
100 x 32 FRAME
CORNER DETAIL OF TWO SCREENS
DETAILS

2 FRONT ELEVATION
TEAK SCREEN
1·500
ELM BANDING
BLACK PLASTIC
QUILTED HIDE
TEAK VENEER
SCREEN

REAR ELEVATION
900
SAFE
FILING
ELM VENEER

5 DRAWER DETAIL
DRAWER RUNNER HOUSING (UNIT WITH NO RAILS)

LEATHER FINISH TO 25 BLOCKBOARD TOP
920
DRAWER
75 x 22 RAILS
TEAK VENEER ELM BAND
PANELS COVERED IN HIDE QUILTED USING ALUMINIUM BUTTONS
900
18 SLIDING VENEERED BLOCKBOARD DOORS
18 SHELF & FRONT IN BLOCKBOARD
100
SECTION A-A COUNTER DETAIL
ELM BAND

3 DETAIL A
38 x 16 TEAK EDGING

4 HIDE QUILTED
PADDING
ALUMINIUM BUTTONS
6 PLYWOOD
PANEL DETAIL

SHEET 109

elm top and bottom and a black plastic inset plinth are shown in the front elevation. The rear elevation shows provision under the counter for drawers, filing systems, storage cupboards fitted with sliding doors, and safe.

A detail showing the finish to the top and edging is shown in Sheet 109.3, with an upholstered panel detailed in Sheet 109.4.

Drawer construction is dealt with under 'desks', a reference to the detail (Sheet 109.5) is made here. Where units constructed from sheet materials, blockboard, and the like, and have no drawer rails or rails set back from the face, in such cases the drawer sides are housed to receive the runner which is fixed to the unit ends or divisions.

Bar counters

Counters in hotels, public houses, and drink bars where a service for both food and drink is given are called bar counters.

In hotel work the bar counter is the focal point in the room, and with the introduction of new materials the designer's aim is to make sure it remains the focal point where the business is done. Attractive back fittings to the bar and surrounding décor add further to the general presentation of hotel bars and rooms.

Sheet 110.1 shows a panelled reception area and bar. Cloakroom and other facilities with bar counter canopy and display units are suggested items in the project for this high-class purpose-made joinery.

Sheet 110.2 shows the elevation and section of the back fitting to the bar counter and is designed to accommodate wines, spirits, etc. on adjustable glass shelves in the upper portion and glass and bottle storage in the lower portion. A refrigeration unit is also housed in the lower area below the actual level of the bar counter top.

The whole unit is constructed using 25 mm blockboard with laminated plastic facings. It is suggested that contrasting colours may be used; the ends and shelves in white and the top and fascias dove grey with the plinth in black.

A detail at the top of the unit to show the lighting arrangement is shown in Sheet 110.3.

Sheet 110.4 shows the details of the bar counter in the project design above.

In hotel work bar counters range between 1·050 m and 1·175 mm in height for the convenience of the serving of drinks to customers standing or sitting on high stools at the bar counter.

The counter front finish is of teak veneered blockboard facings with all remaining facings and oversailing top finished in laminated plastic on blockboard to enhance the rich colour and teak grain finish to the front.

The projecting top is formed by boxing-out at the front edge and at the returns to strengthen and increase the width of the top for serving. The working top is set at a convenient working height behind the counter for staff use, and in order to support the main counter top the ends and divisions carry through over the working top and are housed for the required shelves as shown in the rear elevation (Sheet 110.5).

Provision needs to be made for bottle and glass storage with a double stainless steel sink unit and refrigerator unit for staff use.

A section through the bar counter is shown in Sheet 110.6.

Two further designs of bar counters are shown in Sheet 114.1 and 114.2, with typical items of bar furniture in tables and chairs of varying designs shown in Sheet 114.3.

The counter in Sheet 114.1 is of framed construction in softwood to support the top. The face of the counter is panelled in grey quilted leather inclined towards a bracketed bag-shelf below the projecting bar counter top. This is plastic-faced plywood finished with a hardwood tongued nosing to both edges. Provision for the washing of glasses and the storage of bottles, etc. is provided by shelves as in the previous example.

A footrail in brass is carried by metal brackets screwed to the floor as shown.

Sheet 114.2 shows a second bar counter unit constructed mainly from 25 mm blockboard.

The splayed front is formed using 32 mm softwood fins to carry the bag-shelf and provide fixing for the veneered blockboard front. The projecting blockboard top is faced with laminated plastic and edged with hardwood nosings. Support for the top on the staff side is by 25 mm supports at intervals from the work top.

An alternate footrest is shown formed using softwood blockings faced with hardwood and finished with an aluminium nosing.

Shop fittings

The plan showing the layout of a number of shop fittings suitable for varying types of shops is shown on Sheet 111.1.

An island counter unit is sited towards the centre of the floor area and would be used for the display and serving of goods to customers. Further straight counters are positioned at points near to goods displayed along the walls.

Display and wall units lettered D, E, F, and G complete the fittings.

The whole plan with the site dimensions as given may be used by the student as a project and along with the display counters and wall units other items of purpose-made joinery read: shop front, battery of swing doors, window back, rear door, and stairs. All of these items are listed and detailed elsewhere in this book, to which reference should be made.

Sheet 111.2 shows the sketch of the island counter to be provided with a hinged door and counter flap over which has been omitted in the drawing.

A section of the counter, 600 mm wide and 900 mm high, with provision for glass adjustable shelves to the front display areas and storage behind with polished plate-glass sliding doors is shown in Sheet 111.3.

The glass sliding doors are in aluminium polished track with plastic inserts for door removal and fitted with insert finger grips.

Sheet 112.1 shows the plan and elevation of the counter unit from which the setting out will be drawn.

COUNTER & FITTING PROJECT

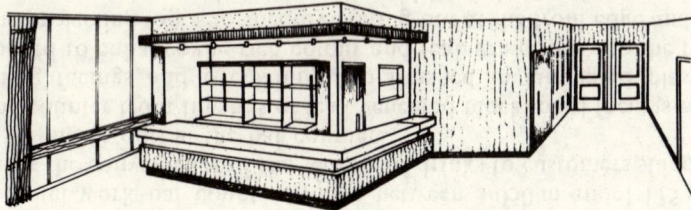

BAR COUNTER & DISPLAY UNITS PROJECT DESIGN — 1

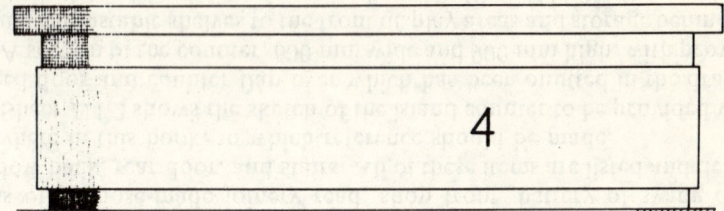

Diagram 2 — BACK FITTING DETAIL

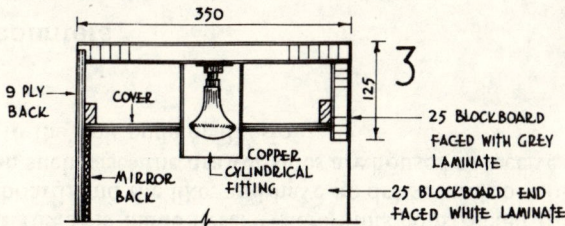

- MIRROR BACK
- POLISHED PLATE GLASS ADJUSTABLE SHELVES
- REFRIGERATION UNIT
- OPEN BOTTLE STORAGE
- GLASS STORAGE
- LIGHTING
- ADJUSTABLE STEEL BRACKETS

ELEVATION
3·700
2·460
1·060
350
500

SECTION A-A — B

BLACK LAMINATED PLASTIC

SPEC^N : 25 BLOCKBOARD LAMINATED PLASTIC FACINGS. SHELVES & ENDS WHITE, TOP & FASCIA'S DOVE GREY. TOP SHELVES 6 POLISHED PLATE GLASS.

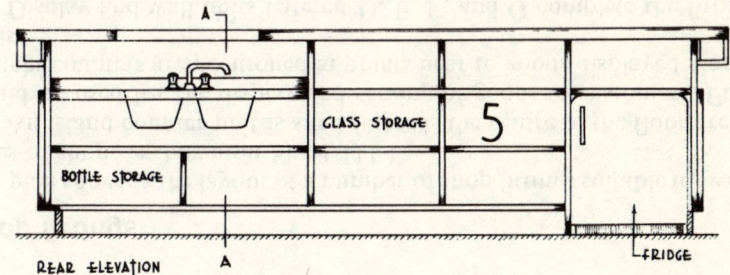

LIGHT FITTING DETAIL B — 3

- 9 PLY BACK
- COVER
- MIRROR BACK
- COPPER CYLINDRICAL FITTING
- 25 BLOCKBOARD FACED WITH GREY LAMINATE
- 25 BLOCKBOARD END FACED WHITE LAMINATE

350
125

FRONT ELEVATION — 4

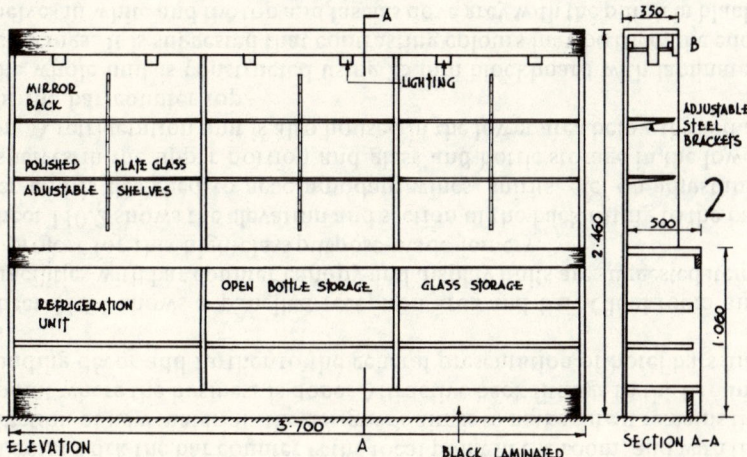

REAR ELEVATION — 5

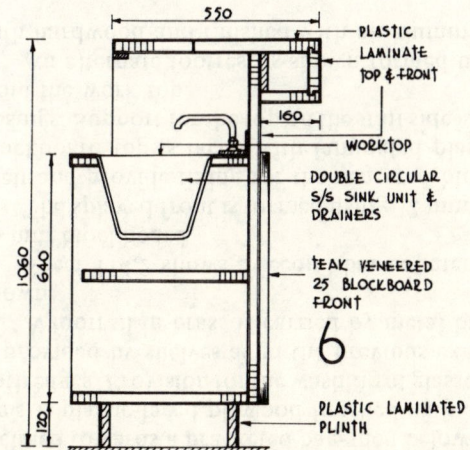

- BOTTLE STORAGE
- GLASS STORAGE
- FRIDGE
- A

550
160
1·060
640
120

SECTION A-A BAR DETAIL — 6

- PLASTIC LAMINATE TOP & FRONT
- WORKTOP
- DOUBLE CIRCULAR S/S SINK UNIT & DRAINERS
- TEAK VENEERED 25 BLOCKBOARD FRONT
- PLASTIC LAMINATED PLINTH

SHEET 110

COUNTER & FITTINGS PROJECT

DESIGN

WALL UNITS
D, E, F, G

DESIGN

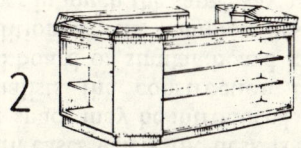

2

DISPLAY AND COUNTER UNIT

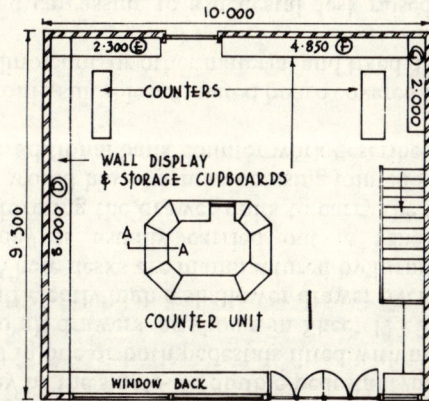

10·000

2·300 (E)   4·850 (F)

COUNTERS

WALL DISPLAY
& STORAGE CUPBOARDS

2·000

9·300

8·000

A — A

COUNTER UNIT   1

WINDOW BACK

PLAN OF LAYOUT

DETAILS

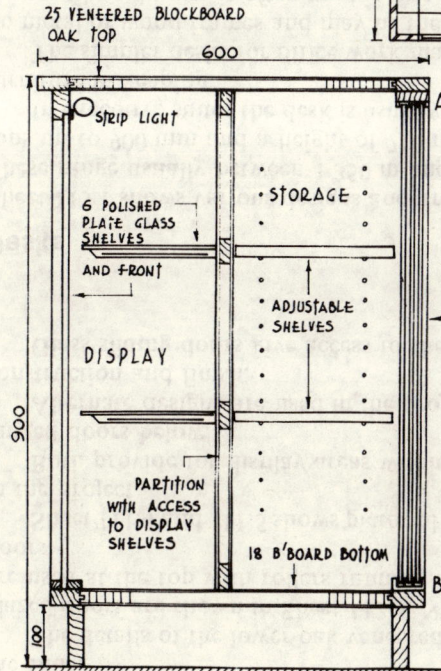

25 VENEERED BLOCKBOARD
OAK TOP

600

STRIP LIGHT

6 POLISHED
PLATE GLASS
SHELVES

AND FRONT

• STORAGE •

• ADJUSTABLE •
SHELVES

DISPLAY

PARTITION
WITH ACCESS
TO DISPLAY
SHELVES

18 B'BOARD BOTTOM

900

100

SECTION A–A

A

B

DETAIL A & B

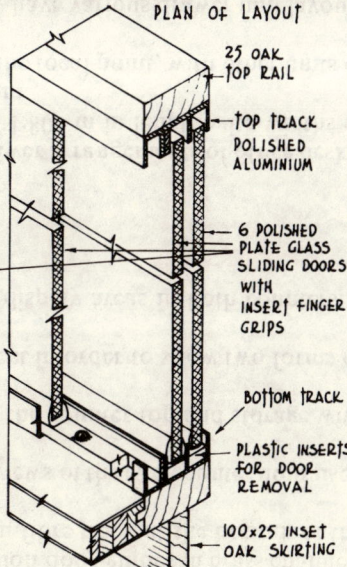

25 OAK
TOP RAIL

TOP TRACK
POLISHED
ALUMINIUM

6 POLISHED
PLATE GLASS
SLIDING DOORS
WITH
INSERT FINGER
GRIPS

BOTTOM TRACK

PLASTIC INSERTS
FOR DOOR
REMOVAL

100×25 INSET
OAK SKIRTING

3

DESIGN

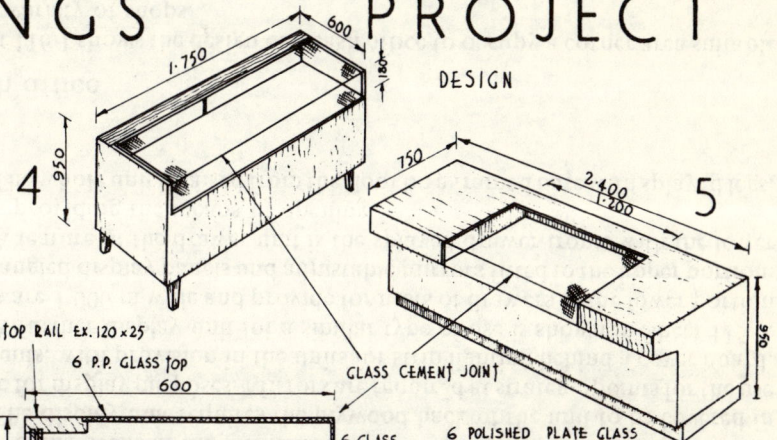

1·750   600   180

950

4

750

2·400
1·200

5

950

DETAILS

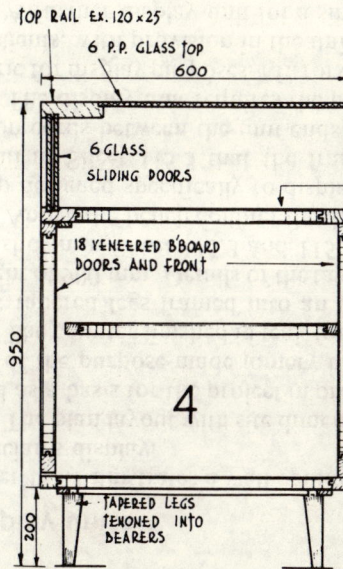

TOP RAIL EX. 120×25

6 P.P. GLASS TOP

600

6 GLASS
SLIDING DOORS

6 GLASS
FRONT

GLASS CEMENT JOINT

6 POLISHED PLATE GLASS
TOP AND DOORS TO DISPLAY

FABRIC COVERED FOR
DISPLAY

SECTION AT
DISPLAY

18 VEN. B'BOARD
TOP WITH OAK
NOSINGS

18 VENEERED B'BOARD
DOORS AND FRONT

950

200

TAPERED LEGS
TENONED INTO
BEARERS

SECTION

4

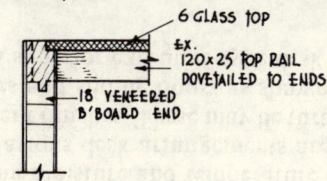

6 GLASS TOP

EX.
120×25 TOP RAIL
DOVETAILED TO ENDS

18 VENEERED
B'BOARD END

TOP AND END JOINTING

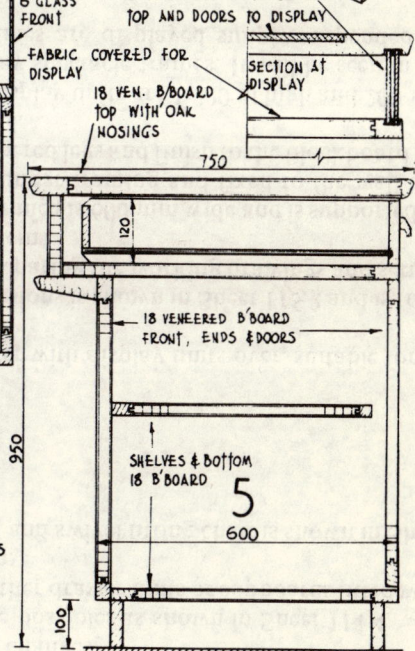

750

120

18 VENEERED B'BOARD
FRONT, ENDS & DOORS

SHELVES & BOTTOM
18 B'BOARD

600

100

SECTION

950

5

SHEET III

Wall units

Sheet 112.2 shows the design for a portion of one of the wall units. The dimensions shown with the lengths of the units and the specification are required by the setter-out.

The unit is constructed from 18 mm blockboard veneered on exposed surfaces and jointed by dovetailing at the top corners with shelves housed into the ends.

The details of the lower oak veneered blockboard sliding doors and the glazed doors are shown in Sheet 112.3. Nylon door guides in brass channels are used at the top with rollers running on fibre track at the bottom of the doors.

Sheet 111.4 and 111.5 shows pictorial views of the two counter units used in the project.

Both provide for display areas within the counter top and storage with hinged doors below.

Alternate designs are used in the project in order to show two forms of construction and finish.

Glass sliding doors give access to the display areas in both counters.

Desks

Sheet 113.1 shows various designs and drawer arrangements of office desks. These range usually between 1·350 m and 1·800 m in length with widths of tops up to 900 mm and a height of 710 mm.

In executive suites the desk is usually the focal point, with other units of furniture in keeping.

The simpler desks for office work may have various drawer unit layouts on metal or wood frames and may be the single- or double-pedestal type. A deeper filing drawer is often used in one or both pedestals fitted with metal runners and automatic locking to all drawers, as shown in Sheet 113.2.

In cases where the desk is sufficiently high a shallower drawer over the knee space may be introduced. Where desks are manufactured by furniture specialists the construction today is usually carried out in veneered blockboard or similar board in forming the drawer units to carry the top. Traditionally, the complete desk would have been made using jointed solid timber in much the same way as traditional bank counter work described on page 264.

Desk tops are now selected from a suitable laminated board covered with a facing of veneer, leather cloth, linoleum, or other material and fixed direct to the pedestals.

Sheet 113.3 shows the framed carcassing to a pedestal desk raised on cross-framed legs.

The ends of the pedestal are framed and finished flush with veneered plywood. The front rails are dovetailed into the end stiles, in the case of the top and bottom rails, and housed for the remainder, and are set back from the face the thickness of the drawer fronts to give an overall flush finish to the unit.

Alternate drawer pulls or handles are shown in Sheet 113.4, with a sunk drawer pull worked in the lower edge of the drawer front, shown in the sketch (Sheet 113.5).

Sheet 113.6 shows a drawer unit fixed to a framed table by screwing to the rails and table legs.

Slotted table plates used for the fixing of unit tops where any movement is likely are shown in Sheet 113.7. It will be seen that the plates have elongated slots in them allowing the screwed top to expand in width when subject to varying moisture and temperature changes.

Various desk arrangements are possible, as shown in Sheet 114.4, with different finishes. These may be further drawer units or cupboards fitted with shelves and sliding doors as shown.

A superior executive-type desk and swivel fitting chair is shown in Sheet 114.5.

Display units

Sheet 115.1 illustrates a wall counter with display units over, suitable for an optician's display.

The plan layout with site dimensions is shown in Sheet 115.2 and is to be used as a basis for the project in preparing the working drawings and setting out of the purpose-made joinery items.

The counter finished in teak formica is 600 mm wide and is supported on teak tapered legs framed into an under-framing and fixed to the wall at a height of 900 mm. Details of the tapered legs and finish to the blockboard top are shown in Sheet 115.3 and 115.4.

Above the bench counter the display units are 1·000 m high and 200 mm deep designed specifically to display spectacle frames. It will be seen in the detail in Sheet 115.5 that the frames are displayed simply suspended on nylon cords between the unit ends.

The display unit requires the plywood back of the unit to be covered in fabric for display purposes. Mirrors are required at strategic points for the use of clients, with provision in the units for strip lighting behind a cover board.

A further display unit for a similar type of use is shown in Sheet 115.6. Units are 1·000 m wide and provide for nests of drawers in the lower portion with angled display panels and adjustable mirrors fitted to the upper portion.

A feature of the drawer unit is the splayed drawer fronts with the lower edges providing the means of opening.

The whole unit is raised from the floor on a framed base with splayed legs.

Cash office

Sheet 116.1 shows the design of a cash office to occupy a corner area suitable for a variety of shops.

The plan, elevation, and section with the main dimensions are shown in Sheet 116.2.

Sheet 116.3 shows the details. The office counter has two heights, a working height of 750 mm for staff and 1·050 m at the public side.

The elevated pay shelf provides the fixing for the plate-glass screen

PROJECT DETAILS

D — 8·000
E ↧ — 2·300
F — 4·850
G — 2·000

DETAILS

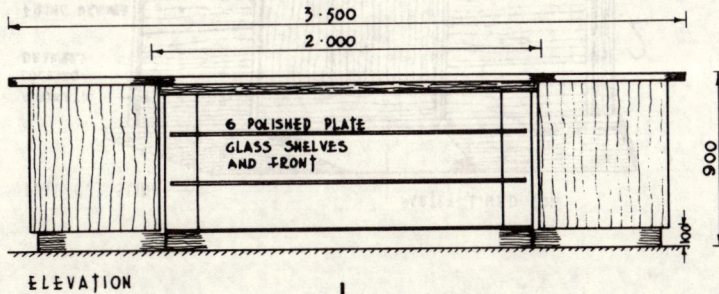

6 POLISHED PLATE
GLASS SHELVES
AND FRONT

3·500
2·000
900
100

ELEVATION

I

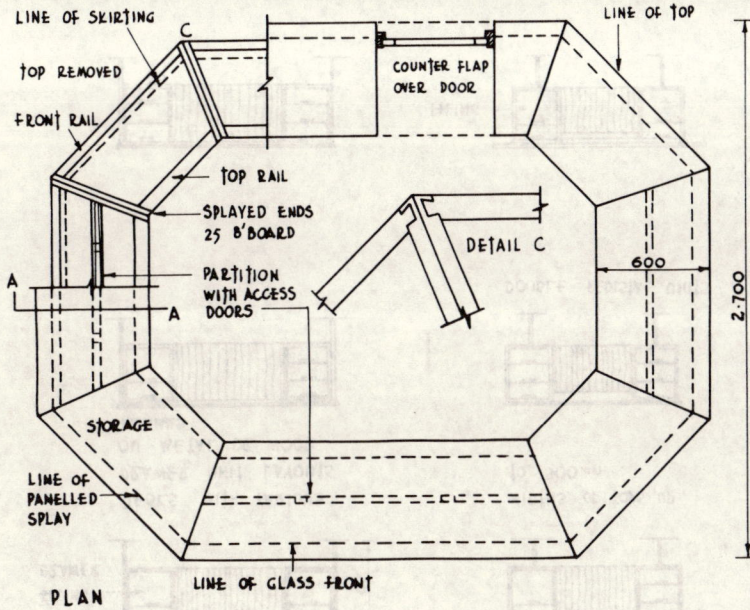

LINE OF SKIRTING
TOP REMOVED
FRONT RAIL
C
COUNTER FLAP OVER DOOR
LINE OF TOP
TOP RAIL
SPLAYED ENDS 25 B'BOARD
DETAIL C
600
2·700
A
A
PARTITION WITH ACCESS DOORS
STORAGE
LINE OF PANELLED SPLAY
LINE OF GLASS FRONT

PLAN

2
DESIGN

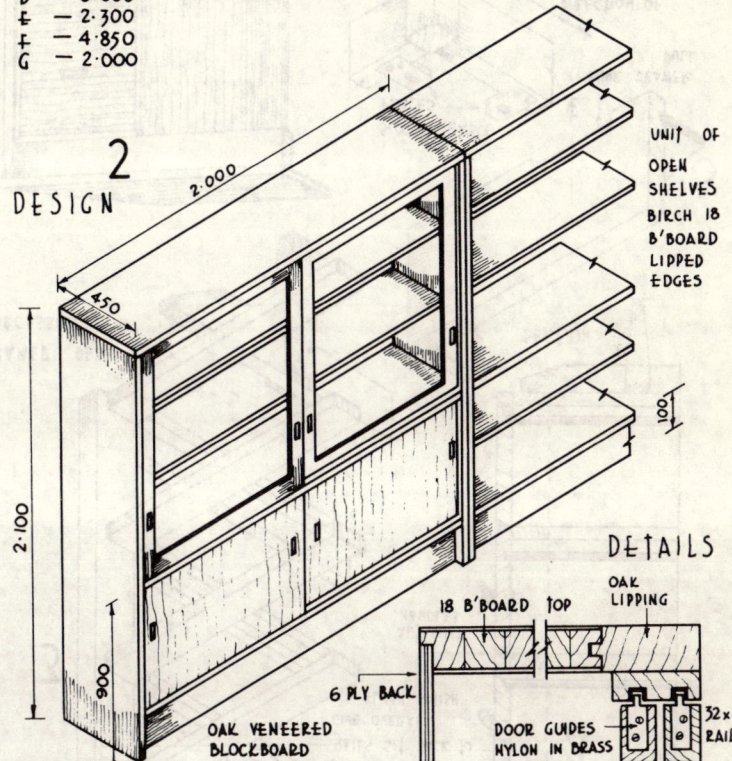

2·000
450
UNIT OF OPEN SHELVES BIRCH 18 B'BOARD LIPPED EDGES
2·100
900
100

OAK VENEERED BLOCKBOARD

DETAILS

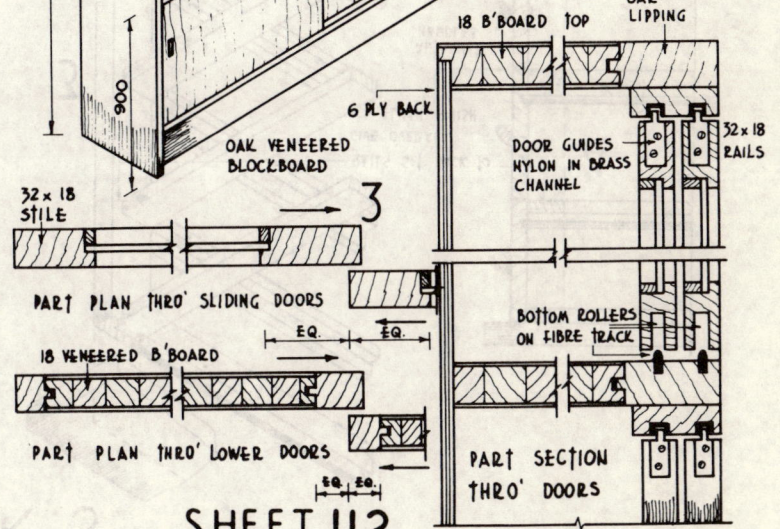

18 B'BOARD TOP
OAK LIPPING
6 PLY BACK
DOOR GUIDES NYLON IN BRASS CHANNEL
32 × 18 RAILS

32 × 18 STILE
3
PART PLAN THRO' SLIDING DOORS
EQ. EQ.
18 VENEERED B'BOARD
PART PLAN THRO' LOWER DOORS
EQ. EQ.

BOTTOM ROLLERS ON FIBRE TRACK
PART SECTION THRO' DOORS

SHEET 112

D E S K S

1·800
710
FILING DRAWER

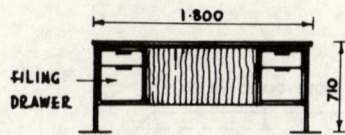

DESKS WITH VARIOUS DRAWER UNIT LAYOUTS ON METAL OR WOOD FRAMES

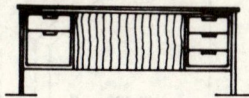

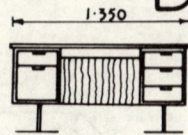

1·350
WIDTHS OF TOPS UP TO 900mm.

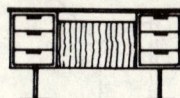

DOUBLE PEDISTAL UNITS

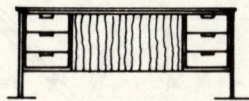

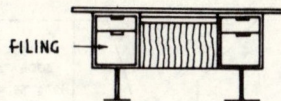

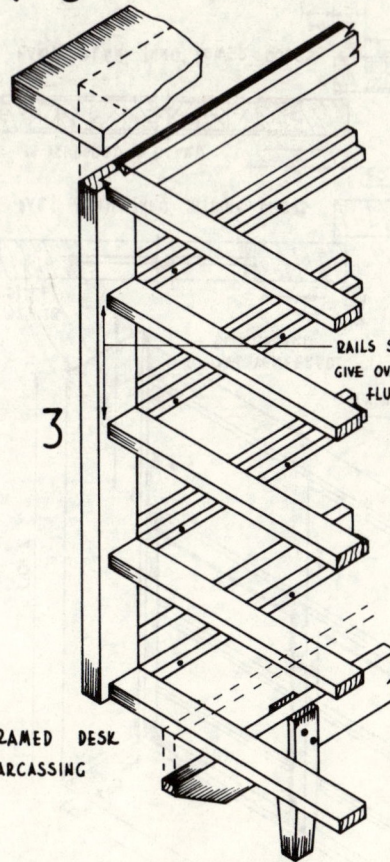

FILING

1

PULL OUT SLIDE
PARTLY LINED TOP

AUTOMATIC LOCKING DRAWERS

FILING DRAWER ON METAL RUNNERS

CROSS FRAMED LEGS

2

RAILS SET BACK TO GIVE OVERALL FLUSH FINISH

ALTERNATE HANDLES

3

FRAMED DESK CARCASSING

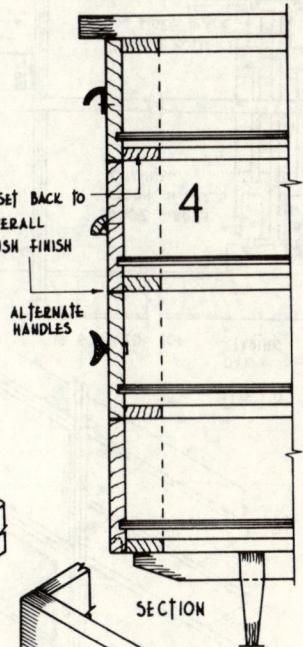

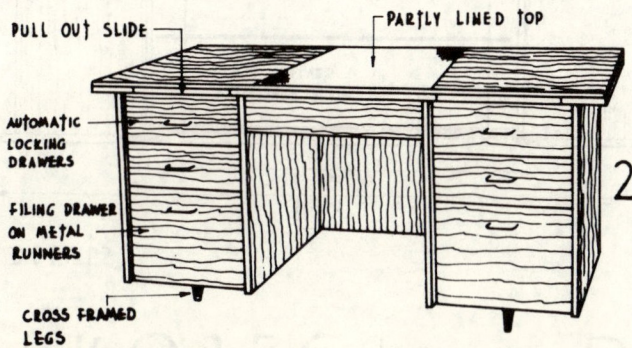

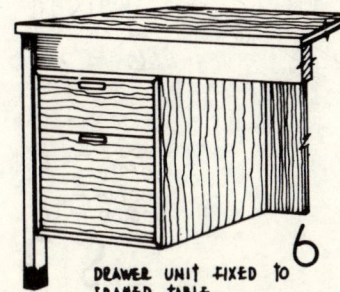

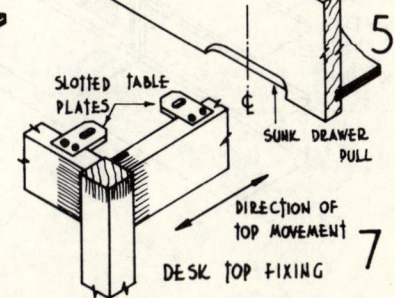

4

SECTION

DRAWER UNIT FIXED TO FRAMED TABLE

6

SLOTTED TABLE PLATES

SUNK DRAWER PULL

DIRECTION OF TOP MOVEMENT

DESK TOP FIXING

5

7

SHEET 113

OFFICE DESKS & BAR FITTINGS

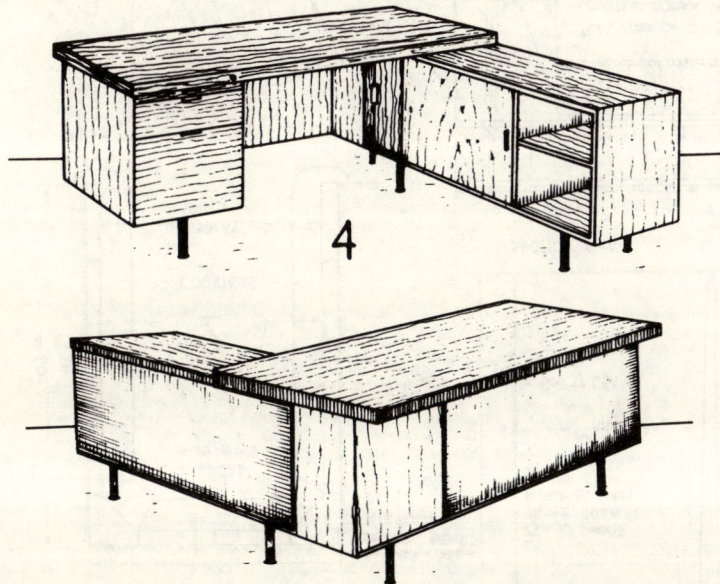

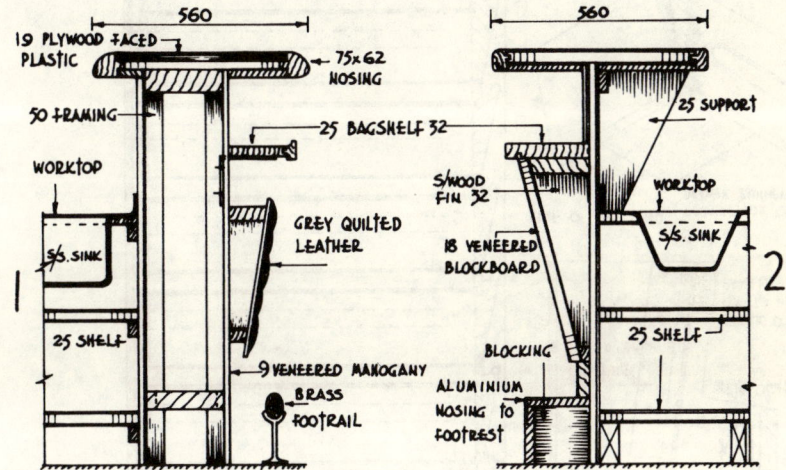

4

5

BAR SECTION

Section 1:
- 19 PLYWOOD FACED PLASTIC
- 75 x 62 NOSING
- 50 FRAMING
- 25 BAGSHELF 32
- WORKTOP
- S/S. SINK
- GREY QUILTED LEATHER
- 25 SHELF
- 9 VENEERED MAHOGANY
- BRASS FOOTRAIL

Section 2:
- 560
- 25 SUPPORT
- S/WOOD FILL 32
- WORKTOP
- 18 VENEERED BLOCKBOARD
- S/S. SINK
- 25 SHELF
- BLOCKING
- ALUMINIUM NOSING TO FOOTREST
- 560

TABLE DESIGNS
750 HIGH
1·200 DIA.

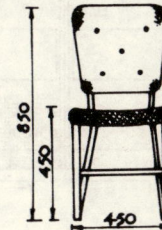

3

CHAIR DESIGNS

850

450

450

FITTINGS

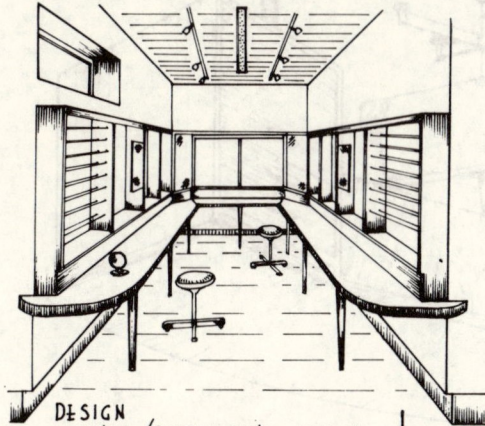

DESIGN
COUNTER / DISPLAY UNITS PROJECT 1

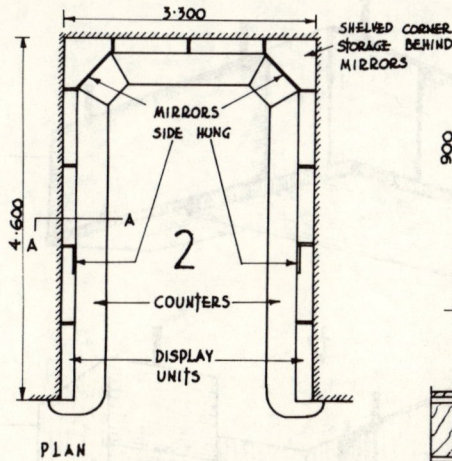

3·300

SHELVED CORNER STORAGE BEHIND MIRRORS

MIRRORS SIDE HUNG

4·600

A

A

2

COUNTERS

DISPLAY UNITS

PLAN

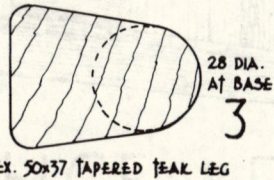

28 DIA. AT BASE

3

EX. 50x37 TAPERED TEAK LEG

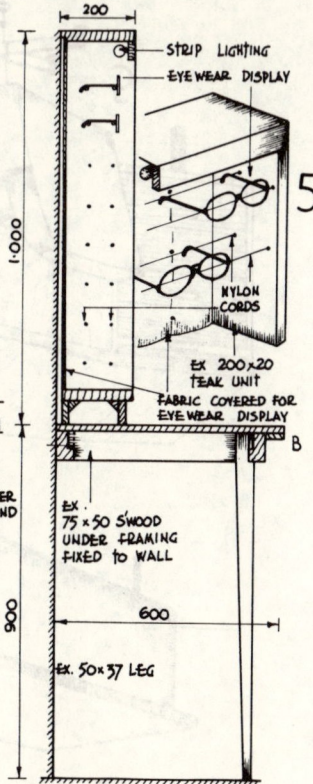

200

1·000

900

STRIP LIGHTING

EYE WEAR DISPLAY

5

NYLON CORDS

EX. 200 x20 TEAK UNIT

FABRIC COVERED FOR EYE WEAR DISPLAY

B

EX. 75 x 50 SWOOD UNDER FRAMING FIXED TO WALL

600

EX. 50x37 LEG

SECTION A-A

DETAILS

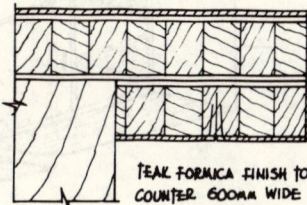

TEAK FORMICA FINISH TO COUNTER 600mm WIDE

DETAIL B 4

DESIGN

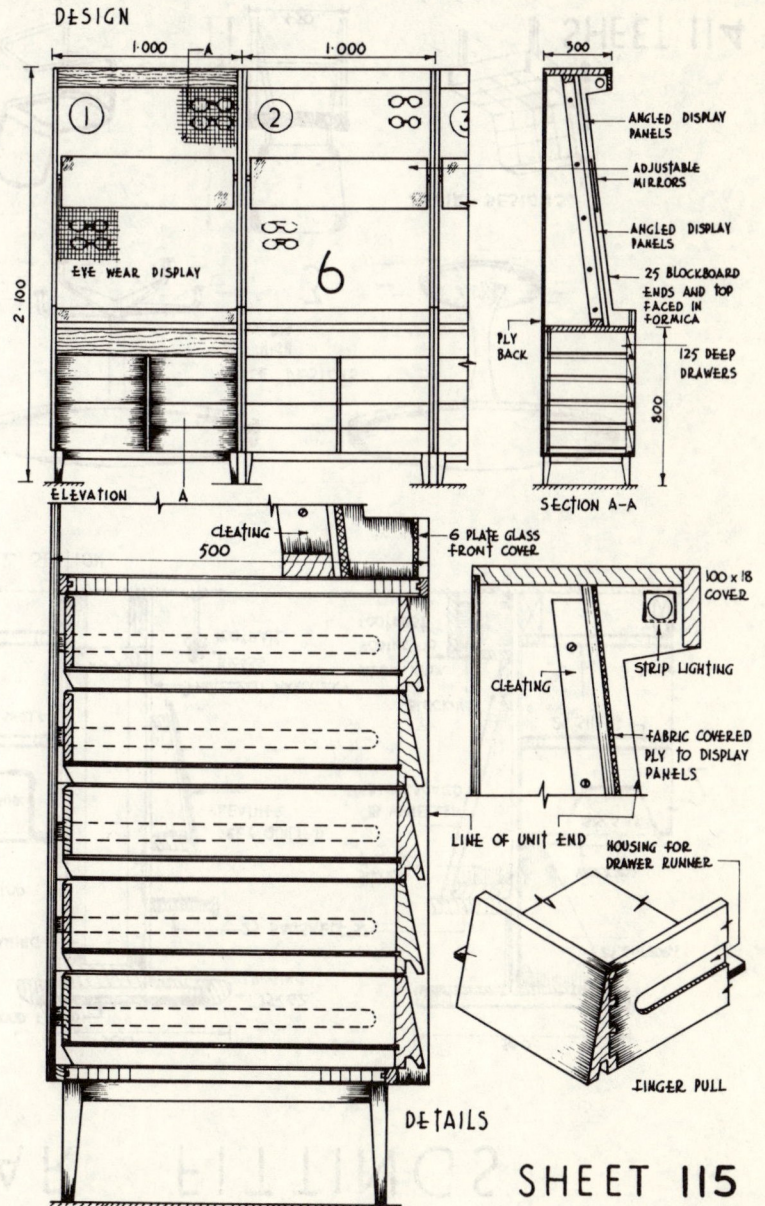

1·000   A   1·000   300

1   2   3

2·100

EYE WEAR DISPLAY   6

ELEVATION   A

ANGLED DISPLAY PANELS

ADJUSTABLE MIRRORS

ANGLED DISPLAY PANELS

25 BLOCKBOARD ENDS AND TOP FACED IN FORMICA

PLY BACK

125 DEEP DRAWERS

800

SECTION A-A

CLEATING 500

6 PLATE GLASS FRONT COVER

100 x 18 COVER

CLEATING

STRIP LIGHTING

FABRIC COVERED PLY TO DISPLAY PANELS

LINE OF UNIT END

HOUSING FOR DRAWER RUNNER

FINGER PULL

DETAILS

SHEET 115

C A S H   D E S K

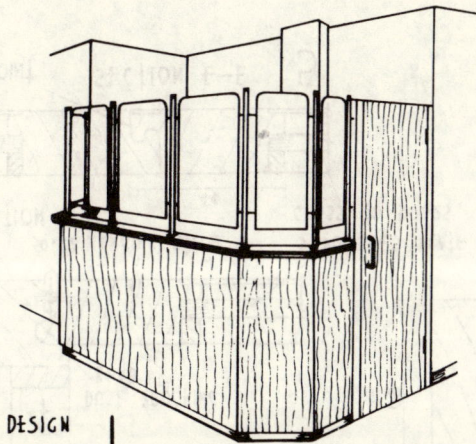

DESIGN 1

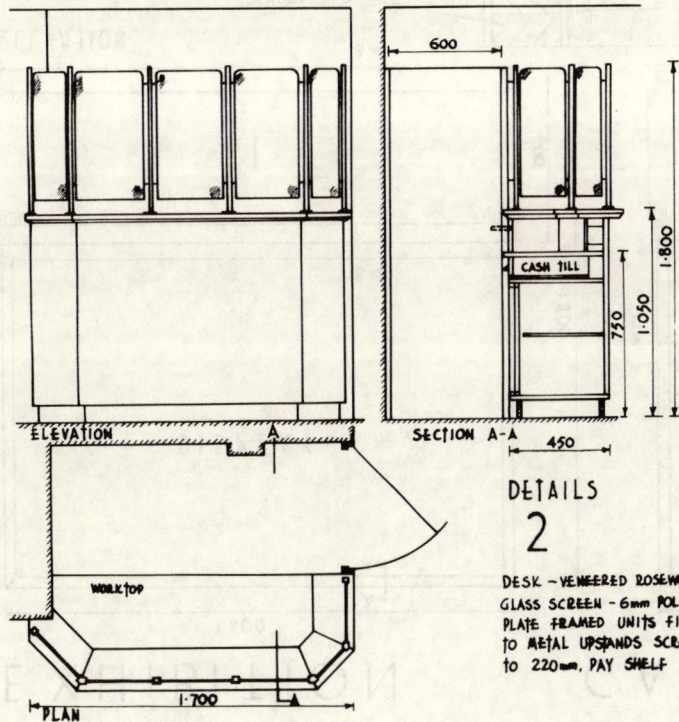

ELEVATION

SECTION A-A

600

1·800

1·050

750

CASH TILL

450

WORKTOP

PLAN

1·700

A

DETAILS 2

DESK - VENEERED ROSEWOOD
GLASS SCREEN - 6mm POLISHED
PLATE FRAMED UNITS FIXED
TO METAL UPSTANDS SCREWED
TO 220mm. PAY SHELF

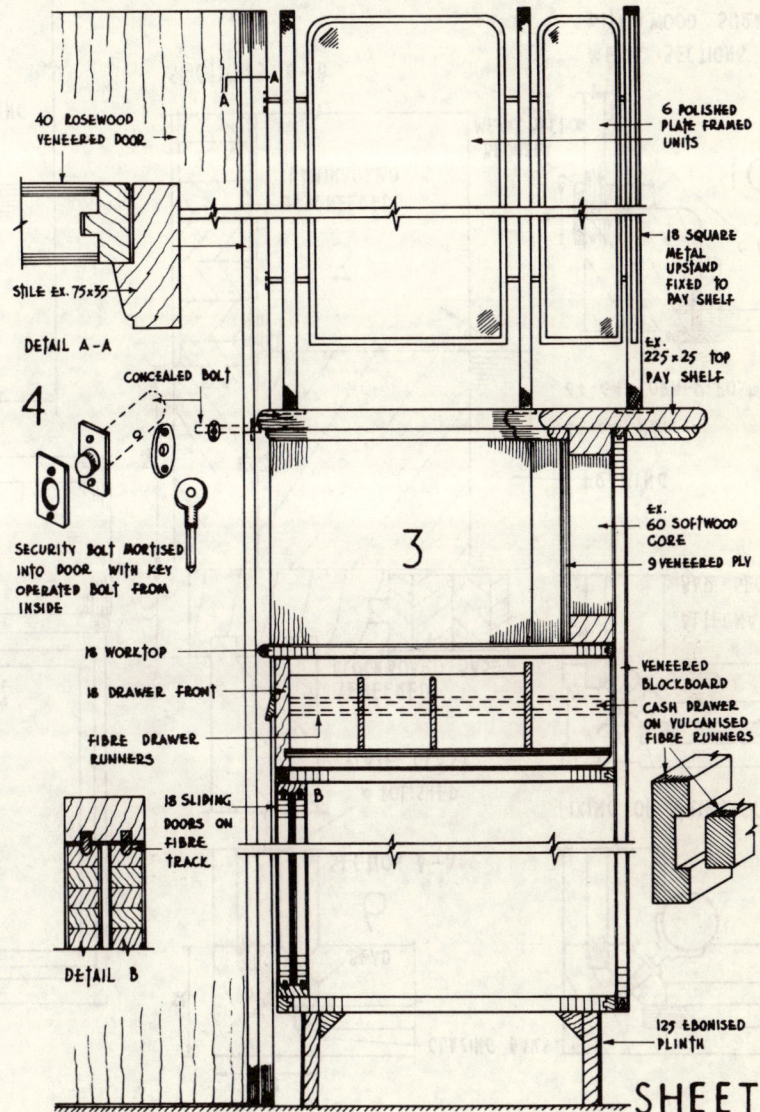

40 ROSEWOOD
VENEERED DOOR

STILE EX. 75x35

DETAIL A-A

A

4

concealed bolt

SECURITY BOLT MORTISED
INTO DOOR WITH KEY
OPERATED BOLT FROM
INSIDE

18 WORKTOP

18 DRAWER FRONT

FIBRE DRAWER
RUNNERS

18 SLIDING
DOORS ON
FIBRE TRACK

B

DETAIL B

6 POLISHED
PLATE FRAMED
UNITS

18 SQUARE
METAL
UPSTAND
FIXED TO
PAY SHELF

EX.
225 x 25 TOP
PAY SHELF

3

EX.
60 SOFTWOOD
CORE

9 VENEERED PLY

VENEERED
BLOCKBOARD

CASH DRAWER
ON VULCANISED
FIBRE RUNNERS

125 EBONISED
PLINTH

SHEET 116

EXHIBITION CASES

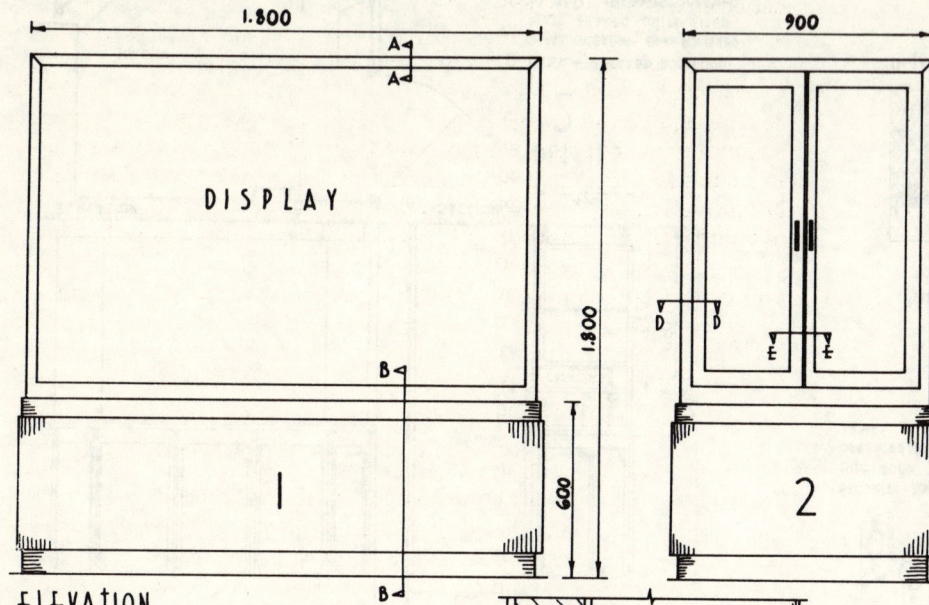

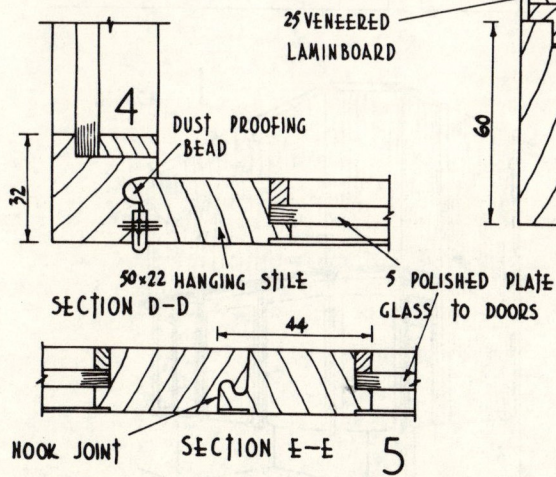

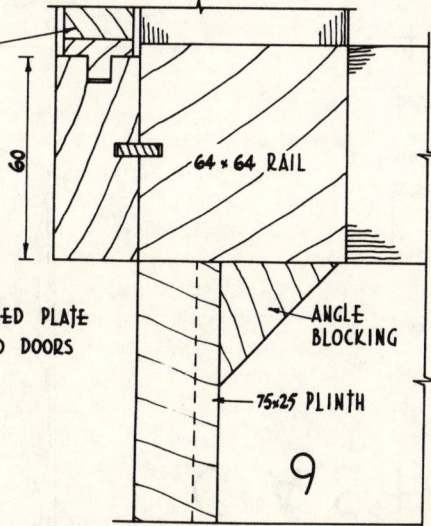

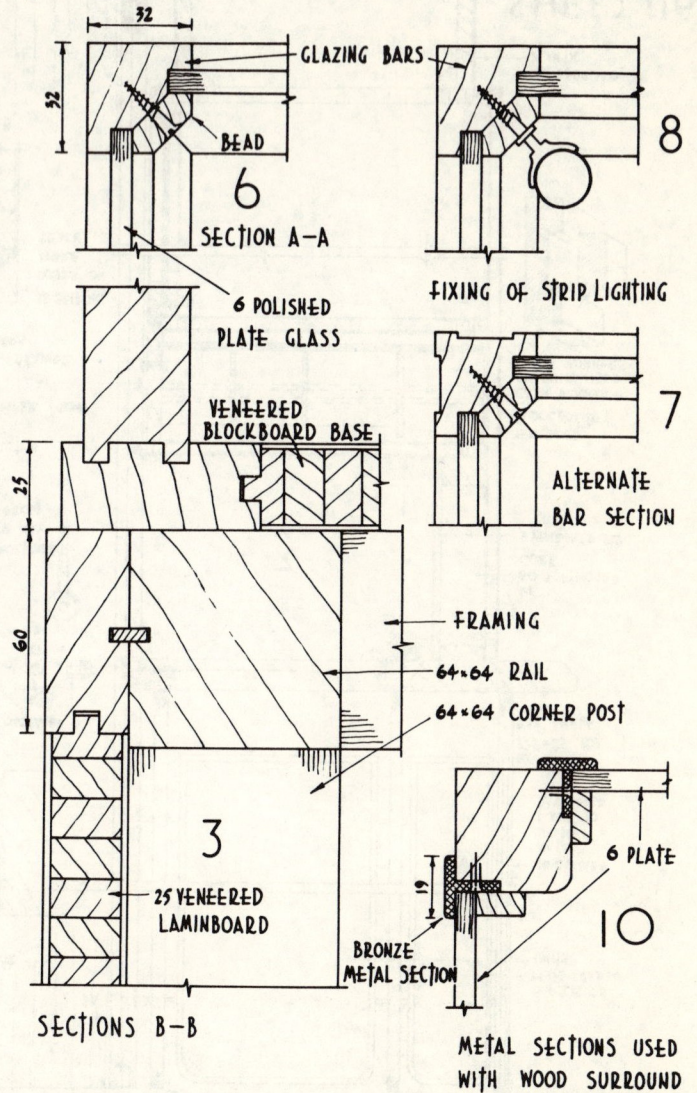

1.800

DISPLAY

1.800

600

ELEVATION

1

900

2

SECTION A-A
32

52

GLAZING BARS

BEAD

6

6 POLISHED PLATE GLASS

FIXING OF STRIP LIGHTING

8

7

ALTERNATE BAR SECTION

VENEERED BLOCKBOARD BASE

25

60

FRAMING

64 × 64 RAIL

64 × 64 CORNER POST

3

25 VENEERED LAMINBOARD

SECTIONS B-B

4

DUST PROOFING BEAD

32

25 VENEERED LAMINBOARD

64 × 64 RAIL

60

ANGLE BLOCKING

75×25 PLINTH

9

50×22 HANGING STILE

SECTION D-D

44

5 POLISHED PLATE GLASS TO DOORS

HOOK JOINT   SECTION E-E   **5**

19

6 PLATE

BRONZE METAL SECTION

10

METAL SECTIONS USED WITH WOOD SURROUND

SHEET 117

supports. Provision is made for cash drawers on fibre drawer runners with sliding doors to cupboard storage areas. The construction follows closely other work on counters previously detailed.

Access to the office is by a flush door opening outwards and fitted with a concealed bolt operated from inside, shown in Sheet 116.4.

Exhibition cases

Sheet 117.1 shows the front elevation and Sheet 117.2 the end elevation of a typical island display case suitable for the display of collections of works of art. The practice illustrated in this example is applicable to shop as well as museum cases. In this design, the case is mounted on a framed stand 600 mm high. This is a simple mortise and tenoned framework in softwood with corner posts which are connected by top and bottom rails tenoned into them and serves a double purpose, providing useful storage space as well as a base. The framework is then faced with 25 mm veneered laminboard edged as shown in Sheet 117.3, mitred at the vertical angles and tongued, glued, and screwed to the framing. The detail of the recessed plinth is shown in Sheet 117.9. This is dovetailed at the corners and secured to the stand by pocket screwing and angle blocking.

Access to the case is provided by a pair of narrow door-frames with hook-jointed meeting stiles. This arrangement is generally better than a single door-frame as the plate-glass panels in wood door-frames impose a great strain on the mortise and tenon joints of the frames.

A method of forming the special joints, which are an essential feature of this class of work in order to prevent the entry of dust, is by the use of double rebates in the construction.

In airtight showcases the protection against the change of air is obtained by a system of beads and grooves. Sheet 117.4 shows a section through a vertical member of the frame which is grooved to receive the bead worked on the edge of the hanging stile of the door. The bead is arranged so that the hinge does not penetrate it when screwed in position. The joining at the centre stiles of the doors is shown in Sheet 117.5, where a hook joint is used.

A further feature of this class of work is that the framework is reduced to the smallest possible dimensions that will keep the plate glass in position, so that an uninterrupted view of the contents may be had. This slenderness of frame requires that the joints at the corners between the rims shall be made with the greatest care and accuracy. Special methods are employed: one method is shown in Sheet 108.12, two members are jointed using a secret dovetail and the third stub-tenoned, with the surfaces mitred. The more common method is to mitre the two top rim members together and stub-tenon the corner vertical bar into these as shown in Sheet 108.13, the whole then being glued together.

Sheet 117.6 shows a section of a glazing member of the frame. This may be left square and rebated for the glass or prepared as Sheet 117.7 with a sinking on each face. The glass is bedded on putty, with a tight joint between the glass and the frame and held in position with beads screwed into the bars. The edges of the glass are usually painted black before glazing to stop any reflection from the cut edges. If tubular strip-lighting is to be fitted (Sheet

117.8), the beads will require grooving on the back side to accommodate the wiring for the lights.

A section through the base of the case is shown in Sheet 117.3. Veneered blockboard is tongued into an outer frame which is mitred and tongued at the corners and screwed from below. The lower member of the case is double tongued into the base frame as shown.

The detail of a corner bar with bronze metal tee sections forming the rebate for the polished plate glass is shown in Sheet 117.10. Metal now plays a large part in the construction of showcases, where bronze is the primary metal used. The angle sections are welded at the corners. Putty is used to bed the glass which is held in position by small metal blocks having a tap screwed into them.

Index